Mycorrhizal Functioning

Mycorrhizal Functioning
An Integrative Plant-Fungal Process

Michael F. Allen, editor

Chapman & Hall
New York   London

First published in 1992 by
Chapman and Hall
an imprint of
Routledge, Chapman & Hall, Inc.
29 West 35th Street
New York, NY 10001-2291

Published in Great Britain by
Chapman and Hall
2-6 Boundary Row
London SE1 8HN

© 1992 Routledge, Chapman & Hall, Inc.

Printed in the United States of America on acid free paper.

All rights reserved. No part of this book may be reprinted or reproduced or utilized in any form or by any electronic, mechanical or other means, now known or hereafter invented, including photocopying and recording, or by an information storage or retrieval system, without permission in writing from the publishers.

Library of Congress Cataloging in Publication Data

Mychorrhizal functioning : an integrative plant-fungal process /
  edited by Michael F. Allen.
    p. cm.
   Includes bibliographical references and indexes.
   ISBN 0-412-01891-8 (cl)
   1. Mycorrhizas. 2. Plant-fungi relationships. I. Allen, Michael
F., 1952–
QK604.2.M92M93 1992
589.2′0452482—dc20   92-20717
                                                      CIP

British Library of Congress Cataloguing in Publication Data also available

Contributors

Edith B. Allen
Department of Biology
Systems Ecology Research Group
San Diego State University
San Diego, CA 92182-0057
USA

Michael F. Allen
Department of Biology
Systems Ecology Research Group
San Diego State University
San Diego, CA 92182-0057

Anne J. Anderson
Department of Biology
Utah State University
Logan, UT 84322-5305
USA

Concepción Azcon-Aguilar
Estación Experimental del Zaidin
Departamento de Microbiología, CSIC
Prof. Albareda 1
18008 Granada
Spain

José Miguel Barea
Estación Experimental del Zaidin
Departamento de Microbiología, CSIC
Prof. Albareda 1
18008 Granada
Spain

Caroline S. Bledsoe
Department of Land, Air and Water Resources
Hoagland Hall
University of California
Davis, CA 95616
USA

Paolo Bonfante-Fasolo
Departimento di Biologia Vegetale
Università di Torino
Viale P. A. Mattioli, 25
I-10125 Torino
Italy

Steven D. Clouse
Department of Biology
San Diego State University
San Diego, CA 92182-0057
USA

Brian Cummings
NPI
417 Wakara Way
Salt Lake City, UT 84108
USA

J. W. Deacon
Institute of Cell and Molecular Biology
University of Edinburgh
School of Agriculture
West Mains Road
Edinburgh EH9 3JG
United Kingdom

Roger Finlay
Department of Microbial Ecology
University of Lund
Helgonavägen 5
S-223 62 Lund
Sweden

A. H. Fitter
Department of Biology
University of York
Heslington, York YO1 5DD
United Kingdom

L. Vincent Fleming
Nature Conservancy Council
106 High Street
Dalbeattie
Kirkcudbrightshire EG5 4HB
United Kingdom

Carl F. Friese
Department of Biology
University of Dayton
Dayton, OH 45469-2320
USA

J. N. Gemma
Department of Botany
University of Rhode Island
Kingston, RI 02881-0812
USA

J. D. Jastrow
Environmental Research Division
Argonne National Laboratory
Argonne, IL 60439
USA

Sherri L. Jeakins
Department of Biology
San Diego State University
San Diego, CA 92182-0057
USA

Richard E. Koske
Department of Botany
University of Rhode Island
Kingston, RI 02881-0812
USA

Hugues Massicotte
Department of Forest Science
Oregon State University
Corvallis, OR 97331
USA

Steven L. Miller
Department of Botany
University of Wyoming
Laramie, WY 82071
USA

R. M. Miller
Argonne National Laboratory
9700 South Cass Avenue
Argonne, IL 60439
USA

Randy Molina
USDA-FS Forestry Research Laboratories
3200 Jefferson Way
Corvallis, OR 97331
USA

D. J. Read
Department of Plant and Animal Sciences
University of Sheffield
Sheffield S10 2TN
United Kingdom

I. R. Sanders
Department of Biology
University of York
Heslington, York YO1 5DD
United Kingdom

Silvano Scannerini
Departimento di Biologia Vegetale
Università di Torino
Viale P. A. Mattioli, 25
I-10125 Torino
Italy

Bengt Söderström
Department of Microbial Ecology
University of Lund
Helgonavägen 5
S-223 62 Lund
Sweden

I. C. Tommerup
CSIRO Division of Forestry
Private Bag
Wembley 6014
Western Australia
Australia

James M. Trappe
Department of Forest Science
Oregon State University
Corvallis, OR 97331
USA

Barbara S. Weinbaum
Department of Biology
San Diego State University
San Diego, CA 92182-0057
USA

J. M. Wilson
Western Australian Department of Agriculture
P.O. Box 110/Marine Ice
Geraldton, Western Australia 6530
Australia

Timothy Wood
NPI
417 Wakara Way
Salt Lake City, UT 84104
USA

Contents

Contributors		v
Preface		xi
Introduction		xiii

Section 1. *Development and Structure*

1.	Fungal reactions to plants prior to mycorrhizal formation R.E. Koske and J.N. Gemma	3
2.	The influence of the plant root on mycorrhizal formation A.J. Anderson	37
3.	The cellular basis of plant–fungus interchanges in mycorrhizal associations. P. Bonfante-Fasolo and S. Scannerini	65
4.	The mycorrhizal mycelium D.J. Read	102
5.	Mycorrhiza and carbon flow to the soil R. Finlay and B. Söderström	134

Section 2. *Interactions*

6.	Interactions between mycorrhizal fungi and other rhizosphere microorganisms C. Azcón-Aguilar and J.M. Barea	163
7.	Interactions between fungal symbionts: VA mycorrhizae J.M. Wilson and I.C. Tommerup	199
8.	Interactions of ectomycorrhizal fungi J.W. Deacon and L.V. Fleming	249
9.	Mycorrhizae, nutrient translocation, and interactions between plants S.L. Miller and E.B. Allen	301
10.	Interactions with the soil fauna A.H. Fitter and I.R. Sanders	333

Section 3. *System Dynamics and Application*

11. Specificity phenomena in mycorrhizal symbiosis: Community-ecological consequences and practical implications — 357
    R. Molina, H. Massicotte and J.M. Trappe
12. Physiological ecology of ectomycorrhizae: Implications for field application — 424
    C.S. Bledsoe
13. The application of VA mycorrhizae to ecosystem restoration and reclamation. — 438
    R.M. Miller and J.D. Jastrow
14. Biotechnology and the future of VAM commercialization — 468
    T. Wood and B. Cummings
15. Mycorrhizae and the integration of scales: From molecules to ecosystems. — 488
    M.F. Allen, S.D. Clouse, B.S. Weinbaum, S. Jeakins, C.F. Friese, and E.B. Allen

Organism Index — 517

Subject Index — 525

Preface

Mycorrhizal associations have generated interest among plant scientists for over a century. The interest continued unabated with ever increasing numbers of published papers, interested scientists, and research questions. This has lead to various subdisciplines through time. Up through the mid-1970's most workers consider themselves as endo-, ecto- or some other specialized order of mycorrhizast. With the advent of interdisciplinary research devoted to topics such as molecular biology or systems ecology in the 1980s, a new form of separation began, that is of scale. Now most scientists are searching for answers to questions that bridge scales and are similar whether one in interested in VA mycorrhizae of grasslands or ectomycorrhizae of coniferous forests.

This book was designed to take this hierarchial approach to understanding the dynamics that regulate mycorrhizae and, in turn, how mycorrhizae affect the systems which they are in. This has been the approach of both the recent 8th North American Conference on Mycorrhizae, held in Jackson, Wyoming, U.S.A. and the 3rd European Symposium on Mycorrhizas in Sheffield, U.K. This book was written to facilitate communication among mycorrhizal researchers working on various scales with one type of mycorrhizae or those working on a single scale with multiple types. It is also designed to introduce the literature of mycorrhizae to ecologists, agronomists, biotechnologists, plant and fungal biologists, and land managers. It is to these scientists that our work relates and whose interest is crucial to the future of the field.

I am grateful for the patience and assistance of Greg Payne of Routledge, Chapman and Hall and to the press for its interest. The support of the following individuals for their assistance as reviewers, proof readers and general support was essential, Edith Allen, Carl Friese, Anne Anderson, Sherri Jeakins, Barbara

Weinbaum, Fred Edwards, Mary De La Cruz, Steve Clouse, Leslie Hickson and Dawn Makis. I would like to dedicate this volume to Martha Christensen, Thomas S. Moore, Jr., Nancy Stanton, and Ty Harrison, who taught me the value of interdisciplinary research.

Michael F. Allen

Introduction
Michael F. Allen

The field of mycorrhizal research, like the study of many aspects of biology, has undergone tremendous changes, especially during the past decade. During the 1970s, most research efforts were devoted to describing physiological or growth responses of various plant species to mycorrhizae and to the application of mycorrhizal fungi to enhance agricultural or forest productivity. The plethora of new research tools during the 1980s from molecular biology to remote sensing provided radically new opportunities to study biology.

In a sense, the study of mycorrhizae was almost uniquely made for taking advantage of the new technologies. Here was a symbiotic relationship between two organisms in which one invaded the tissues and, in some cases, the cells of another. There had to be molecular signals that allowed, promoted, or inhibited the formation of the symbiosis and determined whether it became parasitic, mutualistic, or somewhere in between. At the other extreme, remote sensing was capable of assessing the health of lands that was often correlated with the presence or type of mycorrhizal association present. Moreover, virtually all land plants have been shown to have some reaction to mycorrhizal fungi whether a mycorrhizal symbiosis was established or not. A very large variety of fungi in all three groups of true fungi form mycorrhizae. And these fungi form an important food base for a wide range of animals from mites and nematodes to large herbivores (Allen, 1991). Thus almost any organism or ecosystem that was studied, from bacteria to man and from wetlands to deserts, have some relationship to mycorrhizal associations.

Obviously, this complexity in interactions created a need for heuristic tools to help integrate the sheer volume of information being derived. Hierarchy theory provided a framework for studying various scales and sharing information among these scales. I have tried to follow a generalized hierarchical approach to present what I believe to be a comprehensive overview of mycorrhizal research. This book represents an attempt to view not only what has been done but to give some perspective as to where the field should go. Included are overviews on the molecular signals that regulate formation and activity of mycorrhizal symbioses, interactions among organisms forming mycorrhizae and that depend on the organisms forming mycorrhizae, the roles of mycorrhizae in structuring communities and ecosystems, and, finally, some applied aspects whereby mycorrhizae can be used to enhance the use of human-modified landscapes. Thus, the book is divided into three sections, development and structure, interactions, and system dynamics and application.

As with all fields, not only should information be presented, but the various controversies that fuel the important questions and methods should be enhanced, not buried. It was also my task, with the cooperation of the various authors, to bring out areas of mycorrhizal studies that currently are controversial. It is my view that this is, in part, what makes a field exciting. Several major controversies are presented in the various chapters, such as the nature of the interactions of rhizosphere organisms with mycorrhizal fungi, why or why not do growth responses occur in the field where grazing soil fauna are abundant, and the concepts of "early" and "late" stage mycorrhizal fungi. It is my hope that these controversies will stimulate interest, fuel further discussion, and finally enhance our understanding and predictability of one of the most exciting biological relationships on earth.

In summary, this book is meant to provide an overview of some of the major recent directions in research of the 1980s, to propose some new directions for the 1990s, and to stimulate discussion and scientific interchanges that will lead to new understanding of mycorrhizal symbioses.

References

Allen, M.F. (1991). *The Ecology of Mycorrhizae*. Cambridge University Press, Cambridge.

SECTION 1
Development and Structure

1

Fungal Reactions to Plants Prior to Mycorrhizal Formation

R. E. Koske and J. N. Gemma

SUMMARY. Living roots release a wide variety of soluble, insoluble, and volatile exudates, some of which act as chemical messengers that assist in coordinating the formation of mycorrhizae. Events that may be mediated by these compounds include spore germination, directional growth of germ tubes and hyphae, hyphal growth rate and branching, and positioning of the hyphae near potential entry sites on the root surface. Later stages of mycorrhiza development (formation of appresoria and penetration) in addition may involve surface bound molecules whose role is one of recognition.

Careful observation of preinfection interactions between mycorrhizal fungi and host and nonhost species indicates that the communication is a two-way phenomenon, with each member sending and receiving messages. Whereas previous work has focused on the idea that it is the plant that regulates the extent of involvement of the fungi, new evidence suggests that the fungal partner also may select or reject a particular root. The long evolutionary history and great interdependence (obligate biotrophy) of mycorrhizal fungi and host plant species argue for the existence of an extremely effective means of communication that mediates all aspects of the symbiosis.

Introduction

Communication between the partners of a symbiosis is an essential process that first encourages and later permits the physical association between different species of organisms. As in any intimate association, clear, unambiguous communication between partners is vital for the long-term preservation of the relationship.

The aim of this chapter is to discuss the reactions of some mycorrhizal fungi to plants prior to the formation of mycorrhizae, reactions that are mediated by chemical messages produced by the plants themselves. In addition, the role of messages sent from fungi to roots also will be discussed. Vesicular-arbuscular mycorrhizae (VAM) will be emphasized, although ectomycorrhizae (ECM) also will be included. Only morphological fungal "reactions" to plants will be considered. This topic was well and thoughtfully reviewed by Harley and Smith (1983) and Bowen (1987) and briefly addressed by Anderson (1988). Discussion of some interactions between ECM fungi (ECMF) and plants also are found in Chapter 8 by Deacon and Fleming in this volume.

Although nearly every aspect of the biology of mycorrhizal fungi is affected to some extent by plants, detailed studies of most such interactions are lacking for many areas. As in any young science, correlations abound, but proofs of causation are scarce.

In the majority of mycorrhizal associations, it is routinely acknowledged the fungal partner always benefits, deriving much or all of its carbon nutrition and energy from the host plant. Most of the fungi that form mycorrhizae are physiologically or ecologically obligate biotrophs. Thus, events that result in their locating a host plant prior to formation of the mycorrhiza are vital to their success.

Although the benefits to host plants in mycorrhizal associations are widely acknowledged, the importance of active participation by the plant from the earliest stages of the relationship (including attraction of the fungal partner) often is overlooked. For those obligately mycotrophic plant species, including an estimated 70% of the angiosperms (Trappe, 1987), an early association with their potential fungal partners may be equally as vital to them as to their fungal symbionts.

Considering the high mutual dependency both of mycorrhizal fungi and the majority of vascular plants and the long evolutionary history of the symbiosis (Nicolson, 1975; Pirozynski & Malloch, 1975; Pirozynski & Dalpe, 1989), it is certain that effective communication systems have developed to regulate all aspects of the symbiosis. After many millions of generations in which communication was refined and made selective enough to readily distinguish friend from foe, it is not surprising that the subtle signals exchanged between partners are not easily detected by root biologists.

Whereas specific details differ between the various types of mycorrhizae, all are similar in that fungal structures in the soil (germ tubes, hyphae, hyphal strands/and or rhizomorphs) must contact the roots of potential hosts. Following contact, formation of mycorrhizae occurs as the fungus becomes intimately involved with some portion of the root system. Every phase in the life history of a mycorrhizal fungus is subject to influence by plant roots, including spore germination, rate and directionality of germ tube growth, hyphal branching, recognition of the host, penetration of the root, establishment and extent of colonization of the root, growth of hyphae into the soil, and sporulation.

Root Exudation

Roots produce an immense array of organic chemicals, many hundreds or thousands of which are released to the surrounding soil (Rovira, 1969; Bowen & Theodorou, 1973; Whitfield et al., 1981; Curl & Truelove, 1986). The compounds that are released by roots were classified by Rovira and Davey (1974) according to their mobility in the soils: (1) diffusible-water soluble, (2) diffusible-volatile, and (3) nondiffusible compounds. The latter category includes mucilages and sloughed cells. Absolute amounts of release of the three types of exudates vary with species, stage of development, and cultural conditions. Based on data from studies using $^{14}CO_2$-fed wheat plants, Rovira and Davey (1974) estimated that the relative amounts of the three exudate types released from roots were 1:8–10:3–5, respectively.

The identities of compounds released by roots are poorly known, especially those that are volatile or nondiffusible. The subject of root exudates was recently reviewed by Curl and Truelove (1986), and the reader is referred there and to the excellent reviews by Rovira (1969) and Bowen and Theodorou (1973) for more detailed information than is included in this article. Diffusible water-soluble compounds (hereafter referred to as "water-soluble exudates") include a variety of compounds such as mono- and disaccharides, numerous amino acids and organic acids, flavanones and nucleotides, enzymes, and other substances. A far wider variety of substances released from roots are the diffusible-volatile compounds (referred to as "volatile exudates"). In perhaps the most exhaustive characterization of volatile exudates from roots, Whitfield et al. (1981) isolated 270 distinct organic compounds from the legume *Acacia pulchella* R. Br. Seventy-six were identified, and these included 2 organic acids, 17 alcohols, 7 aldehydes, 12 ketones, 112 esters, 4 phenols, 16 terpenoids, and miscellaneous compounds. Roots would thus appear to have a large chemical vocabulary for communicating with mycorrhizal fungi.

Unfortunately, although the efficacy of volatile compounds as messengers in soil is well documented (Fries, 1966; Stotzky and Schenck, 1976; French, 1985), the realization that volatile compounds may be the most common form of chemical communication between soil organisms has not been generally acknowledged. There appear to be two reasons for this reluctance. First, relatively few studies have been performed to catalog the wide range of responses that are mediated by volatile compounds (mostly due to the technical difficulty in handling volatile substances). Second, the capacity of volatiles to move through soil is unappreciated, although no one disputes the ease with which oxygen diffuses through the soil to roots and CO_2 diffuses away. Only in waterlogged soils is this diffusion slowed, a condition not coincidentally associated with reduced mycorrhizal development. Good aeration often is cited as a requisite for vigorous development of VAM (St. John et al., 1983; Saif, 1981, 1983; T. Wood, personal communication). Interestingly, the chemotropism of the radicle of the parasitic angiosperm *Striga* toward roots of host plants is reduced in wet sites (Sauder, in Williams, 1961).

What characteristics of volatile compounds make them an effective medium for communication? Volatiles are less likely to be inactivated by other soil microorganisms than are water-soluble compounds, a feature that may be of critical importance in enabling volatile root exudates to traverse the rhizosphere, a zone of high microbial activity (Newman & Watson, 1977). In addition, volatiles can be active in two phases, the vapor phase where transfer is rapid, and the liquid phase, in which the volatile compounds are solubilized in water from which they later can revolatilize (Stotzky & Schenck, 1976; French, 1985). Another feature that makes volatile compounds more efficient signals in soil is that they are capable of moving over greater distances more rapidly than are water-soluble compounds. The latter characteristic may be especially important to the VAM symbiosis, since a particular segment of root is susceptible to infection by VAM germ tubes/hyphae for only a short time (e.g., Mosse & Hepper, 1975; Brundrett et al., 1985; Glenn et al., 1985; and see the section on "Initial Stages of Mycorrhizal Formation"). A rapidly transmitted message would appear vital to coordinate the events necessary for infection. Interestingly, very short exposure times (i.e., 20 sec) to a volatile message elicit a variety of developmental responses in fungi, including germination and prolonged directional growth (French, 1985).

Although a variety of water-soluble compounds are exuded from roots, often in significant quantities, such compounds would not diffuse far in soil before they were consumed by bacteria in the rhizosphere (Newman & Watson, 1977). The effects of water-soluble exudates probably are most evident at the root surface and extending a few millimeters into the soil, whereas volatile exudates are probably active in a region extending from the root surface to several centimeters into the soil.

Problems of Method

Various technical problems have slowed our understanding of the reactions of mycorrhizal fungi to their host plants. Some are common to all studies of soil-inhabiting organisms (Curl & Truelove, 1986), some result from the complexity of the symbiosis, and some are anthropogenic.

Opacity of Soil

At the microscopic level the relationship between mycorrhizal fungi and plants is poorly understood for most mycorrhizae, in large part because of the opacity of the soil medium itself. Although microrhizotrons (e.g., Rush et al., 1984; Rygiewicz et al., 1988; Finlay & Read, 1986) have been employed to examine the interactions between fungi and roots, observations generally are limited to a single plane.

Soilless techniques, such as growing plant roots in plastic bags, nutrient films, Petri dish, and test tube cultures (e.g., Mosse, 1962; Mosse & Hepper, 1975; Allen et al., 1979; Hepper, 1981; Koske, 1982; Fortin et al., 1983; Saltveit & Young, 1983, Mosse & Thompson, 1984; Miller-Wideman & Watrud, 1984; Yang & Wilcox, 1984; Chilvers et al., 1986; Becard & Fortin, 1988; Gemma & Koske, 1988a; Strullu et al., 1989) have been very useful in observing the earliest stages of formation of mycorrhizae. The method of Boudarga and Dexheimer (1990) in which VAM were synthesized with seedlings grown on an agar medium in large Petri dishes appears to hold promise. An interesting feature of this technique was the inclusion of activated charcoal (5 g/liter) into the medium. Nevertheless, all of the chambers and containers used in the studies cited above, although useful, are highly artificial. The dynamics of water and gas relations (including CO_2) in soil are probably poorly reproduced in the small culture vessels typically used.

Complexity of the Symbiosis

One of the most frustrating, yet fascinating, aspects of studying mycorrhizal fungi is the interdependence of the symbionts. Fungus, host, soil, climate, and other microorganisms interact in an immensely complex fashion, and the contribution of individual components may be impossible to assess. The effect of environmental factors acting directly on the fungus cannot be separated from factors whose effects are mediated by the host and then indirectly affect fungal behavior.

For example, does the increase in sporulation of various VAMF species at higher temperatures (Schenck & Schroeder, 1974; Schenck & Smith, 1986) result from an enhancement of fungal reproduction at the higher temperatures (direct effect) or from the increased amount of photosynthate made available by the plant at the higher temperatures (indirect effect)? Read et al. (1976) summarized the problem nicely based on their observations in the field, noting that "it is therefore difficult to identify cause and effect." Unfortunately, the situation in the greenhouse or laboratory may be nearly as perplexing.

Laboratory Artifacts

Inhibiting compounds present in culture media also may interfere with attempts to study interactions between the symbionts. The most obvious inhibition, failure of spore germination, is readily recognized (Fries, 1966, 1978; Hepper & Smith, 1976; Bjurman, 1984), although the presence of inhibitors in the medium as the cause of failed germination may not be obvious. More subtle effects of inhibitory compounds in the media on subsequent stages of the symbiosis may not be easily recognized. Watrud et al. (1978) noted that the growth of germ tubes of *Gigaspora margarita* was significantly improved when activated charcoal was added to the medium. Since 1941, Fries (1941) has had success in stimulating germination of basidiospores of some ECMF by streaking media sown with spores with the red yeast *Rhodotorula glutinus*. The yeast may function by inactivating inhibitors in the spores or medium and/or by meeting unknown nutritional requirements.

Other components of the medium may interfere with attempts at synthesizing symbioses, either by directly inhibiting the fungus or by rendering the root less capable of supporting mycorrhiza formation. Thus, Carr et al. (1985) found that concentrations of sucrose exceeding 5 g/liter in the medium reduced growth of hyphae of *Glomus caledonium* and *G. mosseae* in a root–organ culture medium. Becard and Fortin (1988) reported that high concentrations of sucrose (30 g/liter) or KH_2PO_4 (19.1 mg/liter) in a root–organ culture medium prevented the attachment of hyphae of *Gigaspora margarita* to transformed roots of carrot. In contrast, Mugnier and Mosse (1987), using transformed roots of morning glory (*Convolvulus sepium*) and *Glomus mosseae*, found that KH_2PO_4 concentrations to 170 mg/liter had little effect on infection but nitrogen concentration was critical. Growth of germ tubes was prevented when total N in the medium exceeded 2 mM, germ tubes grew but did not attach to roots at 1–2 mM, N, and hyphae showed directional growth and attached to root at <0.2 mM N.

The presence of various carbohydrates in the medium may result in the blocking of receptor/recognition sites on hyphae or host cell walls, which could prevent germ tubes and hyphae from locating host roots. Similar "masking" effects have been demonstrated in several plant pathogenic fungi and in nematodes (e.g., Bushnell & Rowell, 1981; Anderson, 1988). If such a recognition mechanism operates in mycorrhizal symbioses, then, when these sites are blocked, the two organisms would not recognize each other, and the hyphae would not initiate the necessary morphogenetic steps after contacting an apparently suitable host. In such a case, hyphal tips would appear to be "blind" to the presence of the root. This description exactly matches the behavior of VAMF hyphae in several of the attempts at synthesizing mycorrhizae in agar cultures (Mosse, 1962; Mosse & Hepper, 1975; Miller-Wideman & Watrud, 1984) or when hyphae and roots shared a common agar surface (Powell, 1976; Koske & Gemma, personal observation). Such observations have been taken as evidence that contact between symbionts in soil is similarly haphazard (Mosse, 1962; Gianinazzi-Pearson et al., 1988).

Characterization of Root Exudates

Another problem, alluded to earlier, is our inadequate knowledge of the identity of root exudates and of cellular metabolism in roots. Studies with plants grown in greenhouse or laboratory conditions (e.g., "pot cultures" of VAMF) may not be as representative as the investigators wish. Plants grown in greenhouses receive different light quality than do outdoor plants, and fluctuations in temperature, soil moisture, and wind disturbance are less extreme in the greenhouse. Exudates and secondary metabolites of plants grown under greenhouse conditions can differ both qualitatively and quantitatively from those produced by plants grown under field conditions (Rovira, 1969; Bowen & Theodorou, 1973; Powell & Adams, 1973; Curl & Truelove, 1986).

Exudation from plants grown in solution culture may greatly underrepresent levels released from soil-grown plants. Root exudation of carbohydrates and amino acids from barley and maize plants grown in water culture were 2–9 times greater when glass ballotini (small glass spheres) were present in the culture jars (Barber & Gunn, 1974). Contact between roots and the ballotini resulted in mechanical stimulation and increased exudation. In addition, there were qualitative differences between the control and treated (ballotini) plants in the amino acids, both in the exudate and on the root surface. Glycine and lysine, compounds shown by Hepper and Jakobsen (1983) to stimulate growth of germ tubes of *Glomus caledonium*, were present in the exudates from the treated barley plants in concentrations 1.8 and 4.0 times greater than in the control, respectively.

In a parallel situation from plant pathology, it was found that when bean plants were grown in solution culture, they did not exude fatty acids from their roots, although these were released in sterile sand culture (Papavizas & Kovacs, 1972). Since these fatty acids were necessary for spore germination of *Thielaviopsis basicola*, a pathogen of bean roots, the solution culture method of collecting root exudates appears to be inappropriate.

To date, all studies of the effect of root exudates (both water-soluble and volatile) on VAMF (and many studies with ECMF) have employed roots grown in solution culture or under similar conditions (see the section on "VAM and Membrane Permeability Model"). The methods of Schisler and Linderman (1987) for collecting volatile exudates, and of Barber and Gunn (1974) and Papivizas and Kovacs (1972) for collecting water-soluble exudates should be considered by investigators studying the response of mycorrhizal fungi to root exudates.

Another category of exudate, hydrophobic compounds, also may be of biological significance to mycorrhizal fungi. The compounds (some are dihydroquinones) can be recovered easily from rhizosphere soil by extraction with methylene chloride and have been shown to stimulate seed germination of *Striga asiatica*, an obligately parasitic angiosperm (Netzly et al., 1988).

Initially, nearly all investigators appeared to have made the tacit assumption that water-soluble compounds would be the primary compounds on which to focus their attentions. Not surprisingly, the experimental apparatus designed by mycorrhiza investigators could not discriminate between effects resulting from the action of water soluble or volatile exudates. This limitation seldom has been acknowledged, although it may be of importance in interpreting the literature.

Another problem is that the identification of volatile compounds released by roots is technically difficult. Although often active (and present) at low concentrations, many may be easily "lost" to the investigator by adhering to various surfaces during attempts to isolate and identify them (French, 1985), or they may escape as invisible gases.

A final concern, and one that applies to all types of exudates, is that the exuded compounds may be readily modified by microorganisms in the rhizosphere or by contaminants in the assay system (Curl & Truelove, 1986).

Misidentification of Fungi

One major difficulty in interpreting data from published works concerning VAMF is in knowing what species of fungi were used in an experiment, even if the identities were reported. For example, *Glomus fasciculatum* is the most widely cited VAMF in the literature. Hundreds of publications report on this species, yet most isolates identified as *G. fasciculatum* are probably other species such as *G. aggregatum*, *G. invermaium*, *G. intraradix*, and *G. etunicatum* (Walker & Koske, 1987).

Species identification is difficult, and many published descriptions are incomplete, allowing several clearly different biological species to be erroneously included in one taxon. The problem of misidentification is not limited to small-spored species of *Glomus* that can be assigned to *G. fasciculatum*. Equal confusion in identification appears to exist in other "species," especially those most frequently reported in the literature [such as, *G. macrocarpum*, *G. microcarpum*, *G. mosseae*, and *G. etunicatum* (Walker & Koske, unpublished observations)].

The implications of frequent misidentifications in the literature are obvious and enormous. Readers are cautioned that VAMF species whose names appear in this article (and other articles) may be misidentified. From personal experience, we would encourage the greatest skepticism when a fungus identified as *Glomus fasciculatum* is encountered.

A factor that prevents misidentifications in the literature from later being corrected by other investigators is the failure of authors to deposit voucher specimens in recognized herbaria. In 20 randomly selected studies of VAMF published since 1981, only one stated that vouchers had been deposited. Herbaria that are appropriate for lodging voucher specimens are the herbarium at Oregon State University (OSC), Corvallis, and Farlow Herbarium, Cambridge, MA.

Two additional concerns involve the identity and purity of the inoculum used in experiments. Sources of inoculum for use in research have been (1) commercial suppliers, (2) other researchers, and (3) pot cultures initiated by the investigator. Cultures obtained from 1 or 2 (above) may be misidentified and/or contaminated (Walker & Koske, unpublished observations), and once-pure pot cultures of VAMF can readily become contaminated with other VAMF in the greenhouse. Purity and correct identification have been inconsistent with some suppliers of VAMF inocula.

Fungal Reactions Prior to Contact with Roots

In all types of mycorrhizae, the inoculum may be spores or hyphae. VAM infection of roots can be initiated by germ tubes arising from spores, by hyphae growing from other propagules (i.e., vesicles or colonized root fragments), or by external hyphae connected to active VAM. Infection resulting from the first two processes (germination of spores or other propagules) is termed primary infection. Secondary infection results when a hypha deriving its energy from the host plant (active VAM) initiates infection.

Spore Germination

Germination of spores of VAMF has been studied in less than 10% of the described species, with the bulk of the work performed on just two, *G. mosseae* and *G. caledonium*. It appears that the presence of nearby plant roots is not necessary for germination to occur (e.g., Daniels & Trappe, 1980; Koske, 1981a; Tommerup, 1983), although exceptions may be found as more species are examined. Root exudates have been reported to stimulate germination (Graham, 1982; Tommerup, 1985), to inhibit germination (Tommerup, 1984), or to have no effect (Powell, 1976; Daniels & Trappe, 1980; Koske, 1981a; Tommerup, 1985; Gemma & Koske, 1988a; Suriyapperuma, 1990). El-Atrach et al. (1989) found that spores of *G. mosseae* were inhibited by volatile compounds released from the roots of a nonhost species (*Brassica*) but unaffected by water-soluble compounds from the same species.

Since the early work of Mosse (1959), it has been found that the germination of some species of *Glomus* is greatly stimulated by substances present in nonsterile soil (e.g., Daniels & Graham, 1976; Daniels & Trappe, 1980; Azcon-Aguilar et al., 1986a; Azcon, 1987), although under some conditions (including low P) nonsterile soil may be inhibitory (Hetrick & Wilson, 1989; Wilson et al., 1989). Germination of spores of *Glomus versiforme* [=*G. epigaeum*] and *G. mosseae* is stimulated by bacteria associated with roots and spores (Mayo et al., 1986; Azcon, 1989). An unidentified filamentous fungus ("F=4") induced 100% germination in spores of *Glomus mosseae* that were incubated on sterile soil-extract agar (Azcon-Aguilar et al., 1986b). Spores not exposed to F-4 did not germinate. Spores that had remained ungerminated for up to 50 days on the soil-extract agar were rapidly stimulated to 100% germination by F-4. Interestingly, a second fungal isolate (F-3), which earlier had been found to stimulate spore germination of *G. mosseae* on water agar (Azcon-Aguilar et al., 1986a), was not effective when spores were incubated on soil-extract agar.

The stimulatory effects of nonsterile soil and the inhibitory effects of sterile soil on VAMF spore germination in some cases have been variously explained on the basis of substances released by soil microorganisms that are directly stimulatory or that inactive germination inhibitors that are present in the spores or soil (Daniels & Graham, 1976; Daniels & Trappe, 1980; Azcon-Aguilar et al., 1986b). Complicating the interpretation of all studies on the involvement of various factors on spore germination is the phenomenon of spore dormancy (Tommerup, 1983, 1985; Gemma & Koske, 1988b). Obviously, dormant spores may react differently to potential germination stimulators than will spores that have had their dormancy requirements met. In addition, species (and possibly genera) of VAMF may be expected to possess different degrees of dependence on other soil organisms for clues to germination, with the small-spored species (which carry fewer food reserves) being more likely than the large-spored species to require an external stimulus for germination.

Whatever the mode of action of soil microorganisms on spore germination and growth of VAMF hyphae, their effect would be most apparent when young growing roots invade soil that had become fungistatic. Root exudates (water-soluble, volatile, and insoluble) stimulate microbial activity in the immediate vicinity of the roots (the "rhizosphere effect"), and the microorganisms in turn could stimulate the spores of VAMP to germinate and their germ tubes to grow rapidly.

Unlike spores of VAMF, the spores of ECMF frequently germinate poorly or not at all in axenic culture unless exposed to unidentified activating compounds (e.g., Fries, 1966, 1978, 1979). The contrast between the *in vitro* (and, presumably, *in vivo*) germination of spores of VAMF and ECMF apparently indicates different strategies possessed by the fungi. VAMF generally have very wide host ranges and can form mycorrhizae with the vast majority of vascular plant species, while the hosts of ECMF are much fewer. VAMF produce relatively few but unusually large spores, whereas ECMF produce numerous, small spores. The large VAMF spores carry enormous reserves, are capable of germinating several times (Mosse, 1962; Koske, 1981b), and the germ tubes of at least some species are chemotropic toward roots of host plants over distances of greater than 1 cm (Koske, 1982; Gemma & Koske, 1988a; Suriyapperuma, 1990). In addition, the spores of at least some VAMF species are held in a dormancy period after they are formed, preventing them from immediately germinating and reinfecting the host plant (Tommerup, 1983, 1985; Gemma & Koske, 1988b).

After the dormancy requirement has been met, germination of spores of many species of VAMF occurs when temperature and moisture conditions are conducive to root growth (e.g., Daniels & Trappe, 1980; Koske, 1981b). Since nearly all VAMF spores are produced underground in the root zone of host plants they are not routinely dispersed far from the site of their formation, it is highly likely that when spores germinate in Spring that they will be in the vicinity of young host roots. It must be reemphasized that the comments on spore germination of VAMF are based on a small percentage of species. Spores of some (or many) may possess strategies more similar to ECMF fungi (see below).

Spores of ECMF are widely dispersed by wind and animals, and individual spores face a far more risky future than do VAMF spores. Since their chances of being deposited in the root zone of a suitable host are relatively small and their reserves are minute, the spores of most species remain dormant even under otherwise conducive conditions, requiring exposure to exudations from host roots or ECMF fungi to germinate (e.g., Melin, 1963; Fries, 1966, 1978, 1979, 1981; Birraux & Fries, 1981; Fries et al., 1987). Spore reserves thus are conserved for respiration and maintenance and are not squandered on production of germ tubes in the absence of a host plant.

Since interactions between propagules of ECMF and host plants or their mycorrhizae are dealt with in Chapter 8 by Deacon and Fleming (this volume), no mention will be made here.

Growth of Germ Tubes and Hyphae

Following germination, the next critical steps in the preinfection phase are those events that lead to contact between germ tubes or hyphae and roots of a host plant. The stages of development of VAM have been described in detail for several VAMF symbioses that were synthesized in agar cultures (Mosse & Hepper, 1975; Miller-Wideman & Watrud, 1984; Mugnier & Mosse, 1987; Becard & Fortin, 1988; Becard & Piche, 1989a,b) and in soil (Brundrett et al., 1985; Garriock et al., 1989; Brundrett & Kendrick, 1990a,b). Evidence has accumulated in recent years suggesting that plants and mycorrhizal fungi have evolved a complex, efficient system of chemical communication that maximizes the likelihood of contact and that both partners of the symbiosis have at least some control as to the extent of total colonization. A one-sided version of this notion was expressed by Siqueira et al. (1982) who suggested that "plants appear to control [VA] mycorrhiza colonization through root carbohydrate metabolism." Earlier, Krupa and Fries (1971) proposed the similar view that volatile and nonvolatile root exudates regulated the earliest stages in the development of ECM in pine. Nylund and Unestam (1982) echoed the view that overwhelming control over the formation of ECM remains with the host plant, and the earliest stages are regulated by root exudates (see section on "Types of Compounds Functioning as Chemical Messengers.")

Studies of VAMF hyphae and germ tubes generally have shown that infection usually occurs when young lateral roots are contacted near the apical 1.4 cm (Mosse & Hepper, 1975; Brundrett et al., 1985; Glenn et al., 1988; Strullu et al., 1989; Brundrett & Kendrick, 1990a). The observation by Mosse and Hepper (1975) that formation of the first entry point renders the region of the root in the immediate area more liable to subsequent penetrations may be indicative of increased exudation resulting from the initial contact. This is discussed in more detail in the section on "Initial Stages of Mycorrhizal Formulation."

Exudations from plant roots affect growth of germ tubes and hyphae of mycorrhizal fungi in three distinct ways: hyphal growth rate, frequency of branching, and directionality of growth. The first two are combined in the discussion below because they typically occur together.

Hyphal Growth Rate and Branching

Responses of VAMF hyphae and germ tubes to water-soluble exudates emanating from roots include increased growth and branching (Mosse, 1962, 1988; Mosse & Hepper, 1975; Graham, 1982; Glenn et al., 1988). Results from these studies were not unequivocal because not all the assay systems were sterile and, because of the design of the experimental apparatus, it was not possible to separate water-soluble from volatile exudates. Indeed, because of limitations of the assay techniques used in many studies designed to assess the response of hyphae of VAMF to root exudates, it is not possible to distinguish between response to water-soluble and volatile factors or a combination of both.

Several studies have documented increased rates of growth of hyphae of both VAMF and ECMF in response to root exudates prior to actual contact between the members of the symbiosis. In other instances, hyphal growth is stimulated only after contact is made. Precontact responses are discussed below, and the postcontact responses are discussed in the section on "Hyphal Growth, Branching, and Invasion of Roots."

Using an ingenious culturing system developed by Becard and Fortin (1988) employing transformed carrot root cultures, Becard and Piche (1989a,b) demonstrated that volatile exudates are functional during the early phases of VAM formation. Growth of germ tubes of *Gigaspora margarita* was stimulated 20-fold over the control by volatile exudates, a response that occurred prior to contact with roots. This effect was noted only when the hyphae remained connected to the germinating spore and when water-soluble root exudates were present. Water-soluble exudates alone had no effect. The synergistic stimulation could be eliminated by adding KOH-soaked cotton wicks, and the volatile component could be partially replaced with 0.5% CO_2. Since KOH absorbs low molecular weight ketones and aldehydes in addition to CO_2 (Noller, 1965), the identity of all the stimulatory compounds remains undetermined. Earlier, Carr et al. (1985) reported an enhancement of hyphal growth from spores of *Glomus mosseae* and *G. caledonium* in response to volatile compounds released from suspension-cultured cells of lucerne (*Medicago sativa*). The results of Becard and Piche (1989b) also may explain Saif's (1984) observations on the stimulatory effect of low concentrations of CO_2 (0.5–4%) on VAM development in plants grown in soil. The apparently contradictory findings that CO_2 inhibit growth of germ tubes of *G. mosseae* (LeTacon et al., 1983) probably was a result of a high concentration (5%) that was used. The typically reduced development of VAM in waterlogged soils (Gerdemann, 1968; Khan, 1974) may in part be a result of the restriction of rapid movement of volatile root exudates through the soil (see the section on "Types of Compounds Functioning as Chemical Messengers."

There are only a few reports of increased hyphal branching in response to root exudates. Graham (1982) investigated the effects of water-soluble exudates collected from *Citrus* roots immersed in water. He found that germ tubes of *Glomus versiforme* (as *G. epigaeum*) were four times longer and produced more branches in the presence of the exudate. Two difficulties in interpreting the significance of this experiment are that the concentration of exudate to which the spores and germ tubes were exposed could not be directly related to levels found in nature and that the exudates may have contained microbial metabolites. In a similar study, root exudates from P-deficient plants of *Trifolium repens* stimulated growth of germ tubes of *Glomus fasciculatum* up to 37.6 times that induced by exudates from P-sufficient plants (Elias & Safir, 1987). Mosse (1988) reported stimulation of growth and branching of hyphae of *G. intraradices*, *G. versiforme*, and *Gigaspora gigantea* by exudates from transformed roots of carrot (*Daucus carrota*). A substantial literature has amassed concerning the interaction between root exudates (presumed to be water-soluble by most investigators) and formation of VAM. This topic is discussed further in the section on "Fungal Reactions after Contact with Roots."

In a study of the response of germ tubes of *Gigaspora gigantea* and *G. margarita* to roots of *Brassica*, Glenn et al. (1988) found that stimulatory exudates were not released from roots of this nonhost species. The hyphal proliferation (branching and growth) that occurred in the vicinity of the roots of host species was not evident near *Brassica* roots. Earlier, Tommerup (1984) found that exudates of *Brassica napus* slowed the growth rates of germ tubes of two *Glomus* species. The observation that nonhosts do not encourage colonization by VAMF agrees with the discovery that roots of *Brassica* do not release volatile exudates that strongly attract germ tubes of *Gigaspora gigantea* (Gemma & Koske, 1988a) (see the section on "Directional Growth").

As found in studies of spore germination of mycorrhizal fungi, some rhizosphere microorganisms can stimulate the growth of germ tubes of VAMF (Mayo et al., 1986; Azcon, 1987, 1989).

Observations on the behavior of VAMF germ tubes and hyphae when they approach within a few millimeters of roots indicate that some hyphal tips are more receptive than others. Production of aseptate hyphae or germ tubes may be essential for penetration of the root cortex and formation of the mycorrhiza. Mosse (1962) noted that in the absence of roots, germ tubes of species of *Glomus* soon became septate and ceased growth. Much-branched infection fans consisting of numerous septate hyphae arising from an aseptate main hypha were produced by two species of *Glomus* when the germ tubes approached within 1.6 mm of host roots (Powell, 1976). Unlike germ tubes arising from spores of *Glomus* species, germ tubes of *Acaulospora laevis* (as "*Endogone* honcy-colored") began branching at a distance of 3.4 mm from the root, and the branches were aseptate. Germ tubes (apparently aseptate) of *Scutellospora heterogama* branched extensively before contacting roots grown in soil, resulting in numerous penetration points (van Nuffelen & Schenck, 1984). Thus, septate hyphae do not appear to be involved in the infection process (and may in fact be very short lived), although they may have an assimilative function (Mosse & Hepper, 1975).

Infection fans were not formed when hyphae arose from fragments of mycorrhizal roots (Powell, 1976). Instead, an aseptate hypha grew out, branched sparingly when 0.6 mm from the root, and directly penetrated the root. Production of infection fans by species of VAMF prior to contact is inconsistent. Fans may form after contact with roots (see the section on "Initial Stages of Mycorrhizal Formation"), or not be formed. Root organ cultures inoculated with spores of *Gigaspora margarita* and *Glomus mosseae* showed that germ tubes branched little before contacting the roots (Mosse & Hepper, 1975; Miller-Wideman & Watrud, 1984; Becard & Fortin, 1988).

Studies of the effects of root exudates on ECMF greatly predate those on VAMF. The pioneering work of Melin over 75 years ago established that the growth of hyphae of many species of ECMF was stimulated by root exudates, including occasionally, exudates from nonhost species such as tomato. These studies have been reviewed (Melin, 1963; Bowen & Theodorou, 1973; Barea, 1986) and will be only briefly covered here. Based on studies in which exudates from culture roots of *Pinus sylvestris* stimulated the growth of numerous species of ECMF (including species that did not grow in their absence), Melin proposed that the exudates contained a growth-promoting metabolite(s), the "M" factor (Melin, 1963). Other authors confirmed the stimulatory effect of root exudates on ECMF hyphae (e.g., Theodorou & Bowen, 1971; Nylund & Unestam, 1982). Using whole roots of Norway spruce (*Picea abies*), Nylund and Unestam (1982) found that hyphae of the slow-growing ECMF *Piloderma croceum* increased their growth rate when they approached within a few millimeters of the root.

There is some controversy regarding the nature of the "M" factor. In some ECMF, the "M" factor could be replaced by NAD (Melin, 1963), and Bowen and Theodorou (1973) later questioned the uniqueness of such a factor. They noted that the explanation for its effect may be that amino acids in the root exudate make up for deficiencies or inappropriate concentrations of amino acids in the medium. Other compounds also may substitute for live roots as a source of the "M" factor. Low concentrations of CO_2 (0.64%) or malic acid, thymine, and Tween 80 (compounds associated with CO_2 fixation) could replace live roots in stimulating the growth of hyphal fragments of *Cantharellus cibarius* (Straatsma & Bruinsma, 1986; Straatsma et al., 1986).

More recently, Fries et al. (1985) tested lipids in pine root exudate (collected from seedlings grown in sterile vermiculite). Two ECMF species (*Laccaria amethystina* and *L. bicolor*) that are associated with pine showed increased hyphal growth in response to the lipids, while a third species (*Leccinum aurantiacum*) that is mycorrhizal with aspen (*Populus tremuloides*) was unaffected. Three findings from this study were significant. There was an optimum concentration of exudate for maximal response, and higher concentrations were inhibitory, a phenomenon that must be considered in all exudate studies. Second, as indicated above, the effects were selective for certain fungi. Third, the increase in growth rate was not a result of the lipids functioning as the major carbon source, but rather as a trigger on some rate-limiting metabolic pathway.

It is probable that the control of the early stages of mycorrhizal formation (both VAM and ECM) will be found to be mediated by mixtures of compounds acting in concert. Evidence for this hypothesis is found in studies on the attraction of germ tubes of *Gigaspora gigantea* to host roots (see the section on "Directional Growth") and in the diversity of substances that have been found to function as the "M" factor for some ECMF. In addition, it is intuitively apparent (if not proven experimentally) that the clarity and selectivity (for mycorrhizal fungi) of the message that is sent from the roots demand that more than one compound be used.

Directional Growth

Despite the relatively low density of VAMF spores in soil (in contrast to other root-inhabiting fungi) and the nutritionally obligate nature of the symbiosis, it generally has been assumed that contact results from a sequence of passive events, mediated by chance (Mosse, 1962; Miller-Wideman & Watrud, 1984; Gianinazzi-Pearson et al., 1988). However, many, if not all, biotrophic interactions in soil involve attraction of the biotroph to the host organism. Documented instances of such chemotropism are numerous and include many fungal pathogens of roots (e.g. Zentmeyer, 1961; Gooday, 1975; Wildermuth et al., 1984), the nitrogen-fixing rhizobia (Currier & Strobel, 1976) mycophagous collembola (Bengtsson et al., 1988), roots of parasitic angiosperm (Williams, 1961; Kujit, 1969), myco-parasites (Barnett & Binder, 1973), and plant parasitic, mycophagous, and bacteriophagous nematodes (e.g., Townshend, 1964). The earlier, yet persistent view that host and fungus locate each other by chance, passive behavior simply is untenable. Chance contact cannot be ignored as a process that results in the members of the symbiosis locating each other, and it has been effective in agar cultures (e.g., Mosse, 1962; Mosse & Hepper, 1975; Miller-Wideman & Watrud, 1984). However, in the soil milieu, other mechanisms (e.g., volatile and water-soluble attractants) may be much more biologically significant than chance contact.

An unknown volatile exudate(s) from corn (*Zea mays*) and bean (*Phaseolous vulgaris*) roots was responsible for a chemotropic locational response by aerial germ tubes of *Gigaspora gigantea* (Koske, 1982). Germ tubes grew through the air over distances of up to 11 mm to contact roots grown in whole plant culture. Supportive evidence that volatile organic compounds act as messengers in the early events of VAM development (affecting directionality and/or growth rate) was provided by the work of St. John et al. (1983) who found that the extent of VAM colonization in seedlings of clover grown in sand culture with *G. margarita* was reduced by 10 times when the soil atmosphere was exposed to $KMnO_4$, a strong oxidant of organic compounds.

The chemotropic locational response of germ tubes of *G. gigantea* was investigated more intensively using a root – organ culture assay (Gemma, 1987; Gemma & Koske, 1988a). In the system, arterial germ tubes contacted roots in 80% of the tomato cultures and in 55% of corn root cultures. Roots of nonhost plant species [Kohlrabi (*Brassica oleracea* var. *caulo-rapa*), a crucifer, and beet (*Beta vulgaris*), a chenopod were less "attractive" to germ tubes. The chemotropism of germ tubes to roots of host plants was essentially eliminated in the presence of a selective inhibitor of organic volatiles ($KMnO_4$) or in the presence of KOH, which decreases the level of CO_2 and inactivates some aldehydes and ketones (Noller, 1965). It is not clear whether the locational growth response is mediated by a single compound or by a combination of volatiles.

Roots of beet were contacted by germ tubes of *G. gigantea* in 46% of the cultures (Gemma & Koske, 1988). This relatively high value for a nonhost species suggests that some of the herbaceous members of the Chenopodiaceae have retained at least some of the communication system required to form VAM. Members of the family are known to ocassionally form VAM (e.g., Allen, 1983; Miller et al., 1983), and the herbaceous *Chenopodium quinona* forms VAM when treated with the herbicide simazine (see the section on "VAM and Membrane Permeability Model"), a compound that stimulates root exudation (Schwab et al., 1982). If Trappe's (1987) hypothesis that angiosperms are evolving toward autotrophy (lack of mycorrhizae) is correct, then we should expect to find a variety of nonhost species with some (though diminished) abilities to communicate with mycorrhizal fungi, especially in the preinfection stages.

The ability of root systems of four of five varieties of corn (*Zea mays* L.) to attract germ tubes of *G. gigantea* in root–organ culture declined at higher P levels in the medium (Suriyapperuma, 1990). Interestingly, the variety with the lowest efficiency at P uptake did not show a decrease in attraction of germ tubes to its roots even at the highest P concentration tested.

Carr et al. (1985) noted that hyphae of *Glomus mosseae* grew toward cells of wheat (*Triticum aestivum*), and Mugnier and Mosse (1987) also reported a slight rhizotropic response by germ tubes of the same species. Powell (1976) found little evidence that water-soluble exudates guided germ tubes to roots, although germ tubes of *Acaulospora laevis* showed a slight directional growth response toward roots when the distance between them was 3.4 mm or less. In a study of VAM produced from spore inocula of six species of VAMF, van Nuffelen and Schenck (1984) found that germ tubes of four of the species (*G. etunicatum*, *G. intraradices*, *G. mosseae*, and *Entrophospora* sp.) occasionally made a directional change when they approached the root so that the angle of contact was 90°. Mosse and Hepper (1975) similarly reported that germ tubes of *G. mosseae* were oriented 90° to the root surface.

Hyphae of the ECMF *Piloderma croceum* were attracted to roots of *Picea abies* in response to an unidentified root metabolite (Nylund & Unestam, 1982). Yang and Wilcox (1984) also presented evidence that indicated attraction of ECMF hyphae to host roots.

Volatile root exudates could elicit directed hyphal growth by affecting the site of cell wall synthesis at the hyphal tip. This could result from the exudate increasing the permeability of the fungal plasmalemma to protons at a specific location in the apical zone (Harold et al., 1985).

Fungal Reactions after Contact with Roots

Once hyphae or germ tubes of mycorrhizal fungi have contacted a root, surface recognition factors including topography (Manocha & Chen, 1990) probably assume the major regulatory function (Harley & Smith, 1983; Smith & Gianinazzi-Pearson, 1988; Anderson, 1988; Gianinazzi-Pearson et al., 1988), although root exudates and other compounds in the apoplast may still influence some aspects of the attachment, penetration, and infection process. The molecular mechanisms by which members of the symbiosis recognize each other and permit the morphological development necessary to allow synthesis of the mycorrhiza are unknown. Although there is much conjecture on the topic, little information exists even generally characterizing the recognition mechanism (e.g., Gianinazzi et al., 1983; Harley & Smith, 1983; Tester et al., 1987; Massicotte et al., 1987; Gianinazzi-Pearson et al., 1988; Anderson, 1988; Smith & Gianinazzi-Pearson, 1988). The recognition that occurs between fungus and host in ericoid mycorrhizae appears to be mediated by interactions between extracellular carbohydrates located on the fungal wall and on the host wall (Gianinazzi-Pearson et al., 1986; Bonfante-Fasolo et al., 1987; Bonfante-Fasolo, 1988).

The importance of accurate identification by each member of the potential symbiosis and the dangers to each if an error is made argue for an elegant system of confirmatory "check points" at various phases in colonization. It is apparent that numerous plant and fungal genes are required for the appropriate compatibility reactions that must occur at each stage of development (Anderson, 1988).

The difference in the capability of aseptate and septate VAMF hyphae to initiate infection (see the section on "Hyphal Growth Rate and Branching") may result from differential expression of genes that affect the ability of the hyphal tips to exchange identification information with the host. Anderson (1988) suggests that phenolic compounds in root exudates may induce gene expression in hyphae of mycorrhizal fungi prior to their arrival at the root surface. Hyphae unable to respond to such stimulus by synthesizing the right "molecular keys" would not be recognized as potential symbionts at the host surface.

In the nonmycorrhizal species, it has been suggested that, while the fungus may get through the first checkpoints, e.g., at the epidermal surface and in the first few cortical cells), the plant soon recognizes the intruder as such, and further spread of the mycorrhizal fungus is checked (Tommerup, 1984; Tester et al., 1987; Allen et al., 1989). Such a series of checks would explain the limited development of both VAMF and ECMF that sometimes occurs on nonhost plants (Tester et al., 1987; Glenn et al., 1985; Ocampo et al., 1980; Theodorou & Bowen, 1971), the restricted VAM formation that occurs in Gentianaceae (Tester et al., 1987), and the significance of the stimulatory effect of root exudates of such nonhost species as tomato (*Lycoperscium esculentum*) and carrot (*Daucus carrota*) on ECMF (Melin, 1963; Fries & Swedjemark, 1986).

The aborted infections and limited development of VAM in nonhost species suggest that at least one of the partners of the symbiosis did not receive the appropriate chemical message and responded by halting further development. Although it is generally thought that it is the plant that responds to the VAMF by mobilizing a "set" of chemical defenses that limits the spread of the fungus (Harley & Smith, 1983; Tester et al., 1987; Allen et al., 1989), in some cases, it may be the fungus that rejects the plant. The fungus may recognize the nonmycotroph as a plant from which it cannot obtain the intermediate(s) that have made the VAMF the obligate symbionts that they are. There is evidence for this view from the experiment in which limited VAM development occurred in roots of the nonmycotrophic species *Chenopodium quinona* treated with the herbicide simazine (Schwab et al., 1982). Since simazine increases root exudation, the enhancement of VAMF colonization could be a result of the fungus being able to obtain nutrient(s) in the exudate in sufficient quantity that are not normally present. Under such conditions, the fungus may not so quickly reject this normally nonhost plant.

Hyphal Growth, Branching, and Invasion of Roots

There are a few reports indicating that the hyphae of some VAMF increase their growth rate and branching after they contact roots, but prior to the formation of arbuscules (Mosse, 1962; Mosse & Hepper, 1975; Hepper, 1981; Mugnier & Mosse, 1987). In contrast to earlier findings of infection fans formed as VAMF hyphae first approached roots (see above), fan-like growths of septate hyphae were produced by *G. mosseae* after aseptate germ tubes had become attached to the root (Mugnier & Mosse, 1987). The observation that septate VAMF hyphae cannot initiate infection (Powell, 1976; Mugnier & Mosse, 1987) suggests that they lack some feature (perhaps lectin-like recognition factors at the tips of their hyphae) that are present on the aseptate hyphae or are senescent.

Initial Stages of Mycorrhizal Formation

After fungal hyphae or germ tubes contact roots, the communication between the symbionts becomes more refined. In the earliest stages after contact, water-soluble and volatile exudates may still function as chemical messengers to the fungi, perhaps directing them to particular regions of the root where penetration is possible.

The study of VAM formation in leek (*Allium porrum*) by Brundrett et al. (1985) clearly describes many of the events. Seedlings of leek were transplanted to an established pot culture of a "nurse" plant, a mature mycorrhizal leek plant (colonized by *Glomus versiforme*). Seedlings were harvested from the pot at 2, 4, 6, and 8 days after transplanting. The experiment showed that after hyphae contacted the roots of the seedling, they grew over and adhered to the root surface and formed appressoria. The appressoria formed only in the apical 14 mm of root and were located in the depressions between adjacent epidermal cells. Penetration soon followed, and the fungus invaded the cells of the exodermis and cortex, forming inter- and intracellular hyphae. Arbuscules and vesicles were produced within 1–3 days after transplanting. Once formed, infection units spread longitudinally in the root at a rate of 0.75 mm/day. Additional details of VAM formation are found in Harley and Smith (1983), Sanders and Sheikh (1983), Garriock et al. (1989), and Brundrett and Kendrick (1990a).

Hyphae appear to select specific sites on the root surface at which to form appressoria (and penetration hyphae) in response to specific signal molecules bound to the host cell wall and to the exudation of other, more mobile compounds from within the root. Penetration of roots of VAMF occurs in those young areas near the rot tip where extensive suberization has not yet occurred (Brundrett & Kendrick, 1990a). In addition, in those plant species in which the exodermis is dimorphic (composed of both long and short cells), it is only through the short cells that penetration occurs. Long cells become suberized earlier than do the short cells, and hyphae do not enter them (Hussey, 1982; Brundrett & Kendrick, 1990b). Formation of suberin lamellae in the exodermis may exclude VAMF infection not only by the physical/chemical resistance of the wall itself, but also because the nature of exudation (both water-soluble and volatile exudates) will be dramatically altered following complete suberization (Esau, 1965). In the absence of suberin lamellae in exodermal cells, leakage of compounds from the root *in concert* with surface recognition factors (both topography and surface-bound identifier molecules) may stimulate the preinfection morphogenesis by hyphae located directly over the appropriate penetration site.

Brundrett and Kendrick (1990b) suggested that the precursors of suberin (that are present even in cells not yet suberized, such as the short cells in the exodermis) may be used by VAMF to locate regions of the root that are susceptible to penetration. This agrees with Anderson's (1988) hypothesis that phenolics (which include the precursors of suberin) in root exudates affect fungi prior to penetration. Further, synthesis of suberin in the exodermis appear to be stimulated by hyphae of mycorrhizal fungi (Brundrett & Kendrick, 1990b). The enhanced production (and leakage) of suberin precursors after initial contact between roots and VAMF hyphae could thus lead to increased colonization of the root once penetration has occurred, explaining the observation of this very event by Mosse and Hepper (1976) (see the section on "Growth of Germ Tubes and Hyphae").

Two mechanisms are known by which VAMF may increase their ability to respond to roots of host plants. In both phenomena, the fungi induce young (colonizable) roots to approach the spores or hyphae. The first mechanism involves the stimulation of root branching after the roots have been contacted by germ tubes and/or development of the primary infection has occurred (Mosse, 1962; Gemma & Koske, 1988a; Berta et al., 1990). This stimulation in rooting may result from plant hormones produced by the fungus (Barea & Azcon-Aguilar, 1982), or induced by the fungus (Allen et al.; 1980, 1982). The latter may be linked to surface-bound recognition factors on the host cell wall (Gemma & Koske, 1988a) or to transfer of a plasmid from fungus to host. Whatever the cause, the increased branching of roots provides more exuding "target" roots to which the fungus may respond. The production of plant hormones by ECMF that induce root branching may similarly enhance mycorrhiza formation (Gay, 1988). Interestingly, the precursors of the IAA are supplied by the host.

The second mechanism that brings the partners in VAM closer together is the attraction of root tips to spores whose germ tubes have not successfully located roots. When spores of *Gigaspora gigantea* were separated from roots of tomato in root organ culture, lateral roots were attracted to spores by a volatile organic compound(s) released from the spores (Gemma & Koske, 1988a). This phenomenon is remarkably similar to the mating behavior that occurs in the Chytridiomycete, *Allomyces macrogynus* (Pommerville, 1977) and, in fact, may offer some clues to the process in VAM (see the section on "Types of Compound, Functioning as Chemical Messengers"). In *A. macrogynus*, male gametes are rapidly attracted to the larger, less mobile female gametes by the sex hormone sirenin (a terpenoid). Males gametes release a substance that attracts female gametes, although the time required for this effect to take effect is much longer.

Formation of ECM is more complex than is VAM. After contacting host roots, ECMF may undergo extensive branching and form a hyphal envelope or mantle around the portion of the root that is contacted. Thigmotropic reactions may be important at this stage (Thompson et al., 1989), although it is difficult to separate the effects from those elicited by water-soluble exudates occurring on the surface of the epidermal cells of the root. Concentrations of volatile compounds (including oxygen) (Read & Armstrong, 1972) and surface bound recognition factors also affect ECM formation. Bowen (1987) isolated a variety of yeasts, bacteria, and actinomycetes from the mantle of ECM formed between *Rhizopogon luteolus* and *Pinus radiata*. Nearly all of the microorganisms stimulated the growth of the fungal symbiont and formation of the mycorrhiza.

After the envelope or mantle has formed, hyphae then penetrate into the cortex, forcing between cells or entering through larger, natural openings (Warrington et al., 1981). Once inside the cortex, hyphae are stimulated to branch, forming the Hartig net (Nylund & Unestam, 1982). Details of entry and ECM formation vary, and the reader is referred to the works of Harley and Smith (1983), Peterson and colleagues (Piche et al., 1983; Massicotte & Peterson, 1987; Massicotte et al., 1987; Melville et al., 1987) and others (Brown & Sinclair, 1981; Nylund & Unestam, 1982; Kottke & Oberwinkler, 1986).

VAM and Membrane Permeability Model

Over the past 10 years, several lines of investigation were undertaken toward providing an understanding of the relationship between water-soluble exudates emanating from roots and the growth and colonizing vigor of VAMF. In the earliest studies it was generally noted that VAM development (percent of root system colonized, abundance of entry points) was highly correlated with the amount of exudates (reducing sugars and amino acids), and it appeared that some of these compounds served as carbon sources for the fungi (Menge et al., 1978; Ratnayake et al., 1978; Graham et al., 1981; Ferguson & Menge, 1982; Dixon et al., 1988, Mosse, 1988). High levels of exudation were correlated with extensive fungal development in the rhizosphere and rhizoplane, later accompanied by high levels of infection. Further, the amount of exudation appeared to be regulated by the leakiness of plasmalemmas of root cortical cells, which in turn was determined by levels of cellular P. The amount of root P was mediated by VAMF (through their connection to P reserves in soil). These interactions were incorporated into a model by Graham et al. (1981). Under P-deficient conditions, roots become leaky, resulting in high exudation that stimulates the growth of VAMF and subsequent colonization. When mycorrhizal development (both inside and outside the root) is great enough to provide P-sufficiency to roots, membrane leakiness is decreased, and further colonization is halted. In the elegant model, the plant was thought to control the early stages of mycorrhizal formation via the water-soluble exudates. However, as noted in the section on "Directional Growth," the response of VAMF to roots also may be regulated via the exudation of volatile exudates whose release is affected by the P concentration of the roots (Suriyapperuma, 1990).

Further evidence for this model (as it involved water-soluble exudates) was provided by Schwab et al. (1983) who found that under P stress, nonmycorrhizal roots of Sudan grass (*Sorghum vulgare*) released greater quantities of water-soluble compounds, including glycine and lysine, than did P-sufficient plants. Interestingly, Hepper and Jakobsen (1983) found that the growth of germ tubes of *Glomus caledonium* was greatly stimulated by exposure to these same two amino acids and to cysteine.

When plants of the nonhost species *Chenopodium quinona* were treated with the herbicide simazine, water-soluble exudation increased and limited VAM formation also occurred (Schwab et al., 1982). An increase in ECM formation in pine also has been attributed to treatment with the same herbicide (Smith & Ferry, 1979).

Evidence for the membrane leakiness model has been contradictory. Unexpectedly, P-deficient plants were not found to always be leakier than P-sufficient ones (Schwab et al., 1983; Elias & Safir, 1987; Thomson et al., 1986), nonhost species were sometimes leakier than host species (Schwab et al., 1984; Azcon & Ocampo, 1984), and some mycorrhizal roots exuded more than nonmycorrhizal roots (Laheute & Berthelin, 1986). Researchers in Australia (Jasper et al., 1979; Same et al., 1983; Thomson et al., 1986) showed that the increased exudation of soluble carbohydrates from roots in response to low P resulted from higher concentration of carbohydrates in roots rather than from changed membrane permeability. Research by Elias and Safir (1987) with *Trifolium repens* suggested that qualitative rather than quantitative differences between the water soluble exudates of P-deficient and P-sufficient roots were responsible for the differences in effect of the exudates on VAMF. An earlier study by Schwab et al. (1984), however, failed to detect qualitative differences between host and nonhost root water-soluble exudates. In fact, qualitative and quantitative differences in exudation may be of relatively minor significance in controlling the amount of VAMF infection in roots. The higher number of aborted appressoria formed at high P concentrations (Amijee et al., 1989) suggests that regulation of entry occurs well after contact is made between the two potential partners. Root exudation, however, may regulate the rate of spread of the fungus over the root surface.

A partial explanation for the controversy may be in the emphasis on water exudates. If membrane leakiness is a controlling factor, both volatile and water-soluble exudation would be expected to be affected. Assay techniques that permitted variable amounts of the volatile exudates to escape would be expected to give inconsistent results, especially between laboratories. The findings of Becard and Piche (1989b) (see the section on "Hyphal Growth Rate and Branching") on the synergistic interaction between volatile and water-soluble exudates as they affect hyphal growth illustrate the importance of controlling both kinds of exudation when investigating early stages on the symbiosis.

Types of Compounds Functioning as Chemical Messengers

It now appears that formation of mycorrhizae is regulated by water-soluble and volatile exudates and by surface-bound recognition molecules. The involvement of water-insoluble root products and nondiffusible substances in the root is unknown. The general sequence of messages sent to fungi by both VAM and ECM hosts may be remarkably similar, although the specific compounds may differ greatly.

Such a sequence can be formulated by arranging the messenger molecules in order of their capacity to move through the soil (Figs. 2.1 and 2.2). Thus, volatile

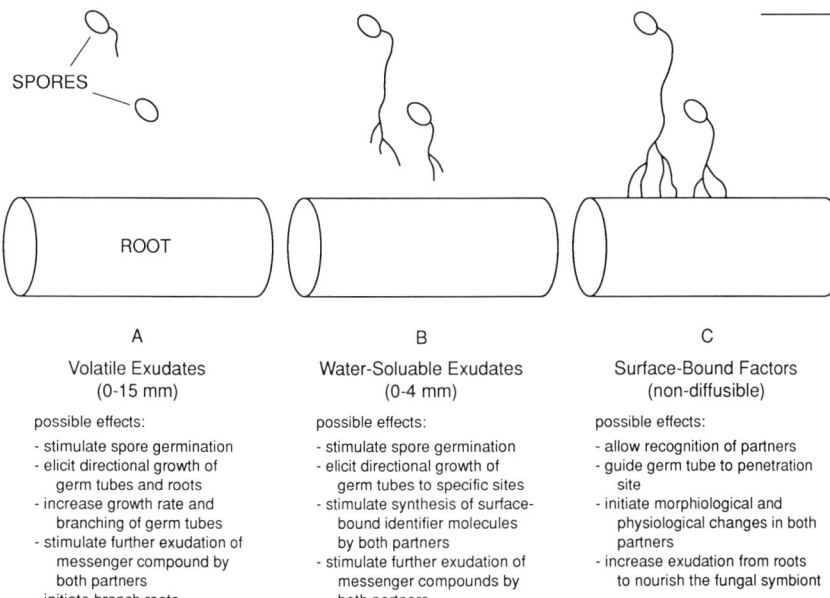

Figure 1.1. Possible communications between mycorrhizal fungi and plant roots prior to formation of mycorrhizae. Spores are shown at different distances from the root surface. Possible effects of volatile compounds (A), of both water-soluble and volatile exudates (B), and of water-soluble and volatile exudates and surface-bound messengers (C) are illustrated.

exudates from the hosts are likely to be involved in stimulation of germination (ECMF), germ tube growth, and directionality. Volatile exudates from the fungus may induce branching or directional growth by the roots and perhaps otherwise alert the root to the presence of a potential symbiont. Although not yet demonstrated, it is likely that synergistic effects between the partners, mediated through their exudates (similar to the manner in which sexual hormones regulate mating in the fungi), will be discovered. In such an interaction, the exudations of one partner stimulate the exudation of stimulatory molecules by the other partner in an ever-increasing cycle of exchange (Gemma & Koske, 1988a).

Indeed, it may be found that the attraction of mycorrhizal fungi to roots (and of roots to the fungi) is mediated by compounds that are analogs of fungal sex hormones. Volatile terpenoids (precursors of trisporic acid) are exchanged between mating strains in members of the Mucorales, and these messenger compounds elicit differentiation, directional growth, and other physiological changes (Ende, 1978).

Water-soluble exudates may function similarly to volatile exudates, but over shorter distances. These substances from the root may lead the hyphae to appropriate locations on the root surface for penetration. In addition, water-soluble

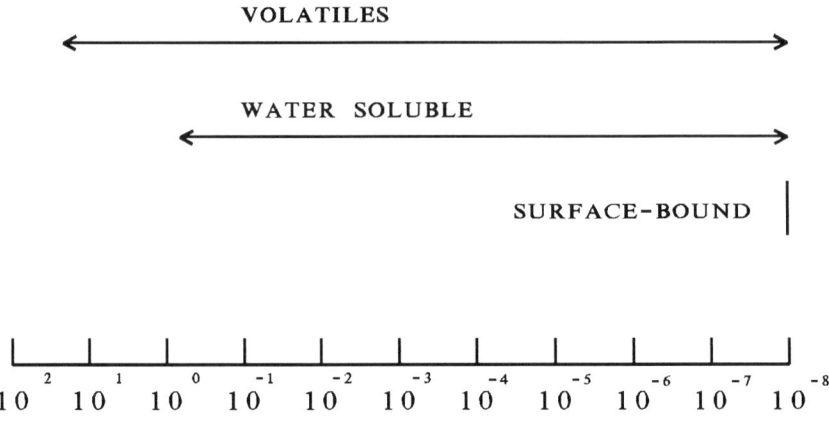

Figure 1.2. Effective distances that different kinds of root exudates and surface-bound molecules can function in preinfection communication with mycorrhizal fungi in a soil environment.

exudates may provide the mycorrhizal fungi with energy-containing compounds or essential nutrients to stimulate branching and allow the fungus to colonize the plant. Water-soluble exudates from the fungus may stimulate the production of signal molecules on the root surface or stimulate further exudation from the roots.

Surface bound recognition molecules both on the fungus and on the host are likely to be important when the fungi encounter plant cell surfaces, including the middle lamella. Different phases of the colonization may be expected to be regulated by different groups of compounds.

Nylund and Unestam (1982) proposed a similar sequence of events based on the release of exudates from roots that lead to formation of ECM. Regulation of development by the plant was proposed to occur at several stages: stimulation of hyphal growth and chemotropism, aggregation of hyphae around the root, penetration, and formation of the Hartig net and mantle. Although the fungal symbiont was recognized to have some effect on the colonization process, Nylund and Unestam (1982) argue that it is the host plant that retains overall control of development of the mycorrhiza. Researchers of VAM also will find this article of interest.

Through their long evolutionary history, plants and mycorrhizal fungi have acquired highly effective chemical communication capabilities whose characterization and identification will challenge mycorrhiza scientists for many years.

References

Allen, M. F. (1983). Formation of arbuscular mycorrhizae in *Atriplex gardneri*, Chenopodiaceae: Seasonal response in a cold desrt. *Mycologia* **75,** 773–776.

Allen, M. F., Moore, Y. S., Christensen, M., & Stanton, N. (1979). Growth of vesicular-arbuscular mycorrhizal and nonmycorrhizal *Bouteloua gracilis* in a defined medium. *Mycologia* **71,** 666–669.

Allen, M. F., Moore, T. S., & Christensen, M. (1980). Phytohormone changes in *Bouteloua gracilis* infected by vesicular-arbuscular mycorrhizae: I. Cytokinin increases in the host plant. *Canadian Journal of Botany* **58,** 371 – 374.

Allen, M. F., Moore, T. S., & Christensen, M. (1982). Phytohormone changes in *Bouteloua gracilis* infected by vesicular-arbuscular mycorrhizae. II. Altered levels of gibberelin-like substances and abscisic acid in the host plant. *Canadian Journal of Botany* **60,** 468–471.

Allen, M. F., Allen, E. B., & Friese, C. F. (1989). Response of the non-mycotrophic plant *Salsola kali* to invasion by vesicular-arbuscular mycorrhizal fungi. *New Phytologist* **111,** 45–49.

Amijee, F., Tinker, P. B., & Stribley, D. P. (1989). The development of endomycorrhizal root systems VII. A detailed study of effects if soil phosphorus on collonization. *New Phytologist* **111,** 435–446.

Anderson, A. J. (1988). Mycorrhizae—host specificity and recognition. *Phytopathology* **78,** 375–378.

Azcon, R. (1987). Germination and hyphal growth of *Glomus mosseae* in vitro: Effects of rhizosphere bacteria and cell free culture media. *Soil Biology and Biochemistry* **19,** 417–419.

Azcon, R. (1989). Selective interaction between free-living rhizosphere bacteria and vesicular-arbuscular mycorrhizal fungi. *Soil Biology and Biochemistry* **21,** 639–644.

Azcon, R., & Ocampo, J. A. (1984). Effect of root exudation on VA mycorrhizal infection at early stages of plant growth. *Plant and Soil* **82,** 133–138.

Azcon-Aguilar, C., Diaz-Rodriguez, R., & Barea, J. M. (1986a). Effect of soil microorganisms on spore germination and growth of the vesicular-arbuscular mycorrhizal fungus *Glomus mosseae*. *Transactions of the British Mycological Society* **86,** 337–340.

Azcon-Aguilar, C., Diaz-Rodriguez, R., & Barea, J. M. (1986b). Effect of free-living fungi on the germination of *G. mosseae* on soil extract. In: *Physiological and Genetical Aspects of Mycorrhizae* (Ed. by V. Gianinazzi-Pearson & S. Gianinazzi), pp. 515–519. INRA, Paris.

Barber, D. A., & Gunn, K. B. (1974). The effect of mechanical forces on the exudation of organic substances by the roots of cereal plants grown under sterile conditions. *New Phytologist* **73,** 39–45.

Barea, J. M. (1986). Importance of hormones and root exudates in mycorrhizal phenomena. In: *Physiological and Genetical Aspects of Mycorrhizae* (Ed. by V. Gianinazzi-Pearson & S. Gianinazzi), pp. 177–187. INRA, Paris.

Barea, J. M., & Aczon-Aguilar, C. (1982). Production of plant growth-regulating substances by the vesicular-arbuscular mycorrhizal fungus *Glomus mosseae*. *Applied and Environmental Microbiology* **43,** 810–813.

Barnett, H. L., & Binder, F. L. (1973). The fungal host-parasite relationship. *Annual Review of Phytopathology* **11,** 273–292.

Becard, G., & Fortin, J. A. (1988). Early events of vesicular-arbuscular mycorrhiza formation on Ri T-DNA transformed roots. *New Phytologist* **108**, 211–218.

Becard, G., & Piche, Y. (1989a). New aspects on the acquisition of biotrophic status by a vesicular-arbuscular mycorrhizal fungus, *Gigaspora margarita*. *New Phytologist* **112**, 77–83.

Becard, G., & Piche, Y. (1989b). Fungal growth stimulation by CO_2 and root exudates in vesicular-arbuscular mycorrhizal symbiosis. *Applied and Environmental Microbiology* **55**, 2320–2325.

Bengtsson, G., Erlandsson, A., & Rundgren, S. (1988). Fungal odor attracts soil collembola. *Soil Biology and Biochemistry* **20**, 25–30.

Berta, G., Fusconi, A., Trotta, A., & Scannerini, S. (1990). Morphogenetic modifications induced by the mycorrhizal fungus *Glomus* strain E3 in the root system of *Allium porrum* L. *New Phytologist* **114**, 207–215.

Birraux, D., & Fries, N. (1981). Germination of *Thelephora terrestris* basidiospores. *Canadian Journal of Botany* **59**, 2062–2064.

Bjurman, J. (1984). An organic acid, inhibitory to spore germination of mycorrhizal fungi, formed from agar during autoclaving. *Microbios* **39**, 109–116.

Bonfante-Fasolo, P. (1988). The role of the cell wall as a signal in mycorrhizal associations. In: *Cell to Cell Signals in Plant, Animal and Microbial Symbiosis* (Ed. by S. Scannerini, D. Smith, P. Bonfante-Fasolo, & V. Gianinazzi-Pearson), pp. 219–235. Springer-Verlag, Berlin.

Bonfante-Fasolo, P., Perotto, S., Testa, B., & Faccio, A. (1987). Ultrastructural localization of cell surface sugar residues in ericoid mycorrhizal fungi by gold labeled lectins. *Protoplasma* **139**, 25–35.

Boudarga, K., & Dexheimer, J. (1990). A simple method for the culture and multiplication of VA mycorrhizal fungus. *Agronomie* **10**, 417–422.

Bowen, G. D. (1987). The biology and physiology of infection and its development. In: *Ecophysiology of VA Plants* (Ed. by G. R. Safir), pp. 27–58. CRC Press, Boca Raton, FL.

Bowen, G. D., & Theodorou, C. (1973). Growth of ectomycorrhizal fungi around seeds and roots. In: *Ectomycorrhizae* (Ed. by G. C. Marks & T. T. Kozlowski), pp. 107–150. Academic Press, New York.

Brown, A. C., & Sinclair, W. A. (1981). Colonization and infection of primary roots of Douglas-fir seedlings by the ectomycorrhizal fungus *Laccaria laccata*. *Forest Science* **27**, 111–124.

Brundrett, M., & Kendrick, B. (1990a). The roots and mycorrhizas of herbaceous woodland plants. I. Quantitative aspects of morphology. *New Phytologist* **114**, 457–468.

Brundrett, M., & Kendrick, B. (1990b). The roots and mycorrhizas of herbaceous woodland plants. II. Structural aspects of morphology. *New Phytologist* **114**, 469–479.

Brundrett, M. C., Piche, Y., & Peterson, R. L. (1985). A developmental study of the early stages in vesicular-arbuscular mycorrhiza formation. *Canadian Journal of Botany* **63**, 184–194.

Bushnell, W. R., & Rowell, J. B. (1981). Suppressors of defense reactions: A model for roles in specificity. *Phytopthology* **71**, 1012–1014.

Carr, G. R., Hinkley, M. A., LeTacon, F., Hepper, C. M., Jones, M. G. K., & Thomas, E. (1985). Improved growth of two species of vesicular-arbuscular mycorrhizal fungi in the presence of suspension-cultured plant cells. *New Phytologist* **101**, 417–426.

Chilvers, G. A., Douglass, P. A., & Lapeyrie, F. F. (1986). A paper sandwich technique for rapid synthesis of ectomycorrhizas. *New Phytologist* **103**, 397–402.

Curl, E. A., & Truelove, B. (1986). The *Rhizosphere*. Springer-Verlag, Berlin.

Currier, W. W., & Strobel, G. A. (1976). Chemotaxis of *Rhizobium* spp. to plant root exudates. *Plant Physiology* **59**, 820–823.

Daniels, B. A., & Graham, S. O. (1976). Effects of nutrition and soil extracts on germination of *Glomus mosseae* spores. *Mycologia* **68**, 108–116.

Daniels, B. A., & Trappe, J. M. (1980). Factors affecting spore germination of the vesicular-arbuscular mycorrhizal fungus, *Glomus epigaeus*. *Mycologia* **72**, 457–471.

Dixon, R. K., Garrett, H. E., & Cox, G. S. (1988). Carbohydrate relationships of *Citrus jambhiri* inoculated with *Glomus fasciculatum*. *Journal of the American Society of Horticultural Science* **113**, 239–242.

El-Atrach, F., Vierheilig, H., & Ocampo, J. A. (1989). Influence of non-host plants on vesicular-arbuscular mycorrhizal infection of host plants and on spore germination. *Soil Biology and Biochemistry* **21**, 161–163.

Elias, K. S., & Safir, G. R. (1987). Hyphal elongation of *Glomus fasciculatus* in response to root exudates. *Applied and Environmental Microbiology* **53**, 1928–1933.

Ende, H. van den. (1978). Sexual morphogenesis in the Phycomycetes. In: *The Filamentous Fungi*, Vol. III (Ed. by J. E. Smith & D. R. Berry), pp. 257–274. Wiley, New York.

Esau, K. 1965. *Plant Anatomy*, 2nd ed. Wiley, New York.

Ferguson, J. J., & Menge, J. A. (1982). The influence of light intensity and artificially extended photoperiod upon infection and sporulation of *Glomus fasciculatus* on sudan grass and on root exudation of sudan grass. *New Phytologist* **92**, 183–191.

Finlay, R. D., & Read, D. J. (1986). The structure and function of the vegetative mycelium of ectomycorrhizal plants. I. Translocation of C-labelled carbon between plants interconnected by a common mycelium. *New Phytologist* **103**, 143–156.

Fortin, J. A., Piche, Y., & Godbout, C. (1983). Methods for synthesizing ectomycorrhizas and their effect on mycorrhizal development. *Plant and Soil* **71**, 275–284.

French, R. C. (1985). The bioregulatory action of flavor compounds on fungal spores and other propagules. *Annual Review of Phytopathology* **23**, 173–199.

Fries, N. (1941). Uber die sporenkeimung bei einigen Gasteromyceten und mykorrhizabildenden Hymenomyceten. *Archivs für Mikrobiologie* **12**, 266–284.

Fries, N. (1968). Chemical factors in the germination of spores of Basidiomycetes. In: *The Fungus Spore* (Ed. by M. F. Madelin), pp. 189–200. Butterworths, London.

Fries, N. (1978). Basidiospore germination in some mycorrhiza-forming hymenomycetes. *Transactions of the British Mycological Society* **70**, 319–324.

Fries, N. (1979). The taxon-specific spore germination reaction in *Leccinum*. *Transactions of the British Mycological Society* **73**, 337–341.

Fries, N. (1981). Recognition reactions between basidiospores and hyphae in *Leccinum*. *Transactions of the British Mycological Society* **77**, 9–14.

Fries, N., & Swedjemark, G. (1986). Specific effects of tree roots on spore germination in the ectomycorrhizal fungus *Hebeloma mesophaeum* (Agaricales). In: *Physiological and Genetical Aspects of Mycorrhizae* (Ed. by V. Gianinazzi-Pearson & S. Gianinazzi), pp. 725–730. INRA, Paris.

Fries, N., Bardet, M., & Serck-Hanssen, K. (1985). Growth of ectomycorrhizal fungi stimulated by lipids from a pine root exudate. *Plant and Soil* **86**, 287–290.

Fries, N., Serck-Hanssen, K., Dimberg, L. H., & Theander, O. (1987). Abietic acid, an activator of basidiospore germination in ectomycorrhizal species of the genus *Suillus* (Boletaceae). *Experimental Mycology* **11**, 360–363.

Garriock, M. L., Peterson, R. L., & Ackerley, C. A. (1989). Early stages in colonization of *Allium porrum* (leek) roots the vesicular-arbuscular mycorrhizal fungus, *Glomus versiforme*. *New Phytologist* **112**, 85–92.

Gay, G. (1988). Role des hormones fongigues dan l'association ectomycorhizienne. *Cryptogamie, Mycologie* **9**, 211–219.

Gemma, J. N. (1987). Physiology and ecology of vesicular-arbuscular mycorrhizal fungi in sand dunes. Ph.D. Thesis, University of Rhode Island, Kingston.

Gemma, J. N., & Koske, R. E. (1988a). Pre-infection interactions between roots and the mycorrhizal fungus *Gigaspora gigantea*: chemotropism of germ tubes and root growth response. *Transactions of the British Mycological Society* **91**, 123–135.

Gemma, J. N., & Koske, R. E. (1988b). Seasonal variation in spore abundance and dormancy of *Gigaspora gigantea* and in mycorrhizal inoculum potential of a dune soil. *Mycologia* **80**, 211–216.

Gerdemann, J. W. (1968). Vesicular-arbuscular mycorrhiza and plant growth. *Annual Review of Phytopathology* **6**, 397–418.

Gianinazzi, S., Gianinazzi-Pearson, V., & Marx, C. (1983). Role of the host-arbuscule interface in the VA mycorrhizal symbiosis: Ultracytological studies of processes involved in phosphate and carbohydrate exchange. *Plant and Soil* **71**, 211–215.

Gianinazzi-Pearson, V., Bonfante-Fasolo, P., Dexheimer, J. (1986). Ultrastructural studies of surface interactions during adhesion and infection by ericoid endomycorrhizal fungi. In: *Recognition in Microbe-Plant Symbiotic and Pathogenic Interactions* (Ed. by B. Lugtenberug), pp. 273–281. Springer-Verlag, New York.

Gianinazzi-Pearson, V., Gianinazzi, S., Dexheimer, J., Morandi, D., Trouvelot, A., & Dumas, E. (1988). Recherche sur les mecanismes intervenant dan les interactions symbiotiques plants-champignons endomycorhizogenes VA. *Crytogamie, Mycologie* **9**, 201–209.

Glenn, M. G., Chew, F. S., & Williams, P. H. (1985). Hyphal penetration of *Brassica* (Cruciferae) roots by a vesicular-arbuscular mycorrhizal fungus. *New Phytologist* **99**, 463–472.

Glenn, M. G., Chew, F. S., & Williams, P. H. (1988). Influence of glucosinolate content of *Brassica* (Cruciferae) roots on growth of vesicular-arbuscular mycorrhizal fungi. *New Phytologist* **110**, 217–225.

Gooday, G. W. (1975). Chemotaxis and chemotropism in fungi and algae. In: *Primitive sensory and Communication Systems* (Ed. by M. J. Carlile), pp. 155–204. Academic Press, New York.

Graham, J. H. (1982). Effects of citrus root exudates on germination of chlamydospores of the vesicular-arbuscular mycorrhizal fungus, *Glomus epigaeum*. *Mycologia* **74**, 831–835.

Graham, J. H., Leonard, R. T., & Menge, J. A. (1981). Membrane-mediated decrease in root exudation responsible for phosphorus inhibition of vesicular-arbuscular mycorrhiza formation. *Plant Physiology* **68**, 548–552.

Harley, J. L., & Smith, S. E. (1983). *Mycorrhizal Symbiosis*. Academic Press, London.

Harold, F. M., Kropf, D. Y., & Caldwell, J. H. (1985). Why do fungi drive electric currents through themselves? *Experimental Mycology* **9**, 183–186.

Hepper, C. M. (1981). Techniques for studying the infection of plants by vesicular-arbuscular mycorrhizal fungi under axenic conditions. *New Phytologist* **88**, 641–647.

Hepper, C. M., & Jakobsen, I. (1983). Hyphal growth from spores of the mycorrhizal fungus *Glomus caledonius*: Effect of amino acids. *Soil Biology and Biochemistry* **15**, 55–58.

Hepper, C. M., & Smith, G. A. (1976). Observations on germination of *Endogone* spores. *Transactions of the British Mycological Society* **66**, 189–194.

Hetrick, B. A. D., & Wilson, G. T. (1989). Suppression of mycorrhizal fungus spore germination in non-sterile soil: relationship to mycorrhizal growth response in big bluestem. *Mycologia* **81**, 382–390.

Hussey, R. B. (1982). Interactions between V-A mycorrhizal fungi and *Asparagus officinalis* L. roots. M.S. Thesis, University of Guelph, Ontario.

Jasper, D. A., Robson, A. D., & Abbott, L. K. (1979). Phosphorus and the formation of vesicular-arbuscular mycorrhizas. *Soil Biology and Biochemistry* **11**, 501–505.

Khan, A. G. (1974). The occurrence of mycorrhizas in halophytes and xerophytes, and of *Endogone* spores in adjacent soils. *Journal of General Microbiology* **81**, 7–14.

Koske, R. E. (1981a). *Gigaspora gigantea*: Observations on spore germination of a VA mycorrhizal fungus. *Mycologia* **73**, 288–300.

Koske, R. E. (1981b). Multiple germination of spores *Gigaspora gigantea*. *Transactions of the British Mycological Society* **76**, 328–330.

Koske, R. E. (1982). Evidence for a volatile attractant from plant roots affecting germ tubes of a VA mycorrhizal fungus. *Transactions of the British Mycological Society* **79**, 305–310.

Kottke, U., & Oberwinkler, F. (1986). Root-fungus interactions observed on initial stages of mantle formation and Hartig net establishment in mycorrhizas of *Amanita muscaria* on *Picea abies* in pure culture. *Canadian Journal of Botany* **64**, 2248–2254.

Krupa, S., & Fries, N. (1971). Studies on ectomycorrhizae of pine. I. Production of volatile organic compounds. *Canadian Journal of Botany* **49,** 1425–1431.

Kujit, J. (1969). *The Biology of Parasitic Flowering Plants.* University of California Press, Berkeley.

Laheurte F., & Berthelin, J. (1986). Influence of endomycorrhizal infection by *Glomus mosseae* on root exudation by maize. In: *Physiological and Genetical Aspects of Mycorrhizae* (Ed. by V. Gianinazzi-Pearson & Gianinazzi), pp. 425–429. INRA. Paris.

LeTacon, F., Skinner, F. A., & Mosse, B. (1983). Spore germination and hyphal growth of a vesicular-arbuscular mycorrhizal fungus, *Glomus mosseae* (Gerdemann and Trappe), under decreased oxygen and increased carbon dioxide concentrations. *Canadian Journal of Microbiology* **29,** 1280–1285.

Manocha, M. S., & Chen, Y. (1990). Specificity of attachment of fungal parasites to their hosts. *Canadian Journal of Microbiology* **36,** 69–76.

Massicotte, H. B., & Peterson, R. L. (1987). Ontogeny of *Eucalyptus pilularis-Pisolithus tinctorius* ectomycorrhizae. I. Light microscopy and scanning electron microscopy. *Canadian Journal of Botany* **65,** 1927–1939.

Massicotte, H. B., Ackerley, C. A., & Peterson, R. L. (1987). Localization of three sugar residues in the interface of ectomycorrhizae synthesized between *Alnus crispa* and *Alpova diplophloeus* as demonstrated by lectin binding. *Canadian Journal of Botany* **65,** 1127–1132.

Mayo, K., Davis, R. E., & Motta, J. (1986). Stimulation of germination of spores of *Glomus versiforme* by spore-associated bacteria. *Mycologia* **78,** 426–431.

Melin, E. (1963). Some effects of forest tree roots on mycorrhizal basidiomycetes. In: *Symbiotic Associations, Proceedings of the 13th Symposium of the Society of General Microbiology* (Ed. by P. S. Nutman & B. Mosse), pp. 124–145. Cambridge University Press, Cambridge.

Melville, L. H., Massicotte, H. B., & Peterson, R. L. (1987). Ontogeny of early stages of ectomycorrhizae synthesized between *Dryas intergrifolia* and *Heboloma cylindrosporum*. *Botanical Gazette* **148,** 332–341.

Menge, J. A., Stierle, A. D., Bagyaraj, D. L., Johnson, E. L. V., & Leonard, R. T. (1978). Phosphorus concentrations in plants responsible for inhibition of mycorrhizal infection. *New Phytologist* **80,** 575–578.

Miller, R. M., Moorman, T. B., & Schmidt, S. K. (1983). Interspecific plant association effects on vesicular-arbuscular mycorrhiza occurrence in *Atriplex confertifolia*. *New Phytologist* **95,** 241–246.

Miller-Wideman, M. A., & Watrud, L. S. (1984). Sporulation of *Gigaspora margarita* on root cultures of tomato. *Canadian Journal of Microbiology* **30,** 642–646.

Mosse, B. (1959). The regular germination of resting spores and some observations on the growth requirements of an *Endogone* sp. causing vesicular-arbuscular mycorrhiza. *Transactions of the British Mycological Society* **42,** 273–286.

Mosse, B. (1962). The establishment of vesicular-arbuscular mycorrhiza under aseptic conditions. *Journal of General Microbiology* **27,** 509–520.

Mosse, B. (1988). Some studies relating to "independent" growth of vesicular-arbuscular endophytes. *Canadian Journal of Botany* **66**, 2533–2540.

Mosse, B., & Hepper, C. M. (1975). Vesicular-arbuscular mycorrhizal infections in root organ cultures. *Physiological Plant Pathology* **5**, 215–223.

Mosse, B., & Thompson, J. P. (1984). Vesicular-arbuscular endomycorrhizal inoculum production. I. Exploratory experiments with beans (*Phaseolus vulgaris*) in nutrient flow culture. *Canadian Journal of Botany* **62**, 1523–1530.

Mugnier, J., & Mosse, B. (1987). Vesicular-arbuscular mycorrhizal infection in transformed root-inducing T-DNA roots grown axenically. *Phytopathology* **77**, 1045–1050.

Netzly, D. H., Riopel, J. L., Ejeta, G., & Butler, L. G. (1988). Germination stimulants of witchweed (*Striga asiatica*) from hydrophobic root exudate of Sorghum (*Sorghum bicolor*). *Weed Science* **36**, 441–446.

Newman, E. I., & Watson, A. (1977). Microbial abundance in the rhizosphere: a computer model. *Plant and Soil* **48**, 17–56.

Nicolson, T. H. (1975). Evolution of vesicular-arbuscular mycorrhizas. In: *Endomycorrhizas* (Ed. by F. E. Sanders, B. Mosse, & P. B. Tinker), pp. 25–34. Academic Press, London.

Noller, C. R. (1965). *Chemistry of Organic Compounds*. W. B. Saunders, Philadelphia.

Nylund, J. E., & Unestam, T. (1982). Structure and physiology of ectomycorrhizae. I. The process of mycorrhiza formation in Norway spruce in vitro. *New Phytologist* **91**, 63–79.

Ocampo, J. A., Martin, J., & Hayman, D. S. (1980). Influence of plant interactions on vesicular-arbuscular mycorrhizal infections. I. Host and non-host plants grown together. *New Phytologist* **84**, 27–35.

Papavizas, G. C., & Kovaks, M. F. (1972). Stimulation of spore germination of *Thielaviopsis basicola* by fatty acids from rhizosphere soil. *Phytopathology* **63**, 688–694.

Piche, Y., Peterson, R. L., Howarth, M. J., & Fortin, J. A. (1983). A structural study of the interaction between the ectomycorrhizal fungus *Pisolithus tinctorius* and *Pinus strobus* roots. *Canadian Journal of Botany* **61**, 1185–1193.

Pirozynski, K. A., & Dalpe, Y. (1989). Geological history of the Glomaceae with particular reference to mycorrhizal symbiosis. *Symbiosis* **7**, 1–36.

Pirozynski, K. A., & Malloch, D. W. (1975). The origin of land plants: a matter of mycotrophism. *BioSystems* **6**, 153–164.

Pommerville, J. (1977). Chemotaxis of *Allomyces* gametes. *Experimental Cell Research* **109**, 43–51.

Powell, C. L. (1976). Development of mycorrhizal infections from *Endogone* spores and infected root segments. *Transactions of the British Mycological Society* **66**, 439–444.

Powell, R. A., & Adams, R. P. (1973). Seasonal variation in the volatile terpenoids of *Juniperus scopulorum* (Cupressaceae). *American Journal of Botany* **60**, 1041–1050.

Ratnayake, M., Leonard, R. T., & Menge, J. A. (1978). Root exudation in relation to supply of phosphorus and its possible relevance to mycorrhizal formation. *New Phytologist* **81**, 543–552.

Read, D. J., & Armstrong, W. (1972). A relationship between oxygen transport and the formation of the ectotrophic mycorrhizal sheath in conifer seedlings. *New Phytologist* **71**, 49–53.

Read, D. J., Koucheki, H. K., & Hodgson, J. (1976). Vesicular-arbuscular mycorrhiza in natural vegetation systems. I. The occurrence of infection. *New Phytologist* **77**, 641–653.

Rovira, A. D. (1969). Plant root exudates. *Botanical Reviews* **35**, 35–58.

Rovira, A. D., & Davey, C. B. (1974). Biology of the rhizosphere. In: *The Plant Root and its Environment* (Ed. by E. W. Carson), pp. 153–204. University Press of Virginia, Charlottesville.

Rush, C. M., Upchurch, D. R., & Gerik, T. J. (1984). In situ observations of *Phymatotrichum omnivorum* with a borescope mini-rhizotron system. *Phytopathology* **74**, 104–105.

Rygiewicz, P. T., Miller, S. L., & Durall, D. M. (1988). A root-mycocosm for growing ectomycorrhizal hyphae apart form host roots while maintaining symbiotic integrity. *Plant and Soil* **109**, 281–284.

Saif, S. R. (1981). The influence of soil aeration on the efficiency of vesicular-arbuscular mycorrhizas. I. Effect of soil oxygen on the growth and mineral uptake of *Eupatorium odoratum* L., inoculated with *Glomus macrocarpus*. *New Phytologist* **88**, 649–659.

Saif, S. R. (1983). The influence of soil aeration on the efficiency of vesicular-arbuscular mycorrhizas. II. Effect of soil oxygen on the growth and mineral uptake of *Eupatorium odoratum* L., *Sorghum bicolor* (L.) Moench and *Guziota abyssinica* (L.F.). Cass. inoculated with vesicular-arbuscular mycorrhizal fungi. *New Phytologist* **88**, 649–659.

Saif, S. R. (1984). The influence of soil aeration on the efficiency of vesicular-arbuscular mycorrhizas. III. Soil carbon dioxide and growth and mineral uptake in mycorrhizal and non-mycorrhizal plants of *Eupatorium odoratum* L., *Guziotia abyssinica* (L.f.) Cass., and *Sorghum bicolor* (L.) Moench. *New Phytologist* **96**, 429–435.

Saltveit, M. E., & Young, E. (1983). A method for studying the three-dimensional distribution of roots grown in an artificial medium. *Journal of the American Society of Horticultural Science* **108**, 1023–1025.

Same, B. I., Robson, A. D., & Abbott, L. K. (1983). Phosphorus, soluble carbohydrates and endomycorrhizal infection. *Soil Biology and Biochemistry* **15**, 593–597.

Sanders, F. E., & Sheikh, N. A. (1983). The development of vesicular-arbuscular mycorrhizal infection on plant root systems. *Plant and Soil* **71**, 223–246.

Schenck, N. C., & Schroeder, V. N. (1974). Temperature response of *Endogone* mycorrhiza on soybean roots. *Mycologia* **66**, 600–605.

Schenck, N. C., & Smith, G. A. (1986). Responses of six species of vesicular-arbuscular mycorrhizal fungi and their effects on soybean at four soil temperatures. *New Phytologist* **92**, 193–201.

Schisler, D. A., & Linderman, R. G. (1987). The influence of volatiles purged from soil around douglas-fir ectomycorrhizae on soil microbial populations. In: *Mycorrhizae in the Next Decade, Proceedings of the 7th NACOM* (Ed. by D. M. Sylvia, L. L. Hung, & J. H. Graham), p. 218. University of Florida, Gainesville.

Schwab, S. M., Johnson, E. L. V., & Menge, J. A. (1982). Influence of simazine on formation of vesicular-arbuscular mycorrhizae in *Chenopodium quinona* Willd. *Plant and Soil* **64**, 283–287.

Schwab, S. M., Menge, J. A., & Leonard, R. T. (1983). Comparison of stages of vesicular-arbuscular mycorrhiza formation in sudangrass grown at two levels of phosphorus nutrition. *American Journal of Botany* **70**, 1225–1232.

Schwab, S. M., Leonard, R. T., & Menge, J. A. (1984). Quantitative and qualitative comparison of root exudates of mycorrhizal and nonmycorrhizal plant species. *Canadian Journal of Botany* **62**, 1227–1231.

Siqueira, J. O., Hubbell, D. H., & Schenck, N. C. (1982). Spore germination and germ tube growth of a vesicular-arbuscular mycorrhizal fungus in vitro. *Mycologia* **74**, 952–959.

Smith, J. R., & Ferry, B. W. (1979). Effects of simazine, applied for weed control, on mycorrhizal development of *Pinus* seedlings. *Annals of Botany* **43**, 93–99.

Smith, S. E., & Gianinazzi-Pearson, V. (1988). Physiological interactions between symbionts in vesicular-arbuscular mycorrhizal plants. *Annual Review of Plant Physiology and Plant Molecular Biology* **39**, 221–244.

St. John, T. V., Hays, R. I., & Reid, C. P. P. (1983). Influence of a volatile compound on formation of vesicular-arbuscular mycorrhizae. *Transactions of the British Mycological Society* **81**, 153–154.

Stotzky, G. & Schenck, S. (1976). Volatile organic compounds and microorganism. *CRC Critical Reviews in Microbiology* **4**, 333–382.

Straatsma, G., & Bruinsma, J. (1986). Carboxylated metabolic intermediates as nutritional factors in vegetative growth of the mycorrhizal mushroom *Cantharellus cibarius* Fr. *Journal of Plant Physiology* **125**, 377–381.

Straatsma, G., Griensven, J. L. D. Van, & Bruinsma, J. (1986). Root influence on in vitro growth of hyphae of the mycorrhizal mushroom *Cantharellus cibarius* replaced by carbon dioxide. *Physiologia Plantarum* **67**, 521–528.

Strullu, D. G., Romand, C., Callac, P., Teoule, E., & Demarly, Y. (1989). Mycorrhizal synthesis in vitro between Glomus spp. and artificial seeds of alfalfa. *New Phytologist* **113**, 545–548.

Suriyapperuma, S. P. (1990). Effect of phosphorus concentration on chemotropism and colonization by vesicular-arbuscular mycorrhizal fungi of five varieties of *Zea mays*. M.S. Thesis, University of Rhode Island, Kingston.

Tester, M., Smith, S. E., & Smith, F. A. (1987). The phenomenon of "nonmycorrhizal" plants. *Canadian Journal of Botany* **65**, 419–431.

Theodorou, C., & Bowen, G. D. (1971). Effects of non-host plants on growth of mycorrhizal fungi of radiata pine. *Australian Forestry* **35**, 17–22.

Thomson, B. D., Robson, A. D., & Abbott, L. K. (1986). Effects of phosphorus on the formation of mycorrhizas by *Gigaspora calospora* and *Glomus fasciculatum* in relation to root carbohydrates. *New Phytologist* **103**, 751–765.

Thompson, J., Melville, L. H., & Peterson, R. L. (1989). Interaction between the ectomycorrhizal fungus *Pisolithus tinctorius* and root hairs of *Picea marina*. *American Journal of Botany* **76**, 632–636.

Tommerup, I. C. (1983). Spore dormancy in vesicular-arbuscular mycorrhizal fungi. *Transactions of the British Mycological Society* **81**, 37–45.

Tommerup, I. C. (1984). Development of infection by a vesicular-arbuscular mycorrhizal fungus in *Brassica napus* L. and *Trifolium subterranean* L. *New Phytologist* **98**, 487–495.

Tommerup, I. C. (1985). Inhibition of spore germination of vesicular-arbuscular mycorrhizal fungi in soil. *Transaction of the British Mycological Society* **82**, 267–278.

Townshend, J. L. (1964). Fungus hosts of *Aphelenchus avenae* Bastian, 1865 and *Bursphelenchus fungivorus* Franklin & Hooper, 1962 and the attractiveness to these nematode species. *Canadian Journal of Microbiology* **10**, 727–732.

Trappe, J. M. (1987). Phylogenetic and ecologic aspects of mycotrophy in the angiosperms from an evolutionary standpoint. In: *Ecophysiology of VA Mycorrhizal Plants* (Ed. by G. R. Safir), pp. 5–26. CRC Press, Boca Raton, FL.

van Nuffelen, M., & Schenck, N. C. (1984). Spore germination, penetration, and root colonization of six species of vesicular-arbuscular mycorrhizal fungi on soybean. *Canadian Journal of Botany* **62**, 624–628.

Walker, C., & Koske, R. E. (1987). Taxonomic concepts in the Endogonaceae: IV. *Glomus fasciculatum* redescribed. *Mycotaxon* **30**, 253–262.

Warrington, S. J., Black, H. D., & Coons, L. B. (1981). Entry of *Pisolithus tinctorius* hyphae into *Pinus taeda* roots. *Canadian Journal of Botany* **59**, 2135–2139.

Watrud, L. S., Heithaus, J. J., & Jaworski, E. G. (1978). Evidence for production of inhibitor by the vesicular-arbuscular mycorrhizal fungus *Gigaspora margarita*. *Mycologia* **70**, 821–828.

Whitfield, F. B., Shea, S. R., Gillen, K. J., & Shaw, K. J. (1981). Volatile components form the roots of *Acacia pulchella* R. Br. and their effect on *Phytophthora cinnamomi* Rands. *Australian Journal of Botany* **29**, 195–208.

Wildermuth, G. B., Warcup, J. H., & Rovira, A. D. (1984). Growth of *Gauemannomyces graminis* var. *tritici* in soil in the presence and absence of wheat roots. *Australian Journal of Botany* **29**, 195–208.

Williams, C. N. (1961). Tropism and morphogenesis of *Striga* seedlings in the host rhizosphere. *Annals of Botany* **25**, 407–415.

Wilson, G. W. T., Hetrick, B. A. D., & Kitt, D. G. (1989). Suppression of vesicular-arbuscular mycorrhizal fungus spore germination by nonsterile soil. *Canadian Journal of Botany* **67**, 18–23.

Yang, C. S., & Wilcox, H. E. (1984). Technique for observation of mycorrhizal development under monoxenic conditions. *Canadian Journal of Botany* **62**, 251–254.

Zentmeyer, G. A. (1961). Chemotaxis of zoospores for root exudates. *Science* **133**, 1595–1596.

2

The Influence of the Plant Root on Mycorrhizal Formation

Anne J. Anderson

SUMMARY. **Symbiosis between fungi and plant roots to form a mycorrhizae involves extensive interactions at the molecular level between both partners. Events that condition whether or not a mycorrhiza form are occurring even outside of the root tissue in the rhizosphere and at the rhizoplane. Key molecules in the root exudate and plant surface factors are important in germination, hyphal growth, and penetration of the fungus. Fungal components in turn may condition plant cells to accept or reject fungal growth through resistance mechanisms. Developing molecular biology skills are helping to reveal the nature of the recognition and signaling mechanisms that turn the keys to permit the mycorrhizal state.**

Introduction

Mycorrhizal formation involves the integrated growth of specific fungi within the plant root. This chapter addresses the way in which the root influences mycorrhizal development. Three zones of interaction will be discussed: the rhizosphere, the rhizoplane, and within the root tissue. At each stage, different communication systems exist between the host and the mycorrhizal fungi to regulate their interdevelopment.

The Rhizosphere

The rhizosphere is the initial soil space in which the root, through the components in the root exudates, influences the mycorrhizal fungus. Within the rhizosphere, fungal spore germination and growth of the germ tube may be nurtured. Increased branching and orientation of the hypha toward the root may enhance the process of subsequent colonization. However, the rhizophere nutrients are used by other microorganisms as well. Consequently, a key issue is the survival of the spore and the hypha in this intensely competitive microbial milieu. It is possible that the production by the mycorrhizal fungus of components antagonistic to other rhizosphere organisms is involved.

Effects of Root Exudates

Effects of root exudates on germination of mycorrhizal spores and hyphal growth are observed. Some of the components in the exudates are primarily used as additional nutrients to promote growth. Other factors may be involved in triggering changes in the fungus which are involved in the colonization process. These molecules constitute the earliest communication system which regulates the process of mycorrhizal formation.

Studies by Melin (1959) suggested the presence of an "M factor" in exudates of pine, which would stimulate the germination of spores of several genera of ectomycorrhizal fungi. Fries and Birraux (1980) support this finding with studies with *Hebeloma*. With vesicular-arbuscular mycorrhizal (VAM) fungi there are differential reports of root exudates on spore germination. Bowen (1969) indicated that exudates were not generally stimulatory to germination but Graham (1982) observed stimulation with exudates of citrus and sudan grass.

The presence of root components may affect the growth character of the germinating hypha. A more branched pattern of hyphal growth for the VAM fungus *Gigaspora margarita* in the presence of root components is reported by Becard and Piche (1989a,b). Both exudates and volatile materials from the root were required for maximum effect. Thus, the communicating systems between the plant and the fungus in the rhizosphere may be quite intricate. Becard and Piche (1989b) indicate that CO_2 concentration appears to be one of the factors involved in the altered morphology. Previously Koske (1982) and Gemma and Koske (1988) had suggested that the attraction of germ tubes of another *Gigaspora* species, *G. gigantea*, toward roots was due to volatiles. They observed that the effect was sensitive to inhibition by $KMnO_4$ or KOH, which would absorb CO_2. A role of CO_2 as a signal in altering the growth pattern of the hypha is not surprising: the root and its microbial associates would provide an elevated source of CO_2 in the soil environment. However CO_2 is not the sole determinant. Gemma and Koske (1988) demonstrated the aerial germ tubes of *G. gigantea* were attracted by roots of host as well as nonhost beet but not by another nonhost kohlrabi. Thus, some differential effects of host and nonhost plants on mycorrhizal fungi may be manifest even without root contact. It is interesting that the fungus may also be influencing the host metabolism early in the colonization process without their intimate contact. Gemma and Koske (1988) report that root elongation was apparent prior to contact with the hosts of *G. gigantea* but not with the nonhost beet.

The nutritional state of the plant is significant in the development of the fungus (Koide & Li, 1990). Elias and Safir (1987) observed hyphal elongation of *G. fasciculatus* was enhanced by exudates from *Trifolium repens* but only when the plants were grown under phosphate limitation. The efficacy of the exudates as a fungal growth promoter decreased as the phosphate-limited plants aged. The concept that the phosphate status of the plant affects mycorrhizal formation is well established (Harley & Smith, 1983). Graham et al. (1981) suggested that it is the quantity of the exudate, rather than the quality as hypothesized by Elias and Safir (1987) that changes on phosphate status. Mechanistically, phosphate deficiency in the plant increases the permeability of the plasma membrane of the root cells thereby affecting exudate levels. In support of this argument, artificially induced, elevated root exudation in *Chenopodium quinona* by treatment with an herbicide, simazine, resulted in a higher incidence of VAM formation (Schwab et al., 1982).

The rate of exudation was implicated as a factor that differentiated three VAM host plants from three VAM nonhost plant species (Schwab et al., 1983). Early in seedling development the rate of exudation was greater from the mycorrhizal rather than the nonmycorrhizal hosts. Analysis of simple sugars, amino acids and organic acids did not reveal consistent differences in composition between the exudates from the host and non-host plants (Thomson et al., 1990). Correlation of extent of mycorrhizal colonization with the amount of exudation also was apparent from the observation of Azcon and Ocampo (1981) using wheat cultivars. Wheat cultivars that were colonized to a lesser extent by VAM fungi released less exudates than the more susceptible cultivars.

Lynn Abbott and co-workers suggest that the amount of root exudate is related to the soluble carbohydrate in the root. Specifically, Thomson et al. (1990) report that as phosphate nutrition of the plant increases the amount of soluble carbohydrate decreases and this limits exudation. The effect of limitations in phosphate to increase free reducing sugars extractable from the root is documented for several plants (Thomson et al., 1990). Thus these observations again connect the potential for mycorrhizal formation with plant nutrition through the regulation of root exudate.

Utilization of components in the root exudates for nutrition by the hypha presumably occurs, but which substrates are preferred is not well understood. A variety of compounds (amino acids, organic acids, and sugars) are demonstrated to improve growth of VAM hypha (Hepper, 1984). Leake and Read (1989) demonstrate that proteases secreted from the hypha of ectomycorrhizal fungi may permit the use of more complex proteinaceous materials as carbon and nitrogen sources. Becard and Piche (1989b) recently demonstrated incorporation of CO_2 by germ tubes of a VAM fungus, suggesting that even the respiratory end product of plant metabolism may be utilized. The use of molecular biology techniques would permit the detection of mRNAs, which are specifically expressed on exposure of the germinating hypha to root components. Judelson and Michelmore (1990) have isolated similar "stage-specific" mRNAs from the germinating hypha of a lettuce pathogen, *Bremia lactucae,* which is also an obligate biotroph. However, the root components may be playing additional roles than just providing nutrients to the microbe. Certain components may act as communication molecules and promote changes in gene expression that are essential or valuable in the process by which the mycorrhizal fungus eventually colonizes the plant.

The idea that root components can regulate gene expression of the colonizing microbe is well supported from other root–microbe interaction studies. Phenolics from the root have been demonstrated to act as triggers for gene expression in bacteria, which are root colonizers (Halverson & Stacey, 1986; Kosslak et al., 1990; Long, 1989; Sprent, 1989; Strange et al., 1990). These phenolics promote the expression of genes involved in colonization. The pathogenic *Agrobacterium tumefaciens* responds with altered gene expression to phenolics that are enhanced after wounding the plant, a factor that may promote its colonization of wound sites. In contrast, rhizobia, the beneficial symbiotic bacteria that fix nitrogen, detect other types of phenolics that are constitutively expressed by roots. Recognition of constitutive phenolics may reflect the fact that rhizobia have evolved a penetration mechanism to enter the plant tissue and, unlike *A. tumefaciens,* do not rely on wounds or natural openings for entry. With *Rhizobium* there is extensive specificity in the range of phenolics which are detected. Different *Rhizobium* species display varied responses and there appears to be some correlation with the specificity of the bacterium for its host and recognition of phenolics from that host. Not all of the responses though are stimulatory; some of the root components are demonstrated to be inhibitory to expression of the bacterial genes. The mechanisms involved in these recognition responses are interesting. They appear to involve two component systems that permit transmembrane signaling and gene activation. One protein appears to act as the sensor of the environmental change and the second partner is the transcriptional activator. Several features of the sensor or the activator are conserved between the systems, which recognize different environmental stimuli (Albright et al., 1989). Similar studies to examine the role of phenolics in altering the expression of genes in mycorrhizal fungi are being initiated. Gianinazzi-Pearson et al. (1990) report enhanced germination of VAM spores and hyphal growth by root exudates and plant flavanoids.

Microbial Interactions

The root may also influence the mycorrhizal fungus in the rhizosphere indirectly through the regulation of the types of microbes that are present (Graham, 1988; Linderman, 1988; McAfee & Fortin, 1988). Roots are associated with a conglomerate of diverse microorganisms. Their effects on their mutual growth and performance range from no interaction, to synergism or antagonism. These relationships are becoming increasingly relevant because of the potential use of microbes, including mycorrhizal fungi (Caron, 1989), as commercial biological soil amendments. The beneficial uses of these microbials are to increase plant performance and to suppress the effects of microbial pathogens in a biocontrol approach. Whether microbials that are antagonists of fungal plant pathogens are also deleterious to mycorrhizal fungi is another key issue. Studies by Garbaye and Bowen (1989) with a biocontrol isolate, a fluorescent *Pseudomonas fluorescens*, found that there was a deleterious effect on ectomycorrhizal formation of *Rizopogon luteous*. However, other bacterial isolates, some of which were isolated from within the mantle, were stimulatory to ectomycorrhizal growth and mycorrhizal formation. The putative mechanisms by which microrganisms stimulate mycorrhizal formation are reviewed by deOliveira and Garbaye (1989) and Duponnois and Garbaye (1990). These researchers observed that the proportion of bacteria with "helper" activity increased with time for the Douglas-fir–*L. laccata* system and positive effects were observed both under greenhouse and bare root nursery conditions (Garbaye et al., 1990).

Variable results of the effects of associated microflora on VAM mycorrhizal formation have been observed. Inhibitory interactions occurred between *Glomus fasciculatus* and a beneficial streptomycete, *Streptomyces cinnamomeous* (Krishna et al., 1982). However, Mugnier and Mosse (1987) observed that *S. orientalis* stimulated germination of spores of *G. mosseae*. Will and Sylvia (1990) report that germination of *G. deserticola* spores and hyphal growth was increased by *Klebsiella pneumonia*. In studies of VAM formation by Paulitz and Linderman (1989), negative effects caused by fluorescent pseudomonads were not observed although there was a short-term inhibition of germination by one strain used. Previous studies by Meyer and Linderman (1986a) revealed synergism between a VAM fungus and a fluorescent pseudomonad in stimulating plant performance. As discussed in the lucid review by Linderman (1988) and elsewhere (Ames et al., 1989), the formation of the mycorrhizae also influences the microbial composition of the rhizosphere to produce the mycorrhizosphere. Consequently, the role of the plant as the supplier of the primary nutritional source for the nurturing of the rhizosphere microbial population needs further evaluation. These data will be especially important if management of the mycorrhizosphere is to be desired at a commercial level in agriculture (Caron, 1989).

The mechanisms involved in the interactions of the root–microbe complex may be quite intricate. The bacteria may alter phosphate availability because many isolates are demonstrated to solubilize phosphate, perhaps by production of organic acids (Duponnois & Garbaye, 1990; Leyval & Bethelin, 1989; Will & Sylvia, 1990). The microbes may also contribute plant growth regulators and enzymes, which degrade plant wall components. Synergism thus may be explained by any of these factors adding to the colonization potential of the mycorrhizal fungus. Antagonism between the microbes in the root environment may be ascribed to several types of interactions also. Important processes in antagonism appear to be the production of antibiotics, of complex structures, as well as the more simple poisons such as cyanide (Davison, 1988; Kloepper et al., 1988, Schippers, 1988; Weller, 1988). Limitation of growth by the secretion of high-affinity iron chelators, siderophores, and competition for nutrients are other mechanisms.

Mycorrhizal fungi also produce antibiotics, siderophores, and utilize nutrients. Thus, they have the potential to be antagonists themselves, although the actual mechanism for each system has not been resolved (Garcia-Garrido & Ocampo, 1988). This potential for antagonism of pathogens mediated by VAM fungi is discussed by Caron (1989). The production of components from ectomycorrhizal fungi antagonistic to tree pathogens is surveyed by Kope and Fortin (1989). One example is suppression of *Fusarium oxysporum* f. sp. *pini* by an ectomycorrhizal fungus *Paxillus involutus* (Duschesne et al., 1989). Antagonism is related partially to the production by this mycorrhizal fungus of oxalic acid. It is interesting that the production of oxalic acid is stimulated by root exudates of *Pinus*. Perhaps this scenario is another example of how gene expression relating to colonization is being regulated in the colonizing microbe by plant factors. Whether the production of antibiotics, cyanide, and siderophores, which are reported from *in vitro* studies of other mycorrhizal fungi, is also stimulated by plant metabolites needs to be examined. A third mechanism involving protection of plants by mycorrhizal formation may be dependent on the mycorrhizal fungi stimulating altered defense responses in the plant tissue. Altered accumulation of plant components that will impair the ability of a pathogen to survive in the host tissues is reported. These plant defensive components include phenolics, phytoalexins, and degradative enzymes such as chitinase (e.g., Strobel and Sinclair, 1991).

The Rhizoplane

The rhizoplane is the trigger for a second set of responses involving colonization by the mycorrhizal fungi. The process may be subdivided into discrete events of (1) binding, (2) additional hyphal growth, (3) differentiation, and (4) penetration. Each of these will be discussed in turn.

Binding

Attachment of the hypha to the root surface is likely to involve specific fungal and root components, as discussed in a review of fungal adhesion mechanisms by Manocha and Chen (1989). Several mechanisms are implicated in these and other systems to account for initial binding between organisms. These mechanisms include (1) charge–charge interactions, (2) lectin–carbohydrate recognition, and (3) hydrophobic bonding. Carbohydrate moieties are likely to be involved in both charge–charge as well as lectin recognition (Hardham & Suzaki, 1989; Sharon & Lis, 1989). Root mucilages are implicated in binding of zoospore of a pathogenic fungus (Longman & Callow, 1987). Proteinaceous plant agglutinins are proposed to function in the recognition of cell surface components of spores and bacteria (Tari & Anderson, 1988). Hamer et al. (1987) implicated glucose or mannose residues in attachment of spores of a rice pathogen to leaves. A role of fucose and galactose is suggested in other fungal binding systems (Hohl

& Balsiger, 1988; Elad et al., 1983; Barak et al., 1986). Consequently, it is interesting that Piche et al. (1983) and Piche and Peterson (1985) report fucose as a major sugar at the interface of pine roots with an ectomycorrhizal fungus *Pisolithus tinctorius*. Hydrophobic binding is implicated in the attachment of rust germlings to the cuticle of leaves (Epstein et al., 1987). Adhesion of beneficial bacteria to root surfaces is also slated to involve hydrophobic binding (Vesper, 1987). Similar attachment processes are likely to occur with mycorrhizal fungi involving exterior cell structures. Whether these features are constitutive or whether there is *de novo* induction or modification of existing structures when the fungus is in the presence of the plant is unknown.

Differentiation and Penetration

The plant root surface clearly influences the growth pattern of the germ tubes that are of importance to colonization. The germ tubes may continue to grow on the root surface or they may undergo morphological differentiation to produce an infection structure termed an appressorium. Enhanced branching of the germ tube prior to appressorium formation is reported by Becard and Fortin (1988) for the VAM fungus *G. margarita* on carrot roots transformed by *A. rhizogenes*. Contacts between different roots are possible because of multiple germ tubes and branching of the hypha, as reported by Koske (1982) for bean and *G. gigantea*. The plant may regulate the process of attachment by its nutritional status. Attachment of *G. margarita* hypha to the transformed carrot roots was inhibited by high concentrations of phosphate, sucrose, and sodium. Age of the roots was also a key factor. Greater adhesion occurred to 20-day-old roots than younger roots (Becard & Fortin, 1988).

How contact with the root influences the timing and location of appressorium formation is unknown for mycorrhizal fungi. Similar questions are being asked with fungal pathogens (Hoch & Staples, 1987). Carver and Ingerson (1987) indicate that contact between rust germlings and barley surfaces enhances appressorium formation from long germ tubes. They indicate that appressorium formation by these long germ tubes was stimulated by contact between the plant and other short germ tubes that do not form appressoria. Perhaps the multiple germ tubes and branching habit of VAM fungi have a similar function, to enhance the chance of finding the appropriate location for appressorium formation. Other studies with the bean rust fungus *Uromyces appendiculatus* indicate that fungal cell surface factors involved in adhesion are essential for the orientation of germ tube growth and appressorium formation (Epstein et al., 1987). A delicate recognition system has been developed by this fungus to trigger appressorium formation to occur over the stomates. First, the germling grows in a specific orientation across the cuticular ridges, perhaps to maximize the chance of contact with a stomate. Second, differentiation to form the appressorium is induced by recognition of the change in steepness of the stomatal lip. This process of "thigmo differentiation" (Hoch & Staples, 1987) appears to be due to a sensing mechanism, which detects the change in steepness of the cuticle at the appressorial lip. However, additional chemical triggers for orientation and differentiation may be involved. Sensitivity to pH is suggested for growth orientation of germlings of *Uromyces vicia fabae* (Edwards & Boling, 1986), and such chemicals as potassium ions (Staples et al., 1983, 1984, 1985), simple sugars (Kaminskyi & Day, 1984), and acrolein (Hoch & Staples, 1987) trigger induction of appressorium formation for other rusts. Presumably these factors trigger altered gene expression in the fungus to permit appressorium formation. Expression of unique genes during appressorium formation in rust fungi has been demonstrated using current molecular biology techniques (Bhairi et al., 1990). Whether mycorrhizal fungi can be tricked into production of appressoria by artificial techniques, such as surface contact or chemical signals, awaits study. If achieved, molecular analysis of gene expression would also be feasible.

What are the chemical and physical signals that dictate the growth modes of the mycorrhizal fungi on the root surface and the location of appressoria? Most frequently penetration is observed maximally over a discrete range along the root, just beyond the growing tip. Appressoria from VAM fungi are sometimes observed on root hairs and intimate growth of ectomycorrhizal fungi with these structures is reported (Massicott et al., 1990). Appressoria of *Glomus verisforma* on leek roots were generally positioned along the cell wall between adjacent epidermal cells (Garriock et al., 1989) and as suggested by Bonfante-Fasolo and Scannerini (Chapter 3) perhaps chemical and physical factors are located in this zone that act as triggers for appressorium formation. In comparison, rhizobial infection is correlated with a critical stage of root hair formation. A role for lectins and Ca^{2+} binding proteins has been implicated in the specificity and location of the infection process for rhizobium (Sharon & Lis, 1989; Smith et al., 1988). Currently, root hair walls are being examined (Mort & Grover, 1988) to search for unique properties that may trigger infection. Studies by Lugtenburg and colleagues indicate that a lectin that will recognize *Rhizobium* has a highly defined location, at the tip of the root hair. The locus on the root hair for reception of the Ca^{2+}-associated ricadhesin structure detected on the rhizobial surface appears to be distinct from that of the lectin (Smith et al., 1988). Similarly, for mycorrhizal development Bonafonte-Fasolo and Scannerini (chapter 3) have suggested that the plant cell wall may have specific properties at the infection zone. Other potential chemical triggers at the root surface could be phenolics, inorganic ions, and suberin components.

Specific molecules on the root surface may be acting to recognize adhesive substances that are subsequently produced by the appressorium to cement its initial binding. Studies with *Salmonella,* a human pathogenic bacteria, have revealed the *de novo* synthesis of bacteria proteins after contact with host epithelial cells that are essential for colonization (Finlay et al., 1989). The attachment of

A. tumefaciens to plant surfaces is also correlated with secretions by the bacteria of glucans that adhere to the plant wall (Matthysse et al., 1982).

Internal Penetration

Internal penetration of the root constitutes the third level of interaction of the mycorrhizal fungi with the root. It is this internal growth pattern of the hypha within the root that permits the exchange of nutrients essential to the symbiotic process to occur (Harley & Smith, 1983). The fungus is believed to benefit by receipt of carbon from the plant and in turn provides mineral nutrients and water to the plant. The infection structure that is formed within the root differs with the various types of mycorrhizal fungi. With ectomycorrhizal fungi, the hypha ramify within the plant cell walls but do not penetrate through the cell wall to contact the plant cell plasmalemma. The extent of colonization of the epidermal and cortical layers to produce the Hartig net is variable and is often used as an indicator of the degree of ectomycorrhizal formation. Penetration through walls of the root cells and invagination of the plasmalemma of the plant cell do occur with ericoid and VAM fungi. Generally with VAM fungi the hypha grow intercellularly in the epidermal layer but intracellularly in the cortex. In the event of hyphal penetration of the epidermal cells the VAM fungus usually forms coils but not the highly branched arbuscule structures. It is in the cortical cells that the arbuscules are formed. These structures are the indicator that is used to assess successful VAM infection.

Mechanisms of Internal Growth

The mechanisms by which the plant cell wall is penetrated by VAM fungal hypha either for inter- or intracellular growth, or intercelluarly by ectomycorrhizae are uncertain. Both physical penetration and controlled degradation of the plant cell walls by enzymes may be involved. A theory espoused by Harley (1985) is that the normal process for plant cell wall polymerization is disturbed by the mycorrhizal fungus and permits a softening that facilitates physical penetration of the hypha. The potential for limited degradation of plant cell wall structures by enzymes produced by the mycorrhizal fungus occurs in certain systems. Synthesis of pectic and cellulose degrading enzymes is reported for some ectomycorrhizal and ericoid fungi (Cervone et al., 1988). Garcia-Romera et al. (1990) have detected pectic, hemicellulose, and cellulose degrading enzyme activities in extracts of germinating spores of VAM fungi. It is also possible that other root-associated microbes contribute enzymes that are involved in penetration of the mycorrhizal hypha.

Regulation of any enzymes involved in plant cell wall penetration must be highly controlled perhaps by both fungal and plant mechanisms. Plant cell walls are associated with proteins that inhibit the pectic-degrading enzymes of plant pathogens (Cervone et al., 1989). Regulation in expression of plant cell wall-degrading enzymes in fungal pathogens involves both induction and repression of synthesis. For example, induction of the cutinase gene in *Fusarium solani* f. sp. *pisi* occurs with only certain unsaturated fatty acid monomers that are derived from cutin (Woloshuk & Kolattukudy, 1986). Discrete sized fragments of pectin supplied at a limited concentration range will induce pectinase (Cooper & Wood, 1975; Keon et al., 1987), whereas other fragments are inactive. Higher concentrations of the inducing oligomers and other sugars suppress expression. These subtle effects of plant products on expression of fungal genes concerned with colonization processes may be reflective of the apparent relationship between root exudate and VAM mycorrhizal formation discussed earlier.

What features of the plant and the VAM fungi account for the differential hyphal penetration of epidermal, hypodermal, and cortical cells? In addressing this question Bonafante-Fasolo et al. (Codignola et al., 1989; Bonafonte-Fasolo et al., 1990b) indicate that the epidermal and cortical cell walls of leek differ in their proportions of two major polysaccharides, cellulose and pectin, and degree of substitution with phenolics. Although cellulose is found in all cell types, this polysaccharide was thicker in the epidermal than the hypodermal or cortical cells. The epidermal cells had a much reduced pectic component than the hypodermal or cortical cell walls. However, both cellulose and pectin were still present in the plant cells wall, which are penetrated inter- and intracellularly by mycorrhizal hypha. Thus any involvement of changes in pectinase or cellulase activities in plant cell wall penetration must be limited in extent and time. Perhaps of more relevance is the location of the wall bound phenolics, which may be involved in generating cross-linkages that are more resistant to enzymatic hydrolysis (Fry, 1986). These phenolics are located only in epidermal and hypodermal cells and were lacking in the cortex.

The signals from the plant that condition the formation of arbuscules in certain plant cells in the mycorrhiza may be similar to the processes that determine the formation of feeding structures with fungal plant pathogens. Heath is studying the formation of haustoria mother cells with *Uromyces vignae*, the cowpea rust pathogen. Li and Heath (1990) observed that extracts from noninoculated host plants and to greater extent infected plants enhance haustorium formation. Mixtures of sugars (arabinose, mannose, xylose, sucrose) were also able to trigger haustorium formation to a limited extent (Heath, 1990). Thus Heath concluded that there are simple factors from the plant that initiate this process of haustorium formation.

Physical and biochemical triggers must also condition the limited growth of the ectomycorrhizal fungi within the plant cell walls of the cortex. Penetration through the cell wall may be restricted by the enzyme complement or the physical strength of the fungus. Nylund (1987) proposes that the special acidic, pectic component of the cortical walls is a key feature in determining the zone at which infection is successful. The lignification of the Casperian strip may also be a limiting factor in the depth of colonization into the root tissue.

Metabolism in the Mycorrhiza

Several different techniques are being used to explore altered functions of both the plant and the fungus in this intricate growth status of the mycorrhiza. Changes in metabolism are evident from structural modifications of the roots as a result of mycorrhizal formation (Berta et al., 1990; Stein & Fortin, 1990; Stein et al., 1990). Ectomycorrhizal associations are characterized by more extensive branching. Radial extension of epidermal cells is observed in the interaction of *Betula alleghaniensis* with *P. tinctorius* (Massicotte et al., 1988). More extensive root branching is observed with VAM fungal interactions (Gemma & Koske, 1988). Spanu and Bonafonte-Fasola (1988) also detected an increase in lateral root formation in VAM leek. Perhaps the increased branching of the root increases the soil volume that is exploited by the plant.

Recent biochemical studies by Hilbert and Martin (1988) with the ectomycorrhizal system involving *Pisolithus* and *Eucalyptus* demonstrated that novel proteins termed "mycorrhizins" are synthesized in these mycorrhizae. These proteins were not detected in extracts of nonmycorrhizal roots or from the extracts of the cultured ectomycorrhizal fungus. Similarly Pacovsky (1989) demonstrated with soybean colonized by the VAM fungus *Glomus fasciculatus* that novel proteins of molecular size 16, 17, 18, 22, and 33 kDa constitute over 5% of the total soluble root proteins. Pacovsky (1989) raises the interesting question of whether any of the novel proteins in the VAM interactions have similar functions to the mycorrhizins that are in the ectomycorrhizae. Wyss et al. (1990) studying VAM formation in soybean observed that certain of the mycorrhizins were recognized as antigens by antibodies observed to nodulins. These observations suggest that the two symbioses of roots with Rhizobia and with mycorrhizal fungi have some common features. Enzymes involved in enhanced carbon and nitrogen metabolism are likely candidates, because in both symbioses nutrient exchanges are of key importance.

Nutrient Exchange

Some of the alterations in protein synthesis in a mycorrhizal root could be related to the nutrient exchanges that are occurring between the fungus and the plant. Aspects of these exchanges, centering on carbon flux from the plant and the fungus and inorganic ions, especially phosphate, from the fungus are reviewed recently in Smith and Gianinazzi-Pearson (1988) and Smith and Smith (1990).

Certain mycorrhizins may be associated with the increased phosphate metabolism that is occurring in the mycorrhiza. Presumably the activity of enzymes, possibly kinases, in the fungi required for depolymerization of stored phosphate is enhanced. Acid phosphatase activity is increased in amounts in several mycorrhizal systems including ectomycorrhizae (Antibus et al., 1981). Dodd et al. (1987) revealed that although enhanced acid phosphates activities were detected with two *Glomus* species, a third did not induce any increased activity, although all three fungal species promoted plant growth. Thus the role of enhanced acid phosphatase activity remains uncertain. Studies by Bae and Barton (1989) with an ectomycorrhizal fungus point out that alkaline as well as acid phosphatase could participate. Further, the extent of contribution of root enzymes to any observed changes in phosphatases is unresolved.

The current techniques of molecular biology should provide more definitive data in this area. For example, a phosphatase isozyme with enhanced activity in mycorrhizal roots may be purified and the N-terminal amino acid sequence obtained. Using this sequence, nucleic acid probes that correspond to the amino acids in this domain could be constructed. These probes would be used in Southern analysis of the genome of the host or the mycorrhizal fungus. These data would indicate whether the phosphatase is of fungal or plant origin.

The regulation of the phosphatase activity in relationship to the development of the mycorrhiza could be further explored. Such studies could involve the isolation of the gene from either the host or the fungi and the construction of fusions of its promoter with a marker construct, usually an enzyme that is not normally expressed by the plant or fungus. The promoter probe would be transformed back into the plant or the fungus. Techniques for such manipulations are feasible with ectomycorrhizal fungi and many plants. Fungal transformation may involve electroporation, chemical treatments to alter permeability of protoplasts and their regeneration, or particle bombardment. These methods were extensively discussed during the recent North American Conference on Mycorrhizae (1990). Electroporation, particle bombardment, and the biological vectors of *Agrobacterium tumefaciens* or *A. rhizogenes* are appropriate for transformation of plant tissues (Benfey & Chua, 1989; Davey et al., 1989). The expression of the marker enzyme would be followed in the transformed tissues. These data would reveal the timing and location of the specified enzyme.

The exchange of nutrients between the fungus and the plant requires passage across the fungal cell wall and fungal plasmalemma and the interfacial matrix and plasmalemma of the plant. Consequently there is interest in the morphology and enzyme activities associated with these structures. This research area is recently reviewed by Smith and Smith (1990). Altered cell wall structures of both the plant and the fungus in the arbuscular cells in VAM roots are well documented. Bonafonte-Fasolo et al. (1990a) indicate that the arbuscular walls are thinner than the walls of the intercellular hypha and possess less polymerized chitin, as detected by staining with specific lectins. This concept is supported by studies of *Glomus clarum* in leek by Jabaji-Hare et al. (1990). These workers also detected reduction in the chitin component of arbuscule walls, compared to intercellular hyphae or vesicles. However, fucose and sialic acid were now observed. Presumably these changes are due to altered accumulations of polysaccharides and perhaps glycosylated proteins, which may have either a structural or enzymatic role. The structure of the interfacial matrix has also been probed. Bonfante-Fasolo et al. (1990b) demonstrated that the typical cell wall polymers cellulose and pectin remain present in the wall matrix adjacent to the fungal arbuscular walls and in the seal that surrounds the penetration site. However, microscopy reveals the interfacial matrix is in other ways disorganized and lacking in the degree of polymerization. Thus the arrangement of other plant cell wall polymers must be altered. Presumably these changes in both fungal and plant cell wall types facilitate the active transport of nutrients.

In VAM interactions increased ATPase activity detected by histochemistry at the host membrane is proposed to relate to the enhanced levels of activated transport (Marx et al., 1982). ATPase activity with a proposed similar function is observed in the plasma membrane of the arbuscules. In ectomycorrhizae there is no intracellular penetration and no arbuscular formation, rather the site of nutrient exchange is proposed to be in the Hartig net where the hypha ramify intercellularly within the plant cell walls. Lei and Dexheimer (1988) observed that abundant ATPase activity was associated with the plasmalemma of the hypha in the Hartig net formed between *Pinus sylvestris* and *Laccaria laccata*. They observed that the activity was manifest only when the hypha was associated with viable host cortical cells. Two different scenarios could explain this pattern. Either the loss of activity is due to inhibitory components that are emanating from the dead plant cell or the enhanced activity of the fungal ATPase is dependent on factors from live host cells.

In reviewing Smith and Smith (1990) there are comparisons between interactions of mycorrhizal fungi and their hosts with plant pathogen–host systems concerning the mechanisms that may account for altered host function. Studies with pathogens have revealed that pathogens produce such factors as toxins, enzymes, and plant growth regulators that engineer the metabolism of the host cell. Certain toxins and plant growth regulators have been demonstrated to affect ATPase activities. For example, the bacterial phytotoxin syringomycin enhances plant ATPase activity (Bidwai & Takemoto, 1987a). The mechanism is complex and involves phosphorylation of plant membrane proteins and altered ion flux (Bidwai & Takemoto, 1987b). Perhaps the mycorrhizal fungi have developed similar compounds to modify the functions of the plant cell plasma membrane to enhance the mutual exchange of nutrients.

Plant growth regulators produced by the fungi are one possible class of modifiers. Plant growth regulators, such as indoleacetic acid (IAA) and cytokinins, are synthesized by ectomycorrhizal fungi (Barea, 1986; Gay et al., 1988). Altered levels of gibberellin-like substances and absciscic acid are noted upon VAM formation in *Bouteloua gracilis* (Allen et al., 1980, 1982). The effect of plant growth regulators on plant metabolism are complex; one consequence is altered gene expression. Other targets are the activities or function of preformed structures. For example, IAA is proposed to alter plant cell wall extensibility (Theologis, 1986) through mechanisms involving decreased wall pH, which could be a consequence of increased APTases activity. Altered gene expression is a second type of effect of increased IAA levels (Theologis, 1986).

Nitrogen metabolism in the plant is also sensitive to mycorrhizal formation. Mycorrhizal fungi appear to readily convert soil ammonium to amino acids, which are transported to the plant. In ectomycorrhizae, glutamine is transferred across the interface (Martin et al., 1987, 1988; Finlay et al., 1988, 1989). Recent studies with *Glomus*-colonized soybean indicate enhanced levels of aspartate and arginine (Pacovsky, 1989). Consequently, the role of proteins involved in altered nitrogen metabolism as mycorrhizins is being analyzed.

Balanced Fungal Growth

The growth pattern between the mycorrhizal fungus and the plant is highly balanced and not so excessive as to lead to pathogenicity. Limited growth rates may be an intrinsic quality of these mycorrhizal fungi. VAM fungi have not been successfully grown in artificial culture suggesting that a highly regulated supply of nutrients is essential. The ectomycorrhizal fungi behave as saprophytes but do not grow as readily in artificial media as many plant pathogens.

The plant may also participate in restricted fungal growth by manifesting limited resistance responses. Altered plant cell wall depositions, similar in some instances to papillae observed in pathogen-induced responses, are observed in mycorrhizae. Garriock et al. (1989) report that epidermal and cortical cells respond to inter- and intracellular penetration with thickening of the plant cell wall.

In incompatible challenges with pathogens, alterations in plant cell walls have been associated with deposition of callose, a β-1, 3-linked glucan, increased accumulation of hydroxyproline-rich glycoproteins, and phenolics (Anderson, 1990). Several studies have examined the role of phenolic defenses in mycorrhizal formation. Ronald and Soderhall (1985) found no enhancement of two enzymes involved in phenolic metabolism, phenylalanine ammonia lyase (PAL) and peroxidase, in the ectomycorrhizal symbiosis between *P. sylvestris* and *L. laccata*. However, other ectomycorrhizal interactions are reported to stimulate the formation of phenolics. Sylvia and Sinclair (1983) show the induction of phenolic deposits in roots inoculated with *L. laccata*. Components present in the culture filtrate of the ectomycorrhizal fungus could duplicate the response. Thus it is tempting to speculate that the ectomycorrhizal fungi do produce compounds that stimulate defense responses of plants. Coleman and Anderson (Anderson, 1988) have demonstrated that extracellular products from the ectomycorrhizal culture filtrates of *Rhizopogon* species elicit phytoalexin production in bean and condensed phenolics in pine. Activity is correlated with carbohydrate-enriched structures that resemble the elicitors that have been characterized from plant pathogenic fungi (Anderson, 1990).

These observations raise interesting questions. One is whether the mycorrhizal fungi are adapted to the defense molecules of their hosts, which for phytoalexins vary in structure according to the species (Ebel, 1986). Recently successful colonization of plants by a fungal pathogen has been correlated to the presence of enzymes that are able to modify the phytoalexins of their host (Van Etten et al., 1989). One of the enzymes extensively studied is inducible by the phytoalexin. Consequently, it is possible that some of the mycorrhizins being detected as novel proteins could be of fungal origin and be related to modification of phytoalexins. A second question is if the mycorrhizal fungi possess components that are active as elicitors, how is the triggering of defense responses in the plant avoided? Perhaps root cortical tissues are less responsive to elicitors than other cell types. Cell type specificity in production of phytoalexins has been reported for other systems (Hahn et al., 1985; Ebel, 1986). A slightly enhanced accumulation of phytoalexins is reported for VAM soybean roots (Morandi et al., 1984).

Codignola et al. (1989) using leek and ginkgo challenged with VAM fungi detected no difference in PAL activity for mycorrhizal and nonmycorrhizal roots. Wood and Anderson (unpublished) also observed no changes in PAL activity in crimson clover on VAM formation. Codignola et al. (1989) observed no changes in level or location of cell wall bound phenolics. These wall bound phenolics were restricted to the epidermal and hypodermal cells and were lacking in the cortical cells. However, peroxidases in roots of leek did show a burst of increased activity on *G. versiforme* infection (Spanu & Bonafonte-Fasolo, 1988). The peak occurred early in penetration corresponding to the intercellular growth period and preceding hyphal growth in the cortical cell walls. The level of peroxidase eventually declined to below that in nonmycorrhizal roots. Because certain peroxidases are proposed to catalyze cross-linkage of the glycoprotein extensin (Fry, 1986), in the plant wall, it will be interesting to see whether the plant walls are modified in this manner, during intercellular hyphal growth.

The potential for other types of defense responses such as enhanced chitinase or glucanase activity exists (Godiard et al., 1990). These enzymes have been demonstrated to act synergistically in the degradation of fungal cell walls (Boller et al., 1988). Extracellular products from the ectomycorrhizal fungus *Amanita muscaria* induced enhanced chitinase activities in roots and suspension cells of its host *Picea abies* (Sauter & Hager, 1989). Chitinase activities higher in VAM than non-VAM plants have been reported (Dehne et al., 1978). Recent studies with chitinase and glucanase reveal that different isozymes exist, some of which are extracellular whereas others are vascular in location (Linthorst et al., 1990). Immunocytology using monoclonals based on the isoforms, which are derived from the plant enzymes that are specifically altered in the mycorrhiza, will provide more precise knowledge of the altered status. It will reveal the location of the cells and the cellular structure that is associated with the change in enzyme levels.

The possibility that chitinase is changed in level may reflect on the reduction of chitin in the cell walls of VAM fungi in arbuscules, although these effects could also be controlled by altered synthesis in the fungus. The degradative enzymes chitinase and glucanase could also be functional in the senescence of the arbuscules that occurs as the mycorrhiza age. The prospect of the release of chitin products from the fungus is interesting because of the documented role of these fragments as elicitors of defense responses in some plants, such as pea and lodgepole pine (Miller et al., 1986).

Effect of Genotype on Mycorrhizal Formation

The plant clearly influences the type and extent of mycorrhizal formation. Some plant species such as Chenopodiaceae and Cruciferae are regarded as nonhosts for VAM fungi. These species rarely form VAM roots, especially under monoculture conditions, although some colonization can be forced when roots of the nonhost are cocultivated with a susceptible, colonized plant (Hirrel et al., 1978). Even in the presence of the companion plant the association of Chenopodiaceae and Cruciferae with VAM fungi was incomplete and arbuscules were not formed (Hirrel et al., 1978). Additional observations with *Brassica* by Glenn et al. (1985) revealed penetration could occur but was limited to regions where the cortical cells were necrotic. Thus on healthy roots of Brassicae, although surface contact with hypha was permitted, appressorium formation and subsequent penetration were thwarted. These events were not related to the production by the Brassicaes of glucosinylates, which are suggested to be antimicrobial chemicals. However, in more recent studies with the chenopod *Salsola kali* (Allen et al., 1989), initial invasion and even arbuscule formation were observed on inoculation with a mixture of *Glomus* and *G. margarita*. Subsequently a dramatic response involving browning and lignification of the invaded area was observed and the damaged tissue was sloughed away. Thus here the rejection response occurred later than in the previously discussed reports. This response of lignification and browning is suggestive of a more typical resistance termed hypersensitivity observed against incompatible pathogens (Lamb et al., 1989; Anderson, 1990). One difference is that hypersensitivity is highly localized and restricted to the contacted cells and does not cause senescence of plant organs. Perhaps, because of the extensive intracellular growth of the mycorrhizal fungi, necrosis of a large number of cells was inevitable.

These limited studies of the mechanisms in nonhosts that prevent mycorrhizal formation suggest that there are different stages throughout the process of interaction between root and mycorrhizal fungus at which the process may be halted. Thus several types of resistance mechanisms may be manifest. An intriguing paper by Duc et al. (1989) indicated that certain, but not all, pea and fava bean cultivars that have a nodulation-deficient (*nod*−) phenotype for *Rhizobium* infection also display aborted mycorrhizae. These plants have been termed myc^-. Genetic analysis of the myc^- plants has revealed at least three loci to be involved. The resistance to mycorrhizal formation is apparent with two VAM species *G. mosseae* and *G. intraradices*. Resistance to the VAM fungi in the nod^- plants generally involved the formation of abnormal appressorium. Consequently, as suggested in an earlier section, the plant must have systems that cue where and when the appressorium is to be formed.

Other more subtle effects of host genotype on VAM formation are apparent. Krishna et al. (1985) detected significant differences in colonization potential of 30 genotypes of pearl millet. Colonization varied from 25 to 56%. The effects were only partially correlated to root length for the different genotypes. Rather, the researchers suggested that differences in such defensive traits as phenolics and phytoalexin formation could be involved. Azcon and Ocampo (1981) previously correlated differences in VAM infection of 13 wheat cultivars, with the level of sugar in exudates from the roots. Heckman and Angle (1987) showed variation in VAM colonization between 57 and 80% for soybean cultivars. Variability in *G. versiforme* colonization was observed with progeny of half sib families of one cultivar of alfalfa but not another (Lackie et al., 1988). In pea cultivars, three *Glomus* species infected similar root lengths but the degrees of stimulated root growth and interaction with phosphate were different (Estaun et al., 1987). These workers concluded that host genotype was a major factor in these variabilities.

Specificity in host range is much more advanced with the ericoid and ectomycorrhizal fungi than with VAM isolates (Bonafonte-Fasolo et al., 1984; Malajczak et al., 1982; Molina & Trappe, 1982). Certain ectomycorrhizal fungi are generalists whereas others are specialists and only form complete mycorrhizae and fruiting structures with discrete hosts. Resistance reactions similar to those observed with incompatible pathogens involving the localized accumulations of phenolics are documented to occur with incompatible ectomycorrhizal fungi.

The displayed specificity is highly intricate for certain species. *Rhizopogon occidentalis* is compatible with lodgepole pine yet is incompatible with other hosts such as Douglas fir. In contrast, *R. vinicolor* is compatible with Douglas fir but not lodgepole pine (Molina & Trappe, 1982). This degree of specificity may be compared to that observed by many plant pathogens. Recent studies with bacterial pathogens have correlated resistance with the expression of discrete genes termed, avirulence genes. These genes were predicted to exist by genetic studies with both bacterial and fungal pathogens, but now the techniques of molecular biology enable the isolation of the segments of DNA that encode the avirulence genes and permit their function to be more easily analyzed (Kobayashi et al., 1989; Whalen et al., 1988). It is proposed that the product of the avirulence gene interacts, either directly or indirectly, with products of genes in the host that encode resistance-involved events. This interaction triggers the onset of the complex resistance mechanism of hypersensitivity. It will be intriguing to use molecular techniques of gene transformation to determine whether the specificity of the system, such as that demonstrated by the *Rhizopogon* species, relies on genes that are involved in resistance initiation or compatibility. Genes for compatibility could involve specific cues that permit penetration and Hartig net formation to occur only in the compatible host by recognition of specific structures. Genes for incompatibility would trigger effective defenses in the nonhost but not the host plants.

Conclusion

Roots and their mycorrhizal fungi have coevolved to form a valuable symbiosis. The development of the mycorrhiza presumably requires lock and key communication systems at many sequential stages. We now have the techniques of molecular biology and the processes of gene manipulation in both plant and fungus to dissect key interactions. Advances may be accelerated by the choice of certain model systems to be explored in depth and used as the basis for examining diversity.

Our goal is to gain answers to the basic questions addressed in this review. What conditions a nonhost? How does the fungus attach and penetrate hosts? What limits spread and development to prevent pathogenesis? How are host defence responses avoided? How is nutrient exchange regulated? Understanding these processes may improve our potential to use these beneficial fungi more effectively in developing agricultural systems.

Acknowledgments

Supported in part by Grants from Utah Agricultural Experiment Station, EPA, and NSF. The author thanks Durango and many colleagues for discussion.

References

Albright, L. M., Huala, E., & Ausubel, F. M. (1989). Prokaryotic signal transduction medicated by sensor and regulator protein pairs. *Annual Review of Genetics* **23,** 311–336.

Allen, M. F., Moore, T. S., & Christensen, M. (1980). Phytohormone changes in *Bouteloua gracilis* infected by vesicular-arbuscular mycorrhizae. I. Cytokinin increases in the host plant. *Canadian Journal of Botany* **58,** 371–374.

Allen, M. F., Moore, T. S., & Christensen, M. (1982). Phytohormone changes in *Bouteloua gracilis* infected by vesicular-arbuscular mycorrhizae. II. Altered levels of givverellin-like substances and abscisic acid in the host plant. *Canadian Journal of Botany* **60,** 468–471.

Allen, M. F., Allen, E. B., & Friese, C. R. (1989). Responses of the non-mycotrophic plant. *Salsola kali* to invasion by vesicular-arbuscular mycorrhizal fungi. *New Phytologist* **111**, 45–49.

Ames, R. N., Mihara, K. L., & Bayne, H. G. (1989). Chitin-decomposing actinomycetes associated with a vesicular-arbuscular mycorrhizal fungus from a calcareous soil. *New Phytologist* **111**, 67–71.

Anderson, A. J. (1988). Mycorrhizae-host specificity and recognition. *Phytopathology* **78**, 375–378.

Anderson, A. J. (1990). Themolecular basis of plant resistance mechanisms. In: *Defense Molecules* (Ed. by J. J. Marchalonis & C. L. Reinisch), pp. 17–32. Wiley-Liss, New York.

Antibus, R. K., Croxdale, J. G., Miller, O. K., & Linkins, A. E. (1981). Ectomycorrhizal fungi of *Salix rotundifolia*. III. Resynthesized mycorrhizal complexes and their surface phosphatase activities. *Canadian Journal of Botany* **59**, 2458–2465.

Azcon, R., & Ocampo, J. A. (1981). Factors affecting the vesicular-arbuscular infection and mycorrhizal dependency of thirteen wheat cultivars. *New Phytologist* **87**, 677–685.

Bae, K., & Barton, L. L. (1989). Alkaline phosphatase and other hydrolyases produced by *Cenococcum graniforme*, an ectomycoirrhizal fungus. *Applied and Environmental Microbiology* **55**, 2511–2516.

Barak, R., Elad, Y., & Chet, I. (1986). The properties of L-fucose binding agglutinin associated with the cell wall of *Rhizoctonia solani*. *Archives of Microbiology* **144**, 346–349.

Barea, J. M. (1986). Importance of hormones and root exudates in mycorrhizal phenomena. In: *Physiological and Genetical Aspects of Mycorrhizae* (Ed. by V. Gianinazzi-Pearson & S. Gianinazzi), pp. 177–187. INRA Paris.

Becard, G., & Fortin, J. A. (1988). Early events of vesicular-arbuscular mycorrhiza formation on Ri T-DNA transformed roots. *New Phytologist* **108**, 211–218.

Becard, G., & Piche, Y. (1989a). New aspects on the acquisition of photrophic status by a vesicular-arbuscular mycorrhizal fungus, *Gigaspora margarita*. *New Phytologist* **112**, 77–83.

Becard, G., & Piche, Y. (1989b). Fungal growth stimulation by CO_2 and root exudates in vesicular-arbuscular mycorrhizal symbiosis. *Applied and Environmental Microbiology* **55**, 2320–2325.

Benfey, P. N., & Chua, N. 1989. Regulated genes in transgenic plants. *Science* **244**, 174–181.

Berta, G., Fusconi, A., Trotta, A., & Scannerini, S. (1990). Morphogenetic modifications induced by the mycorrhizal fungus *Glomus* strain E_3 in the root system of *Allium porrum* L. *New Phytologist* **114**, 206–215.

Bhairi, S. M., Staples, R. C., Freve, P., & Yoder, O. C. (1990). Characterization of a gene induced during development of infection structures in the plant pathogenic fungus *Uromyces appendiculatus*. *Gene* **81**, 237–243.

Bidwai, A. P., & Takemoto, J. Y. (1987a). Stimulation of red beet plasmamembrane ATPase activity. *Plant Physiology* **83**, 39–48.

Bidwai, A. P., & Takemoto, J. Y. (1987b). Bacterial phytotoxin, syringomycin, induces a protein kinase mediated phosphorylation of red beet plasma membrane polypeptides. *Proceedings of the National Academy of Science* **84,** 6755–6759.

Boller, T., Mauch, F. C., Mauch-Mani, B., & Lugwig, A. (1988). Inhibition of fungal growth by the chitinases and β-1, 3-glucanases induced by pathogens in pea tissue. *Journal of Cell Biochemistry,* suppl. **12C,** 262.

Bonfante-Fasolo, P., Gianinazzi-Pearson, V., & Martinengo, L. (1984). Ultrastructural aspects of endomycorrhiza in the Ericaceae: IV. Comparison of infection by *Pezizella ericae* in host and non-host plants. *New Phytologist* **98,** 329 333.

Bonfante-Fasolo, P., Faccio, A., Perotto, S., & Schubert, A. (1990). Correlation between chitin distribution and cell wall morphology in the mycorrhizal fungus *Glomus versiforme*. *Mycological Research* **94,** 157–165.

Bonfante-Fasolo, P., Vian, B., Perotto, S., Faccio, A., & Knox, J. P. (1990). Cellulose and pectin localization in roots of mycorrhizal *Allium porrum:* Labeling continuity between host cell wall and interfacial material. *Planta* **180,** 537–547.

Bowen, G. D. (1969). Nutrient status effects on loss of amides and amino acids from pine roots. *Plant and Soil* **30,** 139–142.

Caron, M. (1989). Potential use of mycorrhizae in control of soil-borne diseases. *Canadian Journal of Plant Pathology* **11,** 177–179.

Carver, T. L. W., & Ingerson, S. M. (1987). Responses of *Erysiphe graminis* germlings to contact with artificial and host surfaces. *Physiological and Molecular Plant Pathology* **30,** 359–372.

Cervone, F., Castoria, R., Spanu, P., & Bonfante-Fasolo, P. (1988). Pectinolytic activity in some ericoid mycorrhizal fungi. *Transactions of the British Mycological Society* **91,** 537–539.

Cervone, F., Hahn, M. G., Lorenzo, G. D., Darvill, A., & Albersheim, P. (1989). Host-pathogen interactions, XXXIII. A plant protein converts a fungal pathogensis factor into an elicitor of plant defense responses. *Plant Physiology* **90,** 542–548.

Codignola, A., Verotta, L., Spanu, P., Maffei, M., Scannerini, S., & Bonfante-Fasolo, P. (1989). Cell wall bound-phenols in roots of vesicular-arbuscular mycorrhizal plants. *New Phytologist* **112,** 221–228.

Cooper, R. M., & Wood, R. K. S. 1975. Regulation of synthesis of cell wall degrading enzymes by *Verticillium albo-atrum* and *Fusarium oxysporum* f. sp. *lycopersici*. *Physiological Plant Pathology* **5,** 135–156.

Davey, M. R., Rech, E. L., & Mulligan, B. J. (1989). Direct DNA transfer to plant cells. *Plant Molecular Biology* **13,** 273–285.

Davison, J. (1988). Plant beneficial bacteria. *Biotechnology* **6,** 282–286.

Dehne, H. W., Schoenbeck, F., & Baltruschat, H. (1978). The influence of endotrophic mycorrhiza on plant diseases. 3. Chitinase-activity and the orthithine-cycle. *Zeitschrift fur Pflanzenkrankheiten und Pflanzenschutz* **85,** 666–678.

DeOliveira, P. V. L., & Garbaye, J. (1989). Les microorganismes auxiliares de l'etablissement des symbioses mycorhiziennes. *European Journal of Forest Pathology* **19,** 54–64.

Dodd, J. C., Burton, C. C., Burns, R. G., & Jeffries, P. (1987). Phosphatase activity associated with the roots and the rhizosphere of plants infected with vesicular-arbuscular mycorrhizal fungi. *New Phytologist* **107**, 163–172.

Duc, G., Trouvelot, A., Gianinazzi-Pearson, V., & Gianinazzi, S. (1989). First report of non-mycorrhizal plant mutants (Myc) obtained in pea (*Pisum sativum I.*) and fababean (*Vicia faba l.*). *Plant Science* **60**, 215–222.

Duchesne, L. C., Ellis, B. E., & Peterson, R. L. (1989). Disease suppression by the ectomycorrhizal fungus *Paxillus involutus:* Contribution of oxalic acid. *Canadian Journal of Botany* **67**, 2726–2730.

Duponnois, R., & Garbaye, J. (1990). Some mechanisms involved in growth stimulation of ectomycorrhizal fungi by bacteria. *Canadian Journal of Botany* **68**, 2148–2152.

Ebel, J. (1986). Phytoalexin synthesis: The biochemical analysis of the induction process. *Annual Review of Phytopathology* **24**, 235–264.

Edwards, M. C., & Bowling, D. J. F. (1986). The growth of rust germ tubes towards stomata in relation to pH gradients. *Physiological and Molecular Plant Pathology* **29**, 185–196.

Elad, Y., Barak, R., & Chet, I. (1983). Possible role of lectins in mycoparasitism. *Journal of Bacteriology* **6**, 1431–1435.

Elias, K. S., & Safir, G. R. (1987). Hyphal elongation of *Glomus fasciculatus* in response to root exudates. *Applied and Environmental Microbiology* **53**, 1928–1933.

Epstein, L., Laccetti, L. B., & Staples, R. C. (1987). Cell-substratum adhesive protein involved in surface contact responses of the bean rust fungus. *Physiological and Molecular Plant Pathology* **30**, 373–388.

Estaun, V., Calvet, C., & Hayman, D. S. (1987). Influence of plant genotype on mycorrhizal infection: Response of three pea cultivars. *Plant and Soil* **103**, 295–298.

Finlay, B. B., Heffron, F., & Falkow, S. (1989). Epithelial cell surfaces induce *Salmonella* proteins required for bacterial adherence and invasion. *Science* **243**, 940–943.

Finlay, R. D., Ek, H., Odham, G., & Soderstrom, B. (1988). Mycelial uptake translocation and assimilation of nitrogen from ^{15}N-labelled ammonium by *Pinus sylvestris* plants infected with four different ectomycorrhizal fungi. *New Phytologist* **110**, 59–66.

Finlay, R. D., Ek, H., Odham, G., & Soderstrom, B. (1989). Uptake, translocation and assimilation of nitrogen from ^{15}N-labelled ammonium and nitrate sources by intact ectomycorrhizal systems of *Fagus sylvatica* infected with *Paxillus involutus*. *New Phytologist* **113**, 47–55.

Fries, N., & Birraux, D. (1980). Spore germination in *Hebeloma* stimulated by living plant roots. *Experientia* **36**, 1056–1057.

Fry, S. C. (1986). Cross linking of matrix polymers in growing cell walls of angiosperms. *Annual Review of Plant Physiology* **37**, 165–186.

Garbaye, J., & Bowen, G. D. (1989). Stimulation of ectomycorrhizal infection of *Pinus radiata* by some microorganisma associated with the mantle of ectomycorrhizas. *New Phytologist* **112**, 383–388.

Garbaye, J., Duponnois, R., & Wahl, J. L. (1990). The bacteria associated with *Laccaria laccata* ectomycorrhizas or sporocarps: Effect on symbiosis establishment on Douglas Fir. *Symbiosis* **9**, in press.

Garcia-Garrido, J. M., & Ocampo, J. A. (1988). Interaction between *Glomus mosseae* and *Erwinia carotovora* and its effects on growth in tomato plants. *New Phytologist* **110**, 551–555.

Garcia-Romera, I., Garcia-Garrido, J. M., Martinez-Molina, E., & Ocampo, J. A. (1990). Possible influence of hydrolytic enzymes on vesicular arbuscular mycorrhizal infection of alfalfa. *Soil Biology and Biochemistry* **22**, 149–152.

Garriock, M. L., Peterson, R. L., & Ackerley, C. A. (1989). Early stages in colonization of *Allium porrum* (leek) roots by the vesicular-arbuscular mycorrhizal fungus, *Glomus versiforme*. *New Phytologist* **112**, 85–92.

Gay, G., Rouillon, R., Bernillon, J., & Favre-Bonvin, J. (1988). IAA biosynthesis by the ectomycorrhizal fungus *Hebeloma hiemale* as affected by different precursors. *Canadian Journal of Botany* **67**, 2235–2239.

Gemma, J. N., & Koske, R. E. (1988). Pre-infection interactions between roots and the mycorrhizal fungus *Gigaspora gigantea:* Chemotropism of germ-tubes and root growth response. *Transactions of the British Mycological Society* **91**, 123–132.

Gianinazzi-Pearson, V., Branzanti, B., & Gianinazzi, S. (1990). In vitro enhancement of spore germination and early hyphal growth of a vesicular-arbuscular mycorrhizal fungus by host root exudates and plant flavonoids. *Symbiosis* **7**, 243–255.

Glenn, M. G., Chew, F. S., & Williams, P. H. (1985). Hyphal penetrations of *Brassica* (cruciferae) roots by vesicular-arbuscular mycorrhizal fungus. *New Phytologist* **99**, 463–472.

Godiard, L., Ragueh, F., Froissard, D., Leguay, J., Grosset, J., Chartier, Y., Meyer, Y., & Marco, Y. (1990). Analysis of the synthesis of several pathogenesis-related proteins in tobacco leaves infiltrated with water and with compatible and incompatible isolates of *Pseudomonas solanacearum*. *Molecular Plant-Microbe Interactions* **3**, 207–213.

Graham, J. H. (1982). Effect of citrus root exudates on germination of chlamydospores of the vesicular-arbuscular fungus *Glomus epigaeum*. *Mycologia* **74**, 831–835.

Graham, J. H. (1988). Interactions of mycorrhizal fungi with soilborne plant pathogens and other organisms: An introduction. *Phytopathology* **78**, 365–366.

Graham, J. H., Leonard, R. T., & Menge, J. A. (1981). Membrane-mediated decrease in root exudation responsible for phosphorus inhibition of vesicular-arbuscular mycorrhiza formation. *Plant Physiology* **63**, 548–552.

Hahn, M. G., Bonhoff, A., & Grisebach, H. (1985). Quantitative localization of the phytoalexin Glyceollin I in relation to fungal hyphae in soybean roots infected with *Phytophthora megasperma* f. sp. *glycinea*. *Plant Physiology* **77**, 591–601.

Halverson, L. J., & Stanley, G. (1986). Signal exchange in plant-microbe interactions. *Microbiological Review* **50**, 193–225.

Hamer, J. E., Howard, R. J., Chumley, F. G., & Valent, B. (1987). A mechanism for surface attachment in spores of a plant pathogenic fungus. *Science* **239**, 288–290.

Hardham, A. R., & Suzaki, E. (1989). Glycoconjugates on the surface of spores of the pathogenic fungus *Phytophthora cinnamomi* studied using fluorescence and electron microscopy and flow cytometry. *Canadian Journal of Microbiology* **36**, 183–192.

Harley, J. L. (1985). Specificity and penetration of tissues by mycorrhizal fungi. *Proceedings of the National Academy of Science* **94**, 99–109.

Harley, J. L., & Smith, S. E. (1983). *Mycorrhizal Symbiosis.* Academic Press, New York.

Heath, M. C. (1990). Influence of carbohydrates on the induction of haustoria of the cowpea rust fungus *in vitro*. *Experimental Mycology* **14**, 84–88.

Heckman, J. R., & Angle, J. S. (1987). Variation between soybean cultivars in vesicular-arbuscular mycorrhiza fungi colonization. *Agronomy Journal* **79**, 428–430.

Hepper, C. M. (1984). Isolation and culture of VA mycorrhizal (VAM) fungi. In: *VA Mycorrhiza* (Ed. by C. L. Powell & D. Bagyaraj), pp. 95–112. CRC Press, Boca Raton, FL.

Hilbert, J. L., & Martin, F. (1988). Regulation of gene expression in ectomycorrhizas. I. Protein changes and the presence of ectomycorrhiza-specific polypeptides in the *Pisolithus-Eucalyptus* symbiosis. *New Phytologist* **110**, 339–346.

Hirrel, M. C., Mehravaran, H., & Gerdemann, J. W. (1978). Vesicular-arbuscular mycorrhizae in the Chenopodiaccae and Cruciferae: Do they occur? *Canadian Journal of Botany* **56**, 2813–2817.

Hoch, H. C., & Staples, R. C. (1987). Structural and chemical changes among the rust fungi during appressorium development. *Annual Review of Phytopathology* **25**, 231–247.

Hohl, H. R., & Balsiger, S. (1988). Surface glycosyl receptors of *Phytophthora megasperma* f. sp. *glycinea* and its soybean host. *Botanica Helvatica* **98**, 271–277.

Jabaji-Hare, S. H., Therien, J., & Charest, P. M. (1990). High resolution cytochemical study of the vesicular-arbuscular mycorrhizal association, *Glomus clarum* × *Allium porrum*. *New Phytologist* **114**, 481–496.

Judelson, H. S., & Michelmore, R. W. (1990). Highly abundant and stage-specific mRNAs in the obligate pathogen *Bremia lactucae*. *Molecular Plant-Microbe Interactions* **3**, 225–232.

Kaminskyi, S. G. W., & Day, A. W. (1984). Chemical induction of infection structures in rust fungi I. Sugars and complex media. *Experimental Mycology* **8**, 53–72.

Keon, J. P. R., Byrde, R. J. W., & Cooper, R. M. (1987). Some aspects of fungal enzymes that degrade plant cell walls In: *Fungal Infection of Plants.* (Ed. by G. F. Pegg & P. G. Ayres, pp. 133–157. Cambridge University Press, Cambridge, UK.

Kloepper, J. W., Lifshitz, R., & Schroth, M. N. (1988). *Pseudomonas* inoculants to benefit plant production. *151 Atlas of Science: Animal and Plant Sciences*, pp. 60–64.

Kobayshi, D. Y., Tamaki, S. J., & Keen, N. T. (1989). Cloned avirulence genes from the tomato pathogen *Pseudomonas syringae* pv. *tomato* confer cultivar specificity on soybean. *Proceedings of the National Academy of Science* **86**, 157–161.

Koide, R. T., & Li, M. (1990). On host regulation of the vesicular-arbuscular mycorrhizal symbiosis. *New Phytologist* **114,** 59–74.

Kope, H. H., & Fortin, J. A. (1989). Inhibition of phytopathogenic fungi *in vitro* by cell free culture media of ectomycorrhizal fungi. *New Phytologist* **113,** 57–63.

Koske, R. E. (1982). Evidence for a volatile attraction from plant roots affecting germ tubes of a VA mycorrhizal fungus. *Transactions of the British Mycological Society* **79,** 305–310.

Kosslak, R. M., Joshi, R. S., Bowen, B. A., & Paaren, H. E. (1990). Strain-specific inhibition of *nod* gene induction in *Bradyrhizobium japonicum* by flavonoid compounds. *Applied and Environmental Microbiology* **56,** 1333–1341.

Krishna, K. R., Balakrishna, A. N., & Bagyaraj, D. J. (1982). Interactions between a vesiular-arbuscular mycorrhizal fungi and *Streptomyces cinnamoneous* and their effects on finger millet. *New Phytologist* **92,** 401–405.

Lackie, S. M., Bowley, S. R., & Peterson, R. L. (1988). Comparison of colonization among half-sib families of *Medicago sativa* L. by *Glomus persiforme* (Daniels and Trappe) berch. *New Phytologist* **108,** 477–482.

Lamb, C. J., Lawton, M. A., Dron, M., & Dixon, R. A. (1989). Signals and transduction mechanisms for activation of plant defenses against microbial attack. *Cell* **56,** 215–224.

Leake, J. R., & Read, D. J. (1989). The biology of mycorrhiza in the Ericaceae. XIII. Some characteristics of the extracellular proteinase activity of the ericoid endophyte *Hymenoscyphus ericae. New Phytologist* **112,** 69–76.

Lei, J., & Dexheimer, J. (1988). Ultrastructural localization of ATPase activity in the *Pinus sylvestris/Laccaria laccata* ectomycorrhizal association. *New Phytologist* **108,** 329–334.

Leyval, C., & Bethelin, J. (1989). Interaction between *Laccaria laccata, Agrobacterium radiobacter,* and beech roots: Influence on P, K. Mg, and Fe mobilization from minerals and plant growth. *Plant and Soil* **117,** 103–110.

Li, A., & Heath, M. C. (1990). Effect of intercellular washing fluids on the interactions between bean plants and fungi nonpathogenic on beans. *Canadian Journal of Botany* **68,** 934–939.

Linderman, R. G. (1988). Mycorrhizal interactions with the rhizosphere microflora: The mycorrhizosphere effect. *Phytopathology* **78,** 366–371.

Linthorst, H. J. M., van Loon, L. C., van Rossum, C. M. A., Mayer, A., Bol, J. F., van Roekel, J. S. C., Meulenhoff, E. J. S., & Cornelissen, B. J. C. (1990). Analysis of acidic and basic chitinases from tobacco and petunia and their constitutive expression in transgenic tobacco. *Molecular Plant-Microbe Interactions* **3,** 252–258.

Long, S. R. (1989). *Rhizobium*-legume nodulation. Life together in the underground. *Cell* **56,** 203–214.

Longman, D., & Callow, J. A. (1987). Specific saccharide residues are involved in the recognition of plant root surfaces by zoospores of *Pythium aphanidermatum. Physiological and Molecular Plant Pathology* **30,** 139–150.

Malajczuk, N., Molina, R., & Trappe, J. M. (1982). Ectomycorrhiza formation in

Eucalyptus: I. Pure culture synthesis, host specificity and mycorrhizal compatibility with *Pinus radiata*. *New Phytologist* **91**, 467–482.

Manocha, M. S., & Chen, Y. (1989). Specificity of attachment of fungal parasites to their hosts. *Canadian Journal of Microbiology* **36**, 69–76.

Martin, F., Stewart, G. R., Genetet, I., & Mourot, B. (1988). The involvement of glutamate dehyfrogenase and glutamine synthetase in ammonia assimilation by the rapidly growing ectomycorrhizal ascomycete, *Cenococcum geophilum* Fr. *New Phytologist* **110**, 541–550.

Martin, R., Ramstedt, M., & Soderhall, K. (1987). Carbon and nitrogen metabolism in ectomycorrhizal fungi and ectomycorrhizas. *Biochimie* **69**, 569–581.

Marx, C., Dexheimer, J., Gianinazzi-Pearson, V., & Gianinazzi, S. (1982). Enzymatic studies on the metabolism of vesicular-arbuscular mycorrhizas. IV. Ultracytoenzymological evidence (ATPase) for active transfer processes in the host-arbuscule interface. *New Phytologist* **90**, 37–43.

Massicotte, H. B., Peterson, R. L., Ackerley, C. A., & Melville, L. H. (1990). Structure and ontogeny of *Betula alleghaniensis–Pisolithus tinctorius* ectomycorrhizae. *Canadian Journal of Botany* **68**, 579–593.

Matthysee, A. G., Holmes, K. V., & Gurlitz, R. H. G. (1982). Elaboration of cellulose fibrils by *Agrobacterium tumefaciens* during attachment to carrot cells. *Journal of Bacteriology* **145**, 583–595.

McAfee, B. J., & Fortin, J. A. (1988). Comparative effects of the soil microflora on ectomycorrhizal inoculation of conifer seedlings. *New Phytologist* **108**, 443–449.

Melin, E. (1959). Mycorrhiza. In: *Handbuch der Pflanzenphysiologic*, Vol. 11 (Ed. by W. Ruhland), pp. 605–638. Springer, Berlin.

Meyer, J. R., & Linderman, R. G. (1986a). Response of subterranean clover to dual inoculation with vesicular arbuscular mycorrhizal fungi and a plant growth-promoting bacterium *Pseudomonas putida*. *Soil Biology and Biochemistry* **18**, 185–190.

Meyer, J. R., & Linderman, R. G. (1986b). Selective influence on populations of rhizosphere or rhizoplane bacteria and actino-mycetes by mycorrhizas formed by *Glomus fasciculatus*. *Soil Biology and Biochemistry* **18**, 191–196.

Miller, R. H., Berryman, A. A., & Ryan, C. A. (1986). Biotic elicitors of defense reactions to lodgepole pine. *Phytochemistry* **25**, 611–612.

Molina, R., & Trappe, J. M. (1982). Patterns of ectomycorrhizal host specificity and potential among Pacific Northwest conifers and fungi. *Forest Science* **28**, 423–458.

Morandi, D., Bailey, J. A., & Gianinazzi-Pearson, V. (1984). Isoflavonoid accumulation in soybean roots infected with vesicular-arbuscular mycorrhizal fungi. *Physiological Plant Pathology* **24**, 357–364.

Mort, A. J., & Grover, P. B. (1988). Characterization of root hair cell walls as potential barriers to the infection of plants by rhizobia. The carbohydrate component. *Plant Physiology* **86**, 638–641.

Mugnier, J., & Mosse, B. (1987). Spore germination and viability of vesicular-arbuscular mycorrhizal fungus *Glomus mosseae*. *Transactions of the British Mycological Society* **88**, 411–413.

Nylund, J. (1987). The ectomycorrhizal infection zone and its relations to acid polysaccharides of cortical cell walls. *New Phytologist* **106,** 505–516.

Pacovsky, R. S. (1989). Carbohydrate, protein and amino acid status of *Glycine–Glomus–Bradyrhizobium* symbioses. *Physiologia Plantarum* **75,** 1–9.

Paulitz, T. C., & Linderman, R. G. (1989). Interactions between fluorescent pseudomonads and VA mycorrhizal fungi. *New Phytologist* **113,** 37–45.

Piche, Y., & Peterson, R. L. (1985). Early events in root colonization by mycorrhizal fungi. In: *Proceedings of the 6th North American Conference on Mycorrhizae* (Ed. by R. Molina), pp. 179–180. Oregon State University, Corvallis.

Piche, Y., Peterson, R. L., Howarth, M. J., & Fortin, J. A. (1983). A structural study of the interface between the ectomycorrhizal fungus *Pisolithus tinctorium* and *Pinus strobus* roots. *Canadian Journal of Botany* **61,** 1185–1193.

Ronald, P., & Soderhall, K. (1985). Phenylalanine ammonia lyase and peroxidase activity in mycorrhizal and nonmycorrhizal short roots of scots pine, *Pinus sylvestris* L. *New Phytologist* **101,** 487–494.

Sauter, M., & Hager, A. (1989). The mycorrhizal fungus *Amanita muscaria* induces chitinase activity in roots and in suspension-cultured cells of its host *Picea abies*. *Planta* **179,** 61–66.

Schippers, B. (1988). Biological control of pathogens with rhizobacteria. *Philosophical Transactions of the Royal Society of London* **318,** 283–293.

Schwab, S. M., Johnson, E. L. V., & Menge, J. A. (1982). Influence of simazine on vesicular-arbuscular mycorrhiza formation in *Chenopodium quinona* Wild. *Plant and Soil* **64,** 283–287.

Schwab, S. M., Leonard, R. T., & Menge, J. A. (1983). Quantitative and qualitative comparison of root exudates of mycorrhizal and nonmycorrhizal plant species. *Canadian Journal of Botany* **62,** 1227–1231.

Sharon, N., & Lis, H. (1989). Lectins as cell recognition molecules. *Science* **246,** 227–234.

Smith, G., Kinjne, J. W., & Lugtenberg, B. J. J. (1987). Involvement of both cellulose fibrils and a Ca^{2+} dependent adhesion in the attachment of *Rhizobium leguminosarum* to pea. *Journal of Bacteriology* **169,** 4294–4301.

Smith, G., Logman, T. J. J., Kijne, J. W., & Lutenberg, B. J. J. (1988). Nitrogen fixation: Hundred years after. In: *Rhizobiaceae Cells Bind to Pea Root Hair Tips through Ca^{2+}-Dependent Adhesins* (ed. by H. Bothe, F. J. De Brijn, & W. E. Newton), p. 456. Gustav Fisher, Stuttgart.

Smith, S. E., & Gianinazzi-Pearson, V. (1988). Physiological interactions between symbionts in vesicular-arbuscular mycorrhizal plants. *Annual Review of Plant Physiology and Plant Molecular Biology* **39,** 221–244.

Smith, S. E., & Smith, F. A. (1990). Structure and function of the interfaces in biotrophic symbioses as they relate to nutrient transport. *New Phytologist* **114,** 1–38.

Spanu, P., & Bonfante-Fasolo, P. (1988). Cell-wall-bound peroxidase activity in roots of mycorrhizal *Allium porrum*. *New Phytologist* **109,** 119–124.

Spanu, P., Boller, T., Ludwig, A., Wiemken, A., Faccio, A., & Bonfante-Fasolo, P. (1989). Chitinase in roots of mycorrhizal *Allium porrum:* Regulation and localization. *Planta* **177,** 447–455.

Sprent, J. I. (1989). Which steps are essential for formation of functional legume nodules? *New Phytologist* **111,** 129–153.

Staples, R. C., Grambow, H. J., & Hoch, H. C. (1983). Potassium ion induces rust fungi to develop infection structures. *Experimental Mycology* **7,** 40–46.

Staples, R. C., Hassouna, S., Lacetti, L., & Hoch, H. C. (1984). Metabolic alterations in bean rust germlings during differentiation induced by the potassium ion. *Experimental Mycology* **8,** 183–192.

Staples, R. C., Hassouna, S., & Hoch, H. C. (1985). Effect of potassium on sugar uptake and assimilation by bean rust germlings. *Mycologia* **77,** 248–252.

Stein, A., & Fortin, A. J. (1990). Pattern of root initiation by an ectomycorrhizal fungus on hypocotyl cuttings of *Larix laricina. Canadian Journal of Botany* **68,** 492–498.

Stein, A., Fortin, A. J., & Vallee, G. (1990). Enhanced rooting of *Picea mariana* cuttings by ectomycorrhizal fungi. *Canadian Journal of Botany* **68,** 468–470.

Strange, K. K., Bender, G. L., Djordjevic, M. A., Rolfe, B. G., & Redmond, J. W. (1990). The *Rhizobium* strain NGR234 *nod*D1 gene product responds to activation by the simple phenolic compounds vanillin and isovanillin present in wheat seedling extracts. *Molecular Plant-Microbe Interactions* **3,** 214–220.

Strobel, N. E., & Sinclair, W. A. (1991). Role of flavonolic wall infusions in the resistance induced by *Laccaria bicolor* to *Fusarium oxysporum* in primary roots of Douglas-fir. *Phytopathology* **81,** 420–425.

Sylvia, D. M., & Sinclair, W. A. (1983). Phenolic compounds and resistance to fungal pathogens induced in primary roots of douglas-fir seedlings by the ectomycorrhizal fungus *Laccaria laccata. Phytopathology* **73,** 390–397.

Tari, P. H., & Anderson, A. J. (1988). Fusarium wilt suppression and agglutinability of *Pseudomonas putida. Applied and Environmental Microbiology* **54,** 2037–2041.

Theologis, A. (1986). Rapid gene regulation by auxin. *Annual Review of Plant Physiology* **37,** 407–438.

Thomson, B. D., Robson, A. D., & Abbott, L. K. (1990). Mycorrhizas formed by *Gigaspora calospora* and *Glomus fasciculatum* on subterranean clover in relation to soluble carbohydrate concentrations in roots. *New Phytologist* **114,** 217–235.

Van Etten, H. D., Matthews, D. E., & Matthews, P. S. (1989). Phytoalexin detoxification: Importance for pathogenicity and practical implications. *Annual Review of Phytopathology* **43,** 143–164.

Vesper, S. J. (1987). Production of pili (fimbriae) by *Pseudomonas fluorescens* and correlation with attachment to corn roots. *Applied and Environmental Microbiology* **53,** 1397–1405.

Weller, D. M. (1988). Biological control of soilborne plant pathogens in the rhizosphere with bacteria. *Annual Review of Phytopathology* **26,** 379–407.

Whalen, M. C., Stall, R. E., & Staskawicz, B. J. (1988). Characterization of a gene from a tomato pathogen determining hypersensitive resistance in non-host species and genetic analysis of this resistance in bean. *Proceedings of the National Academy of Science* **85,** 6743–6747.

Will, M. E., & Sylvia, D. M. (1990). Interaction of rhizosphere bacteria, fertilizer, and vesicular-arbuscular mycorrhizal fungi with sea oats. *Applied and Environmental Microbiology* **56,** 2073–2079.

Woloshuk, C. P., & Kolattukudy, P. E. (1986). Mechanism by which contact with plant cuticle triggers cutinase gene expression in the spores of *Fusarium solani* f. sp. *pisi*. *Proceedings of the National Academy of Science* **83,** 1704–1708.

Wyss, P., Mellor, R. B., & Wiemken, A. (1990). Vesicular-arbuscular mycorrhizas of wild-type soybean and non-nodulating mutants with *Glomus mosseae* contain symbiosis-specific polypeptides (mycorrhizins), immunologically cross-reactive with nodulins. *Planta* **182,** 22–26.

3

The Cellular Basis of Plant–Fungus Interchanges in Mycorrhizal Associations

Paola Bonfante-Fasolo and S. Scannerini

SUMMARY. Higher plants and mycorrhizal fungi interact in many different ways, giving rise to mutual associations ranging from ectomycorrhizae to endomycorrhizae. Nutritional interchanges involving carbon, nitrogen, and phosphorus are brought about by cell walls or by specialized areas (interfaces) that mediate the interactions between the plasma membranes of the two partners. The aim of this review is (1) to analyze the cellular structures, organelles, and compartments involved in plant–fungal interchanges in ecto- and in vesicular-arbuscular (VAM) mycorrhizae and (2) to discuss the functional morphological data in the context of nutritional exchanges. Some common mechanisms can be recognized acting in functioning ecto- and endomycorrhizal associations. First, morphogenetic events leading to an increase of the exchange surface occur in both partners (branching of the hyphae in the Hartig net or in the arbuscular hyphae, increased branching of the root system, and transfer-like structures of laying down of new cell wall material on the part of the host). The second mechanism consists in important changes within the enzymic and electrophysiological features of partner plasma membranes. The third is given by the progressive modifications of the exchange surfaces in their permeability characteristics. The complex bidirectional transport in mycorrhizae requires regulation mechanisms that control the changes observed at organismal, tissue, cellular, and molecular levels and the resulting cellular compatibility. Recent information dealing with gene activation and protein expression changes following fungal colonization are discussed in the context of these complexities.

Introduction

Mycorrhizal plants comprise over 2,500 different species—from liverworts to monocotyledons; their symbiotic fungi may be Zygomycotina, Ascomycotina, and Basidiomycotina. Furthermore, mycorrhizae appear equally successful in

Table 3.1. Summary of Characteristics of Main Types of Mycorrhizae and Their Bidirectional Nutrient Exchange

	Symbiont range	Nutritional interactions	
		Symbiont → host	Host → symbiont
Vesicular-arbuscular mycorrhizae	Four genera of Zygomycotina	Mineral nutrients	Carbohydrate
Ectomycorrhizae	Many fungi, including species from 25 families of Basidiomycotina, 7 families of Ascomycotina, and 1 genus of Zygomycotina (Endogone)	Mineral nutrients	Carbohydrate
Ericoid	Many isolates form steriles in culture: one has been identified as *Hymenoscyphus* (Ascomycotina)	Nutrients especially ammonia and organic nitrogen	Carbohydrate
Arbutoid	Those identified so far are mostly Basidiomycotina, which also form ectomycorrhizae	Probably mineral nutrients	Probably carbohydrate
Monotropoid		Carbohydrates and mineral nutrients	None known
Orchid mycorrhizae	Many isolates form sterile mycelia referable to "form" genus *Rhizoctonia*, induced to form sexual stages, referable to about 8 genera of Basidiomycotina including some pathogens	Juvenile achlorophyllous hosts: carbohydrates and mineral nutrients	None known
		Adult photosynthetic hosts: probably mineral nutrients	None known

	Host range
Vesicular-arbuscular mycorrhizae	Many plants species, including representatives of bryophytes, pteridophytes, gymnosperms, and many angiosperms
Ectomycorrhizae	Many trees and shrubs, especially of temperate regions
Ericoid	Members of Ericales with fine hair roots, especially Ericoideae, Vaccinoideae, Rhododendroideae, Epacridaceae, and Empetraceae
Arbutoid	Members of Ericales with sturdier roots including *Arbutus*, *Arctostaphylos*, and Pyrolaceae
Monotropoid	Achlorophyllous members of Ericales such as *Monotropa*, *Sarcodes*, *Pterospora*
Orchid mycorrhizae	All members of the Orchidaceae

From Smith and Douglas (1987).

colonizing diverse environments, from alpine and boreal zones to tropical forests and grasslands. This ever-changing scenario makes mycorrhizae a fascinating biological phenomenon despite the experimental difficulty and dearth of adequate theory.

However, the high degree of structural and functional complexity in mycorrhizal associations may be simpler when the various mycorrhizal types are observed at the cellular level. All mycorrhizae comprise an intricate system formed by cells belonging to two different eukaryotic organisms interacting and communicating to maintain a durable bidirectional nutrient exchanges (Table 3.1).

Although this may be an oversimplification, it is a key to understanding the cascading events leading to those reactions visible at an organismic level, i.e., growth improvement of the autotroph host and life cycle facilitation of the heterotroph fungus.

The Concept of Interface as a Pathway for Nutritional Exchanges

All plant cells, whether they belong to the same or to different tissues or even to different species, communicate through their cell walls. Thus the cell wall is not only viewed as a passive extracellular matrix that maintains shape and rigidity but also as a dynamic structure, a source of information, involved in self and nonself recognition processes (Bolwell, 1988).

In mycorrhizae, all short-range interactions (i.e., direct communication mediated by molecules influencing cells through physical contact) require walls that enable communication and nutrient exchanges between partner plasma membranes. In fact, channels such as plasmodesmata that are structures related to the junctions of animal cells do not occur between plant and fungal mycorrhizal cells (Bonfante-Fasolo & Gianinazzi-Pearson, 1986). This situation gives rise to the establishment of interface structures, where the apoplast is of paramount importance (Fig. 3.1). A comprehensive review of the interface concept was first given in 1973 by Bracker and Littlefield limited to host–parasite interfaces. Current views hold that a succession of structurally different interfaces is a common feature in walled organisms and is shared by a wide number of plants and fungi involved in different types of symbiotic nutritional relationships from mutualistic to pathogenic (Scannerini & Bonfante-Fasolo, 1983; Jeffries, 1987) (Table 3.2).

Recent reviews (Smith & Gianinazzi-Pearson, 1988) and other chapters of this volume (Bledsoe, Chapter 12; Miller & Allen, Chapter 9) deal with the physiology of mycorrhizae at the organismic level; this chapter will consider the cellular structures that make the signal and nutritional exchanges possible, the molecules involved, if known, and the regulation mechanisms. The morphological structures dealing with the exchanges will be analyzed in two different mycorrhizal systems (ectomycorrhizae and vesicular-arbuscular mycorrhizae) discussing new views gained by using a combination of cytochemical, biochemical, and physiological tools.

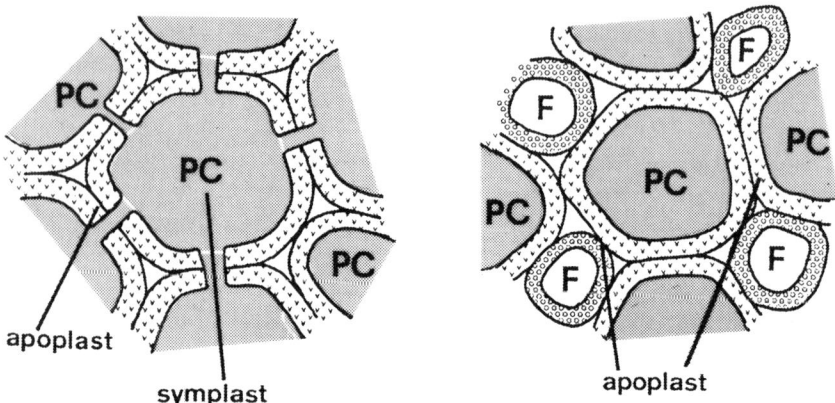

Figure 3.1. Diagram of the different contact zones existing among plant cells (PC) in (A) and between plant and fungal cells (F) in (B). In the latter case, symplastic connections between the partners do not exist. Drawn by Dr. Silvia Perotto.

Table 3.2. Interfaces Occurring in Mycorrhizal Associations

Mycorrhiza	Interface
Vesicular-arbuscular mycorrhizae (VAM)	Fungal wall–host wall
	Fungal wall–host matrix material–host plasmalemma
Ectomycorrhizae	Fungal wall–host wall
	Fungal wall–cementing material–host wall
Ericoid mycorrhizae	Fungal wall–host wall
	Fungal wall–matrix material–host plasmalemma
Ectendomycorrhizae in *Ericales* (arbutoid mycorrhizae)	Fungal wall–host wall
	Fungal wall–matrix material–host plasmalemma
Ectendomycorrhizae in *Ericales* (monotropoid mycorrhizae)	Fungal wall–host wall
	Fungal wall–host wall extensions–host plasmalemma
Endomycorrhizae in orchids	Fungal wall–host wall
	Fungal wall–matrix material–host plasmalemma

From Scannerini and Bonfante-Fasolo (1983).

Structural Bases for Partner Exchanges in Ectomycorrhizae

The host plant has long been held to be the source of carbon for the fungal partner in ectomycorrhizal plants and ectomycorrhizal fungi have been thought to require simple carbohydrates for their symbiotic growth increasing the uptake of any nutrient (Harley & Smith, 1983). More recently the structures involved in uptake and transport have been given greater attention, i.e., extraradical hyphae (Read, Chapter 4, this volume), fungal sheath, host epidermal cells, and the interface zone in the Hartig net. The latter is usually thought to be the most prominent site for transfer of nutrients between host and fungus (Kottke & Oberwinkler, 1987).

Complete descriptions of the morphology and ultrastructure of ectomycorrhizae can be found in Harley and Smith (1983), Scannerini and Bonfante-Fasolo (1983), Strullu (1985), and Kottke and Oberwinkler (1986).

Fungal Structures

The fungal sheath is a pseudoparenchymatous structure, formed by interwoven hyphae, that follows various patterns. When seen at an ultrastructural level, several regions can be identified within the mantle. Generally, the outer regions consist of highly vacuolated hyphae, whereas the inner regions are characterized by hyphae with an active cytoplasm. Cementing material usually occurs between hyphae consisting of remnants of incorporated host cells, including root cap cells or cortical cells and root hairs (Duddridge, 1985) and material of fungal origin (Nylund, 1987). Chemically, these materials are comparable to phenols (Ashford & Allaway, 1985). When an ectomycorrhizal fungus develops a mantle, it stops growing in bundles or in isolated hyphae, and organizes a more complicate structure (Figs. 3.2 and 3.3). Two reasons may account for the dramatically different branching pattern of the fungus during its symbiotic status. First, the root tissues develop a three-dimensional substrate that differs greatly from the flat medium of a Petri dish. Second, the root apex, one of the most important sites for hormone production, could influence fungal morphogenesis. Thus, the formation of an ectomycorrhiza is characterized by the well-known changes in the root morphology (Clowes, 1981) and by less well-known fungal modifications, caused by some potential morphogenetical factor (Nylund, 1988). The strongest impact of such a hypothetical factor can be observed during the Hartig net phase. Here, the longitudinal fungal growth through the intercellular spaces is restricted. Hyphae are oriented transversally to the root axis and branch irregularly, enveloping the cortical cells like fingers of a hand (Kottke & Oberwinkler, 1986) (Fig. 3.2). In addition, the fungal cells of the Hartig net form intimate juxtapositions of their cell walls, forming structures that have been compared to coenocytic transfer cells. This system is thought to act as an amplification of the apoplastic–symplastic surface exchange in *Alnus crispa* and *Alpova diplophloeus* (Massicotte et al., 1987).

Cytologically, the hyphae involved in mycorrhizal formation only partly differ from those growing saprophytically. Glycogen is stored both in the fungus growing in pure culture and in the symbiotic state (Fig. 3.4a). Biochemical analyses demonstrate that mannitol, trealose, and arabitol are the most important storing sugars for ectomycorrhizal fungi (Martin et al., 1988). They probably accumulate in vacuoles. Polyphosphate granules, usually occurring inside vacuoles and easily observable after staining with methachromatic stains act as a storage place for phosphorus inside the fungal cytoplasm (Strullu et al., 1983). Detailed analyses

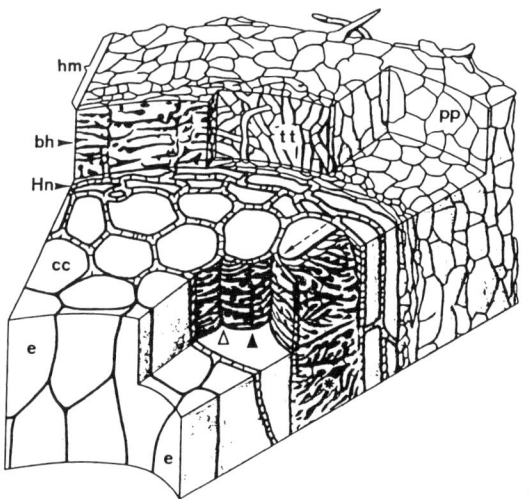

Figure 3.2. Block diagram of hyphal mantle (hm) and Hartig net (Hn): the main growth direction of hyphae is transversal to the root axis (intermitted arrow); branching of hyphae with rare septation (asterix); indentation pattern (black arrow); fountain-like separation of hyphae (open arrow); changes of structures within the mantle. pp, pseudoparenchymatous; tt, tubal cross texture; bh, branched hyphae on the outermost cortex cells; cc, cortex cell; e, endodermis. From Kottke and Oberwinkler (1987) with permission.

show that granules contain P, Cl, K, and Ca, with P and Ca being the most important elements.

Experiments during which *Fagus* mycorrhizae were incubated with KH_2PO_4 demonstrated a strong increase in P content of the mycorrhizal roots (Strullu, 1985). Quantitative microanalysis shows that 2.5–3.7% of the total P is inside the fungal granules. This concentration of P and the presence of Ca and P in the cell walls of the fungal symbiont of *Fagus* mycorrhizae, suggest a transfer mechanism from the cytoplasmic granules toward the interfacial zone. According to Strullu, the granules storing polyanions balanced by cations may regulate the exchanges between the plant and the fungus and this can be considered a feature shared by all the different types of mycorrhizae. However, this mechanism of storing P and Ca is a pattern shared by nonsymbiotic fungi, too. Cairney et al. (1988) recently found that polyphosphate granules are stored in rhizomorphs of saprophytic fungi. The presence of such granules in ectomycorrhizal fungi growing in pure culture strongly suggests that this storage is not related to the symbiotic phase, but it is a metabolic pathway common to many fungi.

A number of experiences have shown that P stored in the fungus subsequently

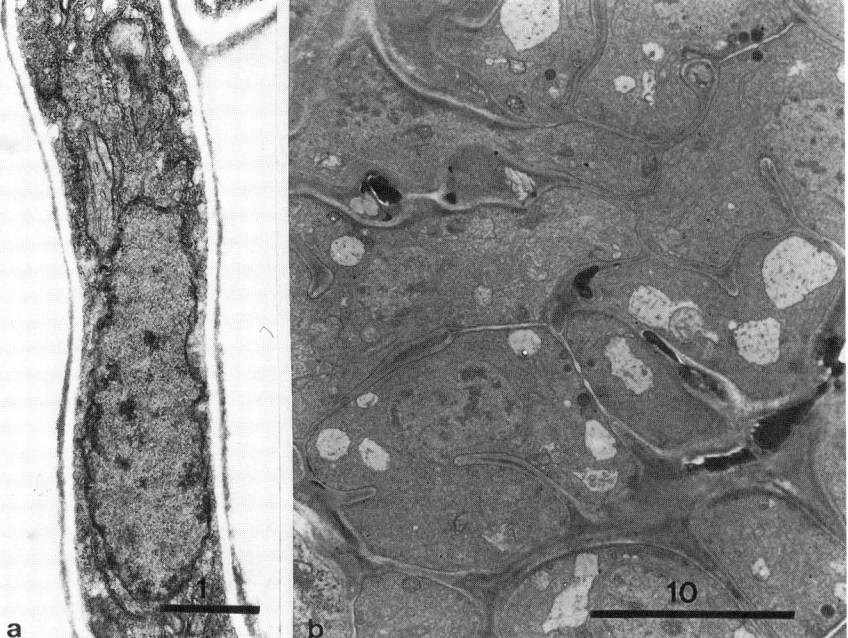

Figure 3.3. Ultrastructural features of *Tuber albidum* growing in pure culture (a) and in association with a host plant (b). In the first case, the fungus shows a typical apical growth; in the second one it develops the mantle as a pseudoparenchymatous structure.

moves slowly into the host by processes that are oxygen dependent and sensitive to metabolic inhibitors. However, no direct evidence of P solubilization, the involvement of specific enzymes or of the direct transfer from fungus to host has been demonstrated (Harley & Smith, 1983; Smith & Smith, 1990).

Fungal hyphae differ in the sheath formation phase both inside the cytoplasm and on the surface. Hartig net hyphae of *Alpova* synthesize large amounts of rough endoplasmic reticulum (Massicotte et al., 1987), while a Basidiomycete colonizing *Pisonia* roots changes its surface features (Ashford et al., 1988). By using an apoplastic tracer, Cellufluor, Ashford and his co-workers found that the fungal hyphae became progressively impermeable to the staining following their process of differentiation. In the tip region, where the sheath is not yet fully differentiated, the apoplastic tracer penetrates, but behind this zone, where the interhyphal spaces are filled with extracellular material, the sheath becomes impermeable. These findings mean that there is a blockage of the sheath apoplast that could provide a sealed compartment at the fungus–soil interface, with a resulting increase of the efficiency of transfer between partners. In fact they suggest that the sealed compartment prevents or at least reduces leakage back into the soil. In this way the symplastic pathway within the fungus may be very

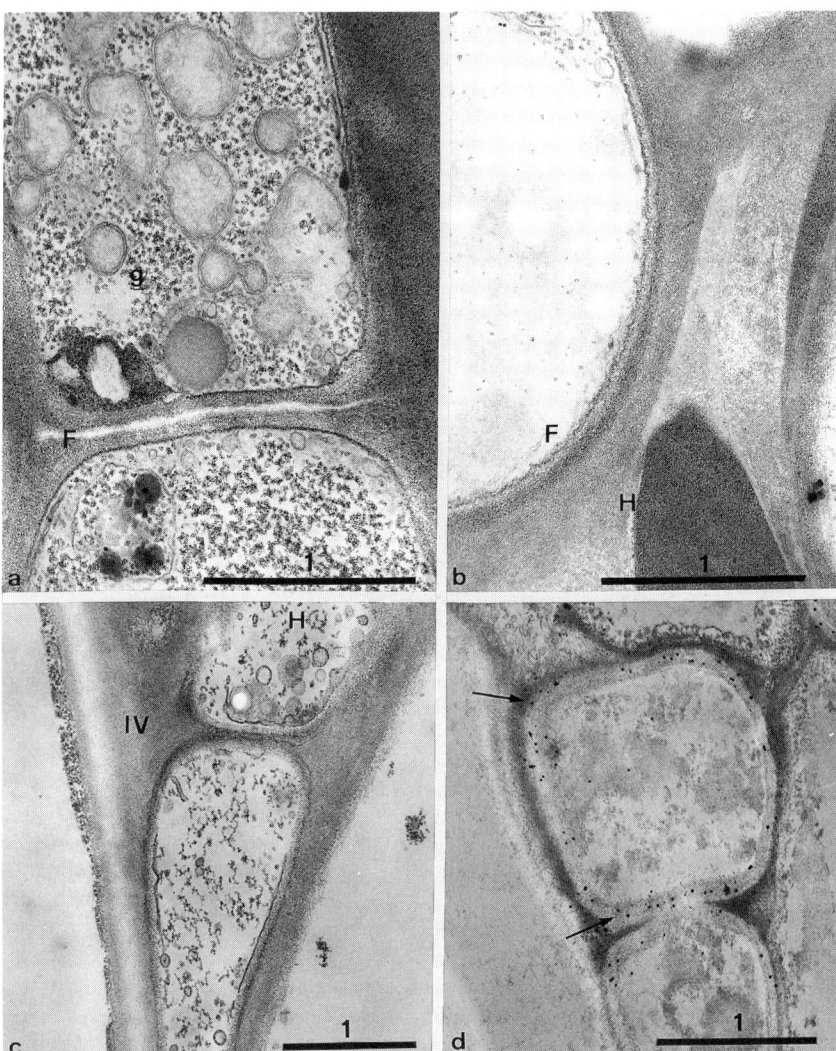

Figure 3.4. Ultrastructural features of *Tuber melanosporum* forming an ectomycorrhiza with *Corylus avellana*. (a) Strong depostion of glycogen granules (g) in a truffle hypha (F) as revealed by the Patag reaction for polysaccharide localization. (b) Interface zone between the mantle hyphae (F) and some remnants of the host surface cells (H): the walls of both partners are recognizable. (c) Interface zone in the Hartig net zone: an *Involving layer* is interposed between host and fungal cell walls. (d) Labeling of thin sections with wheat germ agglutinin linked to colloidal gold to reveal chitin. Gold granules (arrows) occur only on the fungal wall. (Bonfante-Fasolo, unpublished).

important for the radial transfer of material from the soil hyphae to the Hartig net hyphae (Ashford et al., 1988).

Host Partner

Epidermal, cortical, and cap root cells of plants belonging to gymnosperms and dicotyledons are involved in ectomycorrhiza formation.

This great variety of mycorrhizae makes it difficult to present a uniform picture. However, a common feature is that the anatomy of the root changes much more than its cytology. On invasion by mycorrhizal fungi, changes such as the shape and size of the absorbent roots, the formation of dichotomous pine roots, and the disappearance of the cell cap are evident (Clowes, 1981; Fusconi, 1983; Kottke & Oberwinkler, 1987; Pichè et al., 1988), but when seen at the ultrastructural level, many of these differences taper off. For example, epidermal and cortical cells are usually highly vacuolated and the starch deposits are common, even if in the pine mycorrhizae studied so far starch typically disappears (Strullu, 1985).

Although formation of a mycorrhiza always involves the transfer of sugars from the plant to the fungus, there are surprisingly few actual demonstrations of such transfer occurring at the cellular level. As far as we know, the only demonstration of such transfer paradoxically deals with a transfer of C from the fungus to a plant! To demonstrate the occurrence of a direct carbon exchange between plants through the interconnecting mycelium Duddridge et al. (1988) examined the patterns of distribution of ^{14}C-labeled material within the mycelial strands, the sheath, and the root tissues of pine colonized by *Suillus bovinus*. The results of the autoradiographic analysis showed that not only is there a transfer of assimilates from the donor to a receiver plant, but also that the exact path of carbon transfer can be observed. In this pathway, C is found in cytoplasmic hyphae, then in the cytoplasm of the host cortical cells; in addition, extensive labeling is present in the cytoplasm of the endodermal cells. However, there is no direct evidence of the sugar flowing from plant to fungus. The metabolic routes converting sucrose to fungal carbohydrates and other metabolites are not characterized as well as the transport molecules capable of crossing the interfacial zone (see further on in this chapter).

Woolhouse (1975) was the first to suggest that the controlled transfer of solutes required for a functioning mycorrhizal symbiosis is an active transport process. It depends on the membrane activity and may be ATPase dependent (Smith & Smith, 1986). In many different cellular systems, the active transport of cations and anions is related to the activity of the H^+-ATPase (Marrè, 1979) (Fig. 3.5). $H_2PO_4^-$ uptake is suggested to be obtained by a cotransport system with H^+ (Ullrich-Eberius, et al., 1984); other results show that pH lowering improves orthophosphate uptake in corn roots (Sentenac & Grignon, 1985). Moreover, fusicoccin, a fungal toxin that stimulates active H^+ extrusion from the cells, also stimulates the uptake of $H_2PO_4^-$ (Luttge & Clarkson, 1987). In ectomycorrhizae,

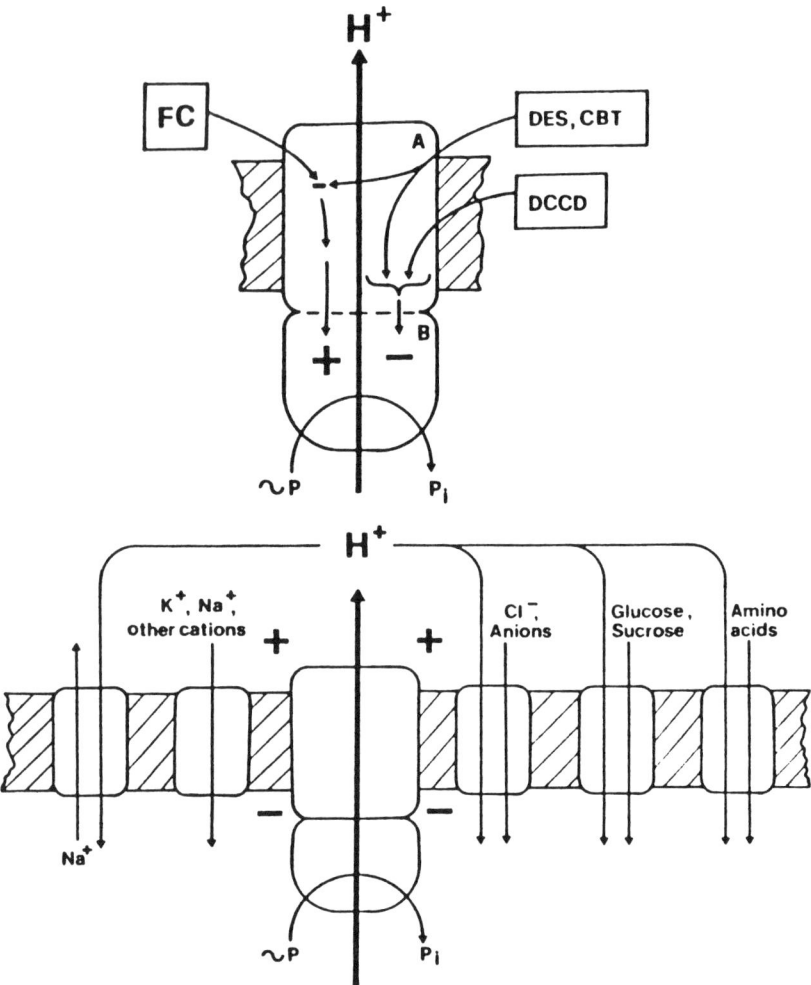

Figure 3.5. Structure of a plasma membrane proton pumping ATPase. Top: model of the pump activated by fusicoccin (FC) with the proposed interpretation of the effects of diethylstilbestrol (DES) and *Cercospora beticola* toxin (CBT) on the ATPase and on the FC binding activities. Bottom: model of utilization of FC-promoted H^+ extrusion for solute transport. From Marré (1979).

the activation of the H^+ pump with a improved activity of the H^+-ATPase of the host plasma membrane may be the mechanism generating the driving energy for P uptake from the fungus toward the host.

The observations by Lei and Dexheimer (1988) are important in order to support the previous hypothesis. In fact, they demonstrated that ATPase activity can be localized in cortical cell plasmalemma, limited to the active steps of the interaction between pine and *Laccaria laccata*. Such activity disappears in the degenerating cortical cells. A symmetrical activity was associated with the fungal plasmalemma, in the external hyphae, in the sheath, and in the Hartig net. Its activity was inhibited by diethylstilbestrol, an inhibitor of plasmalemma ATPase, but not by sodium fluoride, an inhibitor of acid phosphatases. The localization of such activity with the correct inhibitors may offer evidence for the mechanism underlying the transport between the two partners. Smith and Smith (1985, 1990) pointed out that other enzymes with ATP hydrolyzing ability are involved in the cytochemically demonstrated ATPase activity on the plasma membrane: for example, nonspecific phosphatase for which ATP is one substrate without being related to active transport. Moreover, phosphate influx can be under the control of cell wall-bound phosphatases, as recently demonstrated in barley roots and rose cells (Lee, 1988).

Host–Fungus Interface

Duddridge (1985, 1987) considers the interaction between ectomycorrhizal fungi and host plants a two-step process. The first is localized in the early stages of sheath formation when the hyphae make contact with the root surface. The latter is usually covered by mucilage polysaccharides, and the whole of the sheath eventually becomes embedded. Some of these substances stain positively with the Thiery reaction for polysaccharides and have been interpreted as one of the earliest response of both root cells and fungal hyphae during this interaction (Pichè et al., 1988).

The second step of host–fungus interaction occurs when the ectomycorrhiza is fully established. The interface between the fungal sheath and host tissues must change, since remnants of root cap cells, root hairs, and epidermal cells sometimes with phenolic contents are incorporated within the developed sheath (Fig. 3.4b).

Two distinct types of interfaces are usually recognized between the Hartig net hyphae and the host cortical cells (Duddridge, 1985; Kottke & Oberwinkler, 1987). The first type involves direct contact between the host cell wall and the wall of the fungus; the identification of both walls persists. A different situation occurs when an *involving layer* forms between the host and the fungus cell walls, and both symbiont walls lose their integrity (Fig. 3.4c). The development of an involving layer appears to be a feature of mature mycorrhizae, which is remarkably constant, regardless of host and fungus (Duddridge, 1985). However, understanding such a complex interface requires knowledge of the fungus and host cell walls as well as of the hydrolytic enzyme production by the fungus.

Currently, available information on these topics is limited. Even the chemical composition of cell walls in ectomycorrhizal fungi is very scanty. Mangin et al. (1986) reported that *Cenococcum geophilum* cell wall was comprised of high levels of mannose and glucose, a low quantity of chitin, and a high protein content (10%). The same components have been identified using cytochemical affinity techniques. *N*-Acetylglucosamine polymers were localized using the sugar-specific lectin (wheat germ agglutinin, WGA) linked to colloidal gold only in the cell walls of *Tuber melanosporum;* no labeling appearing on the host wall (Fig. 3.4d). Massicotte et al. (1987) found *L*-fucose, mannose, and *N*-acetylglucosamine in the *Alpova diplophleus* walls during its association with *Alnus crispa*. Oddly enough, they observed the same residues in the host walls and hypothesized that such sugar residues bound to a proteinaceous fraction in the host and fungal walls might be used in the synthesis of the elaborate epidermal ingrowths both of the plant and the fungus. From a morphological point of view, fungal cell walls are usually described as electron dense and layered, but their fibrillar organization is not so obvious (Fig. 3.4a).

The discussion of the host cell wall as structure involved in the interface formation is still more complicated. Cell walls are known to be highly heterogeneous systems, so the wide range of the host plants from gymnosperms to angiosperms offers the possibility of listing a large variety of compounds (McNeil et al., 1984). In addition, different cellular types are involved in the establishment of the interface. Thus it is difficult to know the exact components involved at the moment of the interaction. So far cytochemical studies are scanty and limited to gymnosperm mycorrhizae. Nylund (1987) observed that cortical cell wall of pine and spruce contain large amounts of acid polysaccharides, which are absent from all other cell walls studied. The same acid polysaccharides were recognized in the *involving layer* embedding the hyphae of the Hartig net in mycorrhizal roots. In conclusion, because of its staining properties, the involving layer was considered to be of host origin, and it was hypothesized that mycorrhizal formation was linked to the abundance of pectin-like materials. This might increase cell wall flexibility and fungal penetration (Nylund, 1987). Research on the structure of the host cell wall as well as of the interfacial material is greatly needed.

The application of ultrastructural affinity techniques in this field may prove very useful. These techniques make it possible to identify specific molecules inside the cell wall texture (Vian, 1986) and therefore to differentiate the dominions of the two partners clearly.

As with all other symbiotic fungi, the mycorrhizal ones must break down wall polymers locally since they must grow within host cell walls (ectomycorrhizal fungi) or penetrate host cell walls (endomycorrhizal fungi). They also must maintain host cell viability (Keon et al., 1987). The localized breakdown of wall polymers may result from a limited capacity of synthesizing hydrolytic enzymes or from the restriction by host cell walls of fungal enzyme production. The first hypothesis is confirmed by the observation that some ectomycorrhizal fungi such as *Laccaria laccata, Paxillus involutus,* and *Rhizopogon* reduce the viscosity of polygalacturonan. But unlike most necrotrophic parasites a low level of extracellular activity was still detected (Fig. 3.6) (Cooper & Fox, quoted in Keon et al., 1987). Interestingly, the same results were obtained by comparing the pectinolytic

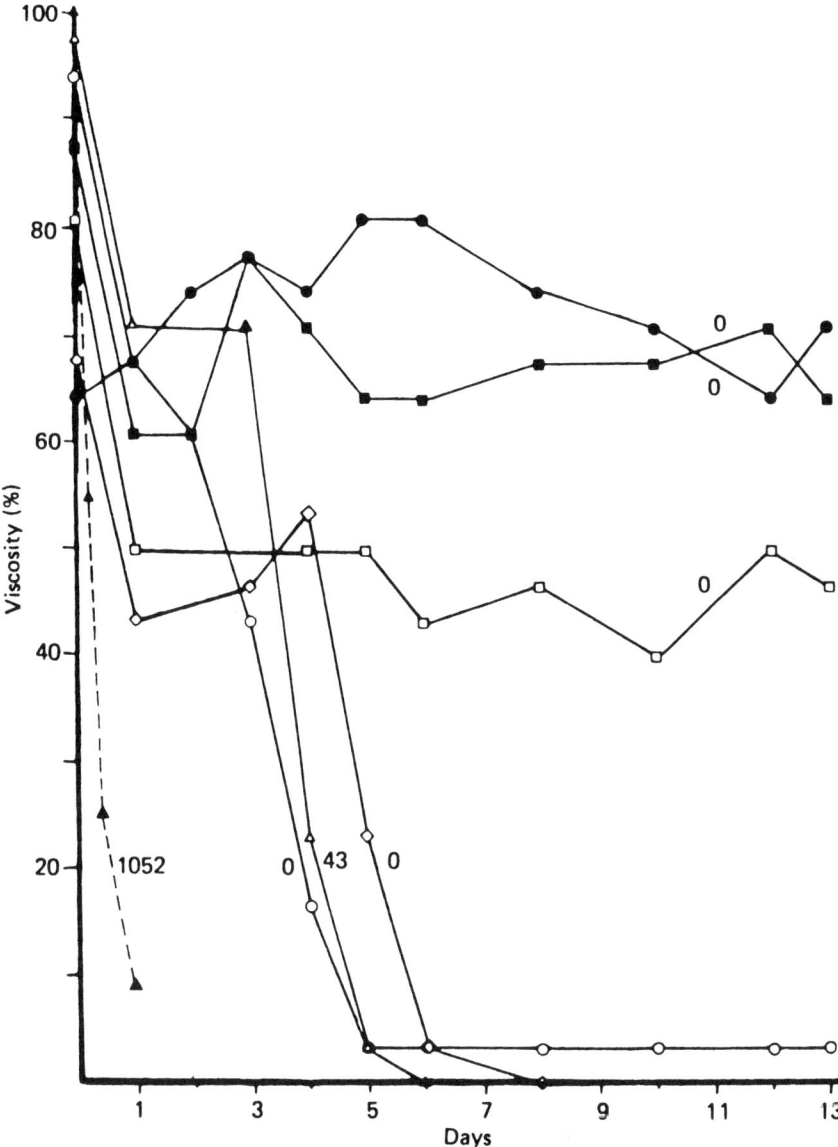

Figure 3.6. Extracellular polygalacturonase (PG) production and effect on polygalacturonan viscosity by mycelium from six mycorrhizal fungi. PG production and viscosity reduction by the necrotrophic *Botrytis fabae* (▲) are included for comparison. Mycorrhizal species were *Laccaria laccata* (△), *Paxillus involutus* (○), *Pisolithus tinctorius* (●), *Rhizopogon* sp. (□), *Rhizopogon roseolus* (isolate A) (■), and *R. roseolus* (isolate B) (◇). Reproduced from Cooper and Fox, in Keon et al. (1987).

activity of some ericoid strains (Cervone et al., 1988), where a very low polygalacturonase activity was detected. In conclusion, mycorrhizal fungi seem capable of producing limited quantities of endopolygalacturonases. On the other hand, the second hypothesis (i.e., fungal enzyme production is under the host control) is supported by the finding of endopolygalacturonase inhibiting proteins (PGIP) isolated from a variety of dicotyledonous plants (Cervone et al., 1989). A PGIP from *Phaseolus* converts fungal endopolygalacturonases into elicitors of plant defense response, acting as a plant resistance factor against fungal invasion. It can be suggested that plant inhibitors of the fungal enzymes may be produced in response to the fungal colonization in ectomycorrhizae, too (Keon et al., 1987). This is a very enthralling hypothesis, as many ectomycorrhiza-specific polypeptides have been found in the *Pisolithus–Eucalyptus* mycorrhiza (Hilbert & Martin, 1988) as a response to fungal colonization.

The Structural Bases for Partner Exchanges in Vesicular-Arbuscular Mycorrhizae

When a vesicular-arbuscular fungus colonizes a root, a complex fungal morphogenesis takes place before the establishment of a nutritional interchange between the two partners. Extraradical hyphae growing from a germinating spore or from infected roots make contact with the surface of the host and may form specialized structures called appressoria. From these, hyphae penetrate the root giving rise to different intraradical structures such as coils, inter-, and intracellular hyphae, vesicles, and highly specialized structures called arbuscules (Fig. 3.7). In addition, morphogenetic modifications of the whole root systems are induced (Berta et al., 1990), probably linked to modifications of the meristematic activity and to a slowing down of the mitotic cycle (Fusconi et al., 1986; Berta et al., 1991). The morphological aspects of the root–fungal interactions have been widely reported elsewhere (Harley & Smith, 1983; Scannerini & Bonfante-Fasolo, 1983; Bonfante-Fasolo, 1984; Strullu, 1985) as well as the complex cellular interactions between the two partners (Bonfante-Fasolo, 1987, 1988). Here, we mostly consider the cellular structures involved in the nutrient exchanges between the fungus and the host.

The Fungus

The complex morphogenesis of the fungus inside the root and the multitude of cellular interactions between the two partners are the means whereby the fungus takes up minerals from the soil and then delivers them to the inner cortical cells of the host near the central cylinder. On the other hand, sugar loading from the host cortical cells toward the fungus is probably located in the inner cortical zone near the phloem elements. Even though evidence purporting to demonstrate that exchanges occur mostly within the inner area of the cortex is weak, many

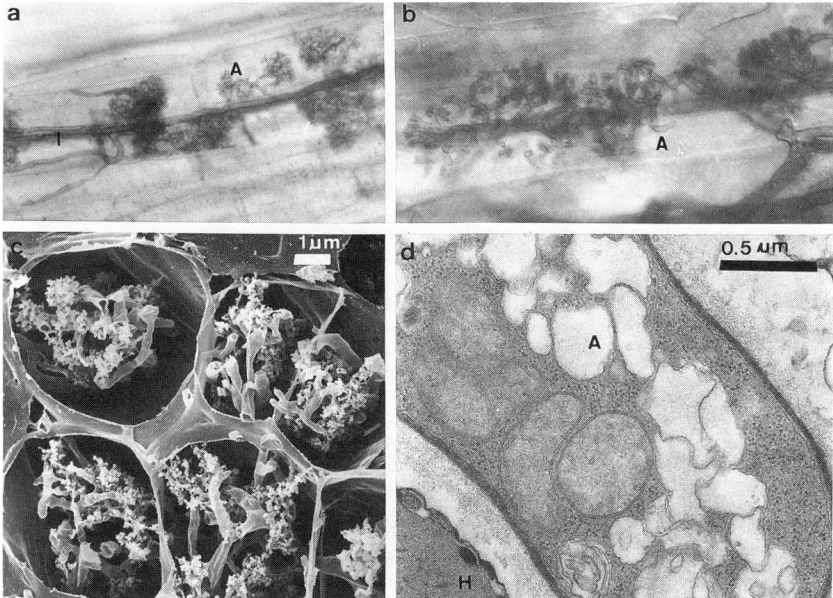

Figure 3.7. Morphological features of VAM fungi living inside roots of different plants. (a) Squash preparation of *Glomus versiforme* forming intercellular hyphae (I) and arbuscules (A) in *Allium porrum*. (b) Detail of a young arbuscule. (c) Arbuscules of *G. caledonium* living in *Ginkgo biloba* roots, as seen at scanning electron microscope. (d) Detail of the ultrastructure of a thin arbuscular hypha (A) of *G. fasciculatum* living in *Vitis vinifera*. (c and d) From Bonfante-Fasolo (1987).

morphological observations (Bonfante-Fasolo, 1988) suggest that the fungus possesses a surface that initially may play an absorbing role, then becomes impermeable. This might avoid a leakage back toward the soil and the root cells of the outer layers. Last, the fungal cell wall becomes thin and plastic to ensure an efficient mineral delivery only to the appropriate host cells.

Detailed observations of the fungal cell wall structure in all its developmental stages performed in some VA fungi and particularly on *Glomus versiforme* demonstrate that the fungal walls during the root colonization are extensively modified. Chlamydospores of many VAM fungi have a thick and complex wall that sometimes possesses a specific architecture called "helicoidal organization" (Bonfante-Fasolo & Vian, 1984) (Figs. 3.8a and 3.9). According to Bouligand's interpretation (1972), this architecture consists of planes formed by parallel and straight fibrils, each plane being laid down rotated by a small angle, with respect to the others, like the steps of a spiral staircase. The whole forms a ply structure with laminae constantly changing in orientation. Oblique sections generate the pattern of arcs seen in the spore, but the arc is an optical illusion, since it consists of short segments of parallel fibrils coming from each plane. This pattern is

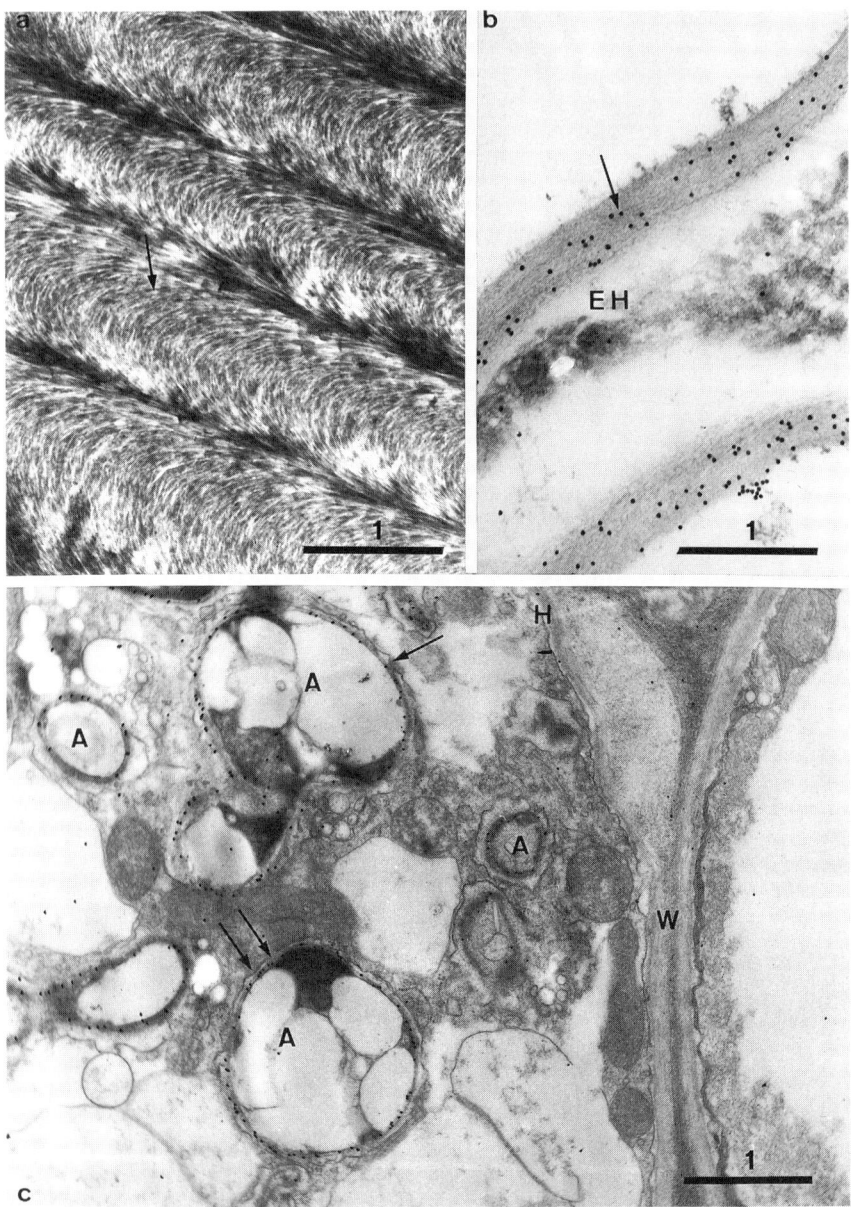

Figure 3.8. Different organizations of the cell wall in VAM fungi. (a) Helicoidal texture of the cell wall in the spore of *Glomus macrocarpum:* arcs of the same width (arrow) are separated by bundles of parallel fibrils. (b) Extraradical hypha (EH) of *G. versiforme* showing a thick fibrillar wall, labeled by gold granules linked to wheat germ agglutinin (WGA). (c) Arbuscular hyphae (A) of *G. versiforme* with a thin and amorphous wall, which is constantly labeled by the WGA lectin, demonstrating the presence of *N*-acetylglucosamine residues. W, host wall.

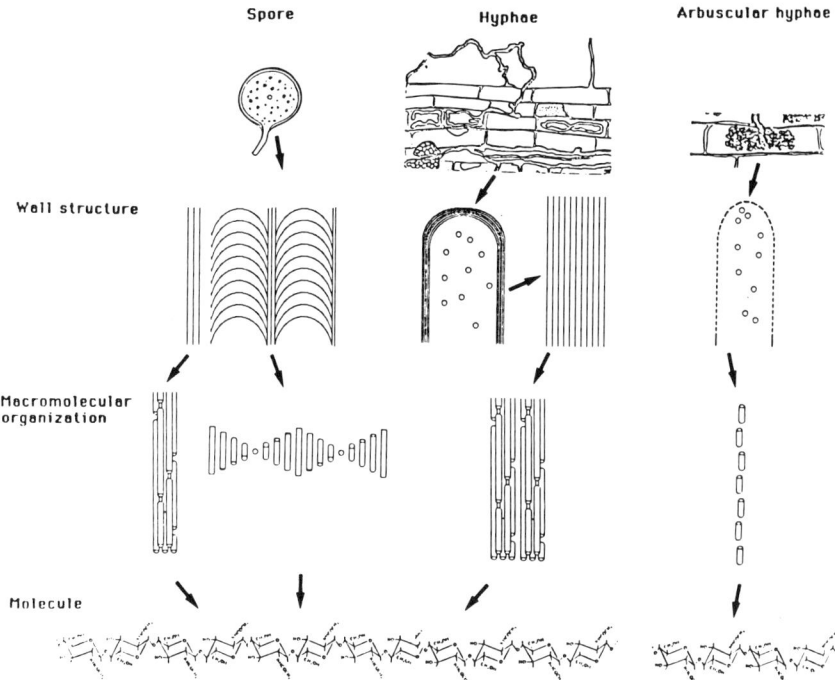

Figure 3.9. Schematic diagram showing the various VAM fungal structures, the wall ultrastructure, and the macromolecular chitin organization, that can be fibrillar (helicoidal or parallel) or amorphous. The last might correspond to a reduced number of N-acetylglucosamine residues in the chain. From Bonfante-Fasolo (1988).

produced by rhythmic activity of the systems involved in the cell wall component deposition and can be interpreted as a time register of cell metabolism (Vian & Roland, 1987). This organization is widespread in the living world; it occurs among crustacea and insects as well as among cell walls of higher plants and algae. However, it had not yet been reported in fungi, with the exception of VAM fungi (see Bonfante-Fasolo, 1987 for a review).

Morphological, biochemical, and cytochemical experiments based on the use of gold probes demonstrated that chitin of the spore cell wall is highly crystalline, with a fibrillar organization, and represents 27% of the cell wall fraction. When the spores germinate in soil, the mycelium has a fungal wall that is very thick in the soil, with a fibrillar texture containing a large amount of chitin (Fig. 3.8b). During the intraradical phase, the fungal wall shows striking changes, which are related to the different types of fungal structures. When the fungus forms coils, in the hypodermal cells, the wall is thick (300–500 nm) with a fibrillar texture rich in alkali-soluble polysaccharides and chitin. The wall thins out when intercellular hyphae run through the root and when the endophytes penetrate cortical host cells. The fungus stretches the host wall and causes the host plasmalemma to invaginate. It then regularly branches forming the arbuscules. The arbuscular wall is very distinctive. It is much thinner, from 30 to 50 nm, and is not fibrillar but amorphous in structure. However, oligomers of *N*-acetylglucosamine can be detected *in situ* (Grandmaison et al., 1988; Bonfante-Fasolo et al., 1991b) after the use of WGA, the lectin considered to possess specific binding sites for GlcNac-linked to FITC or gold (Fig. 3.8c). Moreover, quantification of WGA-gold labeling demonstrated that the granule density changes on the different fungal structures and reached the highest value exactly in the thin amorphous wall of the arbuscular hyphae (Bonfante-Fasolo et al., 1990b). Therefore we can assume that in the apical part of the infection unit, chitin polymerization is not accomplished or chitin does not occur in a crystalline form, while an active synthesis of its oligomers may occur. This agrees with the views on the apices of many filamentous fungi as the specific sites for chitin synthesis and on the plasticity of the fungal wall at these growing points (Wessels, 1986). The conclusion, from such observations, is that chitin or polymers of GlcNac occur throughout fungal morphogenesis, even though their aggregation forms differ (Fig. 3.9).

Recent observations on the fungal wall of *G. versiforme* showed that only the thin arbuscular branches were labeled, when the fluorescent-labeled lectin was used on the thick sections of a hand cut root (Bonfante-Fasolo et al., 1990b). A strong labeling of all the fungal structures was obtained exclusively when the lectin was used after alkali and oxidizing agents. These observations led to the conclusion that the thick-walled hyphae have their chitin partially hidden by nonchitinous cell wall components.

These findings may entail important physiological consequences. Encrusting substances, such melanin and sporopollenin, found primarily in the extraradical phases of the fungus (Grippiolo & Bonfante-Fasolo, 1984), or other unidentified components occurring in the intraradical thick-walled hyphae, modify the chemical physical properties of the fungal wall, thereby altering fungal permeability. In this way, there could be a blockage of the fungal apoplast providing a sealed compartment at the fungal–host interface. Exchanges between the two partners, involving the apoplasts of both organisms, occur only near the apex of the growing hyphae or in the arbuscular branches. In addition, the fact that chitin is embedded within other components suggests that it is sometimes protected from the attack of lytic enzymes, such as chitinases, produced by the plant in response to the first fungal colonization (Spanu et al., 1989). In conclusion, the progressive thinning and simplification of the fungal wall during root colonization may be an important means of ensuring an efficient transfer of nutrients to specific sites of the root and not along the whole contact surface created by the host and fungus.

VAM fungi cannot be grown in pure culture (or this is what our present state of knowledge suggests). We cannot, therefore, compare the cytological features of the symbiotic phase with those of a free phase, which is not influenced by the host. The cytological signs of the nutritional exchanges (organic carbon, nitrogen, and inorganic minerals) between the host and the fungus inside the fungal cytoplasm are very difficult to unveil. Ultrastructural observations show that host sugars are stored as glycogen and lipids in the fungal cells (Fig. 3.10d). Although morphological demonstrations of sugar flux are not available, biochemical studies on the carbohydrate, amino acid, and lipid composition of *Glycine–Glomus* symbioses demonstrated that the balance of these components changes following fungal colonization. Moreover, there is a specific storing of C-20 fatty acids, which are certainly of fungal origin (Pacovsky & Fuller, 1988; Pacovsky, 1989).

Transfer of phosphorus from fungus across the interface toward the host appears to occur in the form of P_i (Smith & Smith, 1990 for references). However, only indirect morphological evidence exists to demonstrate phosphorus flowing from fungal to plant cells in VA mycorrhizae. Scanning electron microscopy and X-ray dispersive microanalysis of onion VA mycorrhizae parenchyma cells detected phosphorus only in hosting-arbuscule cells, whereas sulfur, chlorine, potassium, and calcium were present in both arbuscular and nonarbuscular cells (Schoeknecht & Hatting, 1976). The results suggest an improved phosphorus flux across the plant cell–arbuscule interface.

Electron-dense granules, which are very abundant in the hyphae, represent phosphate storage (Strullu, 1985). They accumulate inside the vacuoles and are probably synthesized via an inducible polyphosphate kinase. Several energy-dispersive X-ray analyses demonstrated high levels of phosphorus and calcium within the electron dense granules of *Glycine max, Allium cepa,* and *Taxus* mycorrhizae (see Strullu, 1985, for a review of results). These electron-dense granules are abundant in the extraradical hyphae, in the intercellular hyphae, and in the arbuscule trunk. Last, they disappear in the thin arbuscular branches (Bonfante-Fasolo, 1984). The mechanisms by which phosphate is lost from hyphae in the presence of host cells are unknown, even though changes in fungal membrane permeability are inferred (Gianinazzi-Pearson & Gianinazzi, 1989). The vacuoles of VA mycorrhizal fungi as well as those from other polyphosphate-storing fungi possess an alkaline phosphatase, which can be localized by using cytochemical techniques (Gianinazzi et al., 1979). The role of this enzyme in phosphate metabolism is still a matter for speculation (Smith & Gianinazzi-Pearson, 1988).

The Host

Over 80% of land plants are hosts for VAM fungi, ranging from bryophytes and pteridophytes to gymnosperms and angiosperms. On the other hand, the species wherein mycorrhizal interactions have been studied in depth are very scanty, making it difficult to generalize as to the behavior of the host cellular structures involved in the exchanges with the fungus. However, some features are constant. Epidermal and cortical cells seem to be the only ones able to trigger fungal penetration, whereas the central cylinder and meristematic cells are not colonized.

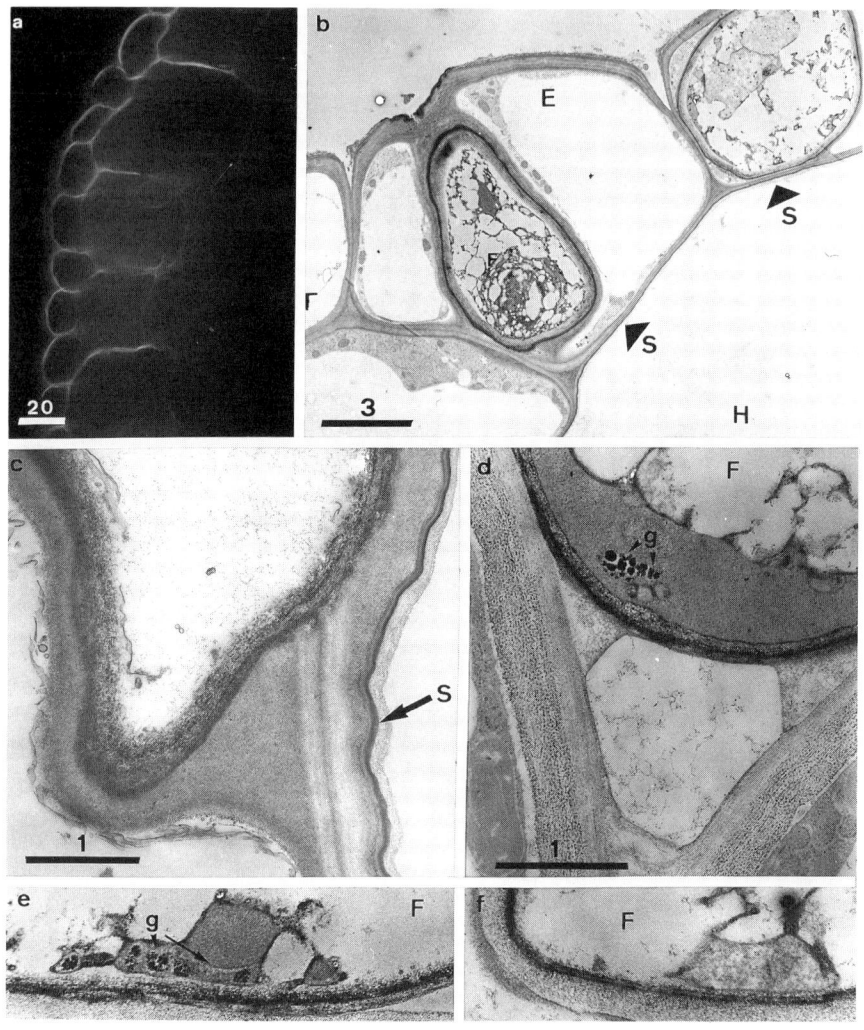

Figure 3.10. Morphological features of the cell walls in mycorrhizal *Allium porrum* roots. (a) Transverse section of the root seen in a fluorescence microscope under UV light: only epidermal and hypodermal cell walls are fluorescent. (b) Epidermal cells (E) crossed by the fungus (F). The host plasmalemma entirely wraps the hypha. A hypodermal cell is lined by a suberin layer (S), while the close cell does not possess such a barrier. (c) Magnification of the suberin layer and the complex hypodermal cell wall. (d) PATAg reaction showing a contact point between an intercellular hypha (F) and a cortical cell wall. Glycogen (g) is strongly stained by the silver grains. (e) Swollen and lamellated texture of the host cell wall when in contact with the fungus. (f) The host wall (H) and the interfacial material (I) show different textures (Vian & Bonfante-Fasolo, unpublished).

Once again, the cell wall plays a very important role in establishing contacts between the two partners. A large body of work shows that each cell type has a specific biochemical composition of cells walls, reflecting large systematic groups (McNeil et al., 1984). A parallel heterogeneity in morphology exists, when the architecture of different cellular types is studied in the presence and in the absence of a VAM fungus. Irrespective of the fungal presence, the epidermal and hypodermal cell walls in *Allium porrum* are reinforced by encrusting substances such as suberin and phenols. Those act as a sort of preformed barrier to the fungal passage (Codignola et al., 1989; Bonfante-Fasolo & Vian, 1989) (Fig. 3.10a–c). The fungus seems to be channeled by them to penetrate only the nonencrusted cells. Similarly, epidermal cells of *Ginkgo biloba* are encrusted by suberin and phenols while inner cortical cells are lignified in specific points, called phi thickenings (Bonfante-Fasolo & Fontana, 1985; Codignola et al., 1989). In this case as well, the VAM fungus does not seem to be capable of crossing the encrusted walls. Similar observations were performed by Brundrett and Kendrick (1990). They found that in five herbaceous woodland plants the way in which fungi entered the roots was determined by endodermal and exodermal cell wall characteristics. By contrast, walls of leek cortical cells where the fungus easily penetrates forming arbuscules do not show autofluorescence under ultraviolet light and are relatively thin and simple in structure (Fig. 3.10e and f). They appear swollen and polylamellate, mostly after the fungal passage (Bonfante-Fasolo & Vian, 1989).

These observations may supply us with new information on two different points: (1) the colonization process followed by the fungus in different host plants is influenced by the composition of cell wall, which is a constraint channeling the fungus; and (2) at least in the case of *Allium porrum,* the progressive thinning of the fungal cell wall is paralleled by a progressive thinning of the host cell wall. These observations suggest that the exchanges between the partners require cell surfaces with appropriate and adjusted characteristics (Bonfante-Fasolo & Vian, 1989). Among the cellular types that could be mycorrhizal, one could suggest that cortical cells are the only ones to possess simple and hydrophilic walls capable of allowing nutrients to cross membranes. The other cell layers probably represent zones in which the apoplastic flow of solution is limited by the presence of impermeable substances. In those zones, the symbiotic fungus may play a relevant role, bypassing the normal apoplastic pathway of solute in roots.

This alternative route for the solute uptake can be very important in, for example, citrus plants, where there is a dimorphic hypodermis regulating solute uptake and where solutes are forced to move symplastically (Smith & Smith, 1990). By using Calcofluor as an apoplastic tracer, we observed that the absorption capability of *Ginkgo biloba* roots is very limited, due to the impermeable characteristics of the outer layer. However, Calcofluor easily penetrates the root in the presence of the fungus, which works as a bridge between the extraradical and intraradical environment (Bonfante-Fasolo & Peretto, 1989). Structural stud-

HOST WALL AND FUNGAL WALL RELATIONSHIPS DURING THE INTRACELLULAR (A) AND INTERCELLULAR (B) PHASE.

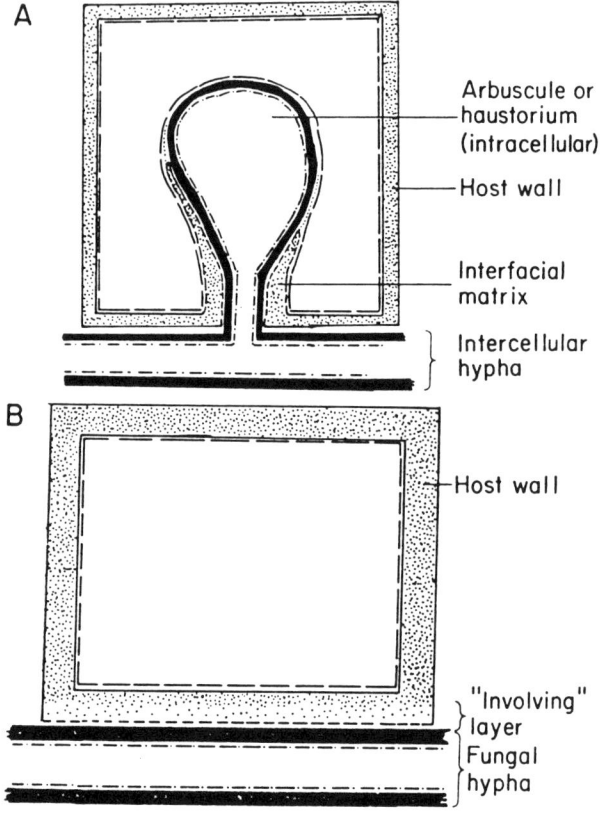

Figure 3.11. Diagram showing host wall and fungal wall relationships during the intracellular (A) and intercellular (B) phase. From Harley and Smith (1983).

ies on host cell wall textures therefore offer a new approach to understanding permeability, of the apoplastic–symplastic pathway, and how VA mycorrhizae actually work.

Interface Zone

As with some biotrophic pathogenic interactions, at the point of the fungal penetration the host plasmalemma invaginates and proliferates around the fungus (Fig. 3.11). In the meantime, material called interfacial material is found between the host plasmalemma and the fungal wall. The interfacial material is morphologically continuous with the host wall, but different in texture. Moreover, it changes following fungal penetration. It is thick and strongly electron-dense around the fungal coils and the penetration points, loose and badly organized around the thin arbuscular branches, and becomes thicker again around the collapsed arbuscular branches (Bonfante-Fasolo, 1984; Bonfante-Fasolo et al., 1981). The material was hypothesized to be of host origin because of its staining properties. In fact, polysaccharides (Dexheimer et al., 1979), cellulose, and pectin-like polysaccharides as well as proteins (Scannerini & Bonfante-Fasolo, 1979) were found in this material by using silver reactions and cytochemical staining. By using specific probes, represented by enzymes or monoclonal antibodies linked to colloidal gold, β-glucans related to cellulose and molecules related to nonesterified galacturonic acid can be detected precisely in the interface zones of mycorrhizal leeks and peas (Bonfante-Fasolo et al., 1990a; Perotto et al., 1990). The molecules are common both to the host cell wall and interfacial material (Fig. 3.12), even though their staining properties differ, suggesting a different assembly pattern. In addition, preliminary ultrastructural observations show that the interfacial material possesses molecules reacting with an antibody raised against melon hydroxyproline-rich glycoproteins (HRGPs) (Mazau et al., 1988). These proteins are structural components of the primary wall and in the meantime they have been demonstrated to increase following fungal attacks (Bolwell, 1988).

Immunogold results, which demonstrate a strong gold labeling in the interface area, suggest that HRGP molecules may occur as structural components of a material similar to a typical primary cell wall or, alternatively, may represent a response of the host plant to the fungal colonization (Bonfante-Fasolo et al., 1991). Observations that some monoclonal antibodies raised against pea membranes cross-react with antigens occurring not only on the perisymbiotic membrane, but also on the interfacial material of mycorrhizal peas (Gianinazzi-Pearson et al., 1990) leads one to look at the interface zone as an apoplastic space of high molecular complexity. These findings and the fact that residues of chitin or N-acetylglucosamine never occur in the interface (Bonfante-Fasolo et al., 1990b) (Fig. 3.8c) are a clear demonstration that many of the molecules occurring in the interfacial material are of host origin.

Some questions are raised as to how the arbuscule-containing cell organizes such an apoplastic compartment, which can be described as a cell wall-like envelope. The material may originate from the peripheral cell wall through a stretching of preexisting cell-wall polymers. Alternatively, the interface material may be laid down as newly synthesized material as a reaction to fungal penetration (Bonfante-Fasolo et al., 1990a). Such new components could be the result of a change in the genome expression of the host during the mycorrhizal state (see the following section). This implies a different cell situation: the new host plasma membrane laid down around the fungus must retain all the enzymatic capabilities typical of a differentiating cell, including β-glucan synthetase activity. In addi-

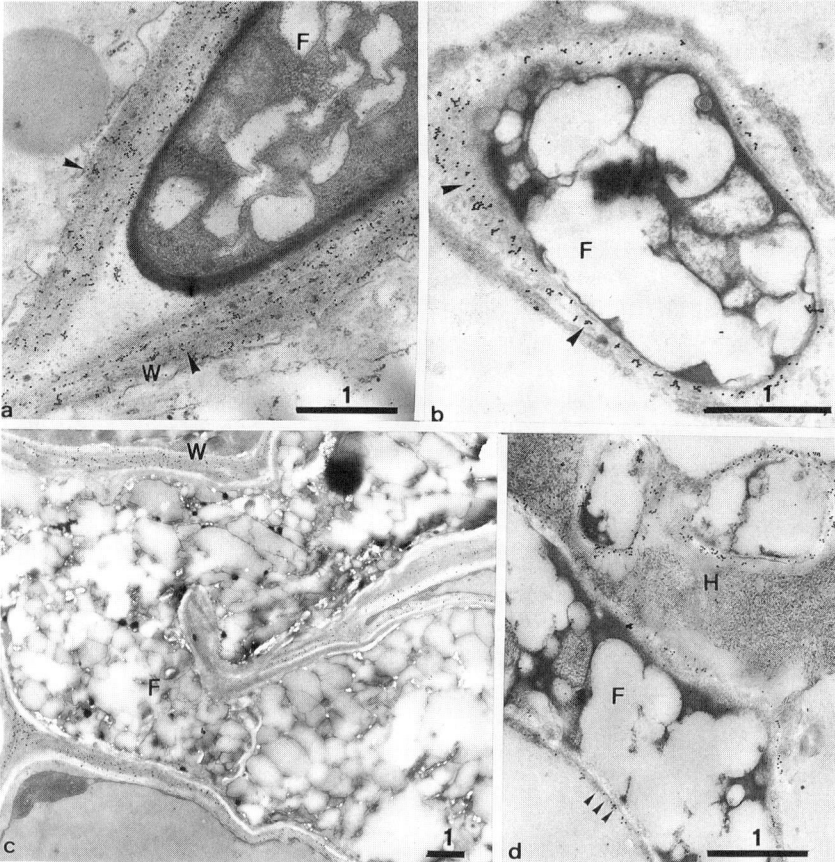

Figure 3.12. Ultrastructural features of the interfacial material after the use of affinity techniques to reveal the origin of the interfacial material. (a) Cellobiohydrolase linked to colloidal gold allows cellulose localization in the primary wall (W) of leek cortical cells. The fungus (F) is not labeled. (b) Loose deposition of gold granules (arrows) after the reaction with CBH complex around the intracellular hyphae (F). (c and d) After an immunogold reaction with a monoclonal antibody capable of recognizing pectin material, gold granules occur on the host wall (W), around the penetrating fungus (F), and around the arbuscular branches. H, host cytoplasm (Bonfante-Fasolo and Perotto, unpublished).

tion, protein and pectin targeting pathways must be changed, going toward the interface space and not toward the peripheral wall.

Looking at the permeability properties of the interfacial material, we can hypothesize a sort of barrier in the external layers, around the coiled hyphae, where other hydrophobic components probably exist in addition to the localized ones. However, plastic and hydrophilic properties of pectin and cellulose may allow the fungal expansion as well as nutrient exchanges between the two partners during the arbuscular phase. In addition, the demonstration of a cell-wall-like envelope around the fungal symbiont allows one to define its topological position clearly: VA fungi are always located in an apoplastic compartment and can be better defined as *extracellular*, according to Smith's terminology (1990), since they are external to the host membrane.

Upon fungal penetration, the host plasmalemma surrounding the fungus proliferates up to 10-fold (Alexander et al., 1988). Compared to the peripheral membrane of a cell belonging to the same tissues in an uninfected root, the host plasmalemma surrounding the hyphal branches shows some modifications of certain enzymatic activities. Neutral phosphatase activity, considered a plasmalemma marker typical of differentiating root cells, was localized on the plasmalemma both around living and collapsed arbuscular branches (Jeanmarie et al., 1985). However, the most important role seems to be played by ATPase activities, which have been cytochemically demonstrated on the interfacial plasma membranes and are consistent with bidirectional transport of solutes involving active uptake by both partners (Smith & Smith, 1990). An Mg-ATPase activity, inhibited by diethylstilbestrol and vanadate, was localized on the plasmalemma around the arbuscule. It disappeared from the membrane surrounding senescent hyphal branches as well as from peripheral plasmalemma (Marx et al., 1982; Gianinazzi-Pearson & Gianinazzi, 1988).

However, as already stated for ectomycorrizae (previous section), the hypothesis of active transport requires the presence of an H^+ pumping activity. According to recent electrophysiological studies, this H^+ pump activity is present in the host cells of mycorrhizal roots of *Allium cepa* and *A. porrum* colonized by *Glomus* sp.E3 (Scannerini, Bellando, & Fieschi, unpublished results; Scannerini et al., 1990). Pump activity has been demonstrated by cellular potentials (PD), the mean values of which are reported in Table 3.3 and by the medium acidification caused by H^+ extrusion. The cellular potential was significantly higher in mycorrhizal than in nonmycorrhizal roots. The reasons for this stable hyperpolarization are not clear. However, the electrophysiological behavior of the host membrane in VA mycorrhizae seems to differ from that occurring in biotrophic pathogenic interactions. Depolarization of the host plasma membrane was demonstrated in different plants treated with elicitors from biotrophic fungal pathogens (Katou et al., 1982; Pelissier et al., 1986). Depolarization is indicative of an increased

Table 3.3. *Cell Potentials in Mycorrhizal and Nonmycorrhizal Roots of Onions (Allium cepa L.).*

	x (mV)	s^2	n	F	p
Nonmycorrhizal	−61	12.2	30		
				341.05	<0.001
Mycorrhizal roots	−79	16.3	30		

permeability of the host membrane and a ion efflux toward the fungus. In the meantime, many ultrastructural investigations proved that there is no ATPase activity in the extrahaustorial membrane (i.e., the host plasma membrane surrounding each fungal haustorium), though such activity is normally present where the plasma membrane lines the host wall (Gay & Woods, 1987). This suggests that the host cell has no metabolic control to oppose solute efflux toward the fungus. Experiments with fluorescein diacetate provide strong support for the hypothesis of a transport coupled with ATPase deficiency in pathogenic interactions (Gay et al., 1987).

In conclusion, cytochemical and electrophysiological data confirm the hypothesis that in VAM an active P uptake by the host is mediated by an H^+-ATPase, allowing a two-way transport, unlike what happens in pathogenic interactions.

Previous morphological analyses have revealed that mycorrhizal interfaces are highly complicated, consisting in two membranes belonging to two different organisms, separated by the two apoplasts. This system is more complex than those usually employed, when ion transport is experimentally studied. Systems can be represented by single organisms, by isolated organs or cells (Lee, 1988) or by isolated plasma membrane vesicles (e.g., Rasi-Caldogno et al., 1985). A model involving membrane activity and cell wall metabolism has been recently presented in order to discuss some factors presumably involved in the process of abscission (Marrè, personal communication, 1988). According to this pattern, cell wall, periplasmic space, and plasmalemma act as a highly unified system via a complex network of molecular signals. Some important points can be listed: cell wall pH may regulate the enzymatic activity and the protein conformation on the plasma membrane; the physicochemical state of the apoplast may be influenced by the membrane activity; the activities and the reciprocal interactions of the H^+ pump, the Ca^{2+} pump, and the plasma membrane redox pump can control transport, turgor pressure, cell wall metabolism, and eventually cell wall lysis.

This pattern may be of theoretical relevance for a better understanding of the transport mechanisms of a plant cell as a whole. A similar complex presence of different factors could also exist in VA mycorrhizae. We can speculate that redox pumps may be affected by cell wall-bound phenols. (Codignola et al., 1989) and may affect the activation of cell wall-bound peroxidases (Spanu & Bonfante-Fasolo, 1988), interfacial Ca^{2+} (Strullu, 1985) could control the protein kinase activity of the partners and their hormonal balance, as well as the observed H^+ extrusion and hyperpolarization (Scannerini et al., 1990). Finally, cell wall acidification could be involved in the host cell wall loosening, fungal penetration, and interfacial material differentiation.

Regulation Mechanisms of the Bidirectional Movements in Mycorrhizae

The analysis of plant and fungal structures in a functioning mycorrhiza demonstrates that (1) the translocation of nutrients to and from both the partners can follow either symplastic or apoplastic routes and (2) the double exchange occurs by means of an active transfer occurring at the interface between the two partners and requiring the crossing of two different membranes and two different apoplastic spaces.

Morphofunctional analyses show the presence of some common mechanisms acting both in ecto- and endomycorrhizal associations. First, all the exchanges are controlled by morphogenetic events leading to an increase of the exchange surface. This is accomplished by deep modifications in the fungal morphology, leading to the branched hyphae of the Hartig net in ectomycorrhizae or to the arbuscular branches of symbiotic fungi in VA mycorrhizae. On the other hand, the host increases its surface with mechanims acting at the organismal and cellular level. Mathematical modeling techniques have been used to demonstrate that the host enhances its root ramification, stimulating the production of adventitious roots in leek (Berta et al., 1990), while at the cellular level there is an enhancement of the cell surface exchange area by producing transfer-like structures or laying down new cell wall material.

The second mechanism consists in changes within the partner plasma membranes. They seem to acquire new enzymatic activities involved in the nutrient flux, or as in endomycorrhizae, there is a new membrane deposition. Membranes can be synthesized *ex novo* or can be derived by fusion of unit membranes from golgi, tonoplast, endoplasmic reticulum, and partner plasma membrane, resulting in a membrane flow modification. The third mechanism consists in progressive modifications of the exchange surfaces in their permeability characteristics. In the only ectomycorrhiza thus studied, the fungus quickly becomes impermeable, avoiding the leakage back of the nutrients (Ashford et al., 1988). Similar mechanisms probably operate in many VAMs, as suggested by preliminary experiments on *Ginkgo biloba* roots. The fungus bypasses the physical barrier represented by the hydrophobic host cell walls of the outer root layers, ensuring an efficient uptake and transfer of nutrients to their delivery to the host. This might be possible with progressive adjustments of the texture and composition of the fungal wall (Bonfante-Fasolo, 1988).

However, the problem is to know how all these events occur. A complex regulation of both the partners is required, since a mycorrhiza functioning from a physiological point of view causes changes at organismal, tissue, cellular, and molecular levels. During development, the mycorrhizal roots undergo anatomical and histological modifications compared with nonmycorrhizal roots (Clowes, 1981; Berta et al., 1990). Simultaneously, the whole host cell organization changes while new membranes and wall materials are laid down at the interfaces (Bonfante-Fasolo, 1987). Recent experiments demonstrate that during ectomycorrhizal formation there is a specific accumulation of 10 different mycorrhiza-specific polypeptides and, in addition to these, hundreds of polypeptides found in uninfected tissues increased or decreased (Hilbert & Martin, 1988). They suggested that the name of *ectomycorrhizins* be used for this specific class of proteins. Similarly, results by Dumas et al. (1990) and by Pacovsky (1989) suggest the presence of a class of new soluble proteins, *Endomycorrhizins,* inside VAM roots. The resulting hypothesis, therefore, is that new mRNA populations appear during mycorrhizal formation, suggesting that the new proteins are synthesized *ex novo* or may appear as new protein species depending on posttranslational modifications.

According to the classical dogma, by definition these findings must be regulated precisely by the host and fungal genomes, giving rise to an intense protein synthesis. Ultrastructural and cytometric observations suggest that the host nuclei in the arbuscule colonized cells of VAM leeks possess active nucleola and a highly decondensed chromatin (Berta et al., 1986, 1990a). The decondensed state has been clearly demonstrated by using static and flow cytofluorimetric evaluations of the chromatin structure with undersaturating concentrations of DNA-specific fluorochromes (Berta et al., 1990b). All these results could suggest the presence of an intense transcriptional activity that is mycorrhiza dependent in the host cells.

The scenario created by the new findings of cell and molecular biology applied to mycorrhizae is much more precise: however, the whole cascading event (fungus–plant interaction, nuclear stimulation, mRNA production, synthesis of new proteins, morphological modifications, changes of the electrophysiological activity) cannot account for the double exchange between the fungus and the host. Interestingly, in plant–pathogenic fungal interactions in which unidirectional transfer occurs, a similar succession of events takes place (Lamb et al., 1989). Many research groups have recently provided evidence of gene activation after fungal contact, the presence of new mRNAs, their localization with *in situ* hybridization techniques, the production of new inducible enzymes, phytoalexins, or the so-called *pathogenesis related* (PR) *proteins.* It is interesting to point out that some of these proteins considered markers of pathogenic interactions have not been found in the presence of VAM fungi, while others have been detected. Glucanase and PR-b_1 protein were not detectable from a quantitative point of view in well-infected VA mycorrhizal tabacco roots (Dumas et al., 1990), while an important increase was found in the presence of the pathogen *Chalara elegans.* High peaks of chitinase and peroxidase activity were observed in leek roots colonized by *G. versiforme* (Spanu & Bonfante-Fasolo, 1988; Spanu et al., 1989), but they appear to be carefully regulated. Their activity peaks are limited to the first instants of the interaction, when the fungus has not yet colonized the whole roots and when probably the nutrient exchanges are not important (Fig. 3.13). When fungal colonization is fully established, the levels of both chitinase and peroxidase fall dramatically.

In addition, when chitinase activity was localized by using an antibody against chitinase with immunogold techniques, it has been demonstrated that the enzymic activity occurs in the intercellular space and in the host vacuole, exactly as in uninfected leaf cells (Mauch & Stahelin, 1989). In the presence of the symbiotic fungus, chitinase activity is localized around the fungus, but it does not seem to

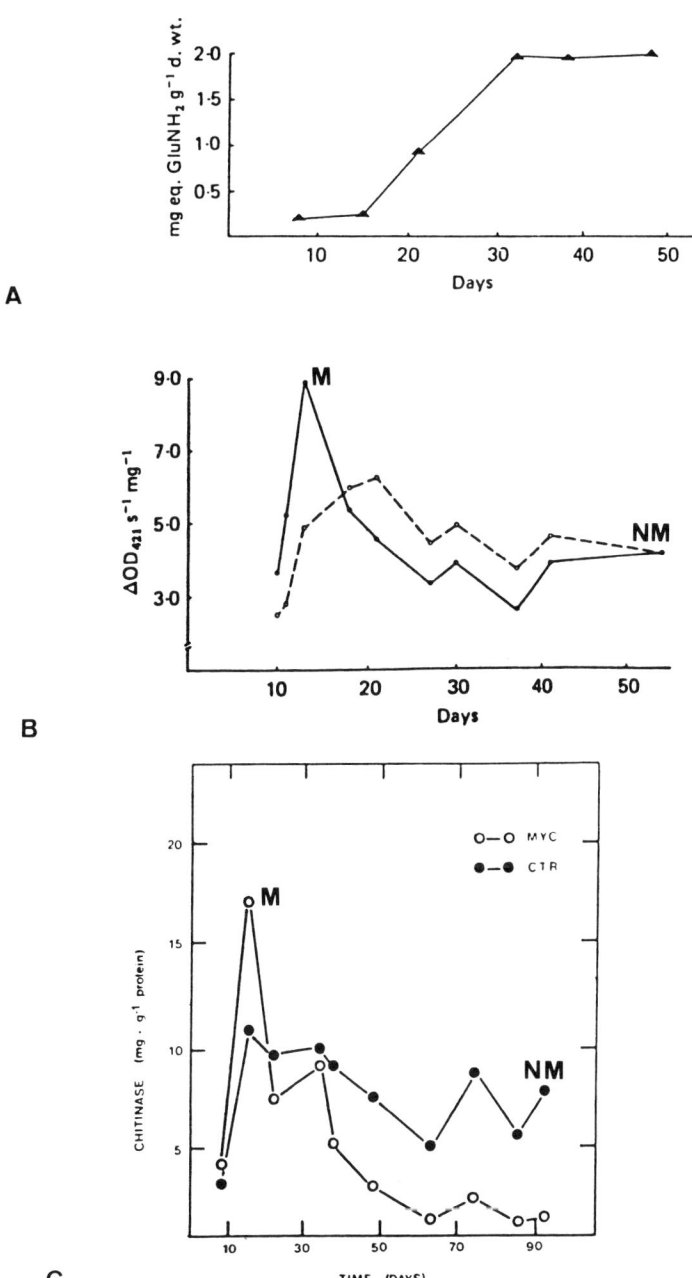

Figure 3.13. Time course of (A) chitin content expressed as glucosamine equivalents per root dry weight during fungal development, (B) cell-wall bound peroxidase, and (C) chitinase activities in mycorrhizal (M) and nonmycorrhizal (NM) roots of leek. The peeks of the two enzymes are limited to the onset of interaction, when the fungal colonization is limited, from Spanu & Bonfante-Fasolo (1988) and Spanu et al. (1989).

get into direct contact with it *in vivo*. On the contrary, chitinase is able to bind to the cell wall of hyphae of *Glomus* treated at 100°C to eliminate soluble polysaccharides (Spanu et al., 1989), confirming that chitin is protected by other cell-wall components, as seen ultrastructurally (Bonfante-Fasolo et al., 1990). A different picture is shown by Benhamou et al. (1990) by studying a plant–pathogenic fungus interaction. Intense chitinase activity accumulates on the fungal surface, suggesting an *in vivo* action. The expression of chitinase and peroxidase in mycorrhizal roots during time course studies suggests a step by step regulation at the level of gene transcription. This control could occur during the different moments of the root colonization, in which aspecific reactions, comparable to those shown by pathogenic associations, are followed by compatible interactions. In the case in which the plant–mycorrhizal fungal interaction is not compatible, as in the nonmycotrophic plant *Salsola kali*, the reaction phase becomes more important leading to a generalized state of bright-yellow autofluorescence, which may indicate a lignification defense process (Allen et al., 1989). These data are in keeping with the current view that parasitic and mutualistic symbionts may shift their nutritional balance, increasing or decreasing the length of their compatibility phase (Scannerini, 1988).

Conclusions

New results obtained as a result of the combined use of cytochemical, biochemical, and physiological experimental approaches have made it possible for us to describe the cellular and—sometimes molecular—bases of nutritional interchanges in plant and mycorrhizal fungi. Exchanges rely on the presence of specialized interfaces, many features of which differ from those identified in pathogenic interactions.

However, we are not yet able to characterize many of the events that lead to cellular and physiological compatibility by acting at the level of genome expression. If the two partners interact, as happens with *Rhizobia* and leguminosae (Nuti et al., 1988), we do not yet know which signals trigger the first cue of their interaction. We are only beginning to understand the conclusion of their relationship.

References*

Alexander, T., Meier, R., Toth, R., & Weber, H. C. (1988). Dynamics of arbuscule development and degeneration in mycorrhizas of *Triticum aestivum* L. and *Avena sativa* L. with reference to *Zea mays*. *New Phytologist* **110**, 363–370.

Allen, M. F., Allen, E. B., & Friese, C. F. (1989). Responses of the non-mycotrophic plant *Salsola kali* to invasion by vesicular-arbuscular mycorrhizal fungi. *New Phytologist* **111**, 45–49.

Ashford, A. E., & W. G. Allaway (1985). Transfers cells and Hartig net in the root epidermis of the sheating mycorrhiza of *Pisonia grandis* R.Br. from Seychelles. *New Phytologist* **100**, 595–612.

Ashford, A. E., Peterson, C. A., Carpenter, J. L., Cairney, J. W. G., & Allaway, W. G. (1988). Structure and permeability of the fungal sheath in the *Pisonia* mycorrhiza. *Protoplasma* **147**, 149–161.

Benhamou, N., Joosten, M. H. A. J., & De Wit, P. J. G. M. (1990). Subcellular localization of chitinase and of its potential substrate in tomato root tissues infected by *Fusarium oxysporum* f.s. *radicis-lycopersici*. *Plant Physiology* **92**, 1108–1120.

Berta, G., Fusconi, A., Sgorbati, S., Trotta, A., & Scannerini, S. (1986). Preliminary results on the ploidy and fine structure of the nuclei of the host cells in a VA mycorrhiza. *Giornale Botanico Italiano* **120**, 84–86.

Berta, G., Fusconi, A., Trotta, A., & Scannerini, S. (1990). Morphogenetic modifications induced by the mycorrhizal fungus *Glomus* strain E_3 on the root system of *Allium porrum* L. *New Phytologist* **114**, 207–215.

Berta, G., Sgorbati, S., Trotta, A., Fusconi, A., & Scannerini, S. (1990a). Correlations between host-endophyte interactions and structural changes in host cell chromatin in a VA mycorrhiza. In: *Endocytobiology IV* (Ed. by P. Nardon, V. Gianinazzi-Pearson, A. M. Grenier, L. Margolis, & D. C. Smith), pp. 145–148. INRA, Paris.

Berta, G., Sgorbati, S., Soler, V., Fusconi, A., Trotta, A., Citterio, M. G., Sparvoli, E., & Scannerini, S. (1990b). Variations in chromatin structure in host nuclei of a vesicular arbuscular mycorrhiza. *New Phytologist* **114**, 199–205.

Berta, G., Tagliasacchi, AM., Fusconi, A., Geriero, D., Trotta, A., & Scannerini, S. (1991). The mitotic cycle in root apial meristems of *Allium porrum* L. is controlled by the endomycorrhizal fungus *Glomus* sp. strain E_3. *Protoplasma* **161**, 12–16.

Bolwell, G. P. (1988). Synthesis of cell wall components: Aspects of control. *Phytochemistry* **27**, 1235–1253.

Bonfante-Fasolo, P. (1984). Anatomy and morphology of VA mycorrhizae. In: *VA Mycorrhizas* (Ed. by C. L. Powell & D. J. Bagyaraj), pp. 5–33. CRC Press, Boca Raton, FL.

Bonfante-Fasolo, P. (1987). Vesicular-arbuscular mycorrhizae fungus-plant interactions at the cellular level. *Symbiosis* **3**, 249–268.

Bonfante-Fasolo, P. (1988). The role of the cell-wall as a signal in mycorrhizal associations. In: *Cell to Cell Signals in Plant, Animal, and Microbial Symbiosis* (Ed. by S. Scannerini, D. C. Smith, P. Bonfante-Fasolo, & V. Gianinazzi-Pearson), pp. 219–235. Springer-Verlag, Berlin.

Bonfante-Fasolo, P., & Fontana, A. (1985). VAM fungi in Gingko biloba roots. Their interactions at cellular level. *Symbiosis* **1**, 53–67.

Bonfante-Fasolo, P., & Gianinazzi-Pearson, V. (1986). Wall and plasmalemma modifications in mycorrhizal symbiosis. In: *Physiological and Genetical Aspects of Mycorrhizae* (Ed. by V. Gianinazzi-Pearson & S. Gianinazzi), pp. 65–74. INRA, Paris.

Bonfante-Fasolo, P., & Peretto, R. (1989). Struttura e permeabilità dell'apoplasto in radici micorrizate di *Allium porrum* L. e *Ginkgo biloba* L. *Giornale Botanico Italiano* **123**, Suppl. 1, 137.

Bonfante-Fasolo, P., & Vian, B. (1984). Wall texture in the spore of a vesicular-arbuscular mycorrhizal fungus. *Protoplasma* **120**, 51–60.

Bonfante-Fasolo, P., & Vian, B. (1989). Cell wall architecture in mycorrhizal roots of *Allium porrum*. *Annales de Botanique* **10**, 97–109.

Bonfante-Fasolo, P., Dexheimer, J., Gianinazzi, S., Gianinazzi-Pearson, V., & Scannerini, S. (1981). Cytochemical modifications in the host-fungus interface during intracellular interactions in vesicular-arbuscular mycorrhiza. *Plant Science Letters* **22**, 13–21.

Bonfante-Fasolo, P., Vian, B., Perotto, S., Faccio, A., & Knox, J. P. (1990a). Cellulose and pectin localization in roots of mycorrhizal *Allium porrum:* Labelling continuity between host cell wall and interfacial material. *Planta* **180**, 537–547.

Bonfante-Fasolo, P., Faccio, A., Perotto, S., & Schubert, A. (1990b). Correlation between chitin distribution and cell wall morphology in the mycorrhizal fungus *Glomus versiforme*. *Mycological Research* **94**, 157–165.

Bonfante-Fasolo, P., Tamagnono, L., Peretto, R., Esquerré-Tugaye, M. T., Mazau, D., Mosiniak, M., & Vian, B. (1991). Immunocytochemical location of hydroxyproline rich glycoproteins at the interface between a mycorrhizal fungus and its host plants. *Protoplasma* **165**, 127–138.

Bouligand, Y. (1972). Twisted fibrous arrangements in biological materials and cholesteric mesophases. *Tissue and Cell* **4**, 189–217.

Bracker, C. E., & Littlefield, L. J. (1973). Structural concepts of host-pathogen interfaces. In: *Fungal Pathogenicity and the Plant's Response* (Ed. by R. J. W. Byrde & C. V. Cutting), pp. 159–317. Academic Press, London.

Brundrett, M., & Kendrick, B. (1990). The roots and mycorrhizas of herbaceous woodland plants. II. Structural aspects of morphology. *New Phytologist* **114**: 469–479.

Cairney, J. W. G., Jennings, D. H., Ratcliffe, R. G., & Southon, T. E. (1988). The physiology of basidiomycete linear organs. II. Phosphate uptake by rhizomorphs of *Armillaria mellea*. *New Phytologist* **109**, 327–333.

Cervone, S., Castoria, R., Spanu, P., & Bonfante-Fasolo, P. (1988). Pectinolytic activity in some ericoid mycorrhizal fungi. *Transactions British Mycological Society* **91**, 537–539.

Cervone, S., Hahn, M. G., De Lorenzo, G., Darvill, A., & Albersheim, P. (1989). Host pathogen interactions XXXIII. A plant protein converts a fungal pathogenesis factor into an elicitor of plant defense responses. *Plant Physiology* **90**, 542–548.

Clowes, F. A. L. (1981). Cell proliferation in ectotrophic mycorrhizas of *Fagus sylvatica* L. *New Phytologist* **87**, 547–555.

Codignola, A., Verotta, L., Maffei, M., Spanu, P., Scannerini, S., & Bonfante-Fasolo, P. (1989). Cell wall bound phenols in roots of vesicular-arbuscular mycorrhizal plants. *New Phytologist* **112**, 221–228.

Dalphe', Y. (1989). Ericoid mycorrhizal fungi in the Myxotrichaceae and Gymnoascaceae. *New Phytologist* **110**, 523–527.

Dexheimer, J., Gianinazzi, S., & Gianinazzi-Pearson, V. (1979), Ultrastructural cytochemistry of the host-fungus interface in the endomycorrhizal association *Glomus mosseae/Allium cepa. Zeitschrift fur Planzenphysiologie* **86,** 189–201.

Duddridge, J. A. (1985). A comparative ultrastructural analysis of the host-fungus interface in mycorrhizal and parasitic associations. In: *Developmental Biology of Higher Fungi* (Ed: by D. Moore et al.), pp. 141–173. Cambridge University Press, Cambridge.

Duddridge, J. A. (1987). Specificity and recognition in ectomycorrhizal associations. In: *Fungal Infection of Plants* (Ed. by G. F. Pegg & P. G. Ayres), pp. 25–44. Cambridge University Press, Cambridge.

Duddridge, J. A., Finlay, R. D., Read, D. J., & Soderstrom, B. (1988). The structure and function of the vegetative mycelium of ectomycorrhizal plants. III. Ultrastructural and autoradiographic analysis of inter-plant carbon distribution through intact mycelial systems. *New Phytologist* **108,** 183–188.

Dumas, E., Tahiri-Alaoui, A., Gianinazzi, S., & Gianinazzi-Pearson, V. (1990). Observations on modifications in gene expression with VA endomycorrhiza development in tobacco: Qualitative and quantitative changes in protein profiles. In: *Endocytobiology IV* (Ed. by P. Nardon, V. Gianinazzi-Pearson, A. M. Grenier, L. Margulis, & D. C. Smith), pp. 153–157. INRA, Paris.

Fusconi, A. (1983). The development of the fungal sheath on *Cistus incanus* short roots. *Canadian Journal Botany* **61,** 2546–2553.

Fusconi, A., Berta, G., Scannerini, S., & Trotta, A. (1986). Meristematic activity in mycorrhizal and uninfected roots of Allium porrum. In: *Physiological and Genetical Aspects of Mycorrhizae.* (Ed. by V. Gianinazzi-Pearson & S. Gianinazzi), pp. 667–672. INRA, Paris.

Gay, J. L., & Woods, A. M. (1987). Induced modifications in the plasma membranes of infected cells. In: *Fungal Infection of Plants* (Ed. by G. F. Pegg & P. G. Ayres), pp. 79–91. Cambridge University Press, Cambridge.

Gay, J. L., Salzberg, A., & Woods, A. M. (1987). Dynamic experimental evidences for the plasmamembrane ATPase domain hypothesis of haustorial transport and for ionic coupling of the haustorium of *Erysiphe graminis* to the host cell (*Hordeum vulgare*). *New Phytologist* **107,** 541–548.

Gianinazzi, S., Gianinazzi-Pearson, V., & Dexheimer, J. (1979). Enzymatic studies on the metabolism of vesicular-arbuscular mycorrhiza. III. Ultrastructural localization of acid and alkaline phosphatase in onion roots infected by *Glomus mosseae* (Nicol. Gerd.). *New Phytologist* **82,** 127–132.

Gianinazzi-Pearson, V., & Gianinazzi, S. (1988). Morphological integration and functional compatibility between symbionts in vesicular arbuscular endomycorrhizal associations. In: *Cell to Cell Signals in Plant, Animal, and Microbial Symbiosis* (Ed: by S. Scannerini, D. C. Smith, P. Bonfante-Fasolo, & V. Gianinazzi-Pearson), pp. 73–84. Springer-Verlag, Berlin.

Gianinazzi-Pearson, V., & Gianinazzi, S. (1989). Phosphorus metabolism in mycorrhizas. In: *Nitrogen, Phosphorus and Sulfur Utilization by Fungi* (Ed: by L. Boddy, R. Marchant, & D. Reed), pp. 227–241. Cambridge University Press, N.Y.

Gianinazzi-Pearson, V., Gianinazzi, S., & Brewin, N. J. (1990). Immunocytochemical localisation of antigenic sites in the perisymbiotic membrane of vesicular-arbuscular endomycorrhiza using monoclonal antibodies reacting against the peribacteroid membrane of nodules. In: *Endocytobiology IV* (Ed. by Nardon, V. Gianinazzi-Pearson, A. M. Grenier, L. Margulis, & D. C. Smith), pp. 127–131. INRA, Paris.

Grandmaison, J., Behamou, N., Furlan, V., & Visser, S. A. (1988). Ultrastructural localization of N-acetilglucosamine residues in the cell wall of *Gigaspora margarita* through its life-cycle. *Biology of the Cell* **63**, 89–100.

Grippiolo, R., & Bonfante-Fasolo, P. (1984). Sporopollenin and melanin-like pigments in the wall of a *Glomus* spore. *Giornale Botanico Italiano* **118**, 88–91.

Harley, J. L., & Smith, S. E. (1983). *Mycorrhizal Symbiosis*. Academic Press, London.

Hilbert, J. L., & Martin F. (1988). Regulation of gene expression in Ectomycorrhizae. I. Protein changes and the presence of ectomycorrhiza specific polypeptides in the *Pisolithus–Eucalyptus* symbiosis. *New Phytologist* **110**, 339–346.

Jeanmairie, C., Dexheimer, J., Marx, C., Gianinazzi, S., & Gianinazzi–Pearson, V. (1985). Effect of vesicular-arbuscular mycorrhizal infection on the distribution of neutral phosphatase activities in root cortical cells. *Journal of Plant Physiology* **119**, 285–293.

Jeffries, P. (1987). Pathways for the exchange of materials in mycoparasitic and plant-fungal interactions. In: *Fungal Infection of Plants* (Ed by G. F. Pegg & P. G. Ayres), pp. 60–78. Cambridge University Press, Cambridge.

Katou, K., Tomiyama, K., & Okamoto, H. (1982). Effects of hyphal wall components of *Phytophthora infestans* on membrane potential of potato tuber cells. *Physiological Plant Pathology* **21**, 311–317.

Keon, J. P. R., Byrde, R. J., Byrde, W., & Cooper, R. M. (1987). Some aspects of fungal enzymes that degrade plant cell walls. In: *Fungal Infection of Plants* (Ed. by G. F. Pegg & P. G. Ayres), pp. 133–157. Cambridge University Press, Cambridge.

Kottke, I., & Oberwinkler, F. (1986). Mycorrhiza of forest trees—structure and function. *Trees* **1**, 1–24.

Kottke, I., & Oberwinkler, F. (1987). The cellular structure of the Hartig net: Coenocytic and transfer cell-like organization. *Nordic Journal Botany* **7**, 85–95.

Lamb, C. J., Lawton, M. A., Dron, M., & Dixon, R. A. (1989). Signals and transduction mechanisms for activation of plant defenses against microbial attack. *Cell* **56**, 215–224.

Lee, R. B. (1988). Phosphate influx and extracellular phosphatase activity in barley roots and rose cells. *New Phytologist* **109**, 141–148.

Lei, J., & Dexheimer, J. (1988). Ultrastructural localization of ATPase activity in the *Pinus sylvestris–Laccaria Laccata* ectomycorrhizal association. *New Phytologist* **108**, 329–334.

Luttge, U., & Clarkson, D. T. (1987). Mineral nutrition: Anions. In: *Progress in Botany* (Ed by H. Dietmar Behnke et al.), pp. 68–86. Springer-Verlag, Berlin.

Mangin, F., Bonaly, R. Botton, B., & Martin, F. (1986). Chemical composition of hyphal walls of the ectomycorrhizal fungus *Cenococcum geophilum*. In: *The Physiological and Genetical Aspects of Mycorrhizae* (Ed by V. Gianinazzi-Pearson & S. Gianinazzi), pp. 451–456. INRA, Paris.

Marré, E. (1979). Fusicoccin: A tool in plant physiology. *Annual Review Plant Physiology* **30**, 273–288.

Martin, F., Ramstedt, M., Soderhall, K., & Canet, D. (1988). Carbohydrate and aminoacid metabolism in the ectomycorrhizal ascomycete *Sphaerosporella brunnea* during glucose utilization. *Plant Physiology* **186**, 935–940.

Marx, C., Dexheimer, J., Gianinazzi-Pearson, V., & Gianinazzi, S. (1982). Enzymatic studies on the metabolism of vesicular-arbuscular mycorrhizas. VI. Ultracytoenzymological evidence (ATPase) for active transfer processes in the host-arbuscule interface. *New Phytologist* **90**, 37–43.

Massicotte, H. B., Ackerley, C. A., & Peterson, R. L. (1987). Localization of three sugar residues in the interface of ectomycorrhizae synthesized between *Alnus crispa* and *Alpova diplophloeus* as demonstrated by lectin binding. *Canadian Journal of Botany* **65**, 1127–1132.

Mauch, F., & Staehelin, L. A. (1989). Functional implications of the subcellular localization of ethylene-induced chitinase and β-1-3-glucanase in bean leaves. *The Plant Cell* **1**, 447–457.

Mazau, D., Rumeau, D., & Esquerre'-Tugaye', M. T. (1988). Two different families of hydroxyprolin-rich glycoproteins in melon callus. *Plant Physiology* **86**, 540–546.

McNeil, M., Darvill, G. A., Fry, C. S., & Albersheim, P. (1984). Structure and function of the primary cell wall of plants. *Annual Review Biochemistry* **53**, 625–663.

Nuti, M. P., Pasti, M. P., & Squartini, A. (1988). Applications of genetic engineering to "symbiontology" in agriculture. In: *Cell to Cell Signals in Plant, Animal, and Microbial Symbiosis* (Eds. by S. Scannerini, D. C. Smith, P. Bonfante-Fasolo, & V. Gianinazzi-Pearson), pp. 347–359. NATO ASI Series Vol. H 17. Springer-Verlag, Berlin.

Nylund, J. E. (1987). The ectomycorrhizal infection zone and its relation to acid polysaccharides of cortical cell walls. *New Phytologist* **106**, 505–516.

Nylund, J. E. (1988). The regulation of mycorrhiza formation. Carbohydrate and hormone theories reviewed. *Scandinavian Journal Forest Research* **3**, 465–479.

Pacovsky, R. S. (1989). Carbohydrate, protein and amino acid status of *Glycine–Glomus–Bradyrhizobium* symbioses. *Physiologia Plantarum* **75**, 346–354.

Pacovsky, R. S., & Fuller, G. (1988). Mineral and lipid composition of *Glycine–Glomus–Bradyrhizobium* symbioses. *Physiologia Plantarum* **72**, 733–746.

Pelissier, B., Thibaud, J. B., Grignon, C., & Esquerre-Tugaye, M. T. (1986). Cell surfaces in plant-microorganism interactions. VII. Elicitor preparations from two fungal pathogens depolarize plant membranes. *Plant Science* **46**, 103–109.

Perotto, S., Vandenbosch, K. A., Brewin, N. J., Faccio, A., Knox, J. P., & Bonfante-Fasolo, P. (1990). Modifications of the host cell wall during root colonization by Rhizobium and VAM Fungi. In: *Endocytobiology IV* (Ed. by P. Nardon, Gianinazzi-Pearson, A. M. Grenier, M. Margulis, & D. C. Smith), pp. 114–117. INRA, Paris.

Piche', Y., Peterson, R. L., & Massicotte, H. B. (1988). Host-fungus interactions in ectomycorrhizae. In: *Cell to Cell Signals in Plant, Animal, and Microbial Symbiosis* (Ed. by S. Scannerini, D. C. Smith, P. Bonfante-Fasolo, & V. Gianinazzi-Pearson), pp. 55–71. NATO ASI Series Vol. H 17. Springer-Verlag, Berlin.

Rasi-Caldogno, F., Pugliarello, M. C., & De Michelis, M. I. (1985). Electrogenic transport of protons driven by the plasma membrane ATPase in membrane vesicles from radish. Biochemical characterization. *Plant Physiology* **77,** 200–205.

Scannerini, S. (1988). The cell structures of plant, animal and microbial symbionts, their differences and similarities. In: *Cell to Cell Signals in Plant, Animal, and Microbial Symbiosis* (Ed. by S. Scannerini, D. C. Smith, P. Bonfante-Fasolo, & V. Gianinazzi-Pearson), pp. 143–157. NATO ASI Series, Vol. H 17. Springer-Verlag, Berlin.

Scannerini, S., & Bonfante-Fasolo, P. (1979). Ultrastructural cytochemical demonstration of polysaccharides and proteins within host-arbuscule interfacial matrix in an endomycorrhiza. *New Phytologist* **83,** 739–744.

Scannerini, S., & Bonfante-Fasolo, P. (1983). Comparative ultrastructural analysis of mycorrhizal associations. *Canadian Journal of Botany* **61,** 917–943.

Scannerini, S., Fieschi, M., Alloatti, G., Sacco, S., & Berta, G. (1990). Cell potential hyperpolarization in *Allium porrum* L. + *Glomus* sp. strain E3 VA mycorrhizae. VIII NACOM, Jackson, Wyoming.

Schoeknecht, J. D., & Hattingh, M. J. (1976). X-ray microanalysis of elements in cells of VA mycorrhizal and non mycorrhizal onions. *Mycologia* **68,** 296–303.

Sentenac, H., & Grignon, C. (1985). Effect of pH on ortophosphate uptake by corn roots. *Plant Physiology* **77,** 136–141.

Smith, D. C. (1990). Symbiosis: An overview, a definition and an examination of the aims of the Colloquium. In: *Endocytobiosis IV* (Ed by P. Nardon, V. Gianinazzi-Pearson, A. M. Grenier, L. Margulis, & D. C. Smith), pp. 29–35. INRA, Paris.

Smith, D. C., & Douglas, A. E., (1987). *The Biology of Symbiosis*. Edward Arnold, London.

Smith, S. E., & Gianinazzi-Pearson, V. (1988). Physiological interactions between symbionts in vesicular-arbuscular mycorrhizal plants. *Annual Review Plant Physiology Plant Molecular Biology* **39,** 221–244.

Smith, F. A., & Smith, S. E. (1986). Movement across membranes: Physiology and biochemistry. In: *Physiological and Genetical Aspects of Mycorrhizae* (Ed. by V. Gianinazzi- Pearson, & S. Gianinazzi), pp. 75–84. INRA, Paris.

Smith, S. E., & Smith, F. A. (1990). Structure and function of the interbiotrophic symbioses as they relate to nutrient transport. *New Phytologist* **114,** 1–38.

Spanu, P., & Bonfante-Fasolo, P. (1988). Cell-wall bound peroxidase activity in roots of mycorrhizal *Allium porrum*. *New Phytologist* **109,** 119–124.

Spanu, P., Boller, T., Ludwig, A., Wiemken, A., Faccio, A., & Bonfante-Fasolo, P. (1989). Chitinase in roots of mycorrhizal *Allium porrum:* Regulation and localization. *Planta* **177,** 447–455.

Strullu, D. G. (1985). *Les mycorhizes. Encyclopedia of Plant Anatomy*. Gebruder Borntraeger, Berlin-Stuttgart.

Strullu, D. G., Harley, J. L., Gourret, J. P., & Garrec, J. P. (1983). A note on the relative phosphorus and calcium contents of metachromatic granules in *Fagus* mycorrhiza. *New Phytologist* **94,** 89–94.

Ulrich-Eberius, C. I., Novacky, A., & Van Bel A. J. E. (1984). Phosphate uptake in *Lemna gibba* G1: Energetics and kinetics. *Planta* **161,** 46–51.

Vian, B. (1986). Ultrastructural localization of carbohydrates. Recent developments in cytochemistry and affinity methods. In: *Biology and Molecular Biology of Plant-Pathogen Interactions* (Ed. by J. A. Bailey), pp. 49–57. NATO ASI Series, Vol. H1. Springer-Verlag, Berlin.

Vian, B., & Roland, J. C. (1987). The helicoidal cell wall as a time register. *New Phytologist* **105,** 345–357.

Wessels, J. G. H. (1986). Cell wall synthesis in apical hyphal growth. *International Review of Cytology* **104,** 37–79.

Woolhouse, H. W. (1975). Membrane structure and transport problems considered in relation to phosphorus and carbohydrate movements and the regulation of endotrophic mycorrhizal associations. In: *Endomycorrhizas* (Ed. by F. E. Sanders, B. Mosse, & P. B. Tinker), pp. 209–240. Academic Press, London.

*Reference list is updated to 1990.

4

The Mycorrhizal Mycelium

D. J. Read

> SUMMARY. The current state of knowledge of the structure and function of the vegetative mycelia produced by ectomycorrhizal, vesicular-arbuscular (VA), and ericoid mycorrhizal roots is analyzed using information gained from laboratory and field-based studies. The extensive nature of the differentiated mycelium of ectomycorrhizal roots is demonstrated and its role in capture of mineral nutrient ions as well as in mobilization of organic nutrients is discussed. The mycelia of VA fungi are also seen to form extensive but relatively undifferentiated networks that facilitate rapid infection of seedlings in natural plant communities, their integration into the network, and the provision of mineral nutrient ions notably phosphate, which are otherwise unavailable to them. The ericoid mycorrhizal mycelium has the dual function of nutrient mobilization, a process achieved through the production of a range of extracellular enzymes, and detoxification, a process involving assimilation of organic acids and sequestration of metal ions. It is concluded that each of the three categories of mycorrhizal mycelium has a distinctive suite of characteristics that provides it with the potential to influence the structure and dynamics of the ecosystem in which it predominates.

Introduction

Of all of the structures of the mycorrhizal root the one that has suffered the greatest neglect is that which is arguably the most important for the nutrition of the host—the external mycelium. Despite being the part of the mutualism that is involved in the processes of mobilization and capture of nutrients from soil, its function, and even its presence, has frequently been ignored. The reason for the lack of understanding of the structure and function of the mycorrhizal mycelium is simply that it is the most difficult part of the system to examine and manipulate experimentally. Whereas the relatively massive mycorrhizal root can be readily extracted from soil and examined in a nondestructive manner, its associated mycelium is not amenable to such treatments. As a result, although we profess to know so much about the function of the mycorrhizal root, we in fact know very little of its vital links with the soil.

Some of the early workers, notably Melin and co-workers, showed awareness of the entire system in experimental analyses of nutrient transfer through intact ectomycorrhizal mycelium. The strength of these experiments was that they demonstrated the potential of the fungal mycelium to provide the pathway for the transfer of the critical plant nutrients nitrogen (Melin & Nilsson, 1952) and phosphorus (Melin & Nilsson, 1950) to the root. Their weakness was that the mycelia were produced under aseptic conditions in media enriched with a soluble exogenous source of carbon. Under these circumstances normal differentiation of the mycorrhizal mycelium is restricted and the possibility of diffusive flow of nutrient ions across hyphae is enhanced. This is especially the case in the absence of a hydrophobic barrier between nutrient sources and sinks (Read & Stribley, 1975a). The view prevailed for many years that the basidiomycete mycelium is little more than animated cotton wool. However, we now know that when grown from their natural substrates in the absence of exogenous carbon the mycelia of many ectomycorrhizal fungi (Read, 1984a; Read et al., 1985) and their wood-rotting counterparts in the Basidiomycotina (Rayner et al., 1985) show very considerable potential for structural differentiation and that their true functional roles can be interpreted only when such differentiation has taken place.

In this chapter our knowledge of the structure and function of the mycorrhizal mycelia of plants with ecto, VA, and ericoid infection is reviewed, gaps in our understanding are highlighted, and possible avenues of advance are explored.

Structure and Function of the Ectomycorrhizal Mycelium

The best known and probably most important fungal species involved in the formation of ectomycorrhizas are basidiomycetes. What knowledge we have of the biology of the mycelium of these fungi in soil has been derived from two different approaches. In one, exemplified by the studies of Ogawa (1977, 1981, 1985), field-based analyses of the distribution of fruit bodies and their vegetative mycelia have been carried out over a number of years. The other approach, first described by Skinner and Bowen (1974) who used sterile media and later developed by Read (1984a) and colleagues, involves laboratory studies using transparent observation chambers containing natural soils in which the development and function of mycorrhizal mycelium can be analyzed in a nondestructive manner. These two types of approach are treated separately below.

Field-Based Studies

Attempts directly to investigate the arrangement of fine mycorrhizal roots and their mycelia in relation to soil particles in the field inevitably involve the risk of disturbance of the structures being examined. Permanently placed root windows (Egli & Kälin, 1990) or minirhizotrons (Taylor, 1987) can enable nondestructive observation at a restricted number of locations but lack flexibility. Alternatively blocks of soil can be fixed *in situ,* cleared, and sectioned (Babel, 1987). One feature that emerges from such studies is that a very high proportion of ectomycorrhizal roots and root clusters are formed in pores between soil particles and not in direct contact with the soil itself. Babel estimated that 90% of fine roots of *Picea abies* and 70% of those of *Fagus sylvatica* were found in this position. Roots situated in soil pores, although being poorly placed to act as nutrient-absorbing organs, are ideally situated to provide effective food bases for their mycorrhizal fungi, the mycelia of which increasingly appear to be the primary absorptive structures. The aerobic environment of the pore will facilitate nutrient transfer from sheath to root, a process involving large oxygen demands (Harley et al., 1953), and enable the proliferation of mycelial strands necessary for the conduction of nutrients to and from the mycorrhiza. Observations such as these highlight the need for more information on the distribution and activity of the ectomycorrhizal mycelium in the field.

Ogawa (1985) described three basic types of mycelial differentiation (Fig. 4.1), the fairy ring, the irregular mat, and the dispersed colony, each of which

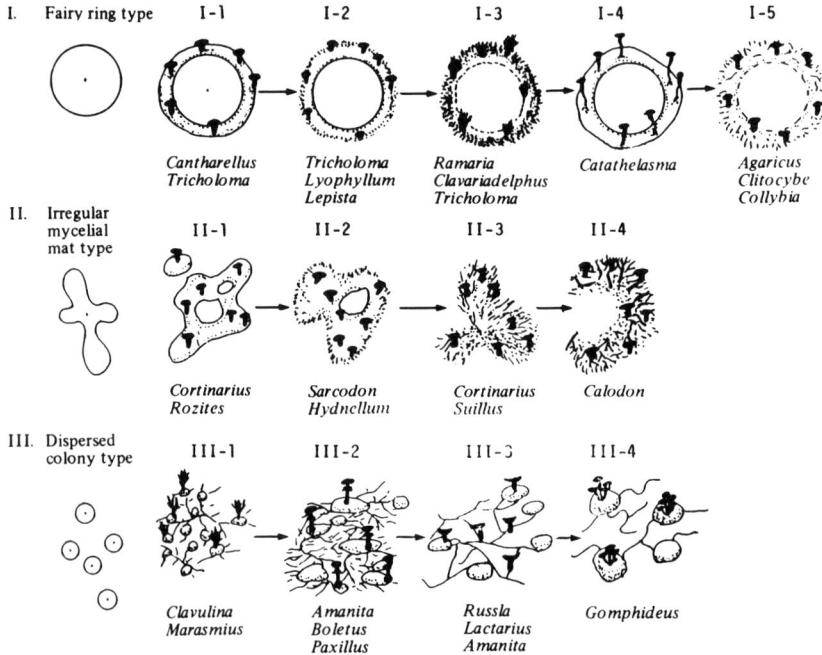

Figure 4.1. The three basic patterns of mycelial distribution, with their variants, as recognized by Ogawa in forests of Japan. From Ogawa (1985), with permission.

has a series of subtypes. Though all of these types are said to be produced by mycorrhizal fungi, closer examination of Ogawa's descriptions reveals that most of the fungi falling into the fairy ring category are, in fact, litter decomposers. Even where "mycorrhizas" are formed by these fungi as in the case of the "shiro" mushroom, *Tricholoma matsutake,* they are said to be of a parasitic kind and to lack a sheath or Hartig net. Diffuse development of undifferentiated mycelium is typical of that seen in fungi, which utilize continuously distributed nutrient resources such as forest litter. It has been called nonunit restricted growth (Rayner et al., 1987). The fungi of the irregular mycelial mat type include true mycorrhiza formers showing differentiation into mycelial strands or chords, but in which the mats are restricted to localized areas of intensive development. Within this category almost certainly falls the well-defined mycelial mats formed by *Hysterangium* and related species of hypogeous fungi (Cromack et al., 1979).

The classic ectomycorrhiza forming fungi of the genera *Amanita, Boletus, Paxillus, Russula,* and *Lactarius* form mycelia of the dispersed colony type. Fungi in all of these genera produce vigorous mycelial strands and widely dispersed fruit bodies. It is not known for certain whether vegetative mycelia interlink the dispersed colonies,, but the extensive nature of the mycelium makes it likely.

The pattern of mycelial growth and differentiation seen in this category is typical of that found in fungi the nutritional resources of which are discontinuously distributed through the soil. In the case of mycorrhizal fungi the resource units are individual short roots and pockets of soil in which nutrient mobilization is particularly active. The mycelial strands provide for rapid transport of nutrients between these localized areas of resource enrichment. A similar growth strategy is employed by wood-rotting fungi, the resource units of which are again irregularly dispersed through the soil (Rayner et al., 1985).

Ogawa's analyses demonstrate that irrespective of "life type" the mycelia of the fungi are concentrated in the "FH" layers of the forest soil as are most of the roots that they colonize. This is evidently the zone in which the crucial processes of nutrient mobilization take place.

More recently interesting information on the population structure and dynamics of the ectomycorrhizal mycelium of *Suillus bovinus* has been obtained by determining the distribution of somatically incompatible clones in the field (Dahlberg & Stenlid, 1990). It was shown that although the number of identifiable clones was greater in younger than in older stands of *Pinus sylvestris,* ca. 800/ha in the former and 125–130/ha in the latter, the diameter of the zone occupied by a single clone increased from 1 to 3 m in the younger, to 30 m in stands over 100 years of age. The reduction in numbers of clones with age suggests that selection pressures operate to eliminate those with the lowest competitive ability. It would be of great importance to known whether these pressures act primarily at the substrate level, where different abilities to mobilize key mineral nutrients may be involved, or at the level of the root where clones may differ in their ability to "capture" uninfected roots and so obtain carbon supplies from an autotroph.

Detailed comparative analysis of the anatomy of mycelial structures produced in the field has been facilitated by microscopic examination of carefully excavated material following, where possible, tracing of connections between roots and identifiable fruit bodies (Agerer, 1988a,b). Agerer (1987) has recognized four levels of complexity in the differentiation of ectomycorrhizal mycelium. The least differentiated, which he regarded as being the most primitive, seen in *Cenococcum geophilum* and in the widespread spruce mycorrhiza *Piceirhiza nigra*, consists simply of a cluster of hyphae emanating from the roots. An advance is seen in those fungi including some species of *Hebeloma* in which anatomosing systems of individual hyphae are produced.

The most advanced types have agglutinated or gelatinous hyphal walls that enable the production of linear organs, the so-called "strands" "chords," or "rhizomorphs." The least well developed of these consists of loose aggregations of hyphae in some cases, for example, *Lactarius deterrimus,* these all being of the same time, whereas in others, such as *Thelephora terrestris,* they are differentiated into thin outer and wide inner structures. The most advanced levels of organisation are seen in the densely packed differentiated strands of genera such as *Suillus, Paxillus, Scleroderma,* and *Pisolithus*. Here outer narrow hyphae, some of them with thick walls, form a compact sheath around wide hyphae forming a central axis. These structures are analogous to the "vessel" hyphae described in the wood-rotting fungus *Serpula lacrymans* (Falck, 1912) and are thought to have the function of water conduction (Duddridge et al., 1980).

Although such studies tell us something of the mycelial structures of individual fungal species they provide no indication of the dynamic interaction between mycelia of different species. It has been estimated (Trappe, 1977) that 2000 species of fungus are potentially able to form mycorrhizas with a single tree species, Douglas fir, and observations such as those of Ogawa (1985) in mature forests indicate that at any point in time, individual trees are hosts to many fungal species, the mycelia of which occupy overlapping domains.

Analyses of fruit body production in plantation forests (Chu Chou, 1979; Deacon et al., 1983; Mason et al., 1983) suggest that temporal as well as spatial factors influence the level of complexity occurring below ground. They reveal that the species composition changes with time, "early stage" fungi are being replaced by "late stage" species as stands age. We can speculate that such species shifts reflect the physiological and biochemical properties of the different fungal mycelia as the quality of the soil resource changes through time. In truth, however, we know very little of the growth dynamics or physiology of individual species of fungi. Our understanding of the interactions between mycelia of different species are even less, and virtually nothing of changes in population structure with time are known. The fact that in a recent study of a Sitka spruce plantation (Taylor & Alexander, 1990) less than 5% of the mycorrhizal root tips were formed by fungi represented in the fruit body population indicates that extrapolation from above ground counts to below ground activity must be carried out with great caution if at all. This is confirmed by the observations of Dahlberg and Stenlid (1990). They showed that although on the basis of fruit body production alone *Suillus bovinus* could be characterised as an "early-stage" fungus, it persists as a mycorrhizal mycelium for long periods, individual clones having a median life span of 36 years.

Faced with the enormous complexity of the soil environment the best hope of progress toward a realistic understanding of the dynamics of the mycorrhizal fungal community lies in the design of experiments that retains the two essential requirements for normal development and differentiation of the fungal mycelium, namely, the host as the source of carbon and natural unsterile soil as the medium for growth. Using this approach some progress, described below, has been obtained.

Laboratory Studies

Observation chambers maintained under controlled environment conditions (20°C day, 15°C night) have been employed to determine the basic kinetics of growth and development of the mycelia of some of the most important ectomycorrhizal fungi (Brownlee et al., 1983; Finlay & Read, 1986a). Seedlings, inoculated under sterile conditions with the required symbiont, are transferred, after infection has developed, to the layers of unsterile acid peat or forest soil between transparent perspex sheets. The progress of mycelial growth can then be followed over months in nondestructive manner. The main ectomycorrhizal species in the genera *Amanita, Boletus, Paxillus, Pisolithus, Rhizopogon, Suillus,* and *Thelephora* form extensive systems in which the mycelia growing from individual roots and even from individual plants merge to form single thalli that ramify extensively through the soil in an integrated manner. Early stages of soil colonization involve the production of a dense hyphal mass in the soil around the infected root (Fig. 4.2). This undifferentiated hyphal mass grows as a diffuse front at a uniform rate through the soil. Fronts forming from individual roots coalesce to form large fan-like structures (Fig. 4.3) advancing through the soil at rates of 2 to 4 mm/day. The density of hyphae at the front may be as many as 250 individual filaments per linear millimeter.

As the front extends differentiation occurs behind it. Hyphal fusions and aggregations lead to the production of the linear organs described above. These form the main linkages between the mycelial front and the root system of the plant and run through territory that, having already been exploited by the advancing front, is left largely unoccupied. The only regions behind the front to retain diffuse mycelial development are localized "patches" in which a particular region of the soil continues to be intensively exploited by a diffuse mycelium. Studies in which $^{14}CO_2$ has been fed to the shoots and $^{32}PO_4$ to the mycelium of mycorrhizal plants in chambers containing patches (Fig. 4.4) have shown that these are sinks in which carbon and phosphate accumulate (Finlay & Read, 1986a,b).

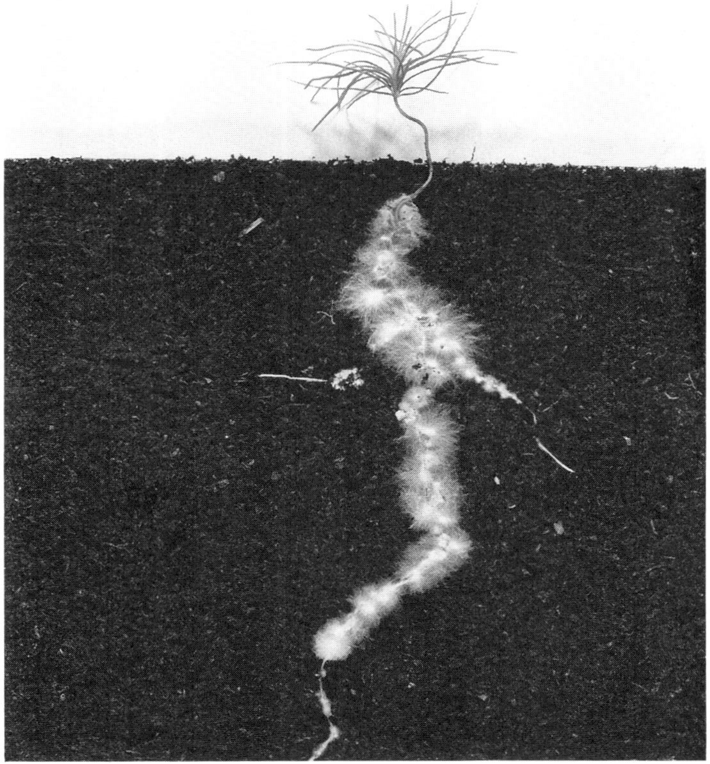

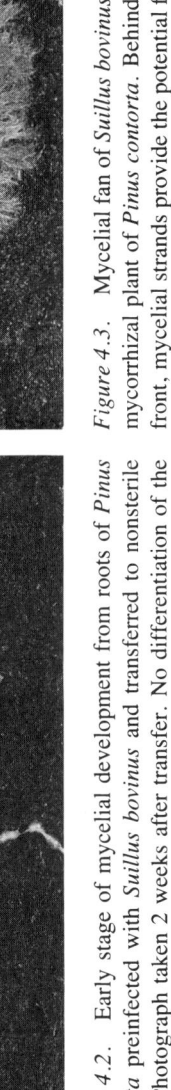

Figure 4.2. Early stage of mycelial development from roots of *Pinus contorta* preinfected with *Suillus bovinus* and transferred to nonsterile peat. Photograph taken 2 weeks after transfer. No differentiation of the mycelium occurs at this stage.

Figure 4.3. Mycelial fan of *Suillus bovinus* growing from a preinfected mycorrhizal plant of *Pinus contorta*. Behind the undifferentiated hyphal front, mycelial strands provide the potential for rapid transfer of captured nutrients to infected roots. Photograph taken 5 weeks after transfer.

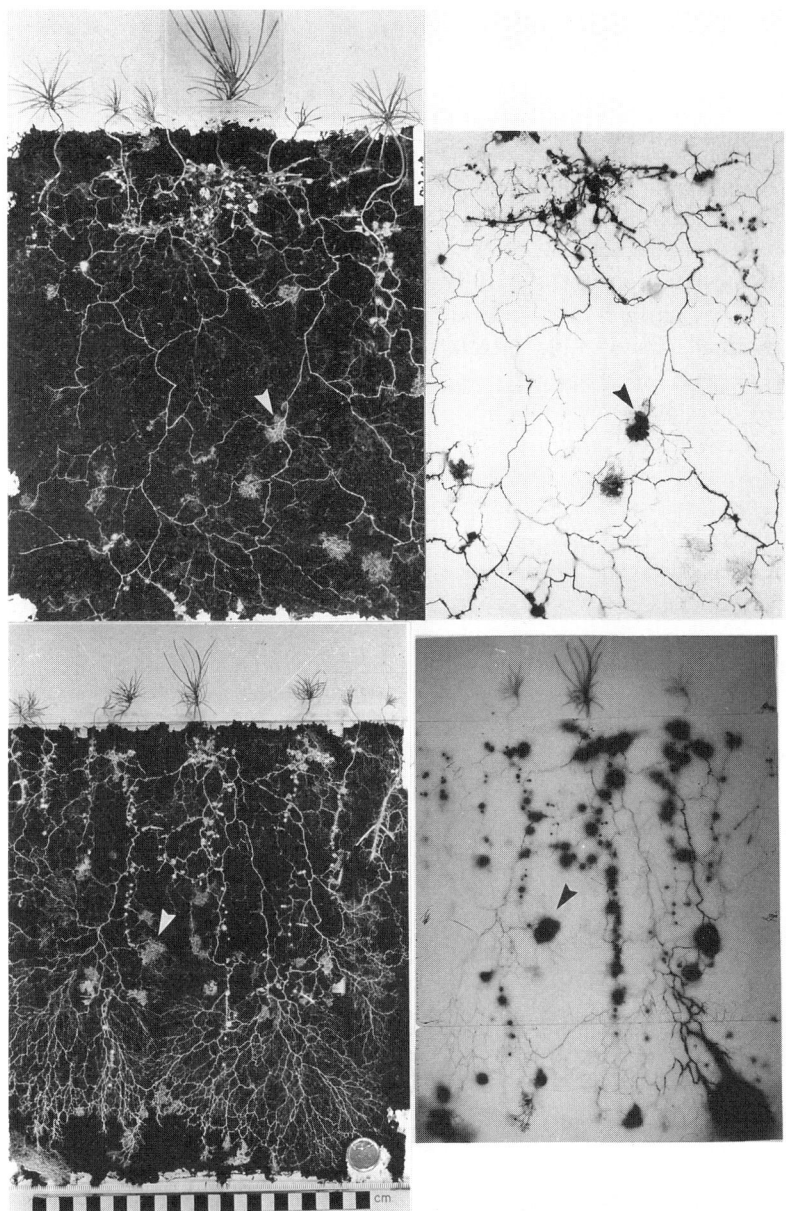

Figure 4.4. Observation chambers containing plants of *Pinus sylvestris* and *Pinus contorta* infected with the ectomycorrhizal fungus *Suillus bovinus,* the mycelium of which has grown from a preinfected central "doner" plant to interconnect all plants in the chambers. Carbon, from $^{14}CO_2$ fed to the donor plant (top left) is shown by autoradiography (top right) to be distributed throughout the mycelium and to accumulate in mycorrhizal roots of "receivers" and in "patches" (arrowed). Phosphorus, fed to the mycelium as $^{32}PO_4$ (bottom left), is shown by autoradiography (bottom right) to be transferred throughout the mycelium, to accumulate in roots of interconnected plants and in patches (arrowed) as well as in the shoots of "receiver" plants.

Clearly it is important to determine the function of the various types of mycelial structure revealed in such studies. On theoretical grounds we can assume that the primary goal of the heterotroph is to capture new food bases. In the case of mycorrhizal fungi, this means infection of new roots. The extensive hyphal front provides an extremely effective structure with which to scavenge for such resources and the rapid onset of infection following contact between uninfected roots and the advancing front confirms that the mycelia have a very high inoculum potential. This vigor, combined with the low host specificity of many ectomycorrhizal fungi, results in the interconnection of individual host plants both within and between fungal species. Recruitment of seedlings into these mycelial networks takes place as the emerging radicle contacts a compatible mycelium (Read et al., 1985).

The growth strategy of the ectomycorrhizal mycelium is comparable with that seen in many wood-decomposing basidiomycetes. The radiating mycelial fans of these fungi forage effectively for irregularly dispersed resource units (Rayner et al., 1985). However, the biology of the mycorrhizal situation differs fundamentally from that of the wood-decomposer in that the requirements of the autotrophic partner must also be considered. Logically, some return on the carbon investment made by the autotroph is required if the association is to be mutualistic. From this perspective the advancing hyphal front is ideally suited to scavenge for mineral nutrients as well as for new roots. Studies in observation chambers have confirmed that when $^{32}PO_4$ (Finlay & Read, 1986b) (Fig. 4.4) or $^{15}NH_4$ (Finlay et al., 1988) is placed near the advancing mycelial front, it is readily absorbed and transported through the anastomosing system of strands to host plants linked by the fungal network.

The requirement for effective mineral nutrient scavenging may, more than any other feature, select for those fungi that develop by diffuse rather than apically dominant rhizomorphic growth as the major ectomycorrhizal associates. Since each individual hyphal element making up the advancing front is an absorptive structure, the resulting increase in surface area for nutrient capture is enormous.

By measuring total lengths of roots and hyphae in representative observation chambers containing pine as host plant and *Suillus bovinus* and *Paxillus involutus* as mycorrhizal fungi, the relative lengths of absorptive surfaces of the fungal and root components of the mutualism was estimated and expressed as a ratio of mycelial length to root length (M:R). From the initial placement of an infected plant into a chamber, hyphal growth outstripped root growth as up to 1000 hyphae emerge from each infected pine root. M:R ratios progressively increase with time as the mycelium colonizes the soil. Using mycelia of *Suillus bovinus,* Read and Boyd (1986) measured mean hyphal lengths of 10–80 m/cm of root giving an M:R ranging from 1000 to 8000:1. These lengths are considerably greater than the maximum values of around 3 m/cm of root length obtained by Jones et al. (1990), who grew cuttings of *Salix* in a sterilized soil mixture inoculated with *Laccaria proxima* or *Thelephora terrestris*. Their lower values may be accounted for by the sterilization treatments that involved a combination of autoclaving, a process known to lead to depression of microbial growth (Salonius et al., 1967), and γ-irradiation, which provokes enhanced release of nitrogenous components from soil organic matter (Bowen & Cawse, 1964; Eno & Popenoe, 1963).

The need to examine mycelial development in the absence of sterilization was further highlighted in a recent study by Erland et al. (1990) who showed that the growth rates of three different ectomycorrhizal fungi extending from infected roots of pine were two to five times greater on unsterilized than on autoclaved peats.

Extrapolation of the most conservative values of hyphal length reported by Read and Boyd to lengths per unit weight and volume of the forest soil employed in their chambers gives values of the order of 200 m/g soil dry weight or 2000 m/cm^3 of fresh soil. Comparing these estimates with the lengths of fine root of the major coniferous host species of these fungi in the field demonstrates the importance of these fronts. Fine roots of *Picea sitchensis* achieve lengths of 1–10 cm/cm^3 of forest soil in plantations (Ford & Deans 1977), whereas those of *Pinus sylvestris* achieve 1–5 cm/cm^3 (Roberts, 1976). Assuming a fine root length of 5 cm/cm^3 and an associated length of mycorrhizal hyphae to be 2000 m/cm^3 the M:R ratio would be 200,000:1.

Although providing understanding of the basic kinetics of growth and differentiation of ectomycorrhizal mycelia, these studies reveal little of their true temporal relationships because they are carried out under conditions of constant day length and controlled temperature. Recently, however, Coutts and Nicoll (1990a) examined the development of mycelial fans of *Thelephora terrestris* and *Laccaria proxima* expanding from roots of Sitka spruce using transparent acrylic tubes filled with peat in the field. A mean extension rate of 3 m/day was recorded for *T. terrestris* mycelium in July coinciding with a peak of soil temperature (Fig. 4.5). This value is comparable with those obtained under controlled environment conditions. Between June and November the fans had grown over a mean distance of 22–25 cm. Rates of extension growth of the longest main roots were generally a little faster than those of the mycelial fans, but the mean distance from the tip of the main root to the nearest emerging short root was normally greater than that to the advancing hyphal front so that these roots were rapidly infected as they emerged. Of particular interest was the observation that the mycelia of *T. terrestris* continued to expand throughout the winter even though the roots were dormant (Fig. 4.6). The rate decreased to a rate of 0.4 mm/day from December to March, but increased again as temperature rose in spring. These results show that the mycelium of *T. terrestris* is a perennial structure. *Laccaria proxima,* in contrast, disappeared as winter progressed. Coutts and Nicoll (1990b) also examined the responses of such mycelia to waterlogging. They observed that the undifferentiated mycelium was killed by the treatment, but regrowth occurred for the mycelial strains when the soil was drained. This is the first indication that the strands also maintain the perennial nature of the mycelium.

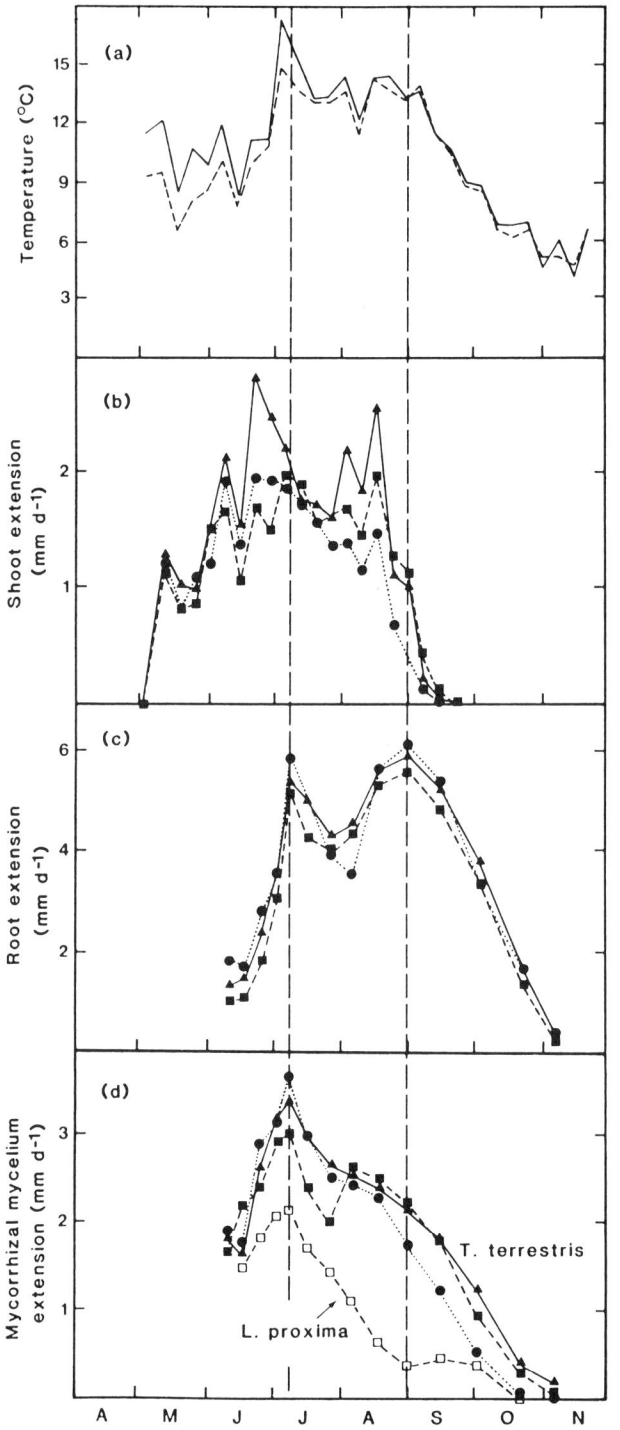

Figure 4.5. Cumulative growth of three Sitka spruce clones and their mycorrhizal fungi during 1987. (a) Mean shoot extension, (b) mean root extension, and (c) mean mycorrhizal mycelium extension. Closed circles, clone 1; closed triangles, clone 2; closed squares, clone 3. In (c) open symbols are *Laccaria proxima*, solid symbols are *Thelephora terrestris*. Vertical bars represent standard errors. From Coutts and Nicoll (1990a), with permission.

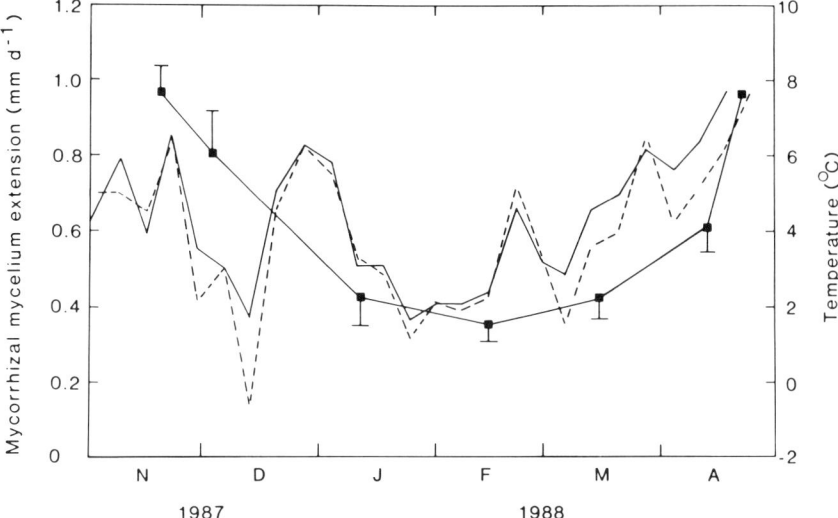

Figure 4.6. Growth of the extramatrical mycelium of *Thelephora terrestris* from dormant Sitka spruce root systems during the winter. Closed squares, mycelial growth; each point is the mean of three plans for each of three clones, growing in freely drained peat. Solid line, mean weekly soil temperature; dashed line, mean weekly air temperature. Vertical bars, standard errors above and below the mean. From Coutts and Nicoll (1990a), with permission.

In view of the large investment in the fungal structure, it is necessary to consider the costs incurred by the plant. Because of the much smaller diameter of the hyphal tube (about 100 times), the provision of a given amount of new absorbing surface in the form of hyphal extension is more cost effective than for the production of the same area by root growth (Harley, 1989). Moreover the disparities between root and hyphal lengths are well in excess of two orders of magnitude.

Based on measurements of fruit body biomass alone, Romell (1939) calculated that the carbon drain imposed by mycorrhizal fungi was equivalent to 10% of annual timber production. Even higher values of carbon demand are suggested by more recent field studies (Fogel & Hunt, 1979; Vogt et al., 1982), which suggest that between 15 and 50% of net primary production may be consumed by the fungi. Studies in pots or observation chambers that facilitate closer control of carbon budgets have largely confirmed the higher values of carbon demand made by the fungi. Reid et al. (1983) estimated that between 20 and 60% of carbon assimilated by *Pinus radiata* was translocated to mycorrhizal roots and considered it possible that half of this may have been consumed to support fungal growth.

Using split chambers in which the respiratory activity of the ectomycorrhizal mycelium was measured independently of that of the roots, Söderström and Read (1987) calculated that up to 30% of total respiration of the below ground system and may be attributable to external mycelium. This mycelium was shown to be entirely dependent on supply of current assimilate for the maintenance of its respiratory activities. Miller et al. (1989) using "mycocosms" containing vermiculite sand and perlite confirmed that considerable quantities of current photosynthate are allocated to the extramatrical mycelium associated with infected seedlings. They showed that mycelia developing naturally from preinfected roots were a significantly larger sink for carbon than were mycelia infecting roots from commercial inoculum. It is evident that if meaningful estimates of the activity of the mycorrhizal mycelium are to be obtained great care must be exercised in selecting realistic experimental conditions. The mycorrhizal status of the host and the environment in which the mycelium develops should be as close as possible to that prevailing in nature.

The size of the discrepancy between the absorptive surface areas provided by the roots and the fungi is itself sufficient to indicate that the roots play an insignificant part in the absorptive process in such systems. Close examination of the structure of the microsites in which ectomycorrhizal roots of pine, larch, spruce, and birch are produced in observation chambers gives strong support for the view that the primary function of these roots is not to absorb nutrients from soil but simply to provide a food base from which the fungus can extend into the soil. As discussed earlier, the roots are primarily formed as individuals or clusters in air pockets between soil particles with which the mycorrhizal sheath itself often has little contact. Individual hyphae extend from the mantle surface to the soil particles, but in almost all cases the main connection between the sheath and the mycelium in the soil takes the form of a mycelial strand emanating from the base of the root or cluster and connecting it with the mycelial network (Fig. 4.7). In such circumstances the mycorrhizal root is not itself an absorbing structure. Rather, it forms a base on which the fungus is dependent for carbon with which to sustain its foraging activity and it provides a facility for storage of products, absorbed and translocated over considerable distance, by the fungi. Such a segregation of absorptive and storage functions is clearly seen in (Fig. 4.4).

The storage function of the ectomycorrhizal sheath has been well documented (Harley & Smith, 1983). The transfer of the absorptive function from roots to mycorrhizal mycelium would be expected to reduce the need for an extensive root system. This may explain the exceptionally low root densities observed in ectomycorrhizal trees such as pine in the field (Kramer & Bullock, 1966).

There are ectomycorrhizal systems in which the sheaths are smooth and appear to lack mycelial contact with the soil or litter around them. Some of the most robust mycorrhizal roots of beech (*Fagus sylvatica*) fall into this category (Harley, 1978). Such roots, when detached and placed into a bathing medium containing phosphate, both absorb and store these ions (Harley & Smith, 1983). Thus, the mechanism of the absorptive process in nature obviously needs to be addressed. Recent studies based on a combination of field excavations and analysis of tissues in the laboratory have shown that even these apparently smooth sheaths can have

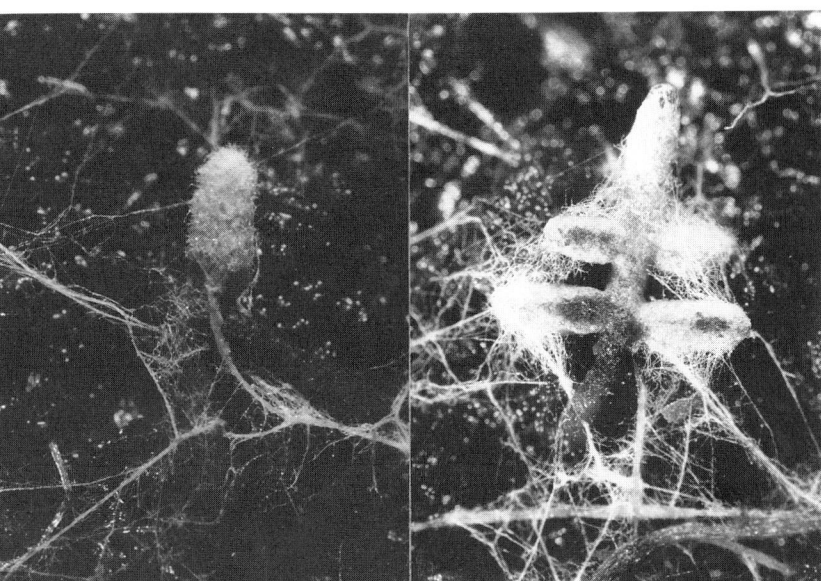

Figure 4.7. Individual mycorrhizal lateral root of *Larix* infected with *Boletinus cauipes* (left) and branched mycorrhizal lateral root of *Picea* infected with *Paxillus involutus* (right), both produced in soil pores, showing the subtending mycelial strands that connect the roots to the mycorrhizal mycelium.

delicate strand like connections with the litter (Brand & Agerer, 1986). These structures have been traced from beech trees to the fruit bodies of *Lactarius subdulcis*. The sheaths of such mycorrhizas, in contrast to those described by Babel (1987), are frequently sandwiched between layers of decaying leaves with which their surfaces have considerable contact. In these situations there may be a combination of direct absorption by the sheath, and translocation by the weak strand-like structures.

Important though the capture of mineral nutrients may be, recent studies suggest that the ectomycorrhizal mycelium may have an even more fundamental function (Read, 1982). Many of the most widely occurring ectomycorrhizal fungi have the ability to degrade polymeric protein molecules and to use the products of proteolysis, which are almost exclusively amino acids (Abuzinadah & Read, 1986a,b; Read et al., 1989), as sole sources of nitrogen. In the field, these fungi preferentially occupy regions of the soil in which practically all of the nitrogen is present in organic form. Subsequent studies have confirmed not only that the nitrogen mobilization seen in pure culture occurs in the mycorrhizal association but also that access of the plant to such resources is entirely dependent on infection. The need now is to determine the role of the mycorrhizal mycelium in the process of nitrogen mobilization in the organic horizons of forest soils occupied by ectomycorrhizal roots.

Observation chambers have again provided valuable insights. Attempts were made to stimulate patch development by adding phosphate and ammonium ions to small areas of the soil in chambers immediately behind an advancing hyphal front. When no response was obtained the experiments were repeated using organically enriched materials. Ground or fragmented samples of leaf litter of beech or larch taken from the "F" horizon of forest soils were added as discrete blocks in advance of the mycelial front. On contact with these materials intensive proliferation of the fungal mycelium took place, and as the front progressed beyond the blocks, the heavily colonized organic material was clearly identifiable as a "patch" (Fig. 4.8).

The first indication that significant amounts of nutrient transfer might be taking place between the patch and the mycorrhizal seedling was obtained when shoots of these plants, which had been chlorotic prior to "patch" formation began to regreen. Subsequent analysis of individual needles over a time course on three separate plants indicated that their revitalization was a result of nitrogen capture (Read, 1990b). Importantly, these patches (formed in homogeneous soil layers) were sinks rather than phosphorus (see Fig. 4.4). Thus, patch development was associated with something other than P release. The need for additional P, as well as carbon, in these patches may now be interpreted as indicating a need for energy to enable the mobilization of nitrogen. We still lack evidence to show that these same fungi are directly involved in mobilizing the organic N of the added litter. However, subsequent analyses have ascertained that the initial levels of ammonium in the source materials are negligible so that some form of attack on the organic residues must be involved.

Among the most labile sources of organic nitrogen in the natural environment of the mycorrhizal root must be the fungal mycelium itself. Fogel and Hunt (1979) calculated that the sheaths of ectomycorrhizal roots contained up to 50% of the below ground N in a Douglas fir forest and Baath and Söderström (1979) working in pine-heaths of Sweden estimated that the fungal biomass contained up to 20% of total soil N. It seems likely that the proteolytic capabilities of the mycorrhizal mycelium will provide the potential for direct recycling of this nitrogen fraction.

Structure and Function of the Vesicular-Arbuscular Mycorrhizal Mycelium

As in the case of the study of ectomycorrhizal mycelium systems, knowledge of the growth and differentiation of VA fungal mycelia has been gained through a combination of destructive field sampling and relatively nondestructive laboratory studies. Examination of the VA mycelium presents even more difficulties than

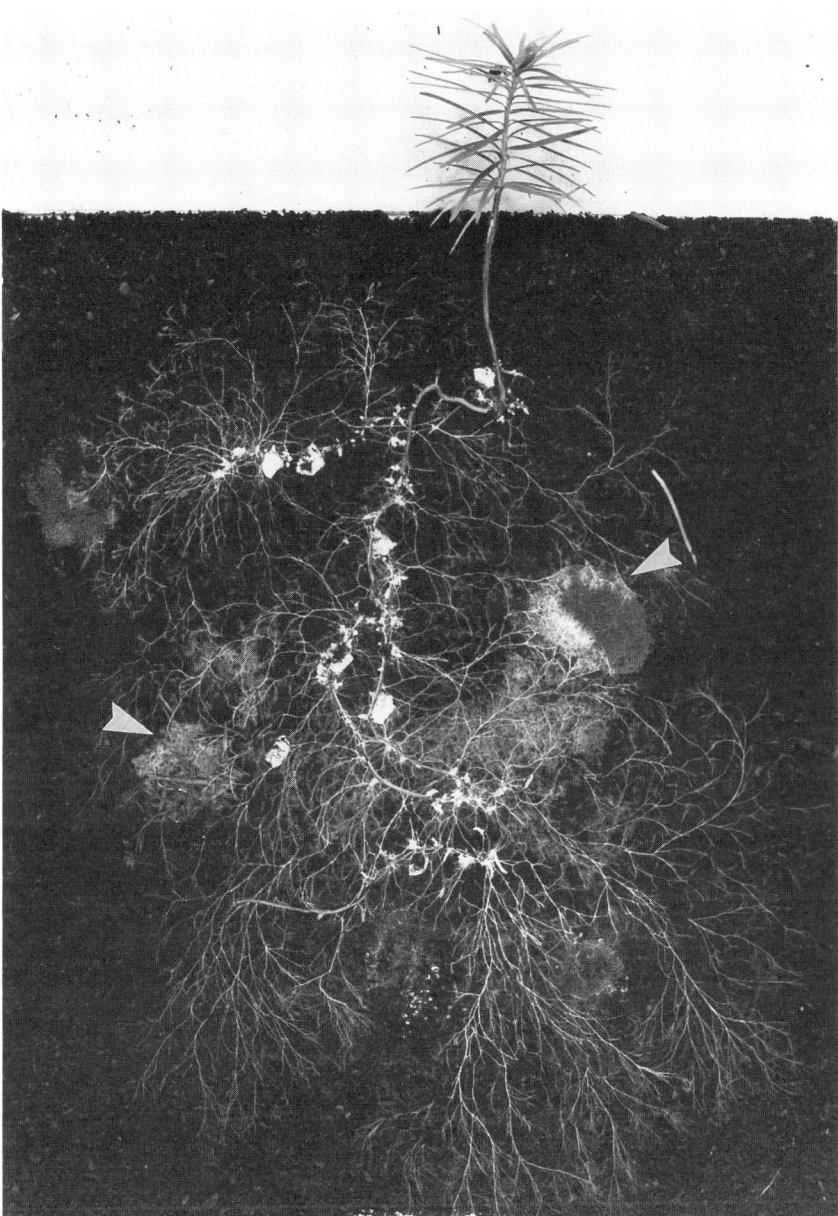

Figure 4.8. Observation chamber with plant of *Larix leptolepis* infected with *Boletinus cauipes* growing on a homogenized matrix of nonsterile humified forest soil to which litter of *Larix* (left arrow) and *Fagus* (right arrow) has been added in discrete blocks. Note "patch" development in association with the enriched material.

does that of ectofungi because it is a more delicate structure. Nonetheless the application of techniques broadly similar to those developed for the study of ectosystems is beginning to reveal that the biology of the two types of mycelium is in many ways comparable.

Field-Based Studies

Although revealing little direct information on the organization of the mycelium in the field, studies of extracted roots and soil samples have provided evidence that the VA mycelium is an extensive structure. Indeed in sand dune ecosystems it has been described as being the main sand binding agency providing for the stabilization of the dunes (Sutton & Sheppard, 1976). Nicolson and Johnston (1979) noted seasonal changes in the quantities of external mycelium recovered from sand dunes at different stages of development, the largest biomass being obtained in summer. A role for the mycelium in sand-binding was again suggested by the observation that the largest quantities, over 4 g mycelial dry weight per liter of sand, were found in the early stages of dune fixation. Measurements of length of VA mycelium in pioneer sand dunes of Florida dominated by sea oats, *Uniola paniculata* (Sylvia, 1986), gave values of 12 m/g/g sand, and 592 m/cm of colonized root length. The former value is similar to those reported by Tisdall and Oades (1979) and Abbott and Robson (1985) for sandy loams.

Some information on structural differentiation of individual hyphal elements can also be gained from field studies. Peyronel (1924) described thick-walled hyphae of diameters between 2 and 30 μm, which gave structure to the mycelium. These, he showed, had characteristic elbow-like bends that were later termed angular projections (Butler, 1939). Mosse (1959) reported that 75% of the external mycelium of apple roots was of the course type, hyphae having a diameter of up to 20 μm. From these course structures finer hyphae grew to penetrate the root. She estimated the number of penetration points to be between 2.6 and 21/mm root of strawberry and between 4.6 and 10.7/mm root in apple.

Recent studies in seminatural grassland (Birch, 1986) indicate similar values. Such large numbers of penetration points suggest the presence of a very vigorous inoculum. Analysis of the development of infection in these grassland ecosystems provides indirect evidence that the extremely high inoculum potential is the result of an extensive and semipermanent mycelial network in undisturbed soil. Birch preimbibed seeds of four of the major herbaceous species of calcareous grassland ecosystem. When placed onto the surface of the soil seedings were infected within 3 to 4 days of radical emergence. The speed and vigor of the infection is itself sufficient to suggest that spores were not the dominant source of inoculum and wet sieving confirmed that viable spore numbers were extremely low in the soil. A factor conducive to the spread of infection and to the maintenance of the hyphal network is the low level of host specificity shown by most of the VA fungi. Allen et al. (1989), examining the response of the nonmycotrophic species *Salsola kali* to the presence of infective mycelium of a VA fungus have highlighted another mechanism whereby a community of compatible mycorrhizal species might be maintained through the activities of the vegetative mycelium. They found that penetration of roots of *S. kali* led to hypertrophic responses at the level of the root and to reduction of vigour of the plants. Responses of this kind would lead to development of plant communities containing guilds of species that were compatible with and interlinked by mycorrhizal mycelium, and to the segregation of these guilds from those containing nonmycotrophic species (Read, 1990a).

Laboratory Studies

Considerable advances in our knowledge of the structure and function of the VA mycelium have been provided by the use of observation chambers similar to those employed for the study of ectomycorrhizal systems. Similar constraints on the design of experimental chambers apply. The substrates in which VA mycelium develops should be as close to those found in nature as possible, free of additional exogenous carbon sources, and unsterile. Dune sand or soil of similar texture has proven to be ideal because it contains little organic matter and so supports only a small population of saprotrophic fungi and has a particle size sufficiently large to enable manipulation. Preinfected plants transferred to shallow transparent dishes of such sand produce an extensive mycorrhizal mycelium, most of which is confined to the base of the dish in association with the roots. Its entire structure can be revealed by careful removal of individual sand grains. Such experiments (Francis & Read, 1984; Read et al., 1985) have shown that the VA mycelium takes the form of an anastomosing network (Fig. 4.9). The differentiation into the hyphal aggregates, seen in ectomycorrhizal systems, is largely lacking, but functionally analogous "arterial hyphae" provide direct connections between resource-rich domains be they roots or soil. From these conduits the finer branching hyphae arise. Application of $^{14}CO_2$ to host plants from which such mycelial systems were growing has confirmed that the hyphal network is a living extension of the root system through which carbon fluxes from root to root and even, in the case of interspecific communities of plants, from species to species (Francis & Read, 1984; Read et al., 1985).

Subsequent analyses (Grime et al., 1987) using larger microcosms of dune sand containing mixtures of plant species have confirmed the hypothesis that the presence of a vigorous mycelial network can have a profound influence on the structure of plant communities. The high inoculum potential and low host specificity of the VA mycelium provide for rapid infection of compatible species as they germinate; subsequently, these individuals are incorporated into an enormous absorptive network the energy costs of which are largely paid by established plants. Species with small seeds and hence small resources benefit most from infection. A consequence of the effectiveness of the mycelium in scavenging for and colonizing new potential food bases in the form of uninfected roots is thus

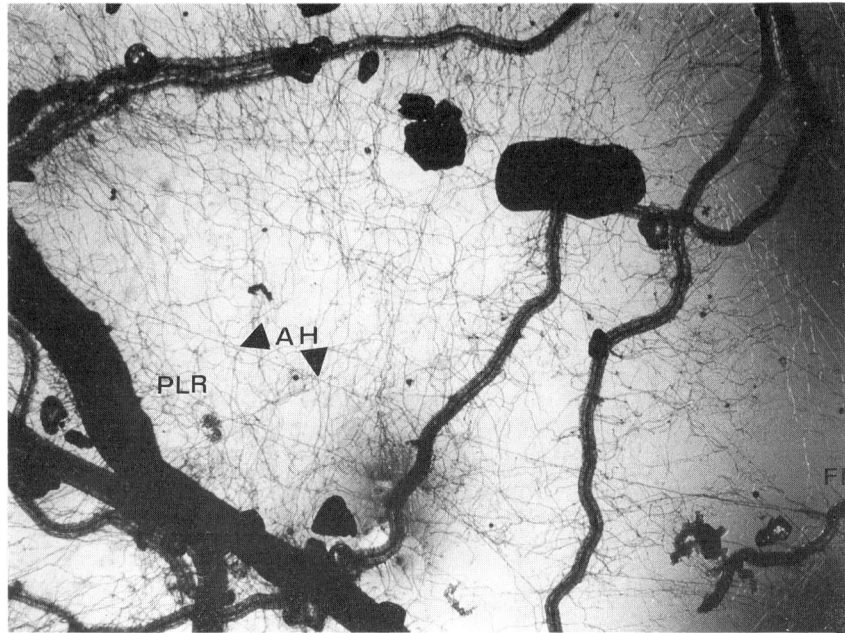

Figure 4.9. Anastomosing mycelial network associated with vesicular-arbuscular infection of roots of *Plantago lanceolata* (PLR) and *Festuca ovina* (FR) grown together in a transparent microcosm containing dune sand. Sand grains have been removed to reveal the network and its "arterial hyphae" (AH), which provide direct connections between roots.

to increase the diversity of species in the higher plant community. The evidence thus points to a major role for the VA mycelium in determining ecosystem dynamics both below and above ground.

This being the case it is important to increase our understanding of the process of development of this mycelium in soil. With this objective in mind, Friese and Allen (1991) employed a modification of the observation chambers described by Finlay and Read (1986a) to carry out a detailed study of hyphal structure and mycelial differentiation. Using glass plates to improve optical clarity they have followed the development of the external hyphal network associated with *Artemisia tridentata* and *Oryzopsis hymenoides* grown in non-sterile sandy soil.

These studies make possible differentiation between hyphae entering and leaving roots (Fig. 4.10). They confirm that runner hyphae (RH), equivalent to the arterial hyphae described earlier, some of which form hyphal bridges (HB) between adjacent roots, can be distinguished structurally from a absorptive hyphal network (AHN). The hyphal architecture of the AHN bears some resemblance, albeit on a much smaller scale, to the ectomycorrhizal hyphal front since it

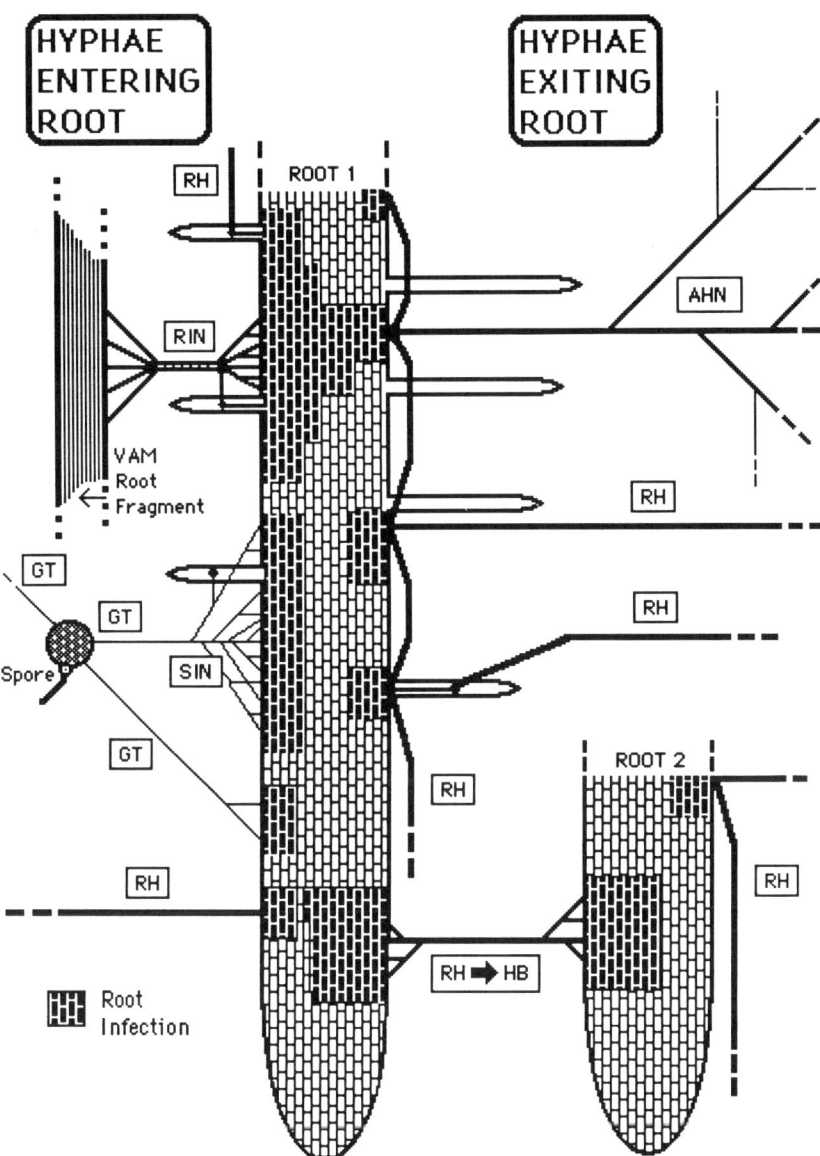

Figure 4.10. Illustration of the various external hyphal architectures recognized growing in the soil of root observation chambers. The primary hyphal types observed in the soil were runner hyphae (RH), hyphal bridges (HB), absorptive hyphal networks (AHN), germ tubes (GT), and infection networks produced by spores (SIN) and root fragments (RIN). Reprinted from Friese and Allen (1991).

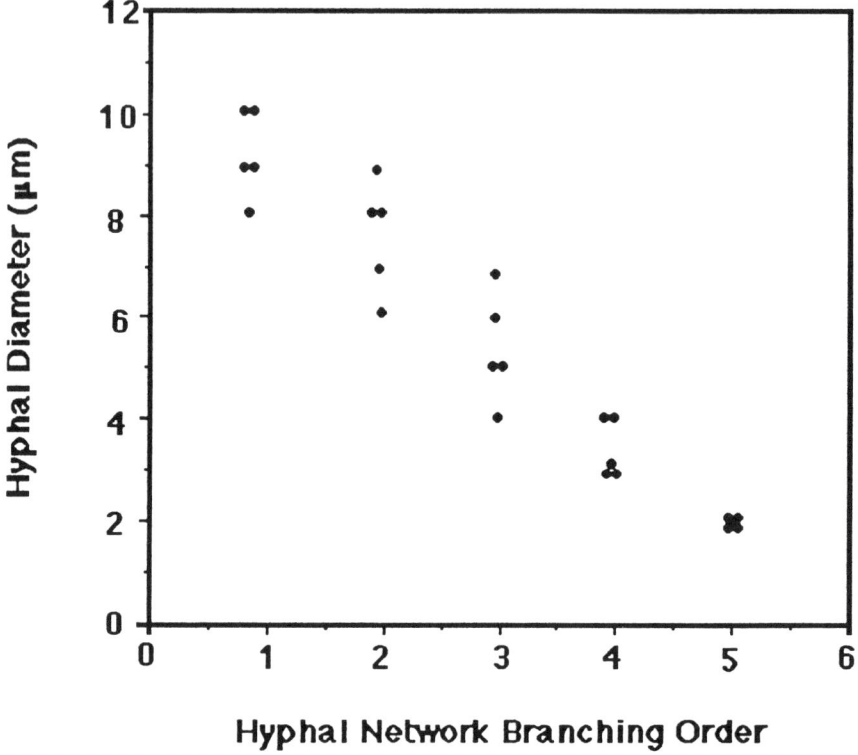

Figure 4.11. The relationship between hyphal network branching order and hyphal diameter (μm). Values represent direct measurements of five hyphal networks out to five consecutive orders of branching, with the first order of branching being the closest to the root and each successive order of branching being further away from the root epidermis. Any subsequent orders of hyphal branching were in the 2 μm diameter range, and no hyphal networks were observed to extend beyond an eighth order of branching. Reprinted from Friese and Allen (1991).

develops into a fan shaped network extending up to 4 cm from the epidermal surface. The branching pattern is always dichotomous and each successive order of branching away from the root leads to successive decrease in hyphal radius (Fig. 4.11). This pattern of division in turn gives consecutive increases in surface area: volume ratio of the hyphal cylinder. Thick hyphae adjacent to the root surface have diameters of approximately 10 μm; those of fifth order branches were around μm. These values are consistent with those reported from the field.

Hyphae entering the root are of three types. These are (1) runner hyphae, which are seen to enter as well as to leave roots and hence to form the hyphal bridges described above; (2) germ tubes (GT), which are initially single points of entry but which proliferate with time to form a branching infection network derived from spores (SIN); and (3) hyphal systems that emerged from dead root fragments.

The latter type is of interest because there is some evidence of aggregation of its hyphal elements to form strand like structures that are the closest approximations in form to the strands of ectomycorrhizal fungi. The elements forming these short linear organs revert to their unitary structures close to the root to form a multiple infection network derived from roots (RIN). Translocation of resources from dead root fragments to living roots through VA mycelium has been reported (Rix & Newman, 1985) and it is possible that these VA "strands" are, like their counterparts in ectomycorrhizal systems, specialized channels for rapid transit of resources between domains.

There is some evidence from laboratory (St. John et al., 1983) and field (Mosse, 1959; Koske et al., 1975) studies that VA hyphae proliferate more extensively on contact with organic particles. St. John et al. (1983) compared lengths of VA hyphae in chambers containing sand with those in adjacent chambers to which organic particles were added. Although finding no indication of orientation of hyphae toward organic matter they observed that having reached such substrates increases of hyphal branching gave rise to greater total lengths of mycelium. This situation is analogous, on a smaller scale, to that seen in the ectomycorrhiza, but there is little evidence to suggest that the VA fungi have comparable abilities to mobilise nutrients. Proliferation of hyphae in association with localized sources of mineral ions has the effect of reducing diffusive resistance to flow and thus of increasing their flux towards the sink (Tinker, 1975).

Since nutrient ions move through the mycelium into the root in solution it is evident that infection must involve supply of water to the plant. Although there are now a number of studies confirming that infection influences the water relations of the host plant there has, until recently, been considerable disagreement about the basis of this effect. One school of thought proposes that any improvement of plant water relations must be only a secondary effect of enhanced phosphorus nutrition (Safir et al., 1971; Nelsen & Safir, 1982; Koide, 1985), and that, in any event, infection can lead to only trivial increases of water flow to the plant (Cooper & Tinker, 1981). However, it is increasingly clear that, providing the VA mycelium is allowed to develop naturally, it can facilitate significant increases of water throughput in infected plants, independently of changes of tissue phosphorus status (Augé et al., 1986).

Direct evidence that the fungal mycelium provides the pathway for water transport has been provided by studies showing that severance of VA hyphae connecting plant roots to the soil leads to a reduction of transpiration rate (Hardie, 1985). Even more convincing is the study of Faber et al. (1991), who used a root chamber containing both a 35-μm mesh screen and an air filled void across which VA hyphae, but no roots, passed to connect plants with moist soil. They found that over a 16-hr period plants in which hyphal links were maintained transpired 35% more water than did those in which the hyphal connections with the moist soil were cut at the start of the period. Since both sets of plants were grown in the mycorrhizal condition under identical circumstances, they were structurally and nutritionally comparable in all respects and differences in transpiration rates can be accounted for only in terms of transport of water through the hyphae. Fluxes of 375 nl of water through each individual hyphal connection per hour were calculated.

The transport, through VA hyphae, of meaningful quantities of water will probably be important not for the provision of amounts sufficient to ensure maximum rates of transpiration, but rather for the maintenance of the minimal supplies required to preserve physiological activity and thus to permit survival. Read and Boyd (1986) pointed out that in any situation in which water supply to the roots is temporarily restricted, the external mycelium will become progressively more beneficial as the hydraulic conductivity of the soil begins to limit uptake at the root surface. A good example of such a benefit was described by Allen and Allen (1986), who observed that VA infection could lead to increases of water potential and reduction of stomatal resistance in a grass of semiarid habitats but that these features were most pronounced under short lived conditions of stress, which they described as representing "ecological crunches."

There is much to suggest that the VA mycelium can fulfill functions comparable with those of the more differentiated ectomycorrhizal systems that facilitate capture and transport of water in sufficient quantity to maintain photosynthetic activity (Boyd et al., 1986) and even survival (Brownlee et al., 1983) of the host plants. Experiments should now be designed to investigate the ability of fully developed VA hyphal systems to increase survivorship of their host plants under conditions of water as well as nutrient stress.

Structure and Function of the Ericoid Mycorrhizal Mycelium

In contrast to the elaborate external mycelial networks associated with most ecto and VA mycorrhizas, the ericoid mycelium does not proliferate far beyond the infected root. The expanded outer cortical cells of the distal parts of ericoid root systems are packed with hyphae of the ascomycete fungus *Hymenoscyphus ericae* (Read) Korf & Kernan, so that up to 80% of the volume of the root can be occupied by vegetative mycelium (Read & Stribley, 1975b). Each infected cortical cell is characteristically penetrated by several hyphae so that communication between the internal and external mycelium is very effective. However, the fine hair roots themselves proliferate so extensively in the decomposition horizons immediately below the litter layer of heathland soil (Gimingham, 1972; Reiners, 1965; Persson, 1980), that a dense root mat readily fulfills the absorptive and transport functions associated with the mycelium of ecto and VA mycorrhizas.

Pure culture studies of the ericoid mycorrhizal fungus have revealed that nutrient mobilization rather than transport is the main function of the extracellular mycelium. The fungus has the ability to degrade many of the recalcitrant polymeric sources of carbon and nitrogen, which are characteristic of mor humus soils produced under ericaceous plants. Lignase (Haselwandter et al., 1990), chitinase (Leake & Read, 1990b), polyphenol oxidase (Leake & Read, 1990c), proteinase (Bajwa et al., 1985; Leake & Read, 1989), and phosphatase (Pearson & Read, 1975; Mitchell & Read, 1981; Straker & Mitchell, 1986) activities have been demonstrated.

Proteolytic activity of the ericoid mycelium provides the infected plant with access to the key growth limiting element of heathland soils, nitrogen, the bulk of which is held an organic combination in raw humus and litter (Stribley & Read, 1974). The importance of lignase, polyphenol oxidase, and chitinase activities probably lies in their abilities to unmask nitrogenous substrates. It has been shown that the ericoid endophyte has access to nitrogen even after the proteinaceous source materials have been precipitated with tannin (Leake & Read, 1990c). Regulation of the activity of this enzyme, which is likely to be crucial for the survival of the plant as well as for the release of nitrogen, appears to be achieved by a simple but ecologically meaningful pH-dependent mechanism (Leake & Read, 1990a). Activity is expressed only under the acidic conditions typical of those prevailing in heathland soils, and is completely repressed under conditions around neutrality of the kind found in the intracellular environment. The products of proteolysis are amino acids (Read et al., 1989), which can be assimilated directly only by mycorrhizal plants (Stribley & Read, 1980). Pure culture studies have shown that ammonium ions are a product of proteolysis only when the fungus is starved of carbon (Read et al., 1989), a situation unlikely to occur normally where the fungus is in the symbiotic state.

The mycelium of *H. ericae* has a very high affinity for cations. This is important in a nutritional sense in that it enhances the ability of mycorrhizal plants to capture essential bases such as calcium (Leake & Read, 1990d), which are present in extremely low concentration in acid heathland soils. This high affinity is also important in a nonnutritional context because toxic metals ions such as Cu and Zn^{2+} can be absorbed and excluded from the plant (Bradley et al., 1981, 1982).

A further attribute of the mycelium is its ability to assimilate and detoxify phenolic and aliphatic acids, which are the natural products of breakdown of the lipid and polyphenol-rich litter of ericaceous plants (Jalal & Read, 1983a,b). Their removal facilitates the penetration of infected ericoid roots into soils that exclude roots of uninfected would-be competitors (Read, 1984b; Leak, 1987). In such newly colonized domains, the fungal endophyte is able to express its biochemical potential, mobilizing nutrients and facilitating their absorption by the host plant.

Conclusion

Each of the three types of mycorrhizal mycelium described here has distinctive features of distribution, structure, and function that provide it with the potential to play a central role in the nutrition, growth, and even survival of its host plants and thus to influence the structure of the ecosystem in which it predominates. Studies of these mycelia reveal an increasingly broad spectrum of physiological properties and of specific adaptations to defined environmental circumstances. In so doing they expose the fact that better understanding of the dynamics of each of the major terrestrial ecosystems of the world will be dependent on a greater awareness of the activities of its fungal mutualists. There is a particular need for further investigation of the involvement of each type of mycorrhizal mycelium in the processes of nutrient mobilization in the natural environment.

References

Abbott, L. K., & Robson, A. D. (1985). Formation of external hyphae in soil by four species of vesicular-arbuscular mycorrhizal fungi. *New Phytologist* **99**, 245–255.

Abuzinadah, R. A., & Read, D. J. (1986a). The role of proteins in the nitrogen nutrition of ectomycorrhizal plants. I. Utilization of peptides and proteins by ectomycorrhizal fungi. *New Phytologist* **103**, 481–493.

Abuzinadah, R. A., & Read, D. J. (1986b). The role of proteins in the nitrogen nutrition of ectomycorrhizal plants. III. Protein utilization by *Betula, Picea,* and *Pinus* in mycorrhizal association with Hebeloma crustuliniforme. *New Phytologist* **103**, 507–514.

Agerer, R. (1987). The ecologically crucial question of ectomycorrhizae: how to make rhizomorphs. In: *Proceedings of the 7th North American Conference on Mycorrhizae* (Ed. by D. M. Sylvia, L. L. Hung, & J. H. Graham), pp. 184–185. University of Florida, Gainesville.

Agerer, R. (1988a). *Colour Atlas of Ectomycorrhizae,* 2nd Fascicle. Einhorn Verlag, Schwabisch Gmund.

Agerer, R. (1988b). Studies on ectomycorrhizae XVII. The ontogeny of the ectomycorrhizal rhizomorphs of *Paxillus involutus* and *Thelephora terrestris*. *Nova Hedwigia* **47**, 311–324.

Allen, E. B., & Allen, M. F. (1986). Water relations of xeri grasses in the field: Interactions of mycorrhizas and competition. *New Phytologist* **104**, 559–571.

Allen, M. F., Allen, E. B., & Friese, C. F. (1989). Responses of the non-mycotrophic plant *Salsola kali* to invasion by vesicular-arbuscular mycorrhizal fungi. *New Phytologist* **111**, 45–49.

Augé, R. M., Schekel, K. A., & Wample, R. L. (1986). Greater leaf conductance of well-watered VA mycorrhizal rose plants is not related to phosphorus nutrition. *New Phytologist* **103**, 10–16.

Baath, E., & Söderström, B. (1979). Fungal biomass and fungal immobilisation of plant nutrients in Swedish coniferous forest soils. *Rev. Ecol. Biol. Sol.* **16**, 477–489.

Babel, U. (1987). *Feinwurzeln und humusmikromorphologie in verschiedenen stark geschädigten Fichtenbestanden*. Schlussbericht. B.M.F.T.

Bajwa, R., Abuarghub, S., & Read, D. J. (1985). The biology of mycorrhiza in the

Ericaceae. X. The utilization of proteins and the production of proteolytic enzymes by the mycorrhizal endophyte and by mycorrhizal plants. *New Phytologist* **101,** 469–486.

Birch, C. P. D. (1986). Development of VA mycorrhizal infection in seedlings in semi-natural grassland turf. In: *Proceedings of the First European Symposium on Mycorrhizas* (Ed. by V. Gianinazzi-Pearson & S. Gianinazzi), pp. 233–239. INRA, Paris.

Bowen, H. J. M., & Cawse, P. A. (1964). Effects of ionising radiations on soil and subsequent crop growth. *Soil Science* **97,** 253–259.

Boyd, R. T., Furbank, R. T., & Read, D. J. (1986). Ectomycorrhiza and the water relations of trees. In: *Proceedings of the First European Symposium on Mycorrhizae* (Ed. by V. Gianinazzi-Pearson & S. Gianinazzi), pp. 689–693. INRA, Paris.

Bradley, R., Burt, A. J., & Read, D. J. (1981). Mycorrhizal infection and resistance to heavy metal toxicity in *Calluna vulgaris*. *Nature (London)* **292,** 335–337.

Bradley, R., Burt, A. J., & Read, D. J. (1982). The biology of mycorrhiza in the Ericaceae. VIII. The role of mycorrhizal infection in heavy metal resistance. *New Phytologist* **91,** 197–209.

Brand, F., & Agerer, R. (1986). Studies on ectomycorrhizae VIII Mycorrhizae formed by *Lactarius subdulcis, L. vellereus* and *Laccaria amethystina* in beech. *Zeitschrift für Mykologie* **52,** 287–320.

Brownlee, C., Duddridge, J. A., Malibari, A., & Read, D. J. (1983). The structure and function of mycelial systems of ecto-mycorrhizal roots with special reference to their role in forming inter-plant connections and providing pathways for assimilate and water transport. *Plant Soil* **71,** 433–443.

Butler, E. J. (1939). The occurrence and systematic position of the vesicular-arbuscular type of mycorrhizal fungi. *Transactions of the British Mycological Society* **22,** 274–307.

Chu-Chou, M. (1979). Mycorrhizal fungi of *Pinus radiata* in New Zealand. *Soil Biology and Biochemistry* **11,** 557–562.

Cooper, K. M., & Tinker, P. B. (1981). Translocation and transfer of nutrients in vesicular-arbuscular mycorrhizas. IV. Effect of environmental variables on movement of phosphorus. *New Phytologist* **88,** 327–39.

Coutts, M. P., & Nicoll, B. C. (1990a). Growth & survival of shoots, roots & mycorrhizal mycelium in clonal Sitka spruce during the first growing season after planting. *Canadian Journal of Forest Research* **20,** 861–868.

Coutts, M. P., & Nicoll, B. C. (1990b). Waterlogging tolerance of roots of Sitka-spruce clones and of strands from *Thelephora terrestris* mycorrhizas. *Canadian Journal of Forest Research* **20,** 1896–1899.

Cromack, K., Sollins, P., Granstein, W. C., Speidel, K., Todd, A. W., Spycher, G., Ching, Y.-Li., & Todd, R. L. (1979). Calcium oxalate accumulation and soil weathering in mats of the hypogeous fungus *Hysterangium crassum*. *Soil Biology and Biochemistry* **11,** 463–468.

Dahlberg, A., & Stenlid, J. (1990). Population structure and dynamics in *Suillus bovinus* as indicated by spatial distribution of fungal clones. *New Phytologist* **115,** 487–493.

Deacon, J. W., Donaldson, S. J., & Last, F. T. (1983). Sequences and interactions of mycorrhizal fungi on birch. *Plant Soil* **71**, 257–62.

Duddridge, J. A., Malibari, A., & Read, D. J. (1980). Structure and function of mycorrhizal rhizomorphs with special reference to their role in water transport. *Nature (London)* **287**, 834–6.

Egli, S., & Kälin, I. (1990). The root window—a technique for the *in vivo* observation of mycorrhiza in the field. *Agriculture Ecosystems and Environment* **28**, 107–111.

Eno, C. F., & Popenoe, H. (1963). The effect of gamma radiation on the availability of nitrogen and phosphorus in soil. *Proceedings of the Soil Science Society of America* **27**, 299–301.

Erland, S., Söderström, B., & Andersson, S. (1990). Effects of liming on ectomycorrhizal fungi infecting *Pinus sylvestris* L. II. Growth rates in pure culture at different pH values compared to growth rates in symbiosis with the host plant. *New Phytologist* **115**, 683–688.

Faber, B., Zasoski, R. J., Munns, D. N., & Shackel, K. (1991). A method for measuring hyphal nutrient and water uptake in mycorrhizal plants. *Canadian Journal of Botany* **69**, 87–94.

Falck, R. (1912). Die Merulius—Faule des Bauholzes. *Hausschwammforschungen* **6**, 1–405.

Finlay, R. D., & Read, D. J. (1986a). The structure and function of the vegetative mycelium of ectomycorrhizal plants. I. Translocation of ^{14}C-labelled carbon between plants interconnected by a common mycelium. *New Phytologist* **103**, 143–156.

Finlay, R. D., & Read, D. J. (1986b). The structure and function of the vegetative mycelium of ectomycorrhizal plants. II. The uptake and distribution of phosphorus by mycelial strands inter-connecting host plants. *New Phytologist* **103**, 157–165.

Finlay, R. D., Ek, H., Odham, G., & Söderström, B. (1988). Mycelial uptake, translocation and assimilation of nitrogen from ^{15}N-labelled ammonium by *Pinus sylvestris* plants infected with four different ectomycorrhizal fungi. *New Phytologist* **110**, 59–66.

Fogel, R., & Hunt, G. (1979). Fungal and arboreal biomass in a western Oregon Douglas fir ecosystem: Distribution patterns and turnover. *Canadian Journal of Forest Research* **9**, 265–256.

Ford, E. D., & Deans, J. D. (1977). Growth of Sitka spruce plantations: Spatial distribution and seasonal fluctuations of lengths, weights and carbohydrate concentrations of fine roots. *Plant Soil* **47**, 463–485.

Francis, R., & Read, D. J. (1984). Direct transfer of carbon between plants connected by vesicular-arbuscular mycorrhizal mycelium. *Nature (London)* **307**, 53–56.

Friese, C. F., & Allen, M. F. (1991). The spread of VA mycorrhizal hyphae in the soil: Inoculum types and external hyphal architecture. *Mycologia* **83**, 409–418.

Gimingham, C. H. (1972). *Ecology of Heathlands*. Chapman & Hall, London.

Grime, J. P., Mackey, J. M. L., Hillier, S. H., & Read, D. J. (1987). Floristic diversity in a model system using experimental microcosms. *Nature (London)* **328**, 420–422.

Hardie, K. (1985). The effect of removal of extraradical hyphae in water uptake by vesicular-arbuscular mycorrhizal plants. *New Phytologist* **101**, 677–684.

Harley, J. L. (1978). Ectomycorrhizas as nutrient absorbing organs. *Proceedings of the Royal Society of London Series B* **203**, 1–21.

Harley, J. L. (1989). The significance of mycorrhiza. *Mycological Research* **92**, 129–139.

Harley, J. L., & Smith, S. E. (1983). *Mycorrhizal Symbiosis*. Academic Press, London.

Harley, J. L., McCready, C. C., & Brierley, J. K. (1953). Uptake of phosphate by excised mycorrhizal roots of beech. IV. The effect of oxygen concentrations upon host and fungus. *New Phytologist* **52**, 124–132.

Haselwandter, K., Bobleter, O., & Read, D. J. (1990). Utilisation of lignin by ericoid and ectomycorrhizal fungi. *Achive für Mikrobiologie* **153**, 352–354.

Jalal, M. A. F., & Read, D. J. (1983a). The organic acid composition of *Calluna* heathland soil with special reference to phyto- and fungi-toxicity. I. Isolation and identification of organic acids. *Plant Soil* **70**, 257–272.

Jalal, M. A. F., & Read, D. J. (1983b). The organic acid composition of *Calluna* heathland soil with special reference to phyto- and fungi-toxicity. II. Monthly quantitative determination of the organic acid content of *Calluna* and spruce dominated soils. *Plant Soil* **70**, 273–286.

Jones, M. D., Durall, D. M., & Tinker, P. B. (1990). Phosphorus relationships and production of extramatrical hyphae by two types of willow ectomycorrhizas at different soil phosphorus levels. *New Phytologist* **115**, 259–267.

Koide, R. (1985). The effect of VA mycorrhizal infection and phosphorus status on sunflower hydraulic and stomatal properties. *Journal of Experimental Botany* **36**, 1087–1098.

Koske, R. E., Sutton, J. C., & Sheppard, B. R. (1975). Ecology of *Endogone* in Lake Huron sand dunes. *Canadian Journal of Botany* **53**, 87–93.

Kramer, P. J., & Bullock, H. G. (1966). Seasonal variations in the proportions of suberised and unsuberised roots of trees in relation to the absorption of water. *American Journal of Botany* **53**, 200–204.

Leake, J. R. (1987). Metabolism of phyto- and fungitoxic phenolic acids by the ericoid mycorrhizal fungus. In: *Proceedings of the Seventh North American Conference on Mycorrhiza* (Ed. by D. M. Sylvia, L. L. Hung, & J. H. Graham), pp. 332–333. University of Florida, Gainesville.

Leake, J. R., & Read, D. J. (1989). The biology of mycorrhiza in the Ericaceae. XIII. Some characteristics of the extracellular proteinase activity of the ericoid endophyte *Hymenoscyphus ericae*. *New Phytologist* **112**, 69–76.

Leake, J. R., & Read, D. J. (1990a). Proteinase activity in mycorrhizal fungi. I. The effect of extracellular pH on the production and activity of proteinase by ericoid endophytes from soils of contrasted pH. *New Phytologist* **115**, 243–250.

Leake, J. R., & Read, D. J. (1990b). Chitin as a nitrogen source for mycorrhizal fungi. *Mycological Research* **94**, 993–995.

Leake, J. R., & Read, D. J. (1990c). The effects of phenolic compounds on nitrogen mobilisation by ericoid mycorrhizal systems. *Agriculture Ecosystems and Environment* **29**, 225–236.

Leake, J. R., & Read, D. J. (1990d). The biology of mycorrhiza in the Ericaceae XV. The effect of mycorrhizal infection on calcium uptake by *Calluna vulgaris* L. Hull. *New Phytologist* **113**, 535–544.

Mason, P. A., Wilson, J., Last, F. T., & Walker, C. (1983). The concept of succession in relation to the spread of sheathing mycorrhizal fungi on inoculated tree seedlings growing in unsterile soils. *Plant Soil* **71**, 247–56.

Melin, E., & Nilsson, H. (1950). Transfer of radioactive phosphorus to pine seedlings by means of mycorrhizal hyphae. *Physiologia Plantarum* **3**, 88–92.

Melin, E., & Nilsson, H. (1952). Transfer of labelled nitrogen from an ammonium source to pine seedlings through mycorrhizal mycelium. *Svensk Botanisk Tidskrift* **46**, 281–5.

Miller, S. L., Durall, D. M., & Rygiewicz, P. T. (1989). Temporal allocation of ^{14}C to extramatrical hyphae of ectomycorrhizal ponderosa pine seedlings. *Tree Physiology* **5**, 239–250.

Mitchell, D. T., & Read, D. J. (1981). Utilization of inorganic and organic phosphates by the mycorrhizal endophytes of *Vaccinium macrocarpon* and *Rhododendron ponticum*. *Transactions of the British Mycological Society* **76**, 255–260.

Mosse, B. (1959). Observations on the extramatrical mycelium of a vesicular arbuscular endophyte. *Transactions of the British Mycological Society* **42**, 439–448.

Nelsen, C. E., & Safir, G. R. (1982). Increased drought tolerance of mycorrhizal onion plants caused by improved phosphorus nutrition. *Planta* **154**, 407–13.

Nicolson, T. H., & Johnston, C. (1979). Mycorrhiza in the Gramineae III. *Glomus fasciculatus* as the endophyte of pioneer grasses in maritime sand dunes. *Transactions of the British Mycological Society* **72**, 261–268.

Ogawa, M. (1977). Ecology of higher fungi in *Tsuga diversifolia* and *Betula ermani*—*Abies mariesii* forests of subalpine zone. *Transactions of the Mycological Society of Japan* **18**, 1–19.

Ogawa, M. (1981). Microbial ecology of Shiro in *T. matsutake* (Ito et Imai) Sing. and its allied species *T. robostum* and *T. zelleri*. *Transactions of the Mycological Society of Japan* **22**, 231–245.

Ogawa, M. (1985). Ecological characters of ectomycorrhizal fungi and their mycorrhizae. *JARQ* **18**, 305–314.

Pearson, V., & Read, D. J. (1975). The physiology of the mycorrhizal endophyte of *Calluna vulgaris*. *Transactions of the British Mycological Society* **64**, 1–7.

Persson, H. (1980). Spatial distribution of fine-root growth, mortality, and decomposition in a young Scots Pine stand. *Oikos* **34**, 77–87.

Peyronel, B. (1924). Prime recherche sulle micorize endotrofiche and sulla microflora radicola normale delle fanergames. *Rivista di Biologia* **5**, 463–85.

Rayner, A. D. M., Powell, K. A., Thompson, W., & Jennings, D. H. (1985). Morphogenesis of vegetative organs. In: *Developmental Biology of Higher Fungi* (Ed. by D. Moore,

L. A. Casselton, D. A. Wood, & J. C. Frankland), pp. 249–279. Cambridge University Press, Cambridge.

Rayner, A. D. M., Boddy, L., & Dowson, C. G. (1987). Genetic interactions and developmental versatility during establishment of decomposer basidiomycetes in wood and tree litter. In: *Ecology of Microbial Communities* (Ed. by T. R. G. Gray, M. Fletcher, & G. Jones), pp. 83–122. Cambridge University Press, Cambridge.

Read, D. J. (1982). In support of Franks organic nitrogen theory. *Angewandte Botanik* **61**, 25–37.

Read, D. J. (1984a). The structure and function of the vegetative mycelium of mycorrhizal roots. In: *The Ecology and Physiology of the Fungal Mycelium* (Ed. by D. H. Jennings & A. D. M. Rayner, pp. 215–240. Cambridge University Press, Cambridge.

Read, D. J. (1984b). Interactions between ericaceous plants and their competitors with special reference to soil toxicity. In: *Weed Control and Vegetation Management in Forests and Amenity Areas. Aspects of Applied Biology* **5**, 195–209. Association of Applied Biologists, Wellesbourne, UK.

Read, D. J. (1990a). Ecological integration by mycorrhizal fungi. In: *Endocytobiology IV* (Ed. by P. Nardon), pp. 99–107. INRA, Paris.

Read, D. J. (1990b). Mycorrhizas in ecosystems. *Experimentia* (in press).

Read, D. J., & Boyd, R. (1986). Water relations of mycorrhizal fungi and their host plants. In: *Water, Fungi and Plants* (Ed. by P. Ayres & L. Boddy), pp. 287–303. Cambridge University Press, Cambridge.

Read, D. J., & Stribley, D. P. (1975a). Diffusion and translocation in some fungal culture systems. *Transactions of the British Mycological Society* **64**, 381–388.

Read, D. J., & Stribley, D. P. (1975b). Some mycological aspects of the biology mycorrhizas in the Ericaceae. In: *Endomycorrhizas* (Ed. by F. E. Sanders, B. Mosse, & P. B. Tinker), pp. 105–17. Academic Press, London.

Read, D. J., Francis, R., & Finlay, R. D. (1985). Mycorrhizal mycelia and nutrient cycling in plant communities. In: *Ecological Interactions in Soil: Plants, Microbes and Animals* (Ed. by A. H. Fitter, D. Atkinson, D. J. Read, & M. B. Usher), pp. 193–217. British Ecological Society Special Publication 4. Blackwell Scientific Publications, Oxford.

Read, D. J., Leake, J. R., & Langdale, A. R. (1989). The nitrogen nutrition of mycorrhizal fungi and their host plants. In: *Nitrogen, Phosphorus and Sulphur Utilization by Fungi* (Ed. by L. Boddy, R. Marchant, & D. J. Read), pp. 181–204. Cambridge University Press, Cambridge.

Reid, C. P. P., Kidd, F. A., & Ekwebelam, S. A. (1983). Nitrogen nutrition, photosynthesis and carbon allocation in ectomycorrhizal pine. *Plant Soil* **71**, 415–421.

Reiners, W. A. (1965). Ecology of a heath-shrub synusia in the pine barrens of Long Island, New York. *Bulletin of the Torrey Botany Club* **92**, 448–464.

Rix, K., & Newman, E. I. (1985). Evidence for rapid cycling of phosphorus from dying roots to living plants. *Oikos* **45**, 174–180.

Roberts, J. (1976). A study of distribution and growth in a *Pinus sylvestris* L. (Scots pine) plantation in Thetford Chase, East Anglia. *Plant Soil* **44**, 607–621.

Rommell, L. G. (1939). The ecological problem of mycotrophy. *Ecology* **20**, 163–167.

Safir, G. R., Boyer, J. S., & Gerdemann, J. W. (1971). Mycorrhizal enhancement of water transport in soybean. *Science* **172**, 581–583.

Salonius, P. O., Robinson, J. B., & Chase, F. E. (1967). A comparison of autoclaved and gamma-irradiated soils as media for microbial colonisation experiments. *Plant Soil* **27**, 239–248.

Skinner, M. F., & Bowen, G. D. (1974). The uptake and translocation of phosphate by mycelial strands of pine mycorrhizas. *Soil Biology and Biochemistry* **6**, 53–56.

Söderström, B., & Read, D. J. (1987). Respiratory activity of intact and excised ectomycorrhizal mycelial systems growing in unsterilized soil. *Soil Biology Biochemistry* **19**, 231–236.

St. John, T. V., Coleman, D. C., & Read, C. P. P. (1983). Association of vesicular-arbuscular mycorrhizal hyphae with soil organic particles. *Ecology* **64**, 957–959.

Straker, C. J., & Mitchell, D. T. (1986). The activity and characterization of acid phosphatases in endomycorrhizal fungi of the Ericaceae. *New Phytologist* **104**, 243–256.

Stribley, D. P., & Read, D. J. (1974). The biology of mycorrhiza in the Ericaceae. IV. The effects of mycorrhizal infection on the uptake of ^{15}N from labelled soil by *Vaccinium macrocarpon* Ait. *New Phytologist* **73**, 1149–1155.

Stribley, D. P., & Read, D. J. (1980). The biology of mycorrhiza in the Ericaceae. VIII. The relationship between mycorrhizal infection and the capacity to utilize simple and complex organic nitrogen sources. *New Phytologist* **86**, 365–371.

Sutton, J. C., & Sheppard, B. R. (1976). Aggregation of sand-dune soil by endomycorrhizal fungi. *Canadian Journal of Botany* **54**, 326–333.

Sylvia, D. M. (1986). Spatial and temporal distribution of vesicular-arbuscular mycorrhizal fungi associated with *Uniola paniculata* in Florida foredunes. *Mycologia* **78**, 734–740.

Taylor, H. M. (1987). Minirhizotron observation tubes. Methods in application for measuring rhizosphere dynamics. *American Society of Agronomy Special Publication 50.*

Taylor, A. F. S., & Alexander, I. J. (1990). Demography and population dynamics of ectomycorrhizas of Sitka spruce fertilised with nitrogen. *Agriculture Ecosystems and Environment* **28**, 493–497.

Tinker, P. B. H. (1975). The soil chemistry of phosphorus and mycorrhizal effects on plant growth. In: *Endomycorrhizas* (Ed. by F. E. Sanders, B. Mosse, & P. B. Tinker), pp. 353–371. Academic Press, London.

Tisdall, J. M., & Oades, J. M. (1979). Stabilisation of soil aggregates by the root systems of rye grass. *Australian Journal of Soil Research* **17**, 429–441.

Trappe, J. M. (1977). Selection of fungi for ectomycorrhizal inoculation in nurseries. *Annual Review of Phytopathology* **15**, 203–22.

Vogt, K. A., Grier, C. C., Meir, C. E., & Edmonds, R. L. (1982). Mycorrhizal role in net production and nutrient cycling in *Abies amabilis* ecosystems in western Washington. *Ecology* **63,** 370–380.

Vogt, K. A., Moore, E. E., Vogt, D. J., Redlin, M. J., & Edmonds, R. L. (1983). Conifer fine root and mycorrhizal root biomass within the forest floors of Douglas fir stands of different ages and site productivities. *Canadian Journal of Forestry Research* **13,** 429–37.

5

Mycorrhiza and Carbon Flow to the Soil

Roger Finlay and Bengt Söderström

SUMMARY. The loss of energy-rich carbon compounds from plant roots to soil microbial populations constitutes a fundamental supply process to the soil ecosystem and the direct supply of host assimilates from mycorrhizal host plants to their fungal symbionts is of significance, not only to the mycorrhizal associations themselves, but also to the soil ecosystem as a whole. The flow of carbon to mycorrhizal roots, and through mycorrhizal mycelia to different components of the soil ecosystem, can clearly be significant, but further information is required about the amounts and types of compounds involved and the mechanisms regulating their translocation and ultimate partitioning. Existing biomass estimates of vesicular-arbuscular and ectomycorrhizal extramatrical mycelium and fruiting structures are considered in relation to theoretical estimates of necessary carbon flow and available experimental estimates, but the lack of adequate methods for quantifying mycorrhizal mycelium remains a barrier to progress in understanding the dynamics of mycorrhizal mycelia and their interactions with other organisms. The flow of carbon into and through mycorrhizal mycelia has a potentially large range of wider effects in the soil ecosystem since energy-rich substrates are required by most biological processes. Potential effects include interactions with phytopathogens and decomposers, chemical defences against grazing of the mycelium, and stabilization of soil aggregates. Energy-rich substrates are also necessary for the synthesis of enzymes to degrade organic polymers and the significance of this enzymatic activity in different environments is discussed in relation to nutrient availability and possible effects on decomposer populations. The formation of mycelial connections between plants and flow of carbon through these may influence plant community development through effects on regeneration processes or plant competition. All of the above effects may have an impact on energy flow and cycling of nutrients.

Introduction

The loss of energy-rich compounds from plant roots to soil microbial populations constitutes one of the fundamental supply processes to the soil ecosystem. Mycorrhizal fungi are unique within the soil microbial community in that they have direct access to host assimilates and the supply of energy-rich carbon compounds from mycorrhizal host plants to their fungal symbionts is thus of significance, not only to the mycorrhizal associations themselves, but also the soil ecosystem as a whole.

In this chapter we review existing knowledge concerning the supply of carbon compounds to mycorrhizal roots and mycelium, and their subsequent distribution and cycling within the soil ecosystem. The carbon requirements of different mycorrhizal fungi are discussed and carbon transfer from host root tissue to the fungal biomass is considered in relation to available information on the amounts and types of compounds that are translocated, as well as their ultimate partitioning. Problems of estimating mycorrhizal mycelial biomass are discussed and existing biomass estimates of vesicular-arbuscular (VA) and ectomycorrhizal extramatrical mycelium and fruiting structures are considered in relation to theoretical estimates of necessary carbon flow and available experimental measurements. Regulatory and adaptive processes influencing mycorrhizal mycelia are also discussed.

Finally the flow of carbon compounds into and through mycorrhizal mycelia is considered in relation to its possible wider effects in the soil ecosystem. The supply of energy-rich substrates is essential for synthesis of enzymes to degrade organic polymers and the extent and significance of this enzymic capability in different environments are discussed in relation to nutrient availability and possible effects on microbial decomposer populations. The formation of mycelial connections between plants and flow of carbon compounds through these may influence the source–sink relationships of differently illuminated mycorrhizal plants, with potential effects on regeneration and community development. The possible existence and significance of mycorrhizosphere effects are discussed in connection with processes such as exudation, death, decomposition and recycling of mycelial nutrients, and production of antibiotics. Possible interactions with other microbial populations are discussed.

Carbon Flow from the Host to the Mycobiont

The processes involved in the development of mycorrhizal infection in root systems frequently lead to increased rates of photosynthesis and translocation of carbon compounds to the root systems of host plants. Information concerning the full extent of this translocation, including that to the extramatrical mycelium, the nature of the compounds involved, and the processes regulating their partitioning between different sinks is, however, still limited.

Carbon Requirements of Mycorrhizal Fungi

Different types of mycorrhizal fungi show wide variation in the extent to which they appear able to grow independently of their normal autotrophic hosts. The

Glomeales of vesicular-arbuscular mycorrhizal (VAM) fungi appear to be ecologically obligate symbionts with little or no capacity for independent growth or production of enzymes to degrade complex carbohydrate polymers such as cellulose or pectin. On the other hand, many species of ectomycorrhizal fungi have been isolated and grown on a wide range of culture media. They are generally able to utilize simple sugars such as the monosaccharides glucose, mannose, and fructose, and show intra- and interspecific variation in the degree to which they are able to use disaccharides such as sucrose and trehalose, the simpler oligosaccharides, and compounds such as starch, glycogen, and inulin. Some strains of some species appear to be able to utilize pectic substances to support growth, but the ability of ectomycorrhizal fungi to use complex carbohydrates such as lignin and cellulose is generally considered to be limited. Ericoid mycorrhizal fungi can also be grown on a wide range of carbon compounds such as monosaccharides (including the pentose xylose), disaccharides such as maltose, sucrose, and cellobiose and complex carbohydrates such as pectin and starch. Their cellulolytic activity in pure culture appears to be restricted.

The fungi isolated from orchid roots differ from most other mycorrhizal fungi in that they possess a considerable capacity for independent saprophytic or parasitic growth. In addition to using soluble sugars, most of the tested isolates appear capable of utilizing starch and pectin as well as complex insoluble polymers such as cellulose and some even degrade lignin (Hadley & Ong, 1978). While fungi such as *Rhizoctonia solani* and *Armillaria mellea* may be widespread as parasites, the distribution and importance of these and other fungi as mycorrhizal symbionts represent a more specialised form of association and orchid mycorrhizas are therefore not considered further in this chapter.

The ease with which many ectomycorrhizal fungi can be cultured and the fact that some of them can spread to the rhizosphere of nonhost plants suggest that some may not be obligately symbiotic. However, under natural ecological conditions, the degree of facultative symbiosis may be low and the production of fruit bodies, in some species at least, has been shown to depend on the presence of living host roots (Romell, 1938, 1939). The distribution of many ericoid endophytes also appears to be wider than that of their host plants, suggesting that they may have some ability to survive as soil saprophytes, or as resting structures in the vicinity of roots of nonhost plants.

Pure culture studies do not fully reflect the natural conditions pertaining when growth is in symbiotic association with a host plant. The nutrient concentrations frequently used in pure culture studies are artificially high and the ability or inability of fungi to grow on single carbon sources may not be an accurate reflection of how these are used in situations in which a natural spectrum of compounds is available. Enzymes for the utilization of certain substances may be adaptive and a supplementary carbon source in addition to other nutrients is often needed to allow enzyme synthesis during the period of adaptation. Problems of catabolite repression have often been overlooked and it is important to distinguish between production of enzymes for restricted and localized hydrolysis and enzymic activity that is sufficient to support growth using the products as the sole carbon source.

Generalizations from *in vitro* studies of activity must be made with caution. The classical view is that ectomycorrhizal fungi are able to use only simple sugars as carbon sources, but a number of interesting exceptions have been discussed by Lindeberg (1986). There is now a growing concensus that the proteolytic activity of some ecto- and ericoid mycorrhizal fungi may be greater than previously appreciated (Abuzinadah et al., 1986; Abuzinadah & Read, 1989; Leake & Read, 1989) and carbon released from such proteolytic activity may reduce the carbon drain to the symbiont from host plants. The ability of the ericoid endophyte *Hymenoscyphus ericae* to utilize organic acids as carbon sources (Leake, 1988) may also be important in this and other respects. The significance of these findings is complicated by the fact that our knowledge of the availability and dynamics of possible organic nutrients is less well developed than for mineral nutrients.

While the ability to grow mycorrhizal fungi in pure culture is of great practical importance to laboratory experimentation, studies of intact mycorrhizal associations have the advantage in that a more natural carbon balance can be achieved when both symbiotic partners are grown together.

Carbon Translocation to Mycorrhizal Roots

It has been assumed for many years that mycorrhizal fungi obtain their carbon compounds from host plants and that there is a net flow of these from autotroph to heterotroph. No direct evidence of this was available until 1957, when Melin and Nilsson (1957) demonstrated the translocation of ^{14}C-labeled photosynthate to roots and fungal sheaths of *Pinus sylvestris* seedlings infected with the ectomycorrhizal fungi *Suillus bovinus* and *Rhizopogon roseolus*. Direct evidence of such translocation was initially more difficult to acquire in VAM systems because of the difficulty of separating the symbionts and the rather scant production of external mycelium in artificial systems. Ho and Trappe (1973) performed the earliest experiments showing transfer of photosynthetically incorporated ^{14}C into external mycelium and spores. In later experiments Bevege et al. (1975) and Cox et al. (1975) demonstrated rapid translocation of photosynthate to infected root systems and subsequent incorporation of labeled carbon into both intracellular hyphae and external mycelium. These initial experiments were followed by many more (see Harley & Smith, 1983) that refined our understanding of the type and relative amounts of different compounds translocated to the fungal symbionts.

Ectomycorrhizas contain fungal-specific carbohydrates such as mannitol, trehalose, and glycogen. The rapid conversion of absorbed plant sugars to metabolic intermediates and fungal storage compounds is one way by which ascomycetous and basidiomycetous fungi are thought to maintain a metabolic sink for photosynthate derived from their hosts (Lewis & Harley, 1965). Although mannitol and trehalose were originally identified as being important carbon sinks, Söderström et al. (1988) also found the alditols, arabitol and erythritol in the mycelia of *Suillus bovinus, Pisolithus tinctorius,* and *Paxillus involutus.*

Ericoid mycorrhizas show the same general carbon incorporation patterns as ectomycorrhizas. Stribley and Read (1974) showed accumulation of ^{14}C into glucose, sucrose, and fructose in nonmycorrhizal roots of *Vaccinium* but in mycorrhizal roots the label was incorporated into the fungal carbohydrates mannitol and trehalose and into polymers of glucose (glycogen) and mannose. These observations are consistent with a general transfer of carbohydrate from the host to the heterotroph, a metabolic sink being maintained by establishment of a concentration gradient in favor of transport to the heterotroph through rapid conversion of plant assimilates to fungal-specific carbohydrates.

There have been few detailed studies of the carbohydrate physiology of VA mycorrhizas. The carbohydrate composition of infected and uninfected roots appears similar, and attempts to detect fungal specific sugars such as the mannitol and trehalose have been largely unsuccessful (Hayman, 1974; Bevege et al., 1975). Lewis (1975) suggested that mannitol was unlikely to be present but that failure to detect trehalose might be due to the low ratio of fungal tissue relative to root tissue. Amijee and Stribley (1987) found only glucose, sucrose, and fructose in mycorrhizal roots of *Allium porrum* infected with *Glomus mosseae* but detected two fungal specific sugars in the external mycelium. Trehalose was found mostly in spores while another unidentified carbohydrate appeared in the mycelium. Lipid synthesis in the fungal component of VA mycorrhizas may have an analogous storage role. Increased total lipid levels in VA mycorrhizal roots of onion, clover, and ryegrass were demonstrated by Cooper and Lösel (1978), although the lipid fractions of infected and uninfected roots did not differ qualitatively. Deposition of lipid droplets and increased amounts of membrane lipids associated with arbuscule formation would contribute to this increase and Cox et al. (1975) demonstrated the incorporation of photosynthetically derived ^{14}C in lipid droplets using autoradiography. Nagy et al. (1980) found increases in the amounts of triglycerides and phospholipids associated with mycorrhizal Citrus roots. Three unidentified fatty acids constituted 31–44% of the total lipids in mycorrhizal roots and were not present in uninfected roots.

Partitioning of Host-Derived Assimilates

Host assimilates transferred to mycorrhizas can be distributed in a number of ways and there has been an increasing number of studies of their partitioning and ultimate distribution, both in experimental microcosms and in natural ecosystems. Experimental study of these processes has been complicated by the difficulty of separating and quantifying mycorrhizal mycelia in natural ecosystems (see below) with the consequence that many field studies have been restricted to fruiting structures and may thus underestimate the full extent of carbon input to the soil ecosystem. Experimental microcosms facilitate the specific study of mycorrhizal mycelia but are generally restricted to studies of early seedling stages since it is not practical to work with mature trees.

The products of photosynthetic assimilation may be distributed in a number of ways. Mycorrhizal infection often results in increased allocation of C to the root system and this may be incorporated into increased root biomass, increased root respiration, mycelial biomass (both within the root and as external mycelium), and mycelial respiration (internal and external), or lost in the form of exudation from the roots or decomposition and leakage of dying mycorrhizal hyphae. Losses may also occur to mycophagous grazer populations.

Many experiments have failed to quantify one or more of these components. Increases in respiration of symbiotic roots may not be entirely due to the respiration of the symbionts themselves. Pate et al. (1979) found higher rates of CO_2 evolution in host tissue subtending *Rhizobium* nodules than in tissue not associated with nodules, and Cox and Tinker (1976) demonstrated that cells containing arbuscules contained 22 times more cytoplasm than adjacent cells not containing fungal structures. Distinction of symbiont respiration from increases in the respiration of host tissue itself is often difficult or impossible, however, and the two are often considered jointly as part of the overall carbon cost of infection. Changes in the size and nutritional status of mycorrhizal plants often complicate direct comparison and fertilizer supplements are sometimes necessary in nonmycorrhizal controls. Comparison of results is also often difficult because of different experimental species and conditions, but some general patterns emerge.

Many investigators have found increased allocation of C to mycorrhizal roots. This is often associated with greater respiratory losses and increased C fixation rates, which may or may not compensate for the increased carbon drain. Pang and Paul (1980) found additional allocation of C to mycorrhizal roots of *Vicia faba* equivalent to about 10% of total photosynthate. Nonmycorrhizal plants allocated 37% of fixed ^{14}C below ground whereas the corresponding figure for mycorrhizal plants was 47%. This difference was due to increased respiration from the roots of infected plants. Mycorrhizal plants had similar yields to uninfected ones, indicating that respiratory losses were compensated for by increased C fixation. Kucey and Paul (1982) found that N_2-fixing *Rhizobium* nodules of mycorrhizal *V. faba* plants utilized 12% of total photosynthate, whereas those of nonmycorrhizal plants only used 6%. The C-fixation rate was also increased by 8% in plants supporting mycorrhizal symbionts. Snellgrove et al. (1982) also found that *Allium porrum* plants translocated 7% more C to mycorrhizal roots than to nonmycorrhizal roots of similar sized plants. The increased C allocation was associated with a decrease in specific leaf mass and increased leaf hydration. They suggested that this adaptation could enable mycorrhizal plants to maintain a greater photosynthetic capacity without increasing plant C requirements. Allocation of carbon to *Rhizobium* nodules of mycorrhizal *Glycine max* plants in experiments by Harris et al. (1985) was increased to 12% of total photosynthate compared with 9% in nonmycorrhizal plants. Photosynthetic fixation rates were increased by up to 47% in plants with both symbionts, compared with nonsymbiotic, fertilizer-treated plants. Below ground CO_2 evolution was similarly increased from 9 to 29% in 6-week-old plants, the two symbionts together accounting for 82% this figure. In a recent, detailed, study Jakobsen and Rosendahl (1990) examined the distribution of ^{14}C in the mycorrhizal plant–soil system of cucumber plants fed with $^{14}CO_2$. Control plants exhibited stress symptoms so detailed comparison of mycorrhizal and nonmycorrhizal plants was not possible, although total root activity and below ground respiration were five times higher in the mycorrhizal plants than the uninfected controls, representing 13.2 and 27%, respectively, of the total photoassimilated ^{14}C. Altogether 43% of the total assimilated ^{14}C was translocated to the root system and 70% of this was lost as CO_2 or extraradical C (exudates plus mycelium).

Estimation of the actual proportion of assimilate allocated to the mycorrhizal symbionts themselves is more difficult since it requires the separation of external mycelium from plant roots and soil, and necessitates assumptions concerning the efficiency of substrate incorporation. Nevertheless, strikingly consistent figures have been suggested. Kucey and Paul (1982) estimated that the mycorrhizal fungi of both nodulated and nonnodulated hosts utilized approximately 4% of the total C fixed by their hosts and that they constituted 5% of the root mass. Harris et al. (1985) estimated that the allocation to 6-week old plants infected with both *Glomus fasiculatum* and *Rhizobium japonicum* was 16.4% of the total assimilated carbon, of which 2.7% was attributed to biomass and 13.7% to respiration. These estimates were 2.8 and 4.6%, respectively, in 9-week-old plants, assuming an equal distribution of activity in intra- and extraradicle hyphae. In the study by Jakobsen and Rosendahl (1990) activity in the external hyphae represented 0.8% of total fixed ^{14}C. These authors estimated total allocation to the external mycelium as 4% and total allocation to the internal hyphae as 16% (assuming a growth yield of 0.2 mg hyphal C per mg substrate C and assuming the internal infection to be 10% of root dry weight).

There are few studies which relate losses of C compounds via root exudation to assimilate allocation to external mycorrhizal mycelium. Kucey and Paul (1982) were unable to account for 3.2% of the ^{14}C loss from roots, which could be accounted for by exudation. Jakobsen and Rosendahl (1990) noted that the activity of soluble and insoluble fractions of extraradical C in mycorrhizal systems was double that found in control systems and, together with the activity in external hyphae, represented 3.1% of total fixed ^{14}C. The hyphae represented 26% of this total so that 2.3% of the total C fixation could be accounted for by exudation, a figure similar to those of Whipps and Lynch (1985). Losses from decomposition, leakage of the hyphae, and internal recycling of carbon compounds remain unquantified although their possible significance is discussed later in this chapter.

Studies of ectomycorrhizal carbon allocation show a similar pattern. Nelson (1964) demonstrated that allocation of carbon to mycorrhizal roots of *Pinus strobus* was almost four times that to nonmycorrhizal roots on a dry weight basis. Bevege et al. (1975) demonstrated an 8-fold increase in the roots of the same *Pinus radiata* plants infected with the ectomycorrhizal fungus *Rhizopogon luteolus* compared with nonmycorrhizal roots. More recent studies by Cairney et al. (1989) of carbon distribution within ectomycorrhizal root systems of *Eucalyptus pilularis* infected with *Pisolithus tinctorius* indicate that the amount of assimilate transferred to mycorrhizal roots is between 42.6 and 4.2 times that transferred to nonmycorrhizal root tips in the same root system. The same study showed that young roots acted as stronger sinks for activity and that there was a progressive reduction in translocation of photosynthate with age. In studies of carbon allocation in *Pinus taeda* and *Pinus contorta* infected with *Pisolithus tinctorius* and *Suillus granulatus* Reid et al. (1983) found that mycorrhizal plants assimilated more CO_2, allocated a greater proportion to their root systems, and lost a greater percentage of ^{14}C by root respiration than did nonmycorrhizal plants.

Regulation of Carbon Flow

The relative importance of different processes regulating carbon flow to mycorrhizal roots is still not clear. It is apparent from the studies mentioned above that mycorrhizal infection is frequently associated with raised photosynthetic rates, increased carbon translocation to infected roots, and higher root respiration rates. Several explanations have been put forward. One suggestion is that irreversible conversion of plant assimilates to fungal-specific carbohydrates occurs (Lewis & Harley, 1965), creating a fungal sink that may indirectly increase the rate of photosynthesis. This is consistent with the idea that photosynthesis is under some control by sink demand. Mycorrhizal fungi may create such a fungal sink by converting carbohydrates to storage products, utilizing assimilates for fungal biomass production, or by using carbohydrates as energy for maintenance metabolism. Although labelled fungal carbohydrates can be measured, the latter two components of this sink have been difficult to quantify, especially with respect to the extramatrical mycelium. Other factors may also have a direct or indirect influence on photosynthesis and translocation. These include improved mineral nutrition, fungal-produced hormones, or changes in the balance of root hormones (Reid et al., 1983). Nylund and Wallander (1988) found circumstantial evidence of hormone effects, supporting the theory of Slankis (1973) that the mycobiont strongly affects its host via auxin action. In their experiments they grew *Pinus contorta* plants under steady-state nutrient conditions in semihydroponic culture. Auxin (IBA and IAA) treatments increased carbohydrate concentrations, but nutrients had no demonstrable effects on root or shoot carbohydrate contents. Free access to balanced, low concentrations of nutrients did not inhibit mycorrhiza formation and photosynthesis was considerably increased in mycorrhizal seedlings even though concentrations of N and Mg were similar in infected and uninfected plants. Growth of non-nutrient-limited mycorrhizal plants was reduced, suggesting that the mycorrhizal fungi imposed a carbon drain on their hosts, obtaining carbohydrates by means of an active process. This contradicts the notion that mycorrhizal fungi are simply the passive recipients of "surplus" plant carbohydrates (Björkman, 1942; see also Nylund, 1988). Similar reductions have been recorded in the growth of non-nutrient-limited plants infected with VA mycorrhizal fungi (Douds et al., 1988). Finlay (1989) speculated that, in certain situations, a deleterious carbon drain on the host plant may also result from infection by a poorly compatible ectomycorrhizal fungus. Studies of incompatible ectomycorrhizal associations between *Pinus sylvestris* and *Boletinus cavipes* (Finlay, 1989) demonstrated apparently normal translocation of ^{14}C-labeled assimilates to the fungal mycelium even though infection of incompatible hosts was poor and translocation of ^{32}P-labeled phosphate to them was severely reduced.

The cost–benefit relations of carbon drain, on one hand, versus growth stimulation through improved nutrient uptake, on the other, deserve further attention and are important when considering the "efficiency" of different mycobionts. Consideration of the relative size of structural and maintenance sinks has been complicated by the difficulty of measuring biomass and respiration of the external mycelial phase of mycorrhizal associations. The first studies of respiratory activity specific to the mycelial phase of ectomycorrhizal associations (Söderström & Read, 1987) demonstrated that approximately 30% of total respiration was due to that of the mycorrhizal mycelium. Mycelial respiration was shown to be highly dependent on the supply of current assimilate and severance of mycelial connections at the roots led to greater than 50% decreases in respiration rate within 24 hr. Such studies show the potential importance of the mycelial phase in carbon cycling.

Recent studies in our laboratory by Erland et al. (1991) show that up to 6% of assimilated carbon is present in the external mycelium of mycorrhizal *Pinus contorta* seedlings within 96 hr of supplying $^{14}CO_2$ to the shoot systems. In pulse labeling experiments the ^{14}C content of mycorrhizal roots was only 50% of that in nonmycorrhizal roots within 48 hr of supplying a 1-hr pulse of $^{14}CO_2$, although the relative amounts of C respired by the roots and the fungus could not be measured separately.

Mycorrhizal Mycelia

Methods of Measurement

Fungal biomass in and on the roots can be estimated by direct sampling and analysis of the roots: fruiting structures are often macroscopic and their biomass can be estimated. The mycelium growing in the soil is more difficult to quantify and our estimates of its biomass are therefore restricted. Much effort has been put into the development of methods for estimating soil fungal biomass in general and a number of more or less imperfect methods exist that can be applied to mycorrhizal systems.

The most commonly used methods for measuring fungal mycelium in soil are still based on microscopic estimations of hyphal length, many of them based on modifications of the agar-film technique originally described by Jones and Mollison (1948). A somewhat simplified preparation technique was introduced by Hanssen et al. (1974) and their membrane filter method is now probably the most commonly used technique for direct microscopical estimation of hyphal length and width. Abbott et al. (1984) modified this technique for studies of vesicular-arbuscular mycorrhizal (VAM) external mycelium. One disadvantage of simple microscopical methods is that they do not distinguish between total biomass (including dead microbial cells), live biomass, and physiologically active biomass. Other methods, which distinguish these components, have been described, and one commonly used principle is to employ fluorogenic substances such as fluoroscein diacetate (FDA) (Söderström, 1977). Each of these methods has specific drawbacks, but one common disadvantage is that distinguishing mycorrhizal hyphae from those of other fungi is not realistically possible.

Other methods of estimating fungal biomass that have been applied to mycorrhizal systems involve the analysis of fungal-specific metabolites such as ergosterol (Salmanowicz & Nylund, 1988) or chitin (Plassard et al., 1982; Whipps, 1987). These methods have been successfully applied to estimate fungal biomass, primarily within host tissue (Hepper, 1977; Bethlenfalvay et al., 1982; Vignon et al., 1986), but their efficacy in soil samples remains questionable as they do not differentiate mycorrhizal fungi from saprophytes or soil animals.

Analyses of respiration rates, or of substances that reflect microbial metabolic activity (ATP, enzyme activity), have often been used to provide relative biomass estimates. These indicators are of restricted use in mycorrhizal research, however, since they are totally unspecific and do not allow distinction of fungal activity, much less mycorrhizal activity.

The use of immunological methods provides a potentially more attractive approach in that these may allow distinction of specific mycorrhizal species. There is a vast literature available on the use of these techniques in medical mycology, but the application of immunological techniques to analysis of fungal systems in soil has been less successful. A number of qualitative studies have been published (e.g., Malajczuk et al., 1975; Aldwell et al., 1983, 1985; Wright et al., 1987; Frankland, 1975) but few attempts have so far been made to apply the technique quantitatively in mycorrhizal studies (Kough & Linderman, 1986). With wider use of DNA probes for identification of microorganisms (e.g., Festl et al., 1986; Stull et al., 1988) we can hopefully look forward to their application in mycorrhizal research.

A general problem in almost all microbial biomass estimations is the conversion of numbers of organisms, lengths of hyphae, or amounts of a specific measured substance to actual biomass. The conversion factors used by different workers and in different soils vary greatly, restricting the general applicability of the methods.

Estimates of Mycorrhizal Mycelial Biomass

Vesicular-Arbuscular Mycorrhiza

The majority of measurements of mycorrhizal mycelium have been made in VAM systems. Measuring fungal biomass in a solid medium such as soil is difficult, and the problem becomes even worse when only a specific proportion, such as the external mycelium of mycorrhizal fungi, is to be estimated. Sanders et al. (1977) analyzed the fungal biomass gravimetrically, thereby avoiding all conversion problems. Using pure sand as a growth support they removed the mycelium attached to the root manually and weighed it to obtain a direct biomass estimate. They found 3.6 μg dry weight of mycelium per cm root using onion plants. In an earlier paper Sanders and Tinker (1973) calculated a length of 80 cm mycelium per cm of infected root (45 cm/cm total root) in onion plants inoculated with a *Glomus* species.

Other studies of external mycelium have been based on estimates of hyphal length. In two papers Abbott and Robson assessed the external mycelium of subterranean clover *Trifolium subterraneum*. In the first study (Abbott et al., 1984) they investigated the effect of phosphate supply on the development of external mycelium of *Glomus fasiculatum*. In control treatments they found 200 cm of hyphae per g of soil after 6 weeks growth. This corresponded to 450 cm of hyphae per cm of infected root. In phosphate treatments the corresponding values were 1523 cm/g soil and 900 cm/cm infected root. They did not attempt to estimate biomass, but assuming a hyphal diameter of 2 μm and a dry weight conversion factor of 0.2 (Bakken & Olsen, 1983), the control and phosphate treatment biomass would be 2.8 and 5.6 μg/cm infected root, respectively, or 1.2 and 9.6 μg/g fresh weight of plant root. However, it should be emphasized that such crude estimates can be used to estimate biomass only within an order of magnitude. If the mean hyphal diameter were set to 4 μm the estimated biomass figures would be 4 times higher.

In the second study, Abbott and Robson (1985) estimated amounts of external mycelium formed by four different endophytes with subterranean clover. In these experiments, *Glomus fasiculatum* produced only small amounts of external mycelium but *Glomus calospora* produced 200 cm/cm of hyphae per infected root after 4 weeks growth, increasing to 3000 cm/cm after 5 weeks and decreasing again to 1200 cm/cm after 7 weeks. *Glomus fasiculatum* produced 1400 cm/cm after 4 weeks and 300 and 200 cm/cm after 5 and 7 weeks growth, respectively. These authors also found that 30–50% of the mycelium in the pots inoculated with these two species was between 1 and 5 μm in diameter. These data illustrate the complexity of biomass estimates and the need for repeated measurements at different time intervals. However, again assuming a mean hyphal diameter of 2 μm and 20% dry weight, the maximum and minimum biomass values in these studies would be 1.2 and 18.9 μg/cm, respectively. In a more recent study, Jakobsen and Rosendahl (1990) estimated that there was 27 m of external hyphae per g soil dry weight, constituting 2.6% of root dry weight.

Chitin analysis has mainly been used to estimate amounts of internal infection. Hepper (1977) showed that chitin measurements correlated well with amounts of infection and that the regression obtained differed significantly between different endophytes. She estimated the fungal biomass in *Centrosema pubescens* roots to be between 50 and 138 μg/mg dry root in plants with high infection rates. Estimates of internal infection of *Trifolium repens* were as high as 17% root dry weight. The method has also been used to estimate soil VAM mycelial biomass (Pacovsky & Bethlenfalvay, 1982). Bethlenfalvay et al. (1982) found approximately 600 mg dry weight of intraradical fungal mycelium per plant using this method and about 100 mg of extraradical fungus.

Schubert et al. (1987) used FDA too estimate the viable proportion of the extramatrical mycelium in systems of *Trifolium repens* inoculated with *Glomus clarum*. As in other studies, the extraradical biomass increased during the growth of the plants reaching an apparent maximum after 8 weeks. Interestingly however, the FDA active proportion of the hyphal length decreased from 75–90% after 20 days to only 5–10% after 40 and 69 days in two different experiments. The low percentages seem to indicate that a stable equilibrium had been reached since in natural ecosystems the proportion of FDA stained mycelium is always rather low (Söderström, 1979). In another study, using iodonitrotetrazolium and NADH to estimate activity of external VAM hyphae, Sylvia (1988) found that the proportion of active hyphae in soil ranged between 0 and 32% of total length, but that the activity of hyphae attached to roots was much higher.

Ectomycorrhiza

Ectomycorrhizal mycelia are often extensive but estimates of mycelial length are fewer than for VA systems and this probably results from the fact that these fungi typically inhabit soils with a high humus content making the mycelium much more difficult to extract and visualize than in clay or sandy soils. Read and Boyd (1986) estimated hyphal lengths of between 10 and 80 cm/cm root length in laboratory systems containing *Pinus sylvestris* and *Suillus bovinus,* suggesting a similar order of magnitude to that in VA systems. The estimation of hyphal lengths in ectomycorrhizal systems is further complicated by the fact that the fungi often form differentiated structures in which hyphae aggregate to form mycelial strands. In contrast to studies of VAM fungi, which have often been carried out in pot cultures, most information on ectomycorrhizal fungi has come from field studies where the aim has been to estimate the energy demands of the fungi.

Ericoid Mycorrhiza

Available evidence suggests that, in contrast to the above systems, ericoid mycorrhizal systems possess a poorly developed external mycelium (Read, 1984). Although the estimated number of entry points is high, ranging from 250 to 2000/ cm root, the external mycelium only extends 0.5 to 1.0 cm from host roots. The ratio of external to internal fungal biomass may be as low as 0.1 (Read, 1984).

Energy Demands of Mycorrhizal Fungi

In contrast to VA mycorrhizal systems, where only 5–10% of the root weight is thought to be accounted for by internal infection, fungal tissue may constitute up to 40% of the dry weight of ectomycorrhizal roots (Harley & Smith, 1983). It is difficult to calculate an accurate biomass since the fungal mass will vary enormously between both different fungal species as well as host species. However, on the basis of such estimates, Fogel and Hunt (1979) calculated a fungal sheath biomass of 6104 kg/ha and a sclerotial biomass of 2158 kg/ha for *Cenococcum geophilum* in a *Pseudotsuga menziesii* forest in Oregon. They also measured the mycelium in the soil and estimated that 50% of the total stand throughput was accounted for by the fungi. In a later study (Fogel and Hunt, 1983) they reduced this figure to 28%. Vogt et al. (1982) estimated 15% of the net primary production in an *Abies amabilis* forest was consumed by mycorrhizal fungi. This study was based on the biomass of fine roots and it is possible to estimate a corresponding figure for Swedish pine forests. Persson (1978) estimated that fine (<2 mm in diameter) root biomass production in a young pine forest was 2030 kg/ha/year. Assuming that 90% of these tips are mycorrhizal and that 40% of the mycorrhizal tip weight is fungal, 730 kg of fungal sheath may be produced per hectare per year. The forest in which this production was estimated is a heathland pine forest where the dominant fruitbody-forming species is *Lactarius rufus*. There are no reliable data on fruitbody production in this forest but Richardson (1970) estimated the annual production of this species to be 265–460 kg fresh weight/ha. It is thus reasonable to assume a production in the Swedish forest of 30 kg/ha/year. There are no direct estimations of mycorrhizal mycelium in this forest, but Finlay and Söderström (1989) suggested a figure of 200 m/g dry weight on the basis of regression relationships between mycelial respiration and biomass. Using data from Söderström (1979) that figure can be converted to 3.5 kg live mycelium/ha dry weight standing crop. If a turnover rate of 1 week during the 5-month vegetation period is assumed the production will be 70 kg mycelium/ha/year. Together these estimates produce a total fungal production of 830 kg/ha/year. If the carbon content is set to 40% and the productivity to 60%, the carbon demand of the ectomycorrhizal fungi in this forest will be 830 kg C/ha/year. The photosynthetic assimilation of the pines in this forest has been calculated to be 5800 kg C/ha/year (Linder & Axelsson, 1982), which means that the ectomycorrhizal fungi, according to these calculations, will use 14–15% of the assimilated carbon, a figure that is strikingly similar to the ones presented by Vogt et al. (1982), Fogel and Hunt (1979), and also similar to estimates of carbon flow to VA mycorrhiza.

Regulation and Adaptation

Mycorrhizal mycelia are dynamic systems, but we still know rather little about the processes that regulate their behaviour. These can be considered on two levels. First, regulation of mycelial growth may be brought about by short-term effects such as changes in soil conditions and light or plant-induced changes. Second, adaptive changes in mycelial structure and function in response to different environments also occur over evolutionary time.

In experiments by Piché and Fortin (1982) the addition of increasing amounts of ammonium (0–2 mg N per seedling) to growth pouches containing *Pinus strobus* improved the development of mycelial strands, whereas increases in phosphorus concentration caused reduced mycorrhizal infection and poorer mycelial development. Reduced light intensities also reduced mycelial development, showing the importance of current assimilate for mycelial biomass production. Little is known about the longevity and turnover of mycorrhizal roots and mycelia and further experimentation in this area is required.

The adaptive features of different mycorrhizal types that have evolved in response to different environments have been elegantly reviewed by Read (1984, 1987). It is noteworthy from the above discussion that the external mycelium of ericoid mycorrhizae is poorly developed compared with VA and ectomycorrhizas. Read (1984) has pointed out that, owing to the high water-holding capacity of mor humus, the roots of ericaceous heathland plants are normally bathed in a dilute solution of carboxylic acids containing free and adsorbed nutrients. Under such conditions the production of an extensive external mycelium would be a wasteful investment of carbon since no nutritional benefit would arise. Investment of carbon in external mycelia is generally higher in ectomycorrhizal and VA mycorrhizal systems but differences do occur. The transition from the mor humus of boreal coniferous forests to the mull humus of temperate deciduous forests is often associated with a general trend toward decreased production of ectomycorrhizal external mycelium. It has been suggested that efficient absorption and storage of leached nutrients in a thick sheath may be more important in environments where there are seasonal flushes of nutrient release from fallen litter, whereas the maintenance of an extensive mycelium may be better adapted to the acquisition of permanently low levels of mineral nutrients (Read, 1984; Finlay et al., 1989). The consequences of these two strategies in terms of carbon flow remain open to speculation.

Information about the dynamics and turnover of mycorrhizal mycelia in natural soil ecosystems is still severely limited. In the following section we consider the consequences of carbon flow to and through mycorrhizal mycelia in terms of its effects on the soil community.

Consequences of Carbon Flow to Mycorrhizal Mycelia: Effects on the Soil Community

It is clear from the preceding discussion that significant amounts of energy rich carbon compounds are supplied directly to the mycorrhizal mycelium in a range of associations. Estimates of the amounts of carbon involved vary but it seems likely that in many soils the mycelia of VA and ectomycorrhizal associations may form a significant proportion of the total fungal biomass. Discussion of the ecological significance of this mycelial phase has, until recently, been restricted to the consideration of ways in which the supply of mineral nutrients to individual plants or roots is ameliorated. However, the significance of mycorrhizal mycelia has recently been considered in the wider context of its possible influence at the community and ecosystem levels (Read, 1984; Read et al., 1985; Finlay & Söderström, 1989). Some of these wider roles are discussed in other chapters. Here we consider these effects in relation to the consequences of carbon input through mycorrhizal mycelia to the soil ecosystem.

Utilization of Simple and Polymeric Organic Compounds

Evidence concerning the degree of hydrolytic activity of different mycorrhizal fungi and its possible significance in different ecosystems is still incomplete. However, the supply of energy-rich carbon compounds for the synthesis of hydrolytic enzymes is clearly of potential importance. The effects of possible activity on nutrient cycling and interactions with the saprophytic microflora are potentially great in terms of overall mineralization processes and microbial immobilization.

The proteolytic capability of different mycorrhizal fungi differs greatly and more detailed studies are required to assess its importance. Abuzinadah et al. (1986) suggested that direct utilization of organic nitrogen sources by ectomycorrhizal fungi would lead to tighter nutrient cycling by restricting losses to decomposer populations. These authors suggested an alternative view of the forest nitrogen cycle based on proteolytic and peptidolytic capability in some ectomycorrhizal fungi. Mycorrhizal fungi are unique in that they have direct access to a supply of energy-rich substrates in the form of plant assimilates, and this may place them at a competitive advantage with respect to the saprophytic microflora. Finlay and Read (1986a) demonstrated that distribution and activity of ectomycorrhizal mycelia were not uniform and speculated that the mycelium may be capable of selective exploitation of localized sites of nutrient enrichment, in a manner similar to that suggested for VAM fungi by St. John et al. (1983). Finlay and Söderström (1989) pointed out that selective colonization of areas of soil rich in organic material would be of particular importance in species with proteolytic capability. The mycelial patches found by Finlay and Read (1986a; 1986b) were strong sinks for both labelled plant assimilates and phosphorus, and the ability to direct carbon compounds required for the synthesis and excretion of protease enzymes, selectively to sites of organic enrichment would clearly be advantageous.

Other organic compounds containing carbon can be used by mycorrhizal fungi and their use as carbon sources may affect the carbon economy of the symbiosis. Trojanowski et al. (1984) suggested that ectomycorrhizal fungi might be able to utilize lignin but experiments by Haselwandter et al. (1987) showed that the ability of the ectomycorrhizal fungus *Paxillus involutus* was low compared with two ericoid endophytes. Other recalcitrant polymers such as tannic acid can also be degraded by the ericoid endophyte *Hymenoscyphus ericae* (Leake & Read, 1989) and soluble, phytotoxic phenolic acids can also be used as carbon sources. The utilization of organic compounds in the above ways should reduce the carbon drain imposed on the host plant by the mycobiont.

Translocation and Uptake of Mineral Nutrients

The allocation of carbon to mycorrhizal mycelia for biomass production and energy requirements is of fundamental significance to processes of mineral nutrient uptake but is not mentioned further here since this is fully discussed in the previous chapter (Chapter 4, Read).

Mycelial Connections between Plants

One consequence of the extensive growth of external mycorrhizal mycelium and the generally low host specificity of mycorrhizal fungi is that inter- and intraspecific mycelial connections are formed between adjacently growing plants. The full ecological significance of these is still uncertain. Evidence from isotope studies suggests that compounds containing C, N or P can move between plants through these connections (Read et al., 1985; Francis et al., 1986; Finlay and Read, 1986a; Haystead et al., 1988). However, the question of whether net transfer actually occurs and, if so, whether the quantities involved are significant is still a matter of some controversy (see Newman, 1988).

There is presently little evidence to suggest that such movement of compounds is quantitatively important compared with "normal" uptake of phosphate or ammonium, or with photosynthetic C fixation. The transfer of amino acids and amides across the host–fungus interface (Finlay et al., 1988, 1989) allows for bidirectional movement of C in addition to the net carbohydrate flux required for fungal growth. Direct evidence for this reverse flux has been provided by Duddridge et al. (1988). However, the integration of plants into a common mycelial network may be of significance without necessarily implying a large net reverse carbon flux. The capacity for lateral movement of host-derived assimilates along chemical concentration gradients from areas of high substrate availability to areas of low substrate availability, as demonstrated by Francis and Read (1984) and Finlay and Read (1986a), will allow small seedlings to become connected to a much larger mycelium than they could support on the basis of their own photosynthetic products alone. Even small reductions in the carbon drain from small, shaded seedlings by virtue of "lateral" transport of assimilate from larger, better illuminated plants may be of significance to plants at or near their compensation point, and important in processes of regeneration, without implying a large net flux of carbon from fungus to plant. Additionally, as Finlay and Read (1986b) pointed out, seedlings with small root systems and limited seed reserves become connected to a mycelium, which can exploit a larger volume of soil for mineral nutrients than would be possible solely on the basis of allocation of their own assimilates. Again, no arguments of net interplant transfer need to be invoked for this to be so.

Unequivocal demonstration of these processes requires further experimentation, but a number of observations lend support to the idea. Experiments by Fleming (1984) demonstrated that ectomycorrhizal infection of seedlings was reduced in cored or trenching areas suggesting that mycelial connections with mature trees were important for the successful establishment of infection. Recent studies using experimental microcosms (Grime et al., 1987), or manipulated field systems (Gange et al., 1990), also suggest that the presence of vesicular-arbuscular mycelial connections may have important effects on community development, influencing the survivorship and competitive ability of particular species and promoting species diversity. Less experimental evidence exists for ectomycorrhizal systems, but recent experiments by Perry et al. (1989) have provided tentative evidence that ectomycorrhizal fungi may mediate competition between different host species. Ultimate effects of mycelial connections on species diversity may depend on the degree of host specificity existing in the ecosystems under consideration, as pointed out by Finlay (1989).

Aggregation of Sandy Soil

Another consequence of the extensive production of external mycelium by mycorrhizal fungi is that the hyphae help in binding soil particles (Sutton & Sheppard, 1976) to form stable soil aggregates. This may reduce soil erosion and water runoff and be particularly important in loose mine soils (Rothwell, 1984). Tisdall and Oades (1979) recorded mycelial lengths of up to 55 m/g of soil associated with grassland species.

Grazing by Animals

It is clear from the preceding discussion of mycorrhizal mycelial biomass that in many situations it may represent a significant proportion of the total soil mycelial biomass and thus represent a significant resource to the mycophagous soil fauna. The subject of grazing interactions between soil fauna and mycorrhizal fungi is dealt with more fully elsewhere in this volume (Fitter & Sanders, Chapter 10) and is mentioned here only in so far as the cycling of carbon compounds is involved. Many previous studies of the effects of soil have been restricted to the overall effects of grazing on nutrient cycling and there have been few specific studies of the direct effects of grazing of mycorrhizal mycelia (Finlay, 1985; Shaw, 1985; Mcgonigle & Fitter, 1987). More studies are required of interactions between mycophagous soil animals and mycorrhizal fungi to determine the extent of grazing and its possible effect on plant growth and mycorrhizal nutrient cycling. One possible use of carbon compounds transferred to mycorrhizal hyphae is the production of secondary compounds to inhibit grazing.

Mycorrhizosphere Effects

The term "mycorrhizosphere" was suggested by Rambelli (1973) to describe the soil surrounding, and influenced by, mycorrhizas. Evidence for the existence and possible significance of mycorrhizosphere effects has been reviewed by a number of authors (e.g., Fogel, 1988; Linderman, 1988) and will be discussed later in this book (Azcon-Aquilar & Barea, Chapter 6).

A key feature of the mycorrhizosphere is the presence of mycorrhizal hyphae that surround the root and extend out from it in the form of dense extramatrical hyphae. These hyphae may extend the limits of the mycorrhizosphere considerably past those of the normal rhizosphere of uninfected roots. Interactions with other soil microorganisms may be direct or indirect, through effects on the host plant. Mycorrhizal infection may cause changes in the quality and quantity of root exudates and secretions, enhance the nutrient and elemental composition, alter the hormone balance of host roots, and increase respiratory losses of CO_2 from the root surface. Since mycorrhizal hyphae often dramatically influence the distribution and absorptive surface area of root systems they present a considerable surface area across which direct interactions may take place with the microbial flora. A "mycosphere" may thus develop around mycorrhizal hyphae in which enhanced microbial populations of altered species composition may occur.

The extramatrical hyphae themselves may exude substances that cause soil and organic matter to aggregate (Sutton & Sheppard, 1976) providing microsites for growth of bacteria, fungi, actinomycetes, and algae. In cases in which the root is more or less completely surrounded by fungal material, such as the sheathed lateral roots of ectomycorrhizal plants, most or all of the substances entering the soil may do so through fungal hyphae. Interactions may be stimulatory, inhibitory, or neutral (Bowen & Thoedorou, 1979). Stimulatory effects of microorganisms on ectomycorrhizal colonization and development have been reported by Garbaye and Bowen (1989) and MacAfee and Fortin (1988). The *in vitro* growth of different ectomycorrhizal isolates can also be both stimulated or inhibited by actinomycetes isolated from the mycorrhizoplane of *Pinus resinosa* (Richter et al., 1989). It seems that distinct microbial communities may have evolved in response to the presence of specific mycorrhizal associations, and the isolation and selection of the appropriate "helper" organism, for use as coinoculants, may offer scope for additive or synergistic growth stimulation.

While exudation of specific compounds from the roots or mycelium may stimulate microorganisms beneficial to the symbiosis, extracellular metabolites may also have an antibiotic effect on certain phytopathogenic microorganisms (Kope & Fortin, 1989). The continuous supply of carbohydrates that mycorrhizal fungi receive from their hosts is probably important in terms of providing the energy for synthesis of the wide range of compounds that are undoubtedly involved in these microbial interactions. It is often assumed that mycorrhizal fungi are at a competitive advantage with respect to the saprophytic flora because of this direct supply of plant assimilates. Interactions with saprophytic populations could thus influence decomposition processes. Abuzinadah et al. (1986) suggested that the increased rates of pine litter decomposition following exclusion of ectomycorrhizal roots (see Gadgil & Gadgil, 1971, 1975) could be caused by the removal of successful competition for limited organic nitrogen by ectomycorrhizal fungi with proteolytic enzymes. Death, decomposition, and leakage of organic compounds from decomposing mycorrhizal hyphae represent a potential input of carbon into the soil ecosystem about which we know very little. Further information about their dynamics and turnover is required. However, one consequence of the proteolytic activity of certain ectomycorrhizal species is that organic compounds released from dying hyphae could be used directly by living mycorrhizal hyphae and the carbon recycled internally within the ectomycorrhizal association, restricting losses to the soil ecosystem through immobilization.

Concluding Remarks

Carbon flow to mycorrhizal roots and through mycorrhizal mycelia to different components of the soil ecosystem can clearly be significant. In forests it has been estimated that 3–5 times more organic matter is returned to the soil in the form of roots and mycorrhiza than is returned by decomposition of litter (Fogel, 1988). Data from many different sources now suggest that as much as 20% of the total carbon assimilated may be transferred to mycorrhizal fungi in both VA and ectomycorrhizal systems. Fewer data exist for ericoid mycorrhizal systems where the production of external mycelium is less extensive. More information is needed about the amounts and types of carbon compounds involved and the processes that regulate their translocation to mycorrhizal roots and flow through the mycelium. Large gaps remain in our knowledge and further information about these important processes is needed, not least because the flow of carbon to mycorrhizal symbionts under different conditions may have important economic implications to crop and timber yields.

The available methods for quantifying the total and active proportion of the external mycorrhizal mycelium are still inadequate and need further development. Until these are improved, progress in understanding the dynamics of mycorrhizal mycelia and their interactions with other organisms will be restricted. For this reason, concepts of mycorrhizal efficiency, involving the balance between carbon drain and the beneficial effects of increased mineral nutrient uptake, are still poorly developed.

The flow of carbon to mycorrhizal mycelia clearly has a potentially huge range of effects since energy-rich substrates are required by most biological processes. Potential effects include interactions with both phytopathogens and decomposers, chemical defences against grazing of the mycelium, and the stabilization of soil aggregates. These may have an impact on energy flow and cycling of nutrients. Wider effects on plant communities may also occur through influence on regeneration processes and plant competition.

References

Abbott, L. K., & Robson, A. D. (1985). Formation of external hyphae in soil by four species of vesicular-arbuscular mycorrhizal fungi. *New Phytologist* **99,** 245–255.

Abbott, L. K., Robson, A. D., & De Boer, G. (1984). The effect of phosphorus on the formation of hyphae in soil by the vesicular-arbuscular mycorrhizal fungus *Glomus fasiculatum*. *New Phytologist* **97,** 437–446.

Abuzinadah, R. A., & Read, D. J. (1989). The role of proteins in the nitrogen nutrition of ectomycorrhizal plants. V. Nitrogen transfer in birch (*Betula pendula* L.) grown in association with mycorrhizal and non-mycorrhizal fungi. *New Phytologist* **112,** 61–68.

Abuzinadah, R. A., Finlay, R. D., & Read, D. J. (1986). The role of proteins in the nitrogen nutrition of ectomycorrhizal plants. II. Utilization of protein by mycorrhizal plants of *Pinus contorta*. *New Phytologist* **103,** 495–506.

Aldwell, F. E. B., Hall, I. R., & Smith, J. M. B. (1983). Enzyme-linked immunosorbent assay (Elisa) to identify endomycorrhizal fungi. *Soil Biology and Biochemistry* **15,** 377–378.

Aldwell, F. E. B., Hall, I. R., & Smith, J. M. B. (1985). Enzyme-linked immunosorbent assay as an aid to taxonomy of the Endogonaceae. *Transactions of the British Mycological Society* **84,** 399–378.

Amijee, F., & Stribley, D. P. (1987). Soluble carbohydrates of vesicular-arbuscular mycorrhizal fungi. *The Mycologist* **21,** 20–21.

Bakken, L. R., & Olsen, R. A. (1983). Buoyant densities and dry matter contents of microorganisms: Conversion of a measured biovolume into biomass. *Applied and Environmental Microbiology* **45**, 1188–1195.

Bethlenfalvay, G. J., Pacovsky, R. S., & Brown, M. S. (1982). Parasitic and mutualistic associations between a mycorrhizal fungus and soybean: Development of the Endophyte. *Phytopathology* **72**, 894–897.

Bevege, D. I., Bowen, G. D., & Skinner, M. F. (1975). Comparative carbohydrate physiology of ecto- and endomycorrhizas. In: *Endomycorrhizas* (Ed. by F. E. Sanders, B. Mosse, & P. B. Tinker), pp. 149–174. Academic Press, London.

Björkman, E. (1942). Über die Bedingungen der Mykorrhizabildung bei Keifer und Fichte. *Symbolae Botanicae Upsaliensis* **6**, 2.

Bowen, G. D., & Theodorou, C. (1979). Interactions between bacteria and ectomycorrhizal fungi. *Soil Biology and Biochemistry* **11**, 119–126.

Cairney, J. W. G., Ashford, A. E., & Allaway, W. G. (1989). Distribution of photosynthetically fixed carbon within root systems of *Eucalyptus pilularis* plants ectomycorrhizal with *Pisolithus tinctorius*. *New Phytologist* **112**, 495–500.

Cooper, K. M., & Lösel, D. (1978). Lipid physiology of vesicular-arbuscular mycorrhiza. I. Composition of lipids in roots of onion, clover and ryegrass infected with *Glomus mosseae*. *New Phytologist* **80**, 143–151.

Cox, G., & Tinker, P. B. (1976). Translocation and transfer of nutrients in vesicular-arbuscular mycorrhizae. I. The arbuscule and phosphorus transfer: a quantitative ultrastructural study. *New Phytologist* **77**, 371–378.

Cox, G., Sanders, F. E., Tinker, P. B., & Wild, J. A. (1975). Ultrastructural evidence relating to host-endophyte transfer in vesicular-arbuscular mycorrhiza. In: *Endomycorrhizas* (Ed. by F. E. Sanders, B. Mosse, & P. B. Tinker), pp. 149–174. Academic Press, London.

Douds, D. D., Johnson, C. R., & Koch, K. E. (1988). Carbon cost of the fungal symbiont relative to net leaf P accumulation in a split-root VA mycorrhizal symbiosis. *Plant Physiology* **86**, 491–496.

Duddridge, J. A., Finlay, R. D., Read, D. J., & Söderström, B. (1988). The structure and function of the vegetative mycelium of ectomycorrhizal plants. III. Ultrastructural and autoradiographic analysis of inter-plant carbon distribution through intact mycelial systems. *New Phytologist* **108**, 183–188.

Erland, S., Finlay, R. D., & Söderström, B. E. (1991). The influence of substrate pH on carbon translocation in ectomycorrhizal and non-mycorrhizal pine seedlings. *New Phytologist* **119**, 235–242.

Festl, H., Ludwig, W., & Schleifer, K. H. (1986). DNA hybridization probe for the *Pseudomonas fluorescens* group. *Applied and Environmental Microbiology* **52**, 1190–1194.

Finlay, R. D. (1985). Interactions between soil microarthropods and endomycorrhizal associations of higher plants. In: *Ecological Interactions in Soil: Plants, Microbes & Animals* (Ed. by A. H. Fitter), pp. 193–217. Blackwell Scientific Publications, London.

Finlay, R. D. (1989). Functional aspects of phosphorus uptake and carbon translocation in incompatible ectomycorrhizal associations between *Pinus sylvestris* and *Suillus grevillei* and *Boletinus cavipes*. *New Phytologist* **112,** 185–192.

Finlay, R. D., & Read, D. J. (1986a). The structure and function of the vegetative mycelium of ectomycorrhizal plants. I. Translocation of ^{14}C-labelled carbon between plants interconnected by a common mycelium. *New Phytologist* **103,** 143–156.

Finlay, R. D., & Read, D. J. (1986b). The structure and function of the vegetative mycelium of ectomycorrhizal plants. II. The uptake and distribution of phosphorus by mycelial strands interconnecting host plants. *New Phytologist* **103,** 157–165.

Finlay, R. D., & Söderström, B. (1989). Mycorrhizal mycelia and their role in soil and plant communities. In: *Developments in Plant and Soil Sciences*, Vol. 39. *Ecology of Arable Land, Perspectives and Challenges* (Ed. by M. Clarholm & L. Bergström), pp. 139–148. Kluwer Academic Publishers, Dordrecht/London.

Finlay, R. D., Ek, H., Odham, G., & Söderström, B. (1988). Mycelial uptake, translocation and assimilation of nitrogen from ^{15}N-labelled ammonium by *Pinus sylvestris* plants infected with four different ectomycorrhizal fungi. *New Phytologist* **110,** 59–66.

Finlay, R. D., Ek, H., Odham, G., & Söderström, B. (1989). Uptake, translocation and assimilation of nitrogen from ^{15}N-labelled ammonium and nitrate sources by intact ectomycorrhizal systems of *Fagus sylvatica* infected with *Paxillus involutus*. *New Phytologist* **113,** 47–55.

Fleming, L. V. (1984). Effects of soil trenching and corings on the formation of ectomycorrhizas on birch seedlings grown around mature trees. *New Phytologist* **98,** 143–153.

Fogel, R. (1988). Interactions among soil biota in coniferous ecosystems. *Agriculture Ecosystems and Environment* **24,** 69–85.

Fogel, R., & Hunt, G. (1979). Fungal and arboreal biomass in a western Oregon Douglas fir ecosystem: Distribution patterns and turnover. *Canadian Journal of Forest Research* **9,** 245–256.

Fogel, R., & Hunt, G. (1983). Contribution of mycorrhizae and soil fungi to nutrient cycling in a Douglas-fir ecosystem. *Canadian Journal of Forest Research* **13,** 219–232.

Francis, R., & Read, D. J. (1984). Direct transfer of carbon between plants connected by vesicular-arbuscular mycorrhizal mycelium. *Nature (London)* **307,** 53–56.

Francis, R., Finlay, R. D., & Read, D. J. (1986). Vesicular-arbuscular mycorrhiza in natural vegetation systems. IV. Transfer of nutrients in inter- and intra-specific combinations of host plants. *New Phytologist* **102,** 103–111.

Frankland, J. (1975) Estimation of live fungal biomass. *Soil Biology and Biochemistry* **7,** 339–340.

Gadgil, R. L., & Gadgil, P. D. (1971). Mycorrhiza and litter decomposition. *Nature (London)* **233,** 133.

Gadgil, R. L., & Gadgil, P. D. (1975). Suppression of litter decomposition by mycorrhizal roots of *Pinus radiata*. *New Zealand Journal of Forest Science* **5,** 35–41.

Gange, A. C., Brown, V. K., & Farmer, L. M. (1990). A test of mycorrhizal benefit in an early successional community. *New Phytologist* **115,** 85–91.

Garbaye, J., & Bowen, G. D. (1989). Stimulation of ectomycorrhizal infection of *Pinus radiata* by some microorganisms associated with the mantle of ectomycorrhizas. *New Phytologist* **112**, 383–388.

Graham, J. H., Linderman, R. G., & Menge, J. A. (1982). Development of external hyphae by different isolates of mycorrhizal *Glomus* spp. in relation to root colonization and growth of Troyer Citrange. *New Phytologist* **91**, 183–189.

Grime, J. P., Mackey, J. M. L., Hillier, S. H., & Read, D. J. (1987). Floristic diversity in a model ecosystem using experimental microcosms. *Nature (London)* **328**, 420–422.

Hadley, G., & Ong, S. H. (1978). Nutritional requirements of orchid mycorrhiza. *New Phytologist* **81**, 561–569.

Hanssen, J. F., Thingstad, T. F., & Goksøyr, J. (1974). Evaluation of hyphal lengths and fungal biomass in soil by a membrane filter technique. *Oikos* **25**, 102–107.

Harley, J. L., & Smith, S. E. (1983). *Mycorrhizal Symbiosis*. Academic Press, London.

Harris, D., Pacovsky, R. S., & Paul, E. A. (1985). Carbon economy of soybean–*Rhizobium*–*Glomus* associations. *New Phytologist* **101**, 427–440.

Haselwandter, K., Bonn, G., & Read, D. J. (1987). Degradation and utilization of lignin by mycorrhizal fungi. In: *Proceedings of the 7th NACOM, Gainsville, Florida, USA, May 1987* (Ed. by D. M. Sylvia, L. L. Hung, & J. H. Graham). p. 331 Institute of Food & Agricultural Sciences, University of Florida, Gainsville, FL.

Hayman, D. S. (1974). Plant growth responses to vesicular-arbuscular mycorrhiza. VI. Effect of light and temperature. *New Phytologist*. **73**, 71–80.

Haystead, A., Malajczuk, N., & Grove, T. S. (1988). Underground transfer of nitrogen between pasture plants infected with vesicular-arbuscular mycorrhizal fungi. *New Phytologist* **108**, 417–423.

Hepper, C. M. (1977). A colorimetric method for estimating vesicular-arbuscular mycorrhizal infection in roots. *Soil Biology and Biochemistry* **9**, 15–18.

Ho, I., & Trappe, J. M. (1973). Translocation of ^{14}C from *Festuca* plants to their endomycorrhizal fungi. *Nature, New Biology* **244**, 30–31.

Jakobsen, I., & Rosendahl, L. (1990). Carbon flow into soil and external hyphae from roots of mycorrhizal cucumber plants. *New Phytologist* **115**, 77–83.

Jones, P. C. T., & Mollison, J. E. (1948). A technique for the quantitative estimation of soil microorganisms. *Journal of General Microbiology* **2**, 54–69.

Kope, H. H., & Fortin, J. A. (1989). Inhibition of phytopathogenic fungi in vitro by cell free culture media of ectomycorrhizal fungi. *New Phytologist* **113**, 57–63.

Kough, J. L., & Linderman, R. G. (1986). Monitoring extra-matrical hyphae of a vesicular-arbuscular mycorrhizal fungus with an immunofluorescence assay and the soil aggregation technique. *Soil Biology and Biochemistry* **18**, 309–313.

Kucey, R. M. N., & Paul, E. A. (1982). Carbon flow, photosynthesis, and N_2 fixation in mycorrhizal and nodulated Faba beans. (*Vicia faba* L.) *Soil Biology and Biochemistry* **14**, 407–412.

Leake, J. R. (1988). Metabolism of phyto- and fungitoxic phenolic acids by the ericoid mycorrhizal fungus. In: *Proceedings of the 7th NACOM, Gainsville, Florida USA, May*

1987 (Ed. by D. M. Sylvia, L. L. Hung, & J. H. Graham), p. 332. Institute of Food & Agricultural Sciences, University of Florida, Gainsville, FL.

Leake, J. R., & Read, D. J. (1989). The biology of mycorrhiza in the Ericaceae. XIII. Some characteristics of the extracellular proteinase activity of the ericoid endophyte *Hymenoscyphus ericae. New Phytologist* **112,** 69–76.

Lewis, D. H. (1975). Comparative aspects of the carbon nutrition of mycorrhizas. In: *Endomycorrhizas* (Ed. by F. E. Sanders, B. Mosse, & P. B. Tinker), pp. 119–148. Academic Press, London.

Lewis, D. H., & Harley, J. L. (1965). Carbohydrate physiology of mycorrhizal roots of beech. III. Movement of sugars between host and fungus. *New Phytologist* **64,** 256–269.

Lindeberg, G. (1986). Physiology of interactions between mycorrhizae and rhizosphere microorganisms. In: *Physiological and Genetical Aspects of Mycorrhizae, Proceedings of 1st European Symposium on Mycorrhizae* (Ed. by V. Gianinazzi-Pearson, & S. Gianinazzi), pp. 197–206. INRA, Dijon, France.

Linder, S., & Axelsson, B. (1982). Changes in carbon uptake and allocation patters as a result of irrigation and fertilization in a young *Pinus sylvestris* stand. In: *Carbon Uptake and Allocation in Subalpine Ecosystems as a Key to Management* (Ed. by R. H. Waring), pp. 38–44. Forest Research Laboratory, Oregon State University.

Linderman, R. G. (1988). Mycorrhizal interactions with the rhizosphere microflora: The mycorrhizosphere effect. *Phytopathology* **78,** 366–371.

Macafee, B. J., & Fortin, J. A. (1988). Comparative effects of the soil microflora on ectomycorrhizal inoculation of conifer seedlings. *New Phytologist* **108,** 443–449.

Malajczuk, N., McComb, A. J., & Parker, C. A. (1975). An immunofluorescence technique for detecting *Phytophthora cinnamomi* Rands. *Australian Journal of Botany* **23,** 289–309.

McGonigle, T. P., & Fitter, A. H. (1987). Evidence that Collembola suppress plant benefit from vesicular-arbuscular mycorrhizas (VAM) in the field. In: *Proceedings of the 7th NACOM, Gainsville, Florida USA, May 1987* (Ed. by D. M. Sylvia, L. L. Hung, & J. H. Graham), p. 209. Institute of Food & Agricultural Sciences, University of Florida, Gainsville, FL.

Melin, E., & Nilsson, H. (1957). Transport of ^{14}C-labelled photosynthate to the fungal associate of pine mycorrhiza. *Svensk Botanisk Tidskrift* **51,** 166–186.

Nagy, S., Nordby, H. E., & Nemec, S. (1980). Composition of lipids in roots in six citrus cultivars infected with the vesicular-arbuscular mycorrhizal fungus *Glomus mosseae. New Phytologist* **85,** 377–384.

Nelson, C. D. (1964). The production and translocation of photosynthate C-14 in conifers. In: *Formation of Wood in Forest Trees* (Ed. by M. H. Zimmerman), pp. 235–257. Maria Mons Cabot Foundation, New York.

Newman, E. I. (1988). Mycorrhizal links between plants: Their functioning and ecological significance. *Advances in Ecological Research* **18,** 243–270.

Nylund, J.-E. (1988). The regulation of mycorrhiza formation—carbohydrate and hormone theories reviewed. *Scandinavian Journal of Forest Research* **3,** 465–479.

Nylund, J.-E., & Wallander, H. (1989). Effects of ectomycorrhiza on host growth and carbon balance in a semi-hydroponic cultivation system. *New Phytologist* **112**, 389–398.

Pacovsky, R. S., & Bethlenfalvay, G. J. (1982). Measurement of the extraradical mycelium of a vesicular-arbuscular mycorrhizal fungus in soil by chitin determination. *Plant and Soil* **68**, 143–147.

Pang, P. C., & Paul, E. A. (1980). Effects of vesicular-arbuscular mycorrhiza on ^{14}C and ^{15}N distribution in nodulated fababeans. *Canadian Journal of Soil Science* **60**, 241–250.

Pate, J. S., Layzell, D. B., & Atkins, C. A. (1979). Economy of carbon and nitrogen in a nodulated and non-nodulated (NO_3 grown) legume. *Plant Physiology* **64**, 1083–1088.

Perry, D. A., Margolis, H., Choquette, C., Molina, R., & Trappe, J. M. (1989). Ectomycorrhizal mediation of competition between coniferous tree species. *New Phytologist* **112**, 501–511.

Persson, H. (1978). Root dynamics in a young Scots pine stand in central Sweden. *Oikos* **30**, 508–519.

Piché, Y., & Fortin, J. A. (1982). Development of mycorrhizae, extramatrical mycelium and sclerotia on *Pinus strobus* seedlings. *New Phytologist* **91**, 211–220.

Plassard, C., Mousain, D., & Salsac, L. (1982). Estimation of mycelial growth of basidiomycetes by means of chitin determination. *Phytochemistry* **21**, 345–348.

Rambelli, A. (1973). The rhizosphere of mycorrhizae. In: *Ectomycorrhizae* (Ed. by G. L. Marks & T. T. Koslowski), pp. 299–343. Academic Press, New York.

Read, D. J. (1984). The structure and function of the vegetative mycelium of mycorrhizalroots. In: *The Ecology and Physiology of the Fungal Mycelium. Symposium of the British Mycological Society* (Ed. by D. H. Jennings & A. D. M. Rayner), pp. 215–240. Cambridge University Press, Cambridge.

Read, D. J. (1987). Development and function of mycorrhizal hyphae in soil. In: *Proceedings of the 7th NACOM, Gainsville, Florida, USA, May 1987* (Ed. by D. M. Sylvia, L. L. Hung, & J. H. Graham), pp. 178–180. Institute of Food & Agricultural Sciences, University of Florida, Gainsville, FL.

Read, D. J., & Boyd, R. (1986). Water relations of mycorrhizal fungi and their host plants. In: *Water, Fungi and Plants* (Ed. by P. G. Ayres & L. Boddy), pp. 287–303. Cambridge University Press, Cambridge.

Read, D. J., Francis, R., & Finlay, R. D. (1985). Mycorrhizal mycelia and nutrient cycling in plant communities. In: *Ecological Interactions in Soil: Plants, Microbes and Animals* (Ed. by A. H. Fitter), pp. 193–217. Blackwell Scientific Publications, London.

Reid, C. P. P., Kidd, F. A., & Ekwebelam, S. A. (1983). Nitrogen nutrition, photosynthesis and carbon allocation in ectomycorrhizal pine. *Plant and Soil* **71**, 415–431.

Richardson, M. J. (1970). Studies on *Russula emetica* and other agarics in a Scots pine plantation. *Transactions of the British Mycological Society* **55**, 217–229.

Richter, D. L., Zuellig, T. R., Bagley, S. T., & Bruhn, J. N. (1989). Effects of red pine (*Pinus resinosa* Ait.) mycorrhizoplane-associated actinomycetes on in vitro growth of ectomycorrhizal fungi. *Plant and Soil* **115**, 109–116.

Romell, L. G. (1938). A trenching experiment in spruce forest and its bearing on the problems of mycotrophy. *Svensk Botanisk Tidskrift* **32**, 89–99.

Romell, L. G. (1939). The ecological problem of mycotrophy. *Ecology* **20**, 163–167.

Rothwell, F. M. (1984). Aggregation of surface mine soil by interaction between VAM fungi and lignin degradation products of lesperdeza. *Plant and Soil* **80**, 99–104.

Salamanowicz, B., & Nylund, J.-E. (1988). High performance liquid chromatography determination of ergosterol as a measure of ectomycorrhizal infection in Scots pine. *European Journal of Forest Pathology* **18**, 291–298.

Sanders, F. E., & Tinker, P. B. (1973). Phosphate flow into mycorrhizal roots. *Pesticide Science* **4**, 385–395.

Sanders, F. E., Tinker, P. B., Black, R. L. B., & Palmerley, S. M. (1977). The development of endomycorrhizal root systems: I. Spread of infection and growth promoting effects with four species of vesicular-arbuscular endophyte. *New Phytologist* **78**, 257–268.

Schubert, A., Marzachi, C., Mazzetelli, M., Cravero, M. C., & Bonfante-Fasolo, P. (1987). Development of total and viable extraradicle mycelium in the vesicular-arbuscular mycorrhizal fungus *Glomus clarum*, Nicol. & Schenk. *New Phytologist* **107**, 183–190.

Shaw, P. J. A. (1985). Grazing preferences of *Onychiurus armatus* (Insecta: Collembola) for mycorrhizal and saprotrophic fungi of pine plantations. In: *Ecological Interactions in Soil: Plants, Microbes and Animals* (Ed. by A. H. Fitter), pp. 333–337. Blackwell Scientific Publications, Oxford.

Slankis, V. (1973). Hormonal relationships in mycorrhizal development. In: *Ectomycorrhizae* (Ed. by G. C. Marks & T. T. Koslowski), pp. 231–298. Academic Press, New York.

Snellgrove, R. C., Splittstoesser, W. E., Stribley, D. P., & Tinker, P. B. (1982). The distribution of carbon and the demand of the fungal symbiont in leekplants with vesicular-arbuscular mycorrhizas. *New Phytologist* **92**, 75–87.

Söderström, B. (1977). Vital staining of fungi in pure cultures and in soil with fluorescein diacetate. *Soil Biology and Biochemistry* **9**, 59–63.

Söderström, B. E. (1979). Seasonal fluctuations of active fungal biomass in horizons of a podzolized pine-forest soil in central Sweden. *Soil Biology and Biochemistry* **11**, 149–154.

Söderström, B., & Read, D. J. (1987). Respiratory activity of intact and excised ectomycorrhizal mycelial systems growing in unsterilized soil. *Soil Biology and Biochemistry* **19**, 231–237.

Söderström, B., Finlay, R. D., & Read, D. J. (1988). The structure and function of the vegetative mycelium of ectomycorrhizal plants. IV. Qualitative analysis of carbohydrate contents of the mycelium interconnecting host plants. *New Phytologist* **109**, 163–166.

St. John, T. V., Coleman, D. C., & Reid, C. P. P. (1983). Growth and spatial distribution of nutrient-absorbing organs: selective exploitation of soil heterogeneity. *Plant and Soil* **71**, 487–493.

Stribley, D. P., & Read, D. J. (1974). The biology of mycorrhizae in the Ericaceae. III. Movement of carbon-14 from host to fungus. *New Phytologist* **73**, 731–741.

Stull, T. L., LiPuma, J. J., & Edlind, T. D. (1988). A broad spectrum probe for molecular epidemiology of bacteria: Ribosomal RNA. *Journal of Infection Diseases* **157**, 280–288.

Sutton, J. C., & Sheppard, B. R. (1976). Aggregation of sand-dune soil by ectomycorrhizal fungi. *Canadian Journal of Botany* **54**, 326–333.

Sylvia, D. M. (1988). Activity of external hyphae of vesicular-arbuscular mycorrhizal fungi. *Soil Biology and Biochemistry* **20**, 39–43.

Tisdall, J. M., & Oades, J. M. (1979). Stabilization of soil aggregates by the root systems of ryegrass. *Australian Journal of Soil Research* **17**, 429–441.

Trojanowski, J., Haider, K., & Hüttermann, A. (1984). Decomposition of ^{14}C-labelled lignin, holocellulose and lignocellulose by mycorrhizal fungi. *Archives of Microbiology* **139**, 202–206.

Vignon, C., Plassard, C., Mousain, D., & Salsac, L. (1986). Assay of fungal chitin and estimation of mycorrhizal infection. *Physiologie Vegetale* **24**, 201–207.

Vogt, K. A., Grier, C. C., Meier, C. E., & Edmonds, R. L. (1982). Mycorrhizal role in net primary production and nutrient cycling in *Abies amabilis* ecosystems in western Washington. *Ecology* **63**, 370–380.

Whipps, J. M. (1987). Method for estimation of chitin content of the mycelium of ectomycorrhizal fungi grown on solid substrates. *Transactions of the British Mycological Society* **89**, 199–203.

Whipps, J. M., & Lynch, J. M. (1985). Energy losses by the plant in rhizodeposition. *Annual Proceedings of the Phytochemical Society of Europe* **26**, 59–71.

Wright, S. F., Morton, J. B., & Sorobuk, J. E. (1987). Identification of a vesicular-arbuscular mycorrhizal fungus by using monoclonal antibodies in an enzyme-linked immunosorbent assay. *Applied and Environmental Microbiology* **53**, 2222–2225.

SECTION 2

Interactions

6

Interactions between Mycorrhizal Fungi and Other Rhizosphere Microorganisms

C. Azcón–Aguilar and J. M. Barea

SUMMARY. Microbe–microbe interactions are crucial to understand the dynamic processes characteristic of rhizosphere establishment and maintenance and affecting plant growth and health. Mycorrhizal fungi are key components of soil microbiota and certain soil microorganisms are known to regulate mycorrhizal formation and function. Conversely, mycorrhizae affect the establishment of rhizosphere populations. Some interactions between mycorrhizae and soil microorganisms involve nutrient cycling, hence having an impact on plant growth and nutrition. Other interactions concern root pathogen activity, thereby affecting biological control to benefit plant health. The ecophysiological and biochemical studies, as already carried out, are giving way to histochemical, isotope-aided, immunological, genetic, and molecular biology-based approaches for a deeper understanding of the dynamics of rhizosphere populations involving mycorrhizal fungi. These innovative methods appear to give new insight to the manipulation of mycorrhizal fungi as a biotechnological tool to improve plant growth and health.

Introduction

Microbial populations in soil concentrate around plant roots stimulated by root exudates, sloughed-off cells and tissue fragments, cell lysates, secretions, etc., supplied by the plant. This region is defined as the rhizosphere (Bowen, 1980; Foster & Bowen, 1982; Lynch, 1983; Curl & Truelove, 1986). This increase in microbial activity, the "rhizosphere effect," in turn, affects the plant because the stimulated microorganisms, saprophytes, parasitic symbionts ("pathogens"), and mutualistic symbionts ("symbionts") carry out a range of activities important to plant growth.

The development of the rhizosphere is a dynamic process incorporating physical, chemical, and biological modifications occurring at the root–soil interface. Key factors in rhizosphere ecology are the qualitative and quantitative changes in microbial populations with time, and the effects of those populations on root morphology and physiology (Foster & Bowen, 1982). These effects are based on the production of plant growth–regulating substances and enzymes. These compounds modify the geometry, distribution, and size of the root system, and increase the losses of carbon compounds from the root (Bowen, 1980). In addition, these substances are known to influence greatly mycorrhizae development and functioning (Barea, 1986).

Another important consequence of the increase in microbial activity in the rhizosphere derives from the capacity of most microorganisms to alter plant nutrient availability. The increased nutrients are the result of changes induced in the physicochemical properties of the microhabitats, together with biochemical reactions undertaken by the rhizosphere microbes.

In the context of events occurring at the root–soil interface, microbe–microbe interactions are crucial to understanding the dynamic processes characteristic of rhizosphere establishment and maintenance (Newman, 1978; Suslow, 1982). Particularly, interactions taking place in the soil of the rhizosphere influence the microbial colonization of the root surface, the rhizoplane, and, ultimately, root infection by pathogens and mutualists. As will be discussed in this chapter, mycorrhizal fungi, as soil inhabitants, are immersed into the domain of these microbial interactions (Barea & Azcón-Aguilar, 1982). Such interactions between mycorrhizal fungi and other soil microorganisms are critical because they regulate mycorrhizal formation and function, and, conversely, because the mycorrhizal status of a plant can affect the microbial populations in its rhizosphere. These facets must be taken into consideration when trying to manage rhizosphere microorganisms, as, for example, the biological control of plant pathogens, to improve plant nutrition, or to establish a microbial population on the root of axenically propagated plants, current topics in the plant soil ecosystem biotechnology (Linderman, 1986).

Accordingly, the purpose of this review is to summarize the main conceptual principles in microbial interactions affecting mycorrhizal fungi, with special emphasis in the following topics: (1) mycorrhizal fungi as components of microbiota, (2) the role of soil microbiota in mycorrhizal formation, (3) the influence of mycorrhizal establishment on rhizosphere populations (introducing the concept of mycorrhizosphere), (4) the impact on plant growth and nutrition of the interactions between mycorrhizal and microorganisms involved in nutrient cycling and other related processes, and (5) the interactions between mycorrhizal fungi and plant pathogens. Finally, and because of the interest in the management of rhizosphere populations, the available information on this topic will be examined from the point of view of the mycorrhizal fungi–other microbiota interactions.

Mycorrhizal Fungi as Components of Soil Microbiota

As stated by Harley and Smith (1983), "mycorrhizas are active, living components of the soil population having some properties like those of roots and some like those of microorganisms." Both aspects will be considered in this chapter.

First, as is well known, mycorrhizal fungi are usually present in soil, in most ecosystems. Because of the difficulties in culturing them axenically by conventional techniques used in soil microbiology, they were ignored as components of soil populations until recently. Thus, information on the role played by mycorrhizal fungi within the framework of the interactions of soil microbiota is still scant. However, several approaches have been, or are being, made toward a better understanding of these aspects of rhizosphere biology.

The activity and population dynamics of microorganisms in soil, most of them heterotrophs, depend on their saprophytic abilities. This also determines the competitive interactions at the root–soil interface by the limited carbon resources released by plants. Since mycorrhizal fungi have low or negligible saprophytic ability, they do not compete strongly with other microorganisms for nutrient resources during the first stages of propagule germination and growth. Conversely, at these stages they are subject to microbial antagonism as well as beneficial reactions to some soil microorganisms (Barea & Azcón-Aguilar, 1982; Bagyaraj, 1984). When the hyphae expand after the mycorrhizae are well established, the intraradical phase of the fungus is protected against microbial antagonism, but the extramatrical mycelia are still interacting with the microbiota of soil. Some hyphae are attacked and lysed by soil organisms, inducing a decrease in mycorrhizal effectivity.

In general, germination of resting propagules of the mycorrhizal fungi is affected by soil fungistasis, which has, at least partially, a microbial origin (Stotzky, 1972). In the case of ectomycorrhizal fungi, early studies (Davey, 1971; Slankis, 1974; Marx & Krupa, 1978) showed that the survival and development of propagules can be affected by soil microorganisms. However, few studies directly demonstrate how soil fungistasis affects these fungi and the ecological implications of the phenomenon. Theodorou and Bowen (1973) found that only a small percentage (about 10%) of ectomycorrhizal fungal spores inoculated into soil were able to germinate. Conversely, microorganisms such as *Rhodotorula* and *Ceratocystis* stimulated spore germination in soil (Fries, 1978, 1979).

Vesicular-arbuscular mycorrhizal (VAM) fungi are considered physiologically obligate symbionts because of a failure to grow them on synthetic media. However, their spores do have the appropriate genetic information and biochemical machinery needed for germination (MacDonald & Lewis, 1978; Hepper, 1979, 1984; Beilby & Kidby, 1982; Siqueira, 1987; Azcón-Aguilar et al., 1991). In fact, they are able to germinate in the absence of any mineral or organic addition, as for example, in water–agar (Hepper & Smith, 1976; Azcón-Aguilar et al., 1986a; Azcón, 1987). However, a number of factors and conditions are known to change the germination rate and the final proportion of germinated spores (Daniels & Trappe, 1980; Siqueira, 1987), including the presence of some microorganisms (Azcón-Aguilar et al., 1986a).

The few reports on VAM fungal spore germination in soil indicate that only about 50% of the inoculated spores readily germinate (Powell, 1976; Daniels & Duff, 1978; Sylvia & Schenck, 1983; Sanders & Sheikh, 1983). Although these papers did not implicate microbial fungistasis against VAM fungi, the microbial component is common in all cases of "general fungistasis" (Stotzky, 1972). In fact, mycostasis completely prevented the germination of quiescent spores of VAM fungi including species of *Glomus, Gigaspora,* and *Acaulospora* in natural soils (Tommerup, 1985), which was associated with changes in the microbial activity in the suppressive soil. In a recent paper by Wilson et al. (1989), spore germination was significantly suppressed in nonsterile soil, which was attributed to the presence of microorganisms. This inhibition was overcome by addition of P to the nonsterile soil (Wilson et al., 1989; Hetrick & Wilson, 1989), which led the authors to suggest that microbial competition for P may induce fungistasis toward VAM fungi.

There are also some reports describing the colonization of spores of VAM fungi by microorganisms. In many cases, these organisms act as parasites (Daniels & Menge, 1980; Ross & Daniels, 1982; Krishna et al., 1982; Siqueira et al., 1984). Some actinomycetes, particularly chitin decomposers, appear to be associated with spores and act as antagonists on VAM fungi in soils (Bagyaraj, 1984; Ames et al., 1989). Bagyaraj (1984) stated that this interaction must be referred to as "parasitism" and not as "hyperparasitism," since VAM fungi are mutualistic symbionts and not parasites. Absence of any antagonism against VAM fungi by other actinomycete isolates has also been reported (Ross & Daniels, 1982) and Mugnier and Mosse (1987) found that one actinomycete enhanced spore germination.

There is abundant information supporting a beneficial influence of microorganisms on VAM fungal spore germination. Studies using soil extracts were initiated by Mosse (1959) who found that spores showing poor germination on water–agar, could be stimulated to germinate in a nonsterile soil–water–agar. She suggested that a water-soluble, heat-labile, dialysable product of microbial origin was responsible for the enhanced germination effect. Mejstrik (1965) also reported that soil microorganisms enhanced spore germination. Subsequently, Daniels and Graham (1976) reported that a dialysate from an untreated soil allowed ready germination of *G. mosseae* spores whereas the dialysate obtained from the autoclaved soil inhibited the process.

Differing isolates appear to have differing effects on VAM fungal spore germination. Azcón-Aguilar et al. (1986b), using an autoclaved soil extract, showed that neither axenic control spores nor spores placed on plates inoculated with an unidentified free-living fungal isolate (F-4) were able to germinate. But the other isolate (F-3) promoted extensive germination. Interestingly, spores that remained ungerminated for 50 days in control or F-4 plates and were then inoculated with F-3 germinated readily. Díaz-Rodriguez et al. (1986) found that autoclaving reduced VAM fungal spores germination.

These effects of microorganisms on VAM fungal spores in soil or soil extracts were aimed at elucidating the possible mechanisms involved. Time-course experiments by Azcón-Aguilar et al. (1986a) and Azcón (1987) in water–agar indicated that although 100% of *Glomus mosseae* spores germinate, some soil microorganisms hastened the germination rate. Mayo et al. (1986) found that bacteria isolated from the spore surface stimulated germination within 14 days of incubation.

In another experiment (Mugnier & Mosse, 1987), no *G. mosseae* spores germinated spontaneously on water–agar, but many germinated when coinoculated with an actinomycete (*Streptomyces orientalis*). Additionally, when *G. mosseae* and the *Streptomyces* were grown in separate compartments of the plate the stimulatory effect persisted, suggesting that the effect was mediated by volatile products. The stimulatory effect on germination found by Azcón-Aguilar et al. (1986a) during the first 2 weeks of incubation of the VAM fungal spores with the free-living fungi were achieved without any contact between the two types of microorganisms. This also supports the involvement of a water-soluble, diffusible "stimulatory substance" or a volatile product of microbial origin.

In addition to the stimulatory effect on germination, hyphae growing from VAM fungal spores were longer, more branched, and formed more vegetative spores in presence of free-living microorganisms than those mycelia arising from axenically germinated spores (Azcón-Aguilar et al., 1986a; Mayo et al., 1986). This effect was noted when the microorganisms were growing at some distance from *G. mosseae* or when cell-free preparations from bacterial cultures were used (Azcón, 1987, 1989). This suggests that these bacteria produce diffusable substances that are able to stimulate fungal growth.

Interestingly, Calvet et al. (1988) found a stimulation of the growth of the *G. mosseae* mycelium by two isolates of *Trichoderma aureoviride* that were able to antagonize *Fusarium* and *Verticillium* species. *Trichoderma* is frequently described as an efficient antagonist of soil pathogens and it is used for the biological control of root diseases. It would be worth investigating the mechanisms regulating antagonism versus synergism of *Trichoderma* toward pathogenic versus mycorrhizal fungi, and studying the potential of using such fungi in combination with mycorrhizal fungi to improve plant growth and to suppress root diseases.

Different arguments can be used to explain the reported stimulatory effects of soil microorganisms on germination and growth of VAM fungal spores. These include the following:

1. Detoxification of the medium. Microorganisms can act by removing inhibitors of mycorrhizal fungi from the medium. These inhibitors may be either naturally present in soil such as heavy metals (Mn and Zn, for example, inhibit spore germination, Hepper & Smith, 1976) or produced during soil or soil extract sterilization (e.g., autoclaving or γ-ray irradiation may induce soil toxicity).

2. Utilization of self-inhibitors of VAM fungi. These compounds may be formed by the fungi and limit their own growth. This action was proposed by Watrud et al. (1978). Their removal from the medium by other microorganisms would improve VAM fungal growth.

3. Production of stimulatory compounds. Water-soluble, diffusible substances or volatile products (or probably both) could be involved. Biologically active substances such as amino acids, plant hormones, vitamins, and other organic compounds can be produced by soil microorganisms (Lynch, 1976, 1983) and are known to stimulate the growth rates of VAM fungi (Hepper, 1979, 1983; Siqueira et al., 1982; Hepper & Jakobsen, 1983; Barea, 1986; Azcón, 1987). Others including St. John et al. (1983), Mugnier and Mosse (1987), and our own experimental evidence support the participation of volatile substances (see Koske & Gemma, Chapter 1).

As previously indicated, the stimulatory effect on germination and growth has been described for fungi and bacteria, including actinomycetes. It could be considered a generalized influence of soil microbiota. However, plant cell cultures from embryo and leaf or stem tissues also stimulated mycelia development of VAM fungal spores grown axenically (Carr et al., 1985). Particularly, leaf cells from alfalfa stimulated mycelial expansion by mean of volatile substances. Similar results were obtained with cell suspensions from calli obtained from several plant materials (Paula, 1988). These results express the universality of the stimulatory effect beyond the microbial world. Recently, Becard and Piché (1989), studying the effects of root metabolites on hyphal growth from spores of a VAM fungus, found that the addition of root exudates and increased CO_2 concentration stimulated the axenic development of *Gigaspora margarita*. The volatile-based microbial effect could act by altering the CO_2/O_2 rate in the microhabitats where VAM fungi develop.

This last observation supports a general effect, possibly regulated simply via increases in the CO_2 concentration resulting from respiration. This would also explain the diversity of results described in the literature about the subject. Although small increases in CO_2 concentration (0.5%) would be a stimulant, higher increases could have the opposite effect (Le Tacon et al., 1983). Therefore, it could be inferred that a certain level of microbial activity would be positive for the growth of the VAM fungi but higher levels would be detrimental. This agrees with results in our laboratory using compartmental plates where the same microorganism had a beneficial influence when its growth was limited by a poor growing medium and a detrimental influence when grown quickly on rich medium. Thus, factors such as soil nutrient content, organic matter, microbial population, and diversity will modulate the effect of soil microbiota on VAM fungi.

In any case, and whatever the mechanism involved, a distinction should be made between the soil microbiota and VAM fungal spore-associated microorganisms. Tommerup (1985) found that bacteria isolated from the spore surface were able to promote germination while the soil from which they were isolated exhibited a strong microbially mediated fungistasis against those VAM fungi. A similar situation was found by Wilson et al. (1989); the suppression observed in nonsterile soils was probably related to microorganisms other than those associated with VAM fungal spores.

An improvement of the "saprophytic" development of *G. mosseae* in soil, as induced by soil microorganisms, was postulated by Azcón-Aguilar and Barea (1985). The experimental design and treatments they used were as follows: (1) a number of surface sterilized spores of *G. mosseae* were inoculated in tubes containing sterile soil-sand and left 2 weeks for germination and initial mycelial development; (2) a filtrate of soil containing microorganisms, but free from mycorrhizal propagules, was added to half of the tubes and allowed to interact with the *G. mosseae* spores for the 2-week incubation period. To detect a possible effect of the soil microorganisms on the initial development of *G. mosseae*, *Medicago sativa* sterile seedlings were planted in the tubes, after the incubation period, to serve as indicators for infectivity. After a further 2-week period, data on the incidence of mycorrhizal infection and the number of "entry points" of the fungus on the roots strongly suggest a direct effect of the soil microbiota on the saprophytic stages of the VAM fungus.

Effect of Soil Microorganisms on Mycorrhizal Formation

Ectomycorrhizae

Early review articles (Davey, 1971; Bowen & Theodorou, 1973; Rambelli, 1973; Slankis, 1973, 1974; Marx & Krupa, 1978) and experimental work (Tribunskaya, 1955; Levinson, 1957; Theodorou, 1967) showed that the rate of spread of ectomycorrhizal fungi on the root system of an appropriate host was affected by mixed populations of soil microorganisms. In particular, rhizosphere microorganisms such as *Azotobacter* spp., *Trichoderma* spp., or *Pseudomonas* spp. stimulated ectomycorrhizal formation (Davey, 1971). Plant hormones or other growth substances, such as auxins or thiamin, which are produced by rhizosphere microorganisms, were hypothesized to be the cause of the microbial effects on ectomycorrhizal formation (Slankis, 1974).

The systematic work by Bowen and Theodorou (1979) illustrates a range of situations showing either beneficial, neutral, or detrimental effects of bacterial treatments on the colonization of the roots of *Pinus radiata* by the ectomycorrhizal fungi *Rhizopogon luteolus*, *Suillus luteus*, *Thelephthora terrestris*, and *Corticum bicolor*. A strain of *Pseudomonas fluorescens*, for example, was a strong antagonist, whereas a *Bacillus* spp. was stimulant. Depressive effects were associated with competition for nutrients or antibiosis. Undoubtedly the result depends on the origin of the microorganism and the mycorrhizal fungus involved (Garbaye & Bowen, 1987) and it is clear that the management of specific microbes in dual inoculation with mycorrhizal fungi can improve ectomycorrhizal formation (Linderberg, 1986; Linderman, 1988) . Accordingly, McAfee and Fortin (1988) showed that the presence of certain soil microorganisms enhanced ectomycorrhizal formation by conifer seedlings, which benefitted the quality of the seedlings for reforestation purposes. Recent publications (Richter et al., 1989; Garbaye & Bowen, 1989) discuss the potential use of "helper" microorganisms for enhancing the efficiency of ectomycorrhizal fungi inoculation in nurseries. These "helper" microorganisms refer to those isolated from the mantle of an ectomycorrhizae, which proved to stimulate mycelial growth of ectomycorrhizal fungi or enhance ectomycorrhizal formation.

Vesicular-Arbuscular Mycorrhizae (VAM)

To parallel the prior section, it is appropriate to consider "endomycorrhizae." However, as far as we know, there is little published information on interactions between soil microorganisms and endomycorrhizal types other than vesicular-arbuscular. Bacteria isolated from mycorrhizal tissues of terrestrial orchids do appear to affect the germination of the orchid seeds (Wilkinson et al., 1989) but, as this is the only paper on non-VA endomycorrhizae, the rest of this discussion will focus on VAM.

It appears that the formation of the first entry point is a critical stage in VAM development (Mosse & Hepper, 1975) and that changes in the root exudation patterns and in the hormonal balance of the plant are involved in the establishment and development of the symbiosis (Barea, 1986). Soil microorganisms are able to produce compounds that increase root cell permeability (Barber & Martin, 1976), thereby enhancing the rates of root exudation (Bowen, 1980). This enhanced exudation, in turn, would stimulate VAM fungal mycelia in the rhizosphere and might facilitate root penetration by the fungus (Mosse, 1962; Azcón-Aguilar et al., 1980; Azcón-Aguilar & Barea, 1985). This suggests a plant-mediated microbial effect on VAM formation (see Chapters 1 and 2). Plant hormones, as produced by soil microorganisms, are also known to stimulate VAM development (Azcón et al., 1978) and these substances could be involved in the formation of the symbiosis. This topic will be considered further in relation to the interaction with N_2-fixing and P-mobilizing microorganisms.

In several situations, it has been shown that soil microorganisms may benefit the development of the symbiosis. Sutton and Sheppard (1976), for example, found that the addition of a nonsterile soil filtrate increased the development of the extraradical mycelia in VA mycorrhizae. In addition, higher levels of mycorrhizal infection in the presence of actinomycetes (Ames, 1989), spore-associated microbes (Wilson et al., 1989), or phosphate-solubilizing bacteria (Azcón-Aguilar et al., 1986a) have been described.

Effects of Mycorrhizae on Microbial Rhizosphere Population: The Mycorrhizosphere

Mycorrhiza formation induces changes in the mineral composition and physiology of plant tissues, and, as a consequence, in root exudation. This fact, together with the development of the fungal mycelium around roots, will produce some chemical and physical modifications of the environment surrounding the mycorrhizae. These changes affect the microbial populations. Therefore, Oswall and Ferchau (1968) coined the term "mycorrhizosphere" to describe the zone of influence of the mycorrhiza, and consequently the so-called "mycorrhizosphere effect" (Linderman, 1988). For both VAM and ectomycorrhizae, reports are available to show selective changes, which affect, both qualitatively and quantitatively, the microbial groups developing in the rhizosphere (mycorrhizosphere + mycosphere) ecosystem. Thus microbial interactions in mycorrhizosphere are of increasing interest (Rambelli, 1973; Linderman, 1986; 1988).

Actually, mycorrhizae usually concentrate nutrients in the rhizosphere. This is the result of the transport and storage by mycorrhizal fungi when the symbiosis is active and the release of materials when the fungus senesces (Harley & Smith, 1983). This process results in a dynamic nutrient turnover. Perhaps the effect with the greatest influence is that derived from changes in the root exudation patterns in mycorrhizal plants (Ratnayake et al., 1978; Graham et al., 1981; Schwab et al., 1983). These three processes, altered exudation, mineral transport and concentration, and fungal death, contribute both carbon and elemental nutrients in high concentrations (Finlay & Söderström, Chapter 5).

Perhaps it was the paper by Katznelson et al. (1962) that first described the different microbial types developing near ectomycorrhizal and nonmycorrhizal roots. Subsequently, several other reports supported the same conclusions (Neal et al., 1964, 1968; Oswald & Ferchau, 1968; Spedaliere de Nuñez, 1980). The review by Rambelli (1973) and the experimental paper by Li and Castellano (1985) indicated that N_2-fixing free-living bacteria were associated with mycelial strands in an ectomycorrhizal fungus, suggesting that this interaction could enhance the N nutrition of the host plant. The ability of bacteria isolated from the mycorrhizosphere of pine to produce plant hormones (Strzelczyk & Pokojska-Burdziej, 1984) can also influence the biology of the root region. Recent reviews point out the interest in basic to applied research on the ectomycorrhizosphere (Perry et al., 1987; Linderman, 1988), with emphasis on the management of the microbial populations. In particular, the management of mycorrhizal and associated microbiota concerning ecosystem structure and processes seems of interest. The most appreciated effects of rhizosphere bacterial populations are (1) those based on the production of biologically active compounds (hormones, chelators, siderophores, enzymes, etc.), (2) those having implications in the nutrient cycling and on the organic matter turnover, and (3) those able to influence soil structure (Perry et al., 1987). The management strategy proposed by Linderman (1988) emphasizes the identification of the best strains of beneficial microbes and the selection of these for compatibility and combined efficacy regarding a particular plant–soil system. The role of volatiles from ectomycorrhizal seedlings as selectively affecting microbial populations in the ectomycorrhizosphere has recently been discussed (Schisler & Linderman, 1989 a,b).

Concerning VA mycorrhizae, early observations (Clough & Sutton, 1978; Tisdall & Oades, 1979) indicated that the surface of external mycelia was colonized by bacteria developing in the mucilaginous material coating the hyphae. Scanning electron microscopy confirmed this observation (Vancura et al., 1988). Barea et al. (1973) first demonstrated a beneficial effect of VAM on the establishment of an introduced *Azotobacter* spp. Later, Bagyaraj and Menge (1978) and Brown and Carr (1984) found that *Azotobacter* maintain high numbers in the VA mycorrhizosphere compared with the rhizosphere of the nonmycorrhizal controls. A similar situation was also found for phosphate-solubilizing bacteria (Azcón et al., 1976; Raj et al., 1981).

Once established, VA mycorrhizae can decrease exudation rates (Graham et al., 1981). This could be unfavorable to the establishment of the bacteria, as root exudation provides an important source of substrates (Bowen, 1980).

Recent systematic experiments have been carried out aimed at studying the establishment of inoculated, or naturally occurring microorganisms in the rhizosphere, as affected by VA mycorrhiza. In this respect, Ames et al. (1984) found that several species of microorganisms, but not all of them, increase in the VA mycorrhizosphere. Therefore, the species composition can be altered by the influence of VA mycorrhizae (Ames et al., 1984).

This was also investigated for the naturally occurring microbiota by Meyer and Linderman (1986). They concluded that *G. mosseae* altered the microbial equilibrium both in the rhizosphere and in the rhizoplane. Importantly, they found changes in specific functional groups of microorganisms, rather than in the total numbers of the population. Secilia and Bagyaraj (1987) reported that different VAM fungi (*Glomus, Gigaspora, Acaulospora,* and *Sclerocystis*) exerted differing effects on selective microbial groups dependent on the VAM fungal species involved. As a consequence Ames et al. (1987) suggested that to get comparable control treatments to a VAM fungal inoculum, it is important to supplement control plants with filtrated washings obtained from the VAM fungal inoculum to be applied.

Interactions between Mycorrhiza and Soil Microorganisms Involved in Plant Growth and Nutrition

A number of soil microorganisms are known to carry out activities that alter the availability of mineral nutrients. This suggests the possibility that synergistic interactions with mycorrhizal fungi may improve plant growth. Examples include phosphate-solubilizing microorganisms, nitrogen fixers, plant growth-promoting rhizobacteria (PGPR), and siderophore-producing microorganisms.

Perhaps the most important types of interaction involving mycorrhizal fungi and beneficial microorganisms are those involving bacteria able to induce nodules on roots of higher plants, to fix atmospheric N_2. The bacterial genera and plants involved are (1) *Rhizobium,* with legumes, (2) *Frankia* (actinomycetes), with several angiosperm families, and (3) *Nostoc* and *Anabaena* (cyanobacteria), with Cycadaceae (gymnosperms). The consequences of the interaction among mycorrhizal fungi, nodulating bacteria, and host plants influence many industries such as agriculture (including production of food, forage, green manure, etc.), forestry, soil reclamation, and erosion control.

Interactions between VAM Fungi and Rhizobium

A century ago, Janse (1896) reported the coexistence of certain bacteria and fungi colonizing the root system of legume plants. The fungi were later described as VAM fungi (Jones, 1924), and Asai (1944) stated that root nodulation by the bacteria can be greatly dependent on the formation of mycorrhizae. Today both the widespread occurrence of VA mycorrhizae in nodulated legumes and the mycorrhizal role in improving nodulation and *Rhizobium* activity within the nodules are universally recognized (Crush, 1974; Daft & El-Giahmi, 1974, 1976; Smith, 1980; Barea & Azcón-Aguilar, 1983; Hayman, 1986; Barea et al., 1987; Barea et al., 1989). These papers include studies completed using growth, N content, acetylene reduction, and even ^{15}N.

The plant species in the subfamilies Papilionoideae and Mimosoideae have been found to form VA mycorrhizae and rhizobial nodules. The third subfamily, Cesalpinoideae, rarely forms nodules. The two large groups, Amherstieae and Detariae, commonly are not nodulated and are ectomycorrhizal. This is interesting because ectomycorrhizae are able to mineralize organic nitrogen and to transfer it to the plant, thereby acting as an alternative to nodules to supply plants with N (Malloch et al., 1980). Therefore, it seems that VAM are the main mycorrhizal type involved with nodulating bacteria.

Because of the relatively high P demand of the N_2 fixation process, it is obvious that a main cause of the beneficial effect of VAM on the symbiotic role of *Rhizobium* must be the P supplied by the fungus. However, nutrients other than P, such as Zn, Cu, Mo, and Ca, can affect the infectivity or the symbiotic effectiveness of *Rhizobium,* therefore, the enhanced uptake of these elements by the VA mycorrhizae could also be involved in the VAM fungus $\times$ *Rhizobium* interactions. Conversely, N supply by N_2 fixation, as carried out by rhizobial activity, could be critical to maintain a balanced physiological status in the plant, which is important for mycorrhiza formation and functioning (Hayman, 1983). In addition, there is a high requirement for nitrogen by the VAM fungi to synthesize chitin, the main constituent of its walls. Therefore, nodulation and VAM formation seem to be interacting processes. In a recent report by Duc et al. (1989), the expression of the myc^- character (nonmycorrhizal plant mutants) in the tested legumes was associated with that of the nod^- character (nonnodulating plants), suggesting that the establishment of both types of symbiosis depends on the expression of some common gene(s) through which the plant controls early steps in the infection processes.

Several approaches have been made to elucidate the physiological and biochemical basis of VAM fungal × *Rhizobium* interactions. Some have aimed at ascertaining whether VA mycorrhizae enhance *Rhizobium* activity through a generalized stimulation of host nutrition, or whether a more localized effect, at the root or nodule level, occurs. Additionally, interactions not directly mediated via the improvement in nutrition have been argued. These can take place either at the precolonization stages, when both microorganisms interact as rhizosphere inhabitants, or during the development of the tripartite symbiosis. In this context, the influence of VAM on plant water relations, hormonal balance, photosynthesis rate, and source–sink relationships have been hypothesized as underlying mechanisms that can account for the interactions between VA mycorrhiza and nodule activity.

In relation to the preinfection stages, soil microorganisms can stimulate the independent growth of mycorrhizal fungi, as discussed before, or help the fungus at the early events of root colonization. *Rhizobium* spp. are particularly able to improve the mycelial growth from *G. mosseae* spores in axenic conditions (Gonzalez, 1988). Extracellular polysaccharides (EPS) from *Rhizobium meliloti* enhanced the formation of VAM by *Medicago sativa* (Azcón-Aguilar et al., 1980). This can be explained by the fact that *Rhizobium* EPS can increase root exudation in their specific host legume (Olivares et al., 1977) and this can affect the development of the fungus in the rhizosphere (Barea, 1986). Cell-free supernatants of *Rhizobium* cultures, containing plant hormones, were able to increase the colonization levels of *M. sativa* by *G. mosseae* to an extent similar to that obtained by a mixture of auxins, giberellins, and cytokinins applied in doses equivalent to those of the supernatants (Azcón et al., 1978; Azcón-Aguilar & Barea, 1978). These experiments, however, did not distinguish whether the effect was directed at the fungus at the precolonization stages, or whether it was mediated by the plant. However, as previously mentioned, VAM fungi can change, both quantitatively and qualitatively, the microbial composition in the rhizosphere (Mosse, 1977; Ames et al., 1984; Meyer & Linderman, 1986; Barea, 1986) thereby regulating the development of *Rhizobium* and other specific microbes. Systematic studies on this are scarce, but Ames and Bethlenfalvay (1987) did not support a VAM fungal effect on the competitive interactions between rhizobial populations for nodulation sites.

In natural conditions VAM fungi and *Rhizobium* colonize the root almost simultaneously, but the two endophytes do not seem to compete for infection sites (Smith & Bowen, 1979). This is important because previous inoculation with one of the endophytes can depress the development of the other (Bethlenfalvay et al., 1985). This has been mainly attributed to competition for carbohydrates when host photosynthesis is limited. When this occurs, VAM fungi usually show a competitive advantage for carbohydrates over the *Rhizobium* (Bayne et al., 1984;

Bethlenfalvay et al., 1985; Brown & Bethlenfalvay, 1987). However, fungal structures previously established within the root, or of some of their metabolic products, may still interfere with nodulation. Root hair development, for instance, can be affected by VAM formation (Linderman, 1988), which would affect the number of nodulation sites.

In spite of the fragmentary nature of the information suggesting the possibility of some *Rhizobium* × VAM interactions at the preinfection, or early colonization stages, there is no doubt that most of the interactions operate through the plant. Therefore, this could be a case of "indirect mutualism" as claimed by Cluett and Boucher (1983). The nature and mechanisms causing the interactions will now be discussed by analyzing the related literature.

The role of VA mycorrhizal at improving N_2 fixation by the *Rhizobium* legume systems is usually accompanied by an increase in nodule biomass, leghemoglobin content, and nitrogenase activity (Kucey & Paul, 1982). The VAM effects on nodulation are closely related with plant growth responses (Gueye et al., 1987; Badr El-Dim & Moawad, 1988; de la Cruz et al., 1988; Louis & Lim, 1988; Kaur & Singh, 1988). Thus, it can be said that these effects occur mainly through host nutrition. In fact, it is well accepted that a minimum level of P in plant tissues is critical for nodulation, suggesting a generalized VAM effect mediated by the common host plant. However, as nodules usually possess two to three times more P than the root on which they are formed (Mosse, 1977), a "special demand" for P to maintain *Rhizobium* activity is necessary. To investigate the possibility that VAM directly supports that extra P demand, a number of time-course experiments have been carried out (Smith & Daft, 1977; Smith et al., 1979; Asimi et al., 1980). These studies all showed that the effect of VAM on *Rhizobium* activity preceded any effect on plant growth. Comparisons of phosphate–response curves between VAM and nonmycorrhizal plants to obtain matched plants (Smith & Daft, 1977) confirmed this conclusion, and, in addition, when these curves became asymptotic in both the mycorrhizal and the nonmycorrhizal plants, the nitrogenase activity was still higher in VAM roots (Waidyanatha et al., 1979).

The study by Asimi et al. (1980) also provides indirect evidence to support a differential P demand for plant growth and nodule functioning, based on the different P dependencies of the processes. They found that increased P additions initially equalled the VAM effects on plant growth, whereas progressively higher doses were needed to eliminate the effects of VAM on nodulation, and even higher for those on nitrogenase activity.

In all, the above mentioned papers indicate the importance of adequate plant P induced by VAM for *Rhizobium* function. The nodules seem to call first for P because of their higher P requirement. Despite the observation that the VAM fungi do not usually invade the nodular tissues in legumes, the proximity of VAM arbuscules to *Rhizobium* bacteroids results in a steady P supply allowing N_2 fixation (Smith et al., 1979).

In addition to the P-mediated effect, several studies have demonstrated the existence of non-P-mediated effects involved in the VAM × *Rhizobium* interactions. For example, it is well known that the reproduction of *Rhizobium* in the rhizosphere and its symbiotic functioning have a number of other mineral constraints (O'Hara et al., 1988). Munns and Mosse (1980) and Hayman (1986) noted that VAM fungi can help the legume and *Rhizobium* to obtain some of these nutrients, mainly Cu, Zn, or Ca. Rai (1988) found that VAM increased Fe supply to the *Rhizobium* legume symbiosis. It is obvious that a role of VAM, complementary to that on P nutrition, may be involved in the mycorrhizal effect on *Rhizobium* activity.

Other nonnutritional factors also may be involved. Ames and Bethlenfalvay (1987) used a split root system with VAM fungi in one of the sides, and increased P in the other. They found a localized, nonsystemic, non-P-mediated effect of VAM on nodule function. Since VAM root systems usually have higher respiration rates (Pang & Paul, 1980), the source–sink relations can be altered and such differences in the C allocation pattern may explain the greater nodule activity. This stresses the fact that a number of microsymbiont interactions are mediated by the plant, and are therefore complex and dependent on several factors affecting plant metabolism. This can explain the varying results found in assays where the experimental conditions were also different. For example, Harris et al. (1985) found that VAM stimulated the early development of nodules, enhancing nodule dry weight; however, the nitrogenase activity remained unaffected. In another situation, Brown et al. (1988) found that VAM increased nodule number and dry weight, but reduced the P-use efficiency of nitrogenase activity. Also of importance is that the photosynthetic P-use efficiency in VAM plants and the N-use efficiency in *Rhizobium*-inoculated plants were higher than in those plants with added P or N, respectively (Brown & Bethlenfalvay, 1988; Brown et al., 1988). Since soil moisture level affects nodule formation and N_2 fixation (Sprent, 1986), and VAM are known to help plant development under drought stress (Cooper, 1984; Nelsen, 1986), it follows that VAM can improve the development and activity of the *Rhizobium*–legume symbiosis under those conditions (Hardie & Leyton, 1981; Buss & Ellis, 1985; Hardie, 1985; Kwapata & Hall, 1985; Dakessian et al., 1986; Peña et al., 1988; Azcón et al., 1988). In the same way, VAM can help *Rhizobium* legume symbiosis under other stress situations (Mosse et al., 1981). For example, nodulation and N_2 fixation were improved by VAM under salinity conditions, as showed by using ^{15}N (El-Atrach et al., 1989).

Root Nodulating Actinomycetes and Cyanobacteria

Mycorrhizal fungi and *Frankia* spp., the actinomycete forming N_2-fixing nodules, coexist in the root system of a number of nonlegume plant species of great ecological value. The actinomycete-nodulated (actinorrhizal) plants are usually mycorrhizal in nature (Trappe, 1986; Rose, 1980; Rose & Trappe, 1980) and recent reviews by Daft et al. (1985) and Gardner (1986) discussed the significance of the interactions between mycorrhizal fungi and *Frankia* in actinorrhizal plants, commonly shrubs and small trees.

Both ectomycorrhizae and VA mycorrhizae have been found to be involved, but VAM is the most common, existing either exclusively or coexisting with ectomycorrhizae (Daft et al., 1985). Essentially, the interactions are rather similar to those with *Rhizobium* but there are some special features in the actinorrhiza–mycorrhiza–plant tripartite symbiosis. Hyphae of VAM fungi colonize nodular tissues, allowing a closer "communication" than with rhizobial nodules. However, the VAM fungal mycelium was not found in the zone of the actinorrhizal nodules containing the endophyte. An ectomycorrhizal sheath around young nodules has also been described (Godbout & Fortin, 1983). In general, it appears that VAM fungi and *Frankia* colonize first the root system simultaneously (Gardner, 1986), and then further ectomycorrhizal colonization can occur. This timing sequence is important because ectomycorrhizae can cause loss of root hairs (Linderman, 1988), critical to *Frankia* penetration. Alternatively, in a recent paper by Chatarpaul et al. (1989), the root system of *Alnus incana* was infected at the same time by ecto- and VA mycorrhiza, and the growth performance of the plant was highest when the seedlings were inoculated simultaneously with *Frankia*, ecto-, and VA mycorrhizal fungi. The report by Godbout and Fortin (1983) showed a normal ectomycorrhizal colonization pattern (mantle+ Hartig net) over a nodule in which the *Frankia* endophyte was located in the basal cells of the nodule lobes. Once formed, the ectomycorrhiza restricted further spread of the actinorrhiza.

Recent papers (Rose & Youngberg, 1981; Gauthier et al., 1983; Gardner et al., 1984; Daft et al., 1985; Gardner, 1986; Russo, 1989) show an improvement of nodulation and N_2 fixation induced by mycorrhizae and an increased mycorrhizal development in well-nodulated plants. As in legumes, the implications of P- or N-mediated mechanisms and hormonal phenomena have been argued. Nevertheless, more physiological and biochemical studies are needed to gain information on the endophyte interaction.

Species belonging to the gymmosperm order Cycadales are nodulated by the cyanobacteria *Nostoc* or *Anabaena*, endophytes coexisting with VAM fungi (Trappe, 1986). Trappe (1986) also pointed out that the only nonlegume species known to be able to be nodulated by *Rhizobium*, i.e., *Parasponia* sp. (Ulmaceae), also appear to form VA mycorrhiza.

Nitrogen-Fixing Free-Living Microorganisms

Azotobacter appears to enhance the mycorrhizal colonization and increase plant growth for both ectomycorrhizae (Tribunskaya, 1955) and VA mycorrhizae (Bagyaraj & Menge, 1978; Azcón et al., 1978; Manjunath et al., 1981; Brown & Carr, 1984; Mohandas, 1987; Ho, 1988). However, the causes of this synergism seem to be associated with hormone production by the bacteria rather than any direct effect derived from N_2 fixation (Mosse et al., 1981; Bagyaraj, 1984).

Interactions between VAM fungi and *Azospirillum* show a similar trend. Several studies (Barea et al., 1983b; Subba Rao et al., 1985a,b; Pacovsky et al., 1985; Pacovski, 1988) found increases in the colonization levels and, in cases, synergistic or additive effects of dual inoculation on plant growth and nutrition. One hypothesis is that the responses are mediated by the production of plant hormones by the bacteria (Harari et al., 1988).

It is obvious that the effect of these bacteria on the morphology, geometry, and physiology of the root system could be exploited in the future to enhance mycorrhizal formation and response. The use of ^{15}N methods would be useful to ascertain the levels of N_2 fixation by the bacteria on the interaction with mycorrhizal fungi. Other future research directions will be discussed later.

Phosphate-Solubilizing Microorganisms

Many soil microorganisms can mobilize phosphate ions from sparingly soluble inorganic and organic P sources, as shown *in vitro*. However, the effectiveness of this process in soil is unclear because of difficulties inherent in the scarcity of available energy sources in nonrhizospheric microhabitats, the problem of amensalism in soil, and the translocation of the phosphate ions to the root surface following such solubilization (Hayman, 1975; Tinker, 1980). If such microbial phosphate solubilization takes place in discrete microhabitats, the released ions would be taken up by the mycorrhizal hyphae reaching those microenvironments. This would diminish refixation and result in a synergistic microbial interaction. Early work by Barea et al. (1975) and Azcón et al. (1976) suggested this possibility to explain some of the plant responses on the growth and P uptake to a dual inoculation with phosphate-solubilizing microorganisms (PSM) and VAM fungi. Another explanation would be that PSM might be acting by means of the production of plant hormones or plant hormone-induced VAM formation (Azcón et al., 1978), an ability that was substantiated for the tested microorganisms (Barea et al., 1976). Using ^{32}P, Raj et al. (1981) found increased P uptake and dry matter production in plants dually treated with VAM fungi and PSM, but a conclusive P-mediated mechanism was not clearly shown. Related literature on this subject was reviewed by Barea et al. (1983a) and the general conclusion reached was that the "P-solubilizing" mechanism cannot be excluded, but the PSM probably act in the rhizosphere by producing growth regulators.

Recently, Azcón-Aguilar et al. (1986c) studied the interaction between VAM fungi and PSM on the utilization of $^{32}P-^{45}Ca$ tricalcium phosphate by soybean plants. They did not find that PSM improved the fertilizer use by VAM fungi, perhaps because of the high pH and Ca content of the test soil, conditions precluding phosphate solubilization (Khasawneh & Doll, 1978). However, they recorded other reactions of interest. For example, the inoculated bacteria increased the VAM colonization level, the shoot N concentration and content, and the shoot-to-root ratio in plants. It is obvious that the bacterial inoculum behaved as a PGPR, as will be discussed later.

In summary, the action mechanism of PSB and their possibilities in rhizosphere biotechnology warrant further consideration. Assays applying ^{32}P and ^{15}N techniques would be useful to ascertain mechanisms accounting for the above-mentioned effects.

Other Microorganisms

Plant growth-promoting rhizobacteria (PGPR) is the term coined to refer to "root-colonizing bacteria" having the ability to enhance plant growth and, occasionally, a certain capacity to control plant pathogens (Suslow, 1982). The current literature usually makes an artificial division of the bacteria potentially beneficial to be inoculated according to their ability to perform a specific role, i.e., N_2 fixers, P-solubilizers, plant pathogen antagonists, etc., and reserves the term PGPR for those with a less defined role. This situation is accepted here.

Pseudomonas spp. are typical PGPR and there are some published reports on their interactions with VAM fungi. Meyer and Linderman (1986) found that a *Pseudomonas* sp. and *Glomus* spp. mutually enhanced each other's colonization, and achieved additive plant growth enhancement. Azcón (1989) further found that interaction between PGRP and VAM fungi was selective, since the effect was dependent on the bacterial isolate involved. It is commonly accepted that the production of plant hormones is involved in these bacterial effects (Linderman, 1988).

Another mechanism of action of PGPR on plant growth seems to be the production of siderophores (Kloepper et al., 1980). These compounds [microbial iron (III) transport agents] are extracellular low-molecular-weight substances. According to recent information these compounds play an important role in plant growth and disease biocontrol (Neilands & Leong, 1986). The siderophores are produced by most fungi and bacteria, including *Pseudomonas, Rhizobium,* and *Azotobacter* (Neilands & Leong, 1986) and ectomycorrhizal fungi (Szaniszlo et al., 1981). These compounds sequester iron making it unavailable to other soil microorganisms, which are unable to produce the chelator or lack the iron-assimilation system for ferric siderophores. Siderophores can affect plants by chelating Fe, which solubilizes and allows for transport of that Fe to the absorbing surfaces of plant root or mycorrhizae. Some mechanisms have been described to explain how the plant uses the nutrient when it reaches the root surface in a chelated form (Leong, 1986); these can be applied to both roots and mycorrhizae.

We are not aware of any reports of interactions between siderophore-producing PGPR and mycorrhizal fungi, but these are to be expected. Actually, mycorrhizal fungi can be involved in the Fe supply to plants (Rai, 1988) and siderophores can be involved in the transfer of Fe to the zones of mycorrhizal uptake. Recently,

Römheld and Marschner (1986) pointed out a root-based strategy to enhance iron uptake by the plant under conditions of iron deficiency. This is based on the induction of a plasma membrane-bound reductase able to reduce ferric chelates (siderophores), which is associated with a proton extrusion dependent on a plasmalemma-bound ATPase of the root cells.

Interactions between Mycorrhiza and Plant Pathogens

Since mycorrhizae are key components in influencing microbial equilibrium in soil, they can affect microorganisms causing diseases of plant roots. As stated by Marx (1973), both mycorrhizal fungi and pathogens are associated in the same substrate, the fine feeder roots of their hosts. Thus, it was postulated that both types of symbionts, the mutualist mycorrhizal fungus (eusymbiont) and the parasitic microorganisms (dissymbiont), could interact as common components of the rhizosphere soil or when colonizing the root. Evidence overwhelmingly shows that both ecto- and VA mycorrhizae may act in this way. Thus, they could be considered as biological control agents to soil-borne diseases of plants (Harley & Smith, 1983; Bagyaraj, 1984; Perrin, 1985; Tello et al., 1987).

Ectomycorrhizae

Marx (1973, 1975) reviewed early reports on the ectomycorrhizae and proposed the following reasons to justify the protective ability shown by these fungi: (1) competition for nutrients, (2) a mechanical barrier effect against penetration of the pathogen, caused by the sheath encasing the roots, (3) stimulation in the mycorrhizosphere of the microbiota antagonist to the pathogen, and (4) the synthesis of antimicrobial compounds by ectomycorrhizal fungi, or by the root when elicited by the mycorrhizal fungus. A more recent review (Perrin, 1985) emphasized that these general hypotheses are still valid. However, the specific mechanisms of the interactions are still poorly understood. Some efforts have been made to establish the conditions providing the expression of the protective ability of ectomycorrhizae. After Perrin (1985), these conditions refer to the specificity of the protective action, the permissive edaphic factors, and the environmental influences.

Recent studies have attempted to elucidate the mechanisms involved. Sampagni et al. (1986) demonstrated that *Laccaria laccata* protected Norway spruce and Douglas-fir seedlings against *Fusarium oxysporum* simply by enhancing plant growth. Subsequently, Sampangi and Perrin (1986) found that *L. laccata* did not inhibit the growth of *F. oxysporum in vitro*, but cell-free preparations from *L. laccata* cultures did inhibit the germination of *Fusarium* spores. However, the ectomycorrhizal fungi were not effective in controlling the pathogen population in soil or the pathogen's infectivity. Other *Laccaria* isolates stimulated germination and growth. In conclusion, the root protection was apparently mediated by the increased production of inhibitory substances by the host, as elicited by the mycorrhizal fungus. In another instance, Chakravarty and Unestam (1986) assessed the ability of species of *Pisolithus, Laccaria,* and *Hebeloma* to protect *Pinus* seedlings against *Fusarium moniliforme* and *Rhizoctonia solani*. They demonstrated that the ability to induce resistance in the plants was not mediated by antibiosis or by physical protection by the sheath because mycorrhizal fungus and the pathogens did not colonize the same part of the root. They attributed the control to the production of compounds, such as phenols, by the mycorrhizal root. The production of antibiotics by the host plant in response to mycorrhizal colonization is also well documented (Duchesne et al., 1987).

Despite the large data base on the synthesis of antifungal substances by ectomycorrhizal fungi *in vitro* (Marx, 1973), there are few reports documenting this *in situ*. This was investigated by Duchese et al. (1988) in the ectomycorrhiza *Pinus-Paxillus involutus* that increased the plant resistance to *Fusarium oxysporum*. They concluded that the protective effect on root disease could be explained by the biosynthesis of antibiotic compounds by the mycorrhizal fungus stimulated by the root exudates of the host plant.

Later, Duchese et al. (1989) studied the same system using a time-course experiment to assess whether the fungitoxic activity coincides in time with that at which the pathogen is suppressed. They found the suppression occurred shortly after the inoculation of the mycorrhizal fungus and they emphasized that the time of inoculation is critical for disease control. Therefore, biocontrol by mycorrhizae is more effective if the fungus is established prior to the invasion of the pathogen.

In any case, the interactions are obviously affected by plant and ectomycorrhizal fungus. Moreover, it is noteworthy that the protection against the pathogen is influenced by the degree of endogenous resistance of individual plants (Duchese et al., 1989) and by the fungal species involved.

Endomycorrhizae

Considerable attention has recently been directed to the question of whether VAM affect the susceptibility of plants to disease and whether VAM fungi can be managed to control soil-borne pathogens. The early literature (Schonbeck, 1979; Dehne, 1982; Cook & Baker, 1983) indicated that an established VAM can exert protection against pathogens, although the interactions between VAM fungi and phytopathogenic microorganisms are complex and difficult to predict (Harley & Smith, 1983).

As with ectomycorrhizae, prophylactic ability, although characteristic of some VAM fungi, is not shown by all species, cannot be applied for all pathogens, and is not expressed on all the substrates or in all environments (Perrin, 1985). Information on interactions between VAM fungi and phytopatogenic fungi, bacteria, and nematodes can be summarized as follows.

Prior colonization by VAM fungi has been shown to protect plants against pathogenic root-infecting fungi, including both wilt and root-roting pathogens such as *Phytophthora, Gaemannomyces, Fusarium, Thielaviopsis, Pythium, Rhizoctonia, Sclerotium,* and *Verticillum.* However, cases of increases in the severity of the disease have also been reported (Bagyaraj, 1984; Graham, 1986).

Few studies have assessed the interaction of VAM fungi and plant pathogenic bacteria (Bagyaraj, 1984). A recent publication by García-Garrido and Ocampo (1988) found that the VAM fungus *G. mosseae* protected tomato plants against *Erwinia caratovora* and that the colony-forming units of the bacteria were reduced in the rhizosphere of VAM plants. These effects were independent of the P concentration in the plant tissues.

Reviews by Bagyaraj (1984), Graham (1986), and Zambolin (1987) indicate that precolonization with VAM fungi increases plant tolerance to nematodes, especially to sedentary endoparasites such as *Meloidogyne.* VAM fungi do not usually colonize root regions infected by the endoparasitic nematodes and, similarly, nematodes do not infect the region colonized by the fungi. It appears that both organisms are mutually inhibitory and their interactions seem to be highly specific and dependent on different ecological factors (Ingham, 1988). In some cases, the protection was exerted by reducing the reproduction of the pathogen.

However, it is difficult to generalize results because these vary greatly with the particular host–pathogen–VAM fungus combination. Increases of severity, decreases in damage, or no effect are sometimes difficult to predict under field conditions (Dehne, 1988; Smith, 1988). It is also important for studies on VAM × pathogen interactions to use nonmycorrhizal control plants of nutritional status comparable to those with VAM, to ascertain whether the VAM effect is related to improved P nutrition, or carried out by other mechanisms (Smith, 1988).

One hypothesis is that the mechanism whereby VAM protects plants against pathogens is similar to that produced by P additions to the soil (Smith, 1988). Although, in some cases, P additions were not as efficient as VAM in reducing disease symptoms (Caron et al., 1986), P and other nutrients supplied by VAM fungi improve plant vigor and must be responsible for a part of the effect. Changes in rhizosphere populations, produced by altered root exudation induced by VAM, could also lower the pathogen densities, either directly, or via stimulation of microbial antagonists to the pathogen. For instance, bacteria producing siderophores have been found to exhibit fungistatic properties against pathogenic fungi (Misaghi et al., 1982, 1988; Wong & Baker, 1984).

Other mechanisms include competition for host photosynthates, space, or infection sites (Smith, 1988). Dehn and Dehne (1986) studied the development of *Cochliobolus sativus* in roots of Gramineae as affected by VAM colonization. They found that the eusymbiont reduced the infection density and the percentage of damaged roots by the dissymbiont. Histological observations showed host incompatibility with the pathogen when already colonized by the VAM fungus. Although the authors did not delineate the mechanisms involved, they suggested that the beneficial VAM effect was related to physiological changes in the host root.

Decreased carbohydrates in the roots, increased production of secondary metabolites such as lignin, ethylene, and phenols (Dehne, 1986), altered enzymatic activities, or induction of phytoalexins synthesis (Morandi & Gianinazzi-Pearson, 1986) may also reduce susceptibility or increase tolerance of a VAM plant root to pathogen attack (Smith & Gianinazzi-Pearson, 1988). In any case, the increased tolerance may be explained just by means of damage compensation by the VA mycorrhizae.

Conclusions and Perspectives

Experimental evidence demonstrates that microbial interactions involving mycorrhizal fungi appear to play a key role in the soil–plant interface. Without doubt plant growth and health can be very much assisted by suitable mycorrhizal fungi and microorganisms acting in coordination at soil–root interfaces. However, the possibility of manipulating mycorrhizal fungi in combination with beneficial microorganisms depends on a better understanding of the ecosystem and on the suitable selection of inocula to be applied. This must be based on criteria of compatibility and efficiency. Although much work has resulted in significant advances in the understanding of microbial growth around roots, there are still many unanswered questions regarding the dynamics of the microbial colonization of the root. These are especially influenced by processes that are difficult to control such as the flow of substrates in the rhizosphere and continual microbial interactions.

The mechanisms and molecular basis of the root–mycorrhizal fungal–microbial interactions are yet poorly understood, and it is obvious that knowledge of such a basis is crucial to successful manipulation of rhizosphere populations. New techniques will have to be developed or adapted to carry out these studies. The use of electron microscopy (TEM, SEM), double antibiotic marking, and histochemical, immunofluorescent, and isotope-based techniques continues to be a key in rhizosphere biology research. The development of modeling approaches, together with *in situ* studies, using minirrhizotrons, appears to be an important step in advancing knowledge of population dynamics in the mycorrhizosphere. A methodology to be developed is based on enzyme-linked immunoabsorbant assay (ELISA) or newer innovative genetic approaches to characterize existing rhizosphere organisms. Beneficial microorganisms can be genetically engineered to be introduced in the mycorrhizosphere, but the consequences of interactions between mycorrhizae and such microorganisms need to be studied.

In any case, although more studies are needed to increase our understanding of the interactions concerning mycorrhizal fungi, it is clear that mycorrhizal introduction can be improved by "auxiliary" soil microorganisms, and the dual inoculation of mycorrhizal fungi together with helper microorganisms can be useful to ensure mycorrhiza establishment.

References

Ames, R. N. (1989). Mycorrhiza development in onion in response to inoculation with chitin-decomposing actinomycetes. *New Phytologist* **112**, 423–427.

Ames, R. N., & Bethlenfalvay, G. J. (1987). Mycorrhizal fungi and the integration of plant and soil nutrient dynamics. *Journal of Plant Nutrition* **10**, 1313–1321.

Ames, R. N., Mihara, K. L., & Bayne, H. G. (1989). Chitin-decomposing actinomycetes associated with a vesicular-arbuscular mycorrhizal fungus from a calcareous soil. *New Phytologist* **111**, 67–71.

Ames, R. N. Mihara, K. L., & Bethlenfalvay, G. J. (1987). The establishment of microorganisms in vesicular-arbuscular mycorrhizal and control treatments. *Biology and Fertility of Soils* **3**, 217–223.

Ames, R. N., Reid, C. P. P., & Ingham, E. R. (1984). Rhizosphere bacterial population responses to root colonization by a vesicular-arbuscular mycorrhizal fungus. *New Phytologist* **96**, 555–563.

Asai, T. (1944). Die bedeutung der mikorrhiza für das pflanzenleben. *Japanese Journal of Botany* **12**, 359–408.

Asimi, S., Gianinazzi-Pearson, V., & Gianinazzi, S. (1980). Influence of increasing soil phosphorus levels on interactions between vesicular-arbuscular mycorrhizae and *Rhizobium* in soybeans. *Canadian Journal of Botany* **58**, 2200–2206.

Azcon, R. (1987). Germination and hyphal growth of *Glomus mosseae* in vitro: Effects of rhizosphere bacteria and cell-free culture media. *Soil Biology and Biochemistry* **19**, 417–419.

Azcon, R. (1989). Selective interaction between free-living rhizosphere bacteria and vesicular-arbuscular mycorrhizal fungi. *Soil Biology and Biochemistry* **21**, 639–644.

Azcon, R., Barea, J. M., & Hayman, D. S. (1976). Utilization of rock phosphate in alkaline soils by plants inoculated with mycorrhizal fungi and phosphate-solubilizing bacteria. *Soil Biology and Biochemistry* **8**, 135–138.

Azcon, R., Azcon-Aguilar, C., & Barea, J. M. (1978). Effects of plant hormones present in bacterial cultures on the formation and responses to VA mycorrhiza. *New Phytologist* **80**, 359–369.

Azcon, R., El-Atrach, F., & Barea, J. M. (1988). Influence of mycorrhiza vs. soluble phosphate on growth, nodulation, and N_2 fixation (^{15}N) in alfalfa under different levels of water potential. *Biology and Fertility of Soils* **7**, 28–31.

Azcon-Aguilar, C., & Barea, J. M. (1978). Effects of interactions between different culture fractions of phosphobacteria and *Rhizobium* on mycorrhizal infection, growth and nodulation of *Medicago sativa*. *Canadian Journal of Microbiology* **24**, 520–524.

Azcon-Aguilar, C., & Barea, J. M. (1985). Effect of soil micro-organisms on the formation of vesicular-arbuscular mycorrhizas. *Transactions of the British Mycological Society* **84**, 536–537.

Azcon-Aguilar, C., Barea, J. M., & Olivares, J. (1980). Effects of *Rhizobium* polysaccharides on VA mycorrhiza formation. *Second International Symposium on Microbial Ecology*, University of Warwick, Coventry, U.K., Abstr. No. 187.

Azcon-Aguilar, C., Diaz-Rodriguez, R. M., & Barea, J. M. (1986a). Effect of soil microorganisms on spore germination and growth of the vesicular-arbuscular mycorrhizal fungus *Glomus mosseae*. *Transactions of the British Mycological Society* **86**, 337–340.

Azcon-Aguilar, C., Diaz-Rodriguez, R. M., & Barea, J. M. (1986b). Effect of free-living fungi on the germination of *G. mosseae* soil extract. In: *Physiological and Genetical Aspects of Mycorrhizae* (Ed. by V. Gianinazzi-Pearson & S. Gianinazzi), pp. 515–519. INRA, Paris.

Azcon-Aguilar, C., Gianinazzi-Pearson, V., Fardeau, J. C., & Gianinazzi, S. (1986c). Effect of vesicular-arbuscular mycorrhizal fungi and phosphate-solubilizing bacteria on growth and nutrition of soybean in a neutral-calcareous soil amended with $^{32}P-^{45}Ca$-tricalcium phosphate. *Plant and Soil* **96**, 3–15.

Azcon-Aguilar, C., Garcia-Garcia, F., & Barea, J. M. (1991). Germinación y crecimiento axénico de los hongos formadores de micorrizas vesículo-arbusculares. In: *Fijación y Movilización Biológica de Nutrientes. Nuevas Tendencias C.S.I.C.* (Ed. by J. L. Gorgé, J. M. Barea, & J. Olivares), pp. 129–147. C.S.I.C., Madrid.

Badr El-Din, S. M. S., & Moawad, H. (1988). Enhancement of nitrogen fixation in lentil, faba bean, and soybean by dual inoculation with rhizobia and mycorrhizae. *Plant and Soil* **108**, 117–124.

Bagyaraj, D. J. (1984). Biological interactions with VA mycorrhizal fungi. In: *VA Mycorrhiza* (Ed. by C. Ll. Powell & D. J. Bagyaraj), pp. 131–153. CRC Press, Boca Raton, FL.

Bagyaraj, D. J., & Menge, J. A. (1978). Interaction between a VA mycorrhiza and *Azotobacter* and their effects on rhizosphere microflora and plant growth. *New Phytologist* **80**, 567–573.

Barber, D. A., & Martin, J. K. (1976). The release of organic substances by cereal roots into soil. *New Phytologist* **76**, 69–80.

Barea, J. M. (1986). Importance of hormones and root exudates in mycorrhizal phenomena. In: *Physiological and Genetical Aspects of Mycorrhizae* (Ed. by V. Gianinazzi-Pearson & S. Gianinazzi), pp. 177–187. INRA, Paris.

Barea, J. M., & Azcon-Aguilar, C. (1982). Production of plant growth-regulating substances by the vesicular-arbuscular mycorrhizal fungus *Glomus mosseae*. *Applied and Environmental Microbiology* **43**, 810–813.

Barea, J. M., & Azcon-Aguilar, C. (1983). Mycorrhizas and their significance in nodulating nitrogen-fixing plants. *Advances in Agronomy*, **36**, 1–54.

Barea, J. M., Brown, M. E., & Mosse, B. (1973). Association between VA mycorrhiza and *Azotobacter*. *Rothamsted Annual Report* 81–82.

Barea, J. M., Azcon, R., & Hayman, D. S. (1975). Possible synergistic interactions between *Endogone* and phosphate-solubilizing bacteria in low-phosphate soils. In:

Endomycorrhizas (Ed. by F. E. Sanders, B. Mosse, & P. B. Tinker), pp. 409–417. Academic Press, London.

Barea, J. M., Navarro, E., & Montoya, E. (1976). Production of plant growth regulators by rhizosphere-solubilizing bacteria. *Journal of Applied Bacteriology* **40**, 129–134.

Barea, J. M., Azcon, R., & Azcon-Aguilar, C. (1983a). Interactions between phosphate solubilizing bacteria and VA mycorrhiza to improve plant utilization of rock phosphate in non acidic soils. *3rd International Congress on Phosphorus Compounds*, Brussels.

Barea, J. M., DeBonis, A. F., & Olivares, J. (1983b). Interactions between *Azospirillum* and VA mycorrhiza and their effects on growth and nutrition of maize and ryegrass. *Soil Biology and Biochemistry* **15**, 705–709.

Barea, J. M., Azcon-Aguilar, C. & Azcon, R. (1987). Vesicular arbuscular mycorrhiza improve both symbiotic N-2 fixation and N uptake from soil as assessed with a N^{15} technique under field conditions. *New Phytologist* **106**, 717–725.

Barea, J. M., Azcon, R., & Azcon-Aguilar, C. (1989). Time-course of N_2-fixation (^{15}N) in the field by clover growing alone or in mixture with ryegrass to improve pasture productivity, and inoculated with vesicular-arbuscular mycorrhizal fungi. *New Phytologist* **112**, 399–404.

Bayne, H. B., Brown, M. S. & Bethlenfalvay, G. J. (1984). Defoliation effects on mycorrhizal colonization, nitrogen fixation and photosynthesis. *Physiologia Plantarum* **62**, 576–580.

Becard, G., & Piche, Y. (1989). Fungal growth-stimulation by CO_2 and root exudates in vesicular-arbuscular mycorrhizal symbiosis. *Applied and Environmental Microbiology* **55**, 2320–2325.

Beilby, J. P., & Kidby, D. K. (1982). The early synthesis of RNA, protein, and some associated metabolic events in germinating vesicular-arbuscular mycorrhizal fungal spores of *Glomus caledonius*. *Canadian Journal of Microbiology* **28**, 623–628.

Bethlenfalvay, G. J., Brown, M. S., & Stafford, A. E. (1985). *Glycine–Glomus–Rhizobium* symbiosis. II: Antagonistic effects between mycorrhizal colonization and nodulation. *Plant Physiology* **79**, 1054–1058.

Bowen, G. D. (1980). Misconceptions, concepts and approaches in rhizosphere biology. In: *Contemporary Microbial Ecology* (Ed. by D. C. Ellwood, M. J. Latham, J. N. Hedger, J. N. Lynch, & J. M. Slater), pp. 283–304. Academic Press, London.

Bowen, G. D., & Theodorou, C. (1973). Growth of ectomycorrhizal fungi around seeds and roots. In: *Ectomycorrhizae: Their Ecology and Physiology* (Ed. by G. C. Marks & T. T. Kozlowski), pp. 107–150. Academic Press, London.

Bowen, G. D., & Theodorou, C. (1979). Interactions between bacterial and ectomycorrhizal fungi. *Soil Biology and Biochemistry* **11**, 119–126.

Brown, M. S., & Bethlenfalvay, G. J. (1987). *Glycine–Glomus–Rhizobium* symbiosis. VI. Photosynthesis in nodulated, mycorrhizal, or N and P fertilized soybean plants. *Plant Physiology* **85**, 120–123.

Brown, M. S., & Bethlenfalvay, G. J. (1988). The *Glycine–Glomus-Rhizobium* symbiosis. VII. Photosynthetic nutrient-use efficiency in nodulated, mycorrhizal soybeans. *Plant Physiology* **86**, 1292–1297.

Brown, M. E., & Carr, G. R. (1984). Interactions between *Azotobacter chroococcum* and vesicular-arbuscular mycorrhiza and their effects on plant-growth. *Journal of Applied Bacteriology* **56,** 429–437.

Brown, M. S., Thamasurakul, S., & Bethlenfalvay, G. J. (1988). The *Glycine–Glomus–Bradyrhizobium* symbiosis. 8. Phosphorus-use efficiency of CO_2 and N_2 fixation in mycorrhizal soybean. *Physiologia Plantarum* **74,** 159–163.

Busse, M. D., & Ellis, J. R. (1985). Vesicular-arbuscular mycorrhizal (*Glomus fasciculatum*) influence on soybean drought tolerance in high phosphorus soil. *Canadian Journal of Botany* **63,** 2290–2294.

Calvet, C., Pera, J., & Barea, J. M. (1988). Interactions of *Trichoderma* spp. with *Glomus mosseae* and two wilt pathogenic fungi. *Second European Symposium on Mycorrhizae*, Prague, Czechoslovakia.

Caron, M., Richard, C., & Fortin, J. A. (1986). Effect of pre-infestation of the soil by a VAM fungus, *Glomus intraradices*, on *Fusarium* growth and root-rot of tomatoes. *Phytoprotection* **67,** 15–19.

Carr, G. R., Hinkley, M. A., Le Tacon, F., Hepper, C. M., Jones, M. G. K., & Thomas, E. (1985). Improved hyphal growth of two species of vesicular-arbuscular mycorrhizal fungi in the presence of suspension cultured plant cells. *New Phytologist* **101,** 417–426.

Chakravarty, P., & Unestam, T. (1986). Role of mycorrhizal fungi in protecting damping-off of *Pinus sylvestris* L. seedlings. In: *Physiological and Genetical Aspects of Mycorrhizae* (Ed. by V. Gianinazzi-Pearson & S. Gianinazzi), pp. 808–811. INRA, Paris.

Chatarpaul, L., Chakravarty, P., & Subramaniam, P. (1989). Studies in tetrapartite symbioses. I. Role of ecto- and endomycorrhizal fungi and *Frankia* on the growth performance of *Alnus incana*. *Plant and Soil* **118,** 145–150.

Clough, K. S., & Sutton, J. C. (1978). Direct observation of fungal aggregates in sand dune soil. *Canadian Journal of Microbiology* **24,** 333–335.

Cluett, H. C., & Boucher, D. H. (1983). Indirect mutualism in the legume–*Rhizobium*–mycorrhizal fungus interaction. *Oecologia* **59,** 405–408.

Cook, R. J., & Baker, K. F. (1983). *The Nature and Practice of Biological Control of Plant Pathogens*. The American Phytopathological Society, St. Paul, MN.

Cooper, K. M. (1984). Physiology of VA mycorrhizal associations. In: *VA Mycorrhiza* (Ed. by C. Ll. Powell & D. J. Bagyaraj), pp. 156–186. CRC Press, Boca Raton, FL.

Crush, J. R. (1974). Plant growth responses to vesicular-arbuscular mycorrhiza. VII. Growth and nodulation of some herbage legumes. *New Phytologist* **73,** 743–752.

Cruz, R. E., De La Manalo, M. Q., Aggangan, N. S., & Tambalo, J. D. (1988). Growth of three legume trees inoculated with VA mycorrhizal fungi and *Rhizobium*. *Plant and Soil* **108,** 111–115.

Curl, E. A., & Truelove, B. (1986). *The Rhizosphere*. Springer-Verlag, Berlin.

Daft, M. J., & El-Giahmi, A. A. (1974). Effect of *Endogone* mycorrhiza on plant growth. VII. Influence of infection on the growth and nodulation in french bean. (*Phaseolus vulgaris*). *New Phytologist* **73,** 1139–1147.

Daft, M. J., & El-Giahmi, A. A. (1976). Studies on nodulated and mycorrhizal peanuts. *Annals of Applied Biology* **83**, 273–276.

Daft, M. J., Clelland, D. M., & Gardner, I. C. (1985). Symbiosis with endomycorrhizas and nitrogen-fixing organisms. *Proceedings of the Royal Society of Edinburgh* **85**, 283–298.

Dakessian, S., Brown, M. S., & Bethlenfalvay, G. J. (1986). Relationship of mycorrhizal growth enhancement and plant growth with soil water and texture. *Plant and Soil* **94**, 439–443.

Daniels, B. A., & Duff, D. M. (1978). Variation in germination and spore morphology among four isolates of *Glomus mosseae*. *Mycologia* **70**, 1261–1267.

Daniels, B. A., & Graham, S. O. (1976). Effects of nutrition and soil extracts on germination of *Glomus mosseae* spores. *Mycologia* **68**, 108–116.

Daniels, B. A., & Menge, J. A. (1980). Hyperparasitization of vesicular-arbuscular mycorrhizal fungi. *Phytopathology* **70**, 584–588.

Daniels, B. A., & Trappe, J. M. (1980). Factors affecting spore germination of the vesicular-arbuscular mycorrhizal fungus *Glomus epigaeus*. *Mycologia* **72**, 457–471.

Davey, C. B. (1971). Nonpathogenic organisms associated with mycorrhizae. In: *Mycorrhizae* (Ed. by E. Hacskaylo), pp. 114–121. USDA, Washington, D.C.

Dehn, B., & Dehne, H. W. (1986). Development of VA mycorrhizal fungi and interactions with *Cochliobolus sativus* in roots of gramineae. In: *Physiological and Genetical Aspects of Mycorrhizae* (Ed. by V. Gianinazzi-Pearson & S. Gianinazzi), pp. 773–779. INRA, Paris.

Dehne, H. W. (1982). Interaction between vesicular-arbuscular mycorrhizal fungi and plant pathogens. *Phytopathology* **72**, 1115–1119.

Dehne, H. W. (1986). Influence of VA mycorrhizae on host plant physiology. In: *Physiological and Genetical Aspects of Mycorrhizae* (Ed. by V. Gianinazzi-Pearson & S. Gianinazzi), pp. 431–444. INRA, Paris.

Dehne, H. W. (1988). Production and use of inocula of VA mycorrhizal fungi at inorganic carrier materials. Second European Symposium on Mycorrhizae, Prague, Czechoslovakia.

Diaz-Rodriguez, R. M., Azcon-Aguilar, C., & Barea, J. M. (1986). Further studies on the effect of free-living microorganisms on the development of *Glomus mosseae* on soil extracts. *Fourth International Symposium on Microbial Ecology*, Ljubljana, Yugoslavia. Abstr. D13-1.

Duc, G., Trouvelot, A., Gianinazzi-Pearson, V., & Gianinazzi, S. (1989). First report of non-mycorrhizal plant mutants (Myc$^-$) obtained in pea (*Pisum sativum L.*) and fababean (*Vicia faba L.*). *Plant Science* **60**, 214–222.

Duchesne, L. C., Peterson, R. L., & Ellis, B. E. (1987). The accumulation of plant-produced antimicrobial compounds in response to ectomycorrhizal fungi. A review. *Phytoprotection* **68**, 17–27.

Duchesne, L. C., Peterson, R. L., & Ellis, B. E. (1988). Pine root exudate stimulates the synthesis of antifungal compounds by the ectomycorrhizal fungus *Paxillus involutus*. *New Phytologist* **108**, 471–476.

Duchesne, L. C., Peterson, R. L., & Ellis, B. E. (1989). The time-course of disease suppression and antibiosis by the ectomycorrhizal fungus *Paxillus involutus*. *New Phytologist* **111**, 693–698.

El-Atrach, F., Azcon, R., & Barea, J. M. (1989). Effect of vesicular-arbuscular infection on plant growth and nitrogen-fixation as estimated by ^{15}N isotope dilution of alfalfa plants under saline conditions. *35th Meeting of the Societé Française de Phytopathologie*. (International Conference on the Mechanisms of the Relationship between Soil-Plant-Microorganisms in the Rhizosphere). pp. 85.

Foster, R. C., & Bowen, G. D. (1982). Plant surfaces and bacterial growth: The rhizosphere and rhizoplane. In: *Phytopathogenic Prokaryotes*, Vol. 1 (Ed. by R. Mount & C. Lacey), pp. 159–185. Academic Press, New York.

Fries, N. (1978). Basidiospore germination in some mycorrhiza-forming Hymenomycetes. *Transactions of the British Mycological Society* **70**, 319–324.

Fries, N. (1979). The taxon-specific spore germination reaction in *Leccinum*. *Transactions of the British Mycological Society* **73**, 337–341.

Garbaye, J., & Bowen, G. D. (1987). Effect of different microflora on the success of mycorrhizal inoculation of *Pinus radiata*. *Canadian Journal of Forest Research* **17**, 941–943.

Garbaye, J., & Bowen, G. D. (1989). Stimulation of ectomycorrhizal infection of *Pinus radiata* by some microorganisms associated with the mantle of ectomycorrhizas. *New Phytologist* **112**, 383–388.

Garcia-Garrido, J. M., & Ocampo, J. A. (1988). Interaction between *Glomus mosseae* and *Erwinia carotovora* and its effect on the growth of tomato plants. *New Phytologist* **110**, 551–555.

Gardner, I. C. (1986). Mycorrhizae of actinorrhizal plants. *MIRCEN Journal* **2**, 147–160.

Gardner, I. C., Clelland, D. M., & Scott, A. (1984). Mycorrhizal improvement in non-leguminous nitrogen-fixing associations with particular reference to *Hippophae-rhamnoides* L. *Plant and Soil* **78**, 189–199.

Gauthier, D., Diem, H. G., & Dommergues, Y. (1983). Preliminary results of research on *Frankia* and endomycorrhizae associated with *Casuarina equisetifolia*. In: *Casuarina Ecology Management and Utilization* (Ed. by S. J. Midgeley, J. W. Turnbull, & R. D. Johnston), pp. 211–217. CSIRO, Melbourne.

Godbout, C., & Fortin, J. A. (1983). Morphological features of synthesized ectomycorrhizae of *Alnus crispa* and *A. rugosa*. *New Phytologist* **94**, 249–262.

Gonzalez, S. B. (1988). Ecología y biotecnología de micorrizas en leguminosas (soja y alfalfa). Ph.D. Thesis, University of Granada.

Graham, J. H. (1986). Citrus mycorrhiza-potential benefits and interactions with pathogens. *HortScience* **21**, 1302–1306.

Graham, J. H., Leonard, R. T., & Menge, J. A. (1981). Membrane mediated decrease in root exudation responsible for phosphorus inhibition of vesicular-arbuscular mycorrhiza formation. *Plant Physiology* **68**, 548–552.

Gueye, M., Diem, H. G., & Dommergues, Y. R. (1987). Variation in N_2 fixation, N and

P contents of mycorrhizal *Vigna unguiculata* in relation to the progressive development of extraradical hyphae of *Glomus mosseae*. *MIRCEN Journal* **3**, 75–86.

Harari, A., Kigel, J., & Okon, Y. (1988). Involvement of IAA in the interaction between *Azospirillum brasilense* and *Panicum miliaceum* roots. *Plant and Soil* **110**, 275–282.

Hardie, K. (1985). The effect of removal of extraradical hyphae on water uptake by vesicular-arbuscular mycorrhizal plants. *New Phytologist* **101**, 677–684.

Hardie, K., & Leyton, L. (1981). The influence of vesicular-arbuscular mycorrhiza on growth and water relations of red clover. I. In phosphate deficient soil. *New Phytologist* **89**, 599–608.

Harley, J. L., & Smith, S. E. (1983). *Mycorrhizal Symbiosis*. Academic Press, London.

Harris, D., Pacovsky, R. S., & Paul, E. A. (1985). Carbon economy of soybean–Rhizobium–Glomus associations. *New Phytologist* **101**, 427–440.

Hayman, D. S. (1975). Phosphorus cycling by soil micro-organisms and plant roots. In: *Soil Microbiology* (Ed. by N. Walker), pp. 67–91. Butterworths, London.

Hayman, D. S. (1983). The physiology of vesicular-arbuscular endomycorrhizal symbiosis. *Canadian Journal of Botany* **61**, 944–963.

Hayman, D. S. (1986). Mycorrhizae of nitrogen-fixing legumes. *MIRCEN Journal* **2**, 121–145.

Hepper, C. M. (1979). Germination and growth of *Glomus caledonius* spores: The effects of inhibitors and nutrients. *Soil Biology and Biochemistry* **11**, 269–277.

Hepper, C. M. (1983). Limited independent growth of a vesicular-arbuscular mycorrhizal fungus *in vitro*. *New Phytologist* **93**, 537–542.

Hepper, C. M. (1984). Regulation of spore germination of the vesicular-arbuscular mycorrhizal fungus *Acaulospora laevis* by soil-pH. *Transactions of the British Mycological Society* **83**, 154–156.

Hepper, C. M., & Jakobsen, I. (1983). Hyphal growth from spores of the mycorrhizal fungus *Glomus caledonius:* Effect of amino acids. *Soil Biology and Biochemistry* **15**, 55–58.

Hepper, C. H., & Smith, G. A. (1976). Observations on the germination of *Endogone* spores. *Transactions of the British Mycological Society* **66**, 189–194.

Hetrick, B. A. D., & Wilson, G. W. T. (1989). Suppression of mycorrhizal fungus spore germination in non-sterile soil: Relationship to mycorrhizal growth response in big bluestem. *Mycologia* **81**, 382–390.

Ho, I. (1988). Interaction between VA-mycorrhizal fungus and *Azotobacter* and their combined effects on growth of tall fescue. *Plant and Soil* **105**, 291–293.

Ingham, E. R. (1988). Interactions between nematodes and vesicular-arbuscular mycorrhizae. *Agriculture Ecosystems and Environment* **24**, 169–182.

Janse, J. M. (1896). Les endophytes radicaux des quelques plantes Javanaises. *Annales du Jardin Botanique Buitenzorg* **14**, 53–212.

Jones, F. R. (1924). A mycorrhizal fungus in the roots of legumes and some other plants. *Journal of Agriculture Research* **29**, 459–470.

Katznelson, H., Rouatt, J. M., & Peterson, E. A. (1962). The rhizosphere effect of mycorrhizal and nonmycorrhizal roots of yellow birch seedlings. *Canadian Journal of Botany* **40,** 377–382.

Kaur, S., & Singh, O. S. (1988). Response of ricebean to single and combined inoculation with *Rhizobium* and *Glomus* in a P-deficient sterilized soil. *Plant and Soil* **112,** 293–295.

Khasewneh, F. E., & Doll, E. C. (1978). The use of phosphate rock for direct application to soils. *Advances in Agronomy* **30,** 159–206.

Kloepper, J. W., Leong, J., Teintze, M., & Schroth, M. N. (1980). Enhanced plant growth by siderophores produced by plant growth-promoting rhizobacteria. *Nature (London)* **286,** 885–886.

Krishna, K. R., Balakrishna, A. N., & Bagyaraj, A. N. (1982). Interaction between a vesicular-arbuscular mycorrhizal fungus and *Streptomyces cinnamoneous* and their effects on finger millet. *New Phytologist* **92,** 401–405.

Kucey, R. M. N., & Paul, E. A. (1982). Carbon flow photosynthesis, and N-2 fixation in mycorrhizal and nodulated faba beans (*Vicia faba* L.). *Soil Biology and Biochemistry* **14,** 407–412.

Kwapata, M. B., & Hall, A. E. (1985). Effects of moisture regime and phosphorus on mycorrhizal infection, nutrient-uptake, and growth of cowpeas (*Vigna-unguiculata* (L) wal P). *Field Crops Research* **12,** 241–250.

Leong, J. (1986). Siderophores: Their biochemistry and possible role in the biocontrol of plant pathogens. *Annual Review of Phytopathology* **24,** 187–209.

Le Tacon, F., Skinner, F. A., & Mosse, B. (1983). Spore germination and hyphal growth of a vesicular-arbuscular mycorrhizal fungus *Glomus mosseae* (Gerdeman & Trappe), under decreased oxygen and increased carbon dioxide concentrations. *Canadian Journal of Microbiology* **29,** 1280–1285.

Levinson, I. (1957). Antagonistic effects of *Alternaria tenuis* on certain root-fungi of forest trees. *Nature* (London) **179,** 1143–1144.

Li, C., & Castellano, M. (1985). Nitrogen-fixing bacteria isolated from within sporocarps of three ectomycorrhizal fungi. In: *North American Conference on Mycorrhizae* (Ed. by R. Molina), pp. 264. Oregon State University, Corvallis.

Linderberg, G. (1986). Physiology of interactions between mycorrhizal and rhizosphere microorganisms. In: *Mycorrhizae: Physiology and Genetics* (Ed. by V. Gianinazzi-Pearson & S. Gianinazzi), pp. 197–206. INRA, Paris.

Linderman, R. G. (1986). Managing rhizosphere microorganisms in the production of horticultural crops. *HortScience* **21,** 1299–1302.

Linderman, R. G. (1988). Mycorrhizal interactions with the rhizosphere microflora. The mycorrhizosphere effect. *Phytopathology* **78,** 366–371.

Louis, I., & Lim, G. (1988). Differential response in growth and mycorrhizal colonization of soybean to inoculation with two isolates of *Glomus clarum* in soils of different P availability. *Plant and Soil* **112,** 37–43.

Lynch, J. M. (1976). Products of soil micro-organisms in relation to plant growth. *CRC Critical Reviews in Microbiology* **5,** 67–107.

Lynch, J. M. (1983). *Soil Biotechnology. Microbiological Factors in Crop Productivity.* Blackwell Scientific Publications, Oxford.

MacDonald, R. M., & Lewis, M. (1978). The occurrence of some acid-phosphatases and dehydrogenases in the vesicular-arbuscular mycorrhizal fungus *Glomus mosseae. New Phytologist* **80,** 135–141.

Malloch, D. W., Pirozynski, K. A., & Raven, P. H. (1980). Ecological and evolutionary significance of mycorrhizal symbioses in vascular plants. *Proceedings of the National Academy of Sciences U.S.A.* **77,** 2113–2118.

Manjunath, A., Mohan, R., & Bagyaraj, D. J. (1981). Interaction between *Beijerinckia mobilis, Aspergillus niger* and *Glomus fasciculatum* and their effects on growth of onion. *New Phytologist* **87,** 723–727.

Marx, D. H. (1973). Mycorrhizae and feeder root diseases. In: *Ectomycorrhizae: Their Ecology and Physiology* (Ed. by D. H. Marx & T. T. Kozlowski), pp. 351–383. Academic Press, New York.

Marx, D. H. (1975). Mycorrhizae and establishment of trees on strip-mined land. *Ohio Journal of Science* **75,** 288–297.

Marx, D. H., & Krupa, S. V. (1978). Mycorrhizae. In: *Interactions between Non-Pathogenic Soil Microorganisms and Plants* (Ed. by Y. R. Dommergues & S. V. Krupa), pp. 401–442. Elsevier, Scientific Publication, Amsterdam.

Mayo, K., Davis, R. E., & Motta, J. (1986). Stimulation of germination of spores of *Glomus versiforme* by spore associated bacteria. *Mycologia* **78,** 426–431.

McAffee, B. J., & Fortin, J. A. (1988). Comparative effects of the soil microflora on ectomycorrhizal inoculation of conifer seedlings. *New Phytologist* **108,** 443–449.

Mejstrik, V. (1965). Study on the development of endotrophic mycorrhiza in the association of *Cladietum marisci*. In: *Plant Microbes Relationships* (Ed. by J. Macura & V. Vancura), pp. 283–290. Publishing House of the Czechoslovak Academy of Sciences, Prague.

Meyer, J. R., & Linderman, R. G. (1986). Selective influence on populations of rhizosphere or rhizoplane bacteria and actinomycetes by mycorrhizas formed by *Glomus fasciculatum*. *Soil Biology and Biochemistry* **18,** 191–196.

Misaghi, I. J., Stowell, L. J., Grogan, R. G., & Spearmen, L. Cl. (1982). Fungistatic activity of water-soluble fluorescent pigments of fluorescent pseudomonads. *Phytopathology* **72,** 33–36.

Misaghi, I. J., Olsen, M. W., Cotty, P. J., & Donndelinger, C. R. (1988). Fluorescent siderophore-mediated iron deprivation: A contingent biological control mechanism. *Soil Biology and Biochemistry* **20,** 573–574.

Mohandas, S. (1987). Field response of tomato (*Lycopersicum esculentum* mill 'Pusa Ruby') to inoculation with a VA mycorrhizal fungus *Glomus fasciculatum* and with *Azotobacter vinelandii*. *Plant and Soil* **98,** 295–297.

Morandi, D., & Gianinazzi-Pearson, V. (1986). Influence of mycorrhizal infection and phosphate nutrition on secondary metabolite contents of soybean roots. In: *Physiological and Genetical Aspects of Mycorrhizae* (Ed. by V. Gianinazzi-Pearson & S. Gianinazzi), pp. 787–791. INRA, Paris.

Mosse, B. (1959). The regular germination of resting spores and some observations on the growth requirements of an *Endogone* sp. causing vesicular-arbuscular mycorrhiza. *Transactions of the British Mycological Society* **42**, 273–286.

Mosse, B. (1962). The establishment of vesicular-arbuscular mycorrhiza under aseptic conditions. *Journal of General Microbiology* **27**, 509–520.

Mosse, B. (1977). The role of mycorrhiza in legume nutrition on marginal soils. In: *Exploiting the Legume-Rhizobium Symbiosis in Tropical Agriculture* (Ed. by J. M. Vincent, A. S. Witney, & J. Bose), pp. 275–292. College of Tropical Agriculture, University of Hawaii, Miscellaneous Publication 145.

Mosse, B., & Hepper, C. (1975). Vesicular-arbuscular mycorrhizal infections in root organ cultures. *Physiology and Plant Pathology* **5**, 215–223.

Mosse, B., Stribley, D. P., & Le Tacon, F. (1981). Ecology of mycorrhizae and mycorrhizal fungi. *Advances in Microbial Ecology* **5**, 137–210.

Mugnier, J., & Mosse, B. (1987). Vesicular-arbuscular mycorrhizal infection in transformed root-inducing T-DNA roots grown axenically. *Phytopathology* **77**, 1045–1050.

Munns, D. N., & Mosse, B. (1980). Mineral nutrition of legume crops. In: *Advances in Legume Science* (Ed. by R. J. Summerfield, & A. H. Butting), pp. 115–125. H. M. Stationery Office, London.

Neal, J. L., Jr., Bollen, W. B., & Zak, B. (1964). Rhizosphere microflora associated with mycorrhizae of Douglas-fir. *Canadian Journal of Microbiology* **10**, 259–265.

Neal, J. L. Jr., Lu, K. C., Bollen, W. B., & Trappe, J. M. (1968). A comparison of rhizosphere microfloras associated with mycorrhizae of red alder and Douglas-fir. In: *Biology of Alder* (Ed. by J. M. Trappe, J. F. Franklin, R. F. Tarrant, & G. M. Hansen), pp. 57–72. USDA, Oregon.

Neilands, J. B., & Leong, S. A. (1986). Siderophores in relation to plant growth and disease. *Annual Review of Plant Physiology* **37**, 187–208.

Nelsen, Ch. E. (1986). The water relations of vesicular-arbuscular mycorrhizal systems. In: *Ecophysiology of VA Mycorrhizal Plants* (Ed. by G. R. Safir), pp. 71–91. CRC Press, Boca Raton, FL.

Newman, E. I. (1978). Root microorganisms: Their significance in the ecosystem. *Biological Reviews* **53**, 511–554.

O'Hara, G. W., Boonkerd, N., & Dilworth, M. J. (1988). Mineral constraints to nitrogen fixation. *Plant and Soil* **108**, 93–110.

Olivares, J., Montoya, E., & Palomares, A. (1977). Some effects derived from the presence of extrachromosomal DNA in *R. meliloti*. In: *Recent Developments in Nitrogen Fixation* (Ed. by W. Newton, J. R. Postgate, & C. Rodriguez-Barrueco), pp. 375–386. Academic Press, London.

Oswald, E. T., & Ferchau, H. A. (1968). Bacterial associations of coniferous mycorrhizae. *Plant and Soil* **28**, 187–192.

Pacovsky, R. S. (1988). Influence of inoculation with *Azospirillum brasilense* and *Glomus fasciculatum* on sorghum nutrition. *Plant and Soil* **110**, 283–287.

Pacovsky, R. S., Fuller, G., & Paul, E. A. (1985). Influence of soil on the interactions between endomycorrhizae and *Azospirillum* in sorghum. *Soil Biology and Biochemistry* **17**, 525–531.

Pang, P. C., & Paul, E. A. (1980). Effects of vesicular-arbuscular mycorrhiza on C-14 and N-15 distribution in nodulated fababeans. *Canadian Journal of Soil Science* **60**, 241–250.

Paula, M. A. De (1988). Germinaçao e crescimento micelial de esporos de fungos micorrízicos vesículo-arbusculares na presença de calos e suspensao de células vegetais in vitro. Graduate Memoir, University of Lavras, Brasil, pp. 1–128.

Peña, J. I., Sanchez-Diaz, M., Aguirreolea, J. & Becana, M. (1988). Increased stress tolerance of nodule activity in the *Medicago–Rhizobium–Glomus* symbiosis under drought. *Journal of Plant Physiology* **133**, 79–83.

Perrin, R. (1985). Peut-on compter sur les mycorhizes pour lutter contre les maladies des plantes ligneuses? *European Journal of Forest Pathology* **15**, 372–379.

Perrin, R., & Nouveau, M. (1986). L'association mycorhizienne de *Pinus sylvestris* avec *Hebeloma crustuliniforme* et *Laccaria laccata* et les maladies causées por *Pythium* spp. In: *Physiological and Genetical Aspects of Mycorrhizaea* (Ed. by V. Gianinazzi-Pearson & S. Gianinazzi), pp. 793–798. INRA, Paris.

Perry, D. A., Molina, R. & Amaranthus, M. P. (1987). Mycorrhizae, mycorrhizospheres, and reforestation: Current knowledge and research needs. *Canadian Journal of Forestry Research* **17**, 929–940.

Powell, C. Ll. (1976). Development of mycorrhizal infections from *Endogone* spores and infected root segments. *Transactions of British Mycological Society* **66**, 439–445.

Rai, R. (1988). Interaction response of *Glomus-albidus* and *Cicer-Rhizobium* strain on iron uptake and symbiotic N_2 fixation in calcareous soil. *Journal of Plant Nutrition* **11**, 863–869.

Raj, J., Bagyaraj, D. J., & Manjunath, A. (1981). Influence of soil inoculation with vesicular-arbuscular mycorrhiza and a phosphate dissolving bacterium on plant growth and ^{32}P uptake. *Soil Biology and Biochemistry* **13**, 105–108.

Rambelli, A. (1973). The rhizosphere of mycorrhizae. In: *Ectomycorrhizae* (Ed. by G. C. Marks & T. T. Kozlowski), pp. 299–350. Academic Press, New York.

Ratnayake, M., Leonard, R. T., & Menge, J. A. (1978). Root exudation in relation to supply of phosphorus and its possible relevance to mycorrhiza formation. *New Phytologist* **81**, 543–552.

Richter, D. L., Zuellig, T. R., Bagley, S. T., & Bruhn, J. N. (1989). Effects of red pine (*Pinus resinosa* Ait.) mycorrhizoplane-associated actinomycetes on *in vitro* growth of ectomycorrhizal fungi. *Plant and Soil* **115**, 109–116.

Römheld, V., & Marschner, H. (1986). Evidence for a specific uptake system for iron phytosiderophores in roots of grasses. *Plant Physiology* **80**, 175–180.

Rose, S. L. (1980). Mycorrhizal associations of some actinomycete nodulated nitrogen fixing plants. *Canadian Journal of Botany* **58**, 1449–1454.

Rose, S. L., & Trappe, J. M. (1980). Three endomycorrhizal *Glomus* spp. associated with actinorrhizal shrubs. *Mycotaxon* **10**, 412–420.

Rose, S. L., & Youngberg, C. T. (1981). Tripartite associations in snowbrush (*Ceanothus velutinus*): Effect of vesicular-arbuscular mycorrhizae on growth, nodulation and nitrogen fixation. *Canadian Journal of Botany* **59,** 34–39.

Ross, J. P., & Daniels, B. A. (1982). Hyperparasitism of endomycorrhizal fungi. In: *Methods and Principles of Mycorrhizal Research* (Ed. by N. C. Schenck), pp. 55–58. The American Phytopathological Society, St. Paul, Minnesota.

Russo, R. O. (1989). Evaluating alder-endophyte (*Alnus acuminata–Frankia–Mycorrhizae*) interactions. *Plant and Soil* **118,** 151–155.

Sampangi, R., & Perrin, R. (1986). Attempts to elucidate the mechanisms involved in the protective effect of *Laccaria laccata* against *Fusarium oxysporum*. In: *Physiological and Genetical Aspects of Mycorrhizae* (Ed. by V. Gianinazzi-Pearson & S. Gianinazzi), pp. 807–810. INRA, Paris.

Sampangi, R., Perrin, R., & Le Tacon, F. (1986). Disease suppression and growth promotion of norway spruce and Douglas-fir seedlings by the ectomycorrhizal fungus *Laccaria laccata* in forest nurseries. In: *Physiological and Genetical Aspects of Mycorrhizae* (Ed. by V. Gianinazzi-Pearson & S. Gianinazzi), pp. 799–806. INRA, Paris.

Sanders, F. E., & Sheikh, N. A. (1983). The development of vesicular-arbuscular mycorrhizal infection in plant-root systems. *Plant and Soil* **71,** 223–246.

Schisler, D. A., & Linderman, R. G. (1989a). Selective influence of volatiles purged from coniferous forest and nursery soils on microbes of a nursery soil. *Soil Biology and Biochemistry* **21,** 389–396.

Schisler, D. A., & Linderman, R. G. (1989b). Response of nursery soil microbial populations to volatiles purged from soil around Douglas-fir ectomycorrhizae. *Soil Biology and Biochemistry* **21,** 397–401.

Schonbeck, F. (1979). Endomycorrhiza in relation to plant diseases. In: *Soil-Borne Plant Pathogens* (Ed. by B. Schippers & W. Gams), pp. 271–280. Academic Press, New York.

Schwab, S. M., Menge, J. A., & Leonard, R. T. (1983). Quantitative and qualitative effects of phosphorus on extracts and exudates of sudangrass roots in relation to vesicular-arbuscular mycorrhiza formation. *Plant Physiology* **73,** 761–765.

Secilia, J., & Bagyaraj, D. J. (1987). Bacteria and actinomycetes associated with pot cultures of vesicular-arbuscular mycorrhizas. *Canadian Journal of Microbiology* **33,** 1067–1073.

Siqueira, J. O. (1987). Cultura axenica e monoxenica dos fungos micorrizicos vesiculoarbusculares. In: *II Reuniao Brasileira Sobre Micorrizas* pp. 44–70, Sao Paulo, Brasil.

Siqueira, J. O., Hubbell, D. H., & Schenck, N. C. (1982). Spore germination and germ tube growth of a vesicular-arbuscular mycorrhizal fungus *in vitro*. *Mycologia* **74,** 952–959.

Siqueira, J. O., Hubbell, D. H., Kimbrough, J. W., & Schenck, N. C. (1984). *Stachbotrys chartarum* antagonistic to azygospores of *Gigaspora margarita* in vitro. *Soil Biology and Biochemistry* **16,** 679–681.

Slankis, V. (1973). Hormonal relationships in mycorrhizal development. In: *Ectomycor-*

rhizae (Ed. by G. C. Marks & T. T. Kozlowski), pp. 231–298. Academic Press, New York.

Slankis, V. (1974). Soil factors influencing formation of mycorrhizae. *Annual Review of Phytopathology* **12**, 437–457.

Smith, G. S. (1988). The role of phosphorus—nutrition in interactions of vesicular-arbuscular mycorrhizal fungi with soilborne nematodes and fungi. *Phytopathology* **78**, 371–374.

Smith, S. E. (1980). Mycorrhizas of autotrophic higher plants. *Biological Reviews* **55**, 475–510.

Smith, S. E., & Bowen, G. D. (1979). Soil temperature, mycorrhizal infection and nodulation of *Medicago truncatula* and *Trifolium subterraneum*. *Soil Biology and Biochemistry* **11**, 469–473.

Smith, S. E., & Daft, M. J. (1977). Interactions between growth, phosphate content and nitrogen fixation in mycorrhizal and non-mycorrhizal *Medicago sativa*. *Australian Journal of Plant Physiology* **4**, 403–413.

Smith, S. E. & Gianinazzi-Pearson, V. (1988). Physiological interactions between symbionts in vesicular-arbuscular mycorrhizal plants. *Annual Review of Plant Physiology and Plant Molecular Biology* **39**, 221–244.

Smith, S. E., Nicholas, D. J. D., & Smith, F. A. (1979). Effect of early mycorrhizal infection on nodulation and nitrogen fixation in *Trifolium subterraneum* L. *Australian Journal of Plant Physiology* **6**, 305–316.

Spedalieri de Nuñez, N. Y. (1980). Flora fúngica asociada con micorrizas de *Pinus elliottii* Engelm. *Revista de Investigaciones Agropecuarias* **15**, 259–269.

Sprent, J. I. (1986). Nitrogen fixation in a sustainable agriculture. *Biology Agriculture* **1**, 153–165.

St. John, T. V., Hays, R. I., & Reid, C. P. P. (1983). Influence of a volatile compound on formation of vesicular-arbuscular mycorrhizas. *Transactions of the British Mycological Society* **81**, 153–154.

Stotzky, G. (1972). Activity, ecology and population dynamic of microorganisms in soil. *Critical Review in Microbiology* **2**, 59–137.

Strzelczyk, E., & Pokojska-Burdziej, A. (1984). Production of auxins and gibberellin-like substances by mycorrhizal fungi, bacteria and actinomycetes isolated from soil and the mycorrhizosphere of pine (*Pinus silvestris* L.). *Plant and Soil* **81**, 185–194.

Subba Rao, N. S., Tilak, K. V. B. R., & Singh, C. S. (1985a). Effect of combined inoculation of *Azospirillum brasilense* and vesicular-arbuscular mycorrhiza on pearl millet (*Pennisetum americanum*). *Plant and Soil* **84**, 283–286.

Subba Rao, N. S., Tilak, K. V. B. R., & Singh, C. S. (1985b). Synergistic effect of vesicular-arbuscular mycorrhizas and *Azospirillum brasilense* on the growth of barley on pots. *Soil Biology and Biochemistry* **17**, 119–121.

Suslow, T. V. (1982). Role of root-colonizing bacteria in plant growth. In: *Phytopathogenic Prokaryotes*, Vol. 1 (Ed. by R. Mount & C. Lacey), pp. 187–223. Academic Press, New York.

Sutton, J. C., & Sheppard, B. R. (1976). Aggregation of sand-dune soil by endomycorrhizal fungi. *Canadian Journal of Botany* **54**, 326–333.

Sylvia, D. M., & Schenck, N. C. (1983). Application of super-phosphate to mycorrhizal plants stimulates sporulation of phosphorus tolerant vesicular arbuscular mycorrhizal fungi. *New Phytologist* **95**, 655–661.

Szaniszlo, P. J., Powell, P. E., Reid, C. P. P., & Cline, G. R. (1981). Production of hydroxamate siderophore iron chelators by ectomycorrhizal fungi. *Mycologia* **73**, 1158–1174.

Tello, J. C., Vares, F., Notario, A. & Lacassa, A. (1987). Las micorrizas: un potencial a explotar en la lucha contra las enfermedades de las plantas. *ITEA* **73**, 40–64.

Theodorou, C. (1967). Inoculation with pure cultures of mycorrhizal fungi of radiata pine growing in partially sterilised soil. *Australian Journal of Forestry* **31**, 303–309.

Theodorou, C., & Bowen, G. D. (1973). Inoculation of seeds and soil with basidiospores of mycorrhizal fungi. *Soil Biology and Biochemistry* **5**, 765–771.

Tinker, P. B. (1980). The role of rhizosphere microorganisms in phosphorus uptake by plants. In: *The Role of Phosphorus in Agriculture* (Ed. by F. Kwasaneh & E. Sample), pp. 617–654. American Society of Agronomy, Madison, WI.

Tisdall, J. M., & Oades, J. M. (1979). Stabilization of soil aggregates by the root systems of ryegrass. *Australian Journal of Soil Research* **17**, 429–441.

Tomerup, I. C. (1985). Inhibition of spore germination of VAM fungi in soil. *Transactions of the British Mycological Society* **85**, 267–278.

Trappe, J. M. (1986). Phylogenetic and ecologic aspects of mycotrophy in the angiosperms from an evolutionary standpoint. In: *Ecophysiology of VA Mycorrhizal Plants* (Ed. by G. R. Safir), pp. 5–25. CRC Press, Boca Raton, FL.

Tribunskaya, A. (1955). A study of the microflora of the rhizosphere of Scots pine seedlings. *Mikrobiologiya* **24**, 188–192.

Vancura, V., Orozco, M. O., Prikryl, Z., & Grauov, O. (1988). Bacteria in the hyphosphere of vesicular-arbuscular mycorrhizal fungi. *Second European Symposium on Mycorrhizae*, Prague, Czechoslovakia, pp. 108.

Waidyanatha, U. P. S., Yogaratnam, N., & Ariyaratne, W. A. (1979). Mycorrhizal infection effect on growth and nitrogen fixation of *Pueraria* and *Stylosanthes* and uptake of phosphorus from two rock phosphates. *New Phytologist* **82**, 147–152.

Watrud, L. S., Heithaus, J. J. III, & Jaworski, E. G. (1978). Geotropism in the endomycorrhizal fungus *Gigaspora margarita*. *Mycologia* **70**, 449–452.

Wilkinson, K. G., Dixon, K. W., & Sivasithamparam, K. (1989). Interaction of soil bacteria, mycorrhizal fungi and orchid seed in relation of soil germination of Australian orchids. *New Phytologist* **112**, 429–435.

Wilson, G. W. T., Hetrick, B. A. D., & Kitt, D. G. (1989). Suppression of vesicular-arbuscular mycorrhizal fungus spore germination by nonsterile soil. *Canadian Journal of Botany* **67**, 18–23.

Wong, P. T. W., & Baker, R. (1984). Suppression of wheat take-all and ophiobolus patch by fluorescent pseudomonads from *Fusarium*-suppressive soil. *Soil Biology and Biochemistry* **16,** 397–403.

Zambolin, L. (1987). Tolerancia de plantas micorrizadas a fitonematóides. II. Reuniao Brasileira sobre Micorrizas, Sao Paulo, Brasil, pp. 103–125.

7

Interactions Between Fungal Symbionts: VA Mycorrhizae

J. M. Wilson and I. C. Tommerup

SUMMARY. **A single root system is often the host for several different mycorrhizal fungi, not all of them equally effective at improving nutrient uptake. Competitive differences between fungi are primarily related to their infectivity, the inoculum densities and propagule distribution of the interacting fungi, the state of activation of propagules and their ability to produce external hyphae, and the spatial relationships of propagules and roots. Aggression is another useful attribute: this is the ability of a fungus to maintain its level of colonisation in a competitive situation. However, while aggressive fungi are likely to be persistent, they often also exhibit low infectivity. Because of such different attributes, interactions between VA mycorrhizal fungi are important in determining the success of a mycorrhizal symbiosis. Likely outcomes of interactions between mycorrizal fungi include changes in the effectiveness of the symbiosis, changes in total colonisation, and alteration in the relative reproductive ability of the interacting fungi. Such effects have both ecological and agricultural implications, particularly in terms of the likely success of field inoculation. There are methodological problems to overcome before further progress can be made; the most significant is the difficulty of identifying individual fungi, particularly those that are closely related.**

Introduction

Scope

The results of any interactions between vesicular-arbuscular mycorrhizal (VAM) fungi will influence the ecological balance within the soil/root matrix. Since these fungi are important as pioneer colonisers (Forster & Nicolson, 1981), as established residents of cultivated soils (Abbott & Robson, 1982b), and for maximum growth of some plants in treated sterilised soils (Menge et al., 1977), their activities are of special interest to both soil conservationists and agriculturalists. As most of the research on this topic has been based on the significance of VA mycorrhizas in improving plant nutrition, this chapter has been written from an agricultural perspective.

Although improved plant nutrition, particularly phosphorus uptake, is recognized as the main benefit of a mycorrhizal association, this is clearly not the only putative benefit. Enhanced phosphorus uptake is common (Tinker, 1975), but other effects include the increased uptake of various trace elements (Cooper & Tinker, 1978), enhanced water use efficiency (Safir et al., 1972; Allen et al., 1981), hormonal changes (Allen et al., 1980) and the exclusion of root pathogens (Schenck & Kellam, 1978). Although the benefits may have some ecological or competitive spinoffs for the plant, and, to a lesser extent, the fungus, the purpose of this chapter is not to discuss the specific benefits of any particular mycorrhizal association. This review seeks to define and describe those characteristics and behaviors of a VAM fungus that make it more likely than its neighbor to succeed as a symbiont. In this respect we will concentrate on characteristics associated with the formation of root infection.

Desirable Outcomes of Interactions

Although VAM fungi are already widespread in agricultural soils, as with rhizobia (Trinick, 1982), different species (Mosse, 1972; Sanders et al., 1977) and isolates (Abbott & Robson, 1978; Carling & Brown, 1980) differ in their ability to improve plant nutrition. Hence, plant production may be improved by introducing into soils those fungi that are more efficient at increasing resource acquisition than those already present. Alternatively, if the indigenous fungi are effective at increasing plant growth, then an understanding of their behavior will help maximize the benefits derived from them.

From an agricultural viewpoint, the most desirable outcome of any interaction between VAM fungi is that the mycorrhiza that results will improve the growth or survival of a particular plant. In the simplest scenario, an "effective" fungus would form a symbiosis with a plant's roots instead of an "ineffective" fungus. In practice it is very much more complicated than this, partly because there are many different attributes, few of which we fully understand, that contribute to making an endophyte "effective." In addition, a single mycorrhizal root system will usually be the result of a symbiosis with several different fungi, not all of them equally "effective" at improving nutrient uptake.

Attributes of an "Effective" VAM Fungus

The most obvious attribute of an effective VAM fungus is that it has a superior ability to perform an important function, such as the uptake of phosphorus. However, this ability is probably not related to intrinsically superior nutrient absorption, but rather to the result of superior infectivity, or the ability of the particular fungus to colonise roots rapidly and extensively (Abbott & Robson, 1981a). A network of external hyphae that extends considerably beyond the root hair zone may also be important (Rhodes & Gerdemann, 1980), because of its role both in nutrient uptake and in infection spread. In comparing four different endophytes, Sanders et al. (1977) showed that phosphorus flow into roots could be satisfactorily explained on the basis of the length of root infected, irrespective of the fungus. They attributed small differences among the fungi to different rates of infection spread. In a comprehensive discussion of the characteristics of effective VAM fungi, Abbott and Robson (1982a) argued that differences among VAM fungi in their effectiveness at increasing plant growth are directly correlated with their abilities to infect roots and to form an extensive network of hyphae in the surrounding soil. Following successful root colonisation, other factors become significant, particularly in the field. Among these are the ability of the fungus to build up inoculum, to persist in the soil from year to year, to readily activate its propagules, and to rapidly form new mycorrhizas. The various phases of the life cycle, all of which can influence the success of a VAM fungus, are shown in Figure 7.1.

The Relevance of the Concept of Competition

In view of infection as a criterion of effectiveness, the ability of different VAM fungi to form a mycorrhiza on a particular root is important. The empirical data on interactions show that there is a range of differing responses from no interactions to distinct interference (e.g., Powell, 1979b; Abbott & Robson, 1981a; Wilson, 1983, 1984b; Lopez-Aguillon & Mosse, 1987; Hepper et al. 1988). If the interaction is neutral the other fungus does not affect the establishment of the desired isolate. However, when one fungus interferes with the establishment of the "effective" isolate, the competitive interactions are important. As can be seen from the ecological literature, however, the definition of competition is not clear (e.g., Odum, 1971). Clark (1965) described two definitions of competition. In a narrow sense, he defined it as an active demand for a factor that is in short supply. In a broad sense, he considered it to involve many factors such as the struggle for limited resources, direct antagonism, mutual predation, susceptibility to parasitism, or any interaction affecting the growth and survival of an individual. Within a community of microorganisms, such as mycorrhizal fungi, determining the specific factor for which these fungi are competing is difficult. Therefore, we will use the term competition in the broad sense.

Resources for Which VAM Fungi Compete

There are disagreements about what limiting factors are, but any essential resource can be the basis for a competitive interaction. Water and oxygen have been proposed, but both were discounted by Clark (1965). Competition for space and

LIFE–CYCLE OF VAM FUNGI

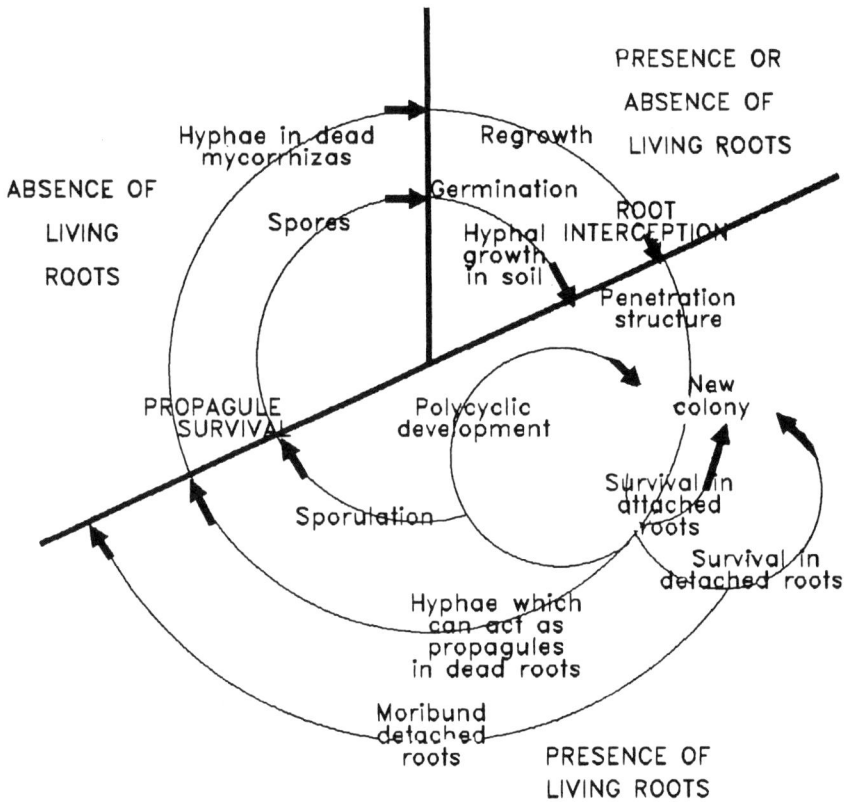

Figure 7.1. The various phases in the life cycle of a VA mycorrhizal fungus. After Tommerup (1985a).

nutrients are the factors which seem most likely to be involved for root-infecting microorganisms such as VAM fungi.

There is doubt as to whether root-infecting microorganisms compete for space. Competition for root volume per se is probably of limited importance, except to the extent that each organism is three dimensional and must necessarily exclude other organisms from the space it occupies. Multiple occupancy of the same part of a root by different VAM fungi can occur (Mosse, 1977; Abbott & Robson,

1982b), but varies according to the fungi concerned, and may be related to their individual aggressiveness (Wilson, 1983, 1984b). Different fungi can occur in the same transverse plane of the root, and can occupy adjacent intercellular spaces (Tommerup, unpublished observation). Whether they can invade adjacent cells, or even the same cell, is unknown. However, it is most common for endophytes to be present singly and for colonisation by different fungi to occur in different parts of the root.

Competition at a site of penetration is an extension of competition for space. Where roots are not uniformly susceptible to infection, the number of penetration sites will be limited, and their positions will vary with root growth and distribution. Localized susceptibility of roots to infection occurs in several plant–microbe interactions. For instance, rhizobia frequently infect through root hairs (Dart, 1977) whereas many pathogens invade the subapical region of the root tip (Huisman, 1982). However, host roots offer very little resistance to infection by mycorrhizal fungi. If all parts of the root are equally susceptible to infection and fungi do not differ in their preferences, then any particular root should provide sites for very large numbers of entry points. This certainly would be enough to allow most active propagules in the average inoculum population to commence colonisation. Nevertheless, data of Wilson (1984a) suggest that in subterranean clover, competition for penetration sites does exist and can occur even between different members of the same population. Smith and Walker (1981) and Hepper (1985) showed that in subterranean clover the area behind the root tip is more likely to become infected than older parts of the root. Hepper (1985) also demonstrated differences between plant genera, with penetration in leek being less affected by root age than subterranean clover.

Ecologically, VAM fungi are obligate biotrophs. The main benefit they derive from the mycorrhizal association is energy in the form of organic nutrients (Rhodes & Gerdemann, 1980; Finley & Söderström, this volume, Chapter 5). Such nutrients also could be the basis of competition among fungi. In fact, competition for sites of penetration might actually reflect competition for nutrients in root exudates within the rhizosphere or at the root surface. The location of the competitive interaction may occur in one or more of the following four places: (1) in the rhizosphere, (2) at the root surface prior to penetration, (3) in the cortex shortly after penetration, or (4) if more than one organism has gained entry, in the root at large. Unfortunately, knowledge of the nutritional requirements of germinating propagules, or of the physiological interactions between the host and fungus, is incomplete, and discussion on the possible nature of any competition for nutrients would be purely speculative.

Root Colonisation: Differences between Fungi

The Concept of "Infectivity"

In comparing interactions between fungi, infectivity becomes critical. However, because it is influenced by many factors, it is difficult to construct a definition that is universally applicable. In a loose sense, infectivity is the amount, usually length, of root colonised by a particular fungus. This is not ideal, because it does not take into account the total active fungal mycelium. However, it is the most practical definition to use until more sophisticated methods of measurement are available. It is important to distinguish this concept of infectivity from that of Plenchette et al. (1987); their definition was used to compare soils rather than specific fungi and was restricted to the initial penetration.

If comparisons of infectivity are to be made between fungi, it is necessary to specify the environment. Relevant parameters include the host, the time frame, the inoculum density, and other factors which affect the result. Abbott and Robson have used the term infectivity as both the total amount of root colonised to a specified time and the rate at which infection was formed (Abbott & Robson, 1981a, b, 1982b). Wilson (1984a) also incorporated inoculum density.

Few comparisons have been made of the infectivities of different fungi and the importance of the uniformity of the inocula used has rarely been acknowledged. Sanders et al. (1977), Abbott and Robson (1981a, 1982b), and Hayman and Tavares (1985) found considerable variability in the infectivities of various fungi, both indigenous and introduced, but their comparisons were probably confounded by differences in inoculum potential. The form, age, and treatment of the inoculum, or the preparation methods of a parent culture can affect the ability of the inoculum to subsequently form infection (Abbott & Robson, 1981b; Wilson & Trinick, 1982; Tommerup, 1983a, 1984a). An even greater component of inoculum potential is propagule density. Infection development exhibits a curvilinear response to inoculum density (Carling et al., 1979; Smith & Walker, 1981; Wilson, 1984a; Khan, 1988) but the maximum level (or plateau) of infection achieved varies among fungi (Wilson & Trinick, 1983), as depicted in Figure 7.2. This led Wilson (1984a) to redefine infectivity as "the plateau level of root length infected, . . . by any given fungus in response to inoculum density." She therefore included a time factor as well as other constraints such as the host, soil pH, temperature, or phosphorus concentration (Jasper et al., 1979; Smith & Bowen, 1979; Rhodes & Gerdemann, 1980; Wilson & Trinick, 1982; Abbott & Robson, 1985a; Hayman & Tavares, 1985; Thomson et al., 1986).

In the investigations which have quantitatively compared the infectivities of different fungi, a wide range of species and conditions have been used. Thus it is very difficult to compare or comment on the superiority of any particular fungal species or isolates. However, close examination of the available data suggests that fungi such as *Glomus fasciculatum, Glomus tenue, Glomus mosseae, Glomus caledonium,* and *Acaulospora laevis* were usually highly infective, whereas species of *Gigaspora* and *Scutellospora* (e.g., *Gigaspora decipiens* and *Scutellospora (Gigaspora) calospora*) usually produced lower levels of infection than other species (e.g. Mosse, 1972; Abbott & Robson, 1977, 1978, 1981a,b, 1982b, 1985b; Sanders et al., 1977; Powell, 1979a; Daniels & Menge, 1981; Wilson, 1983, 1984a,b; Tommerup, 1984a; Hayman & Tavares, 1985; Thomson et al.,

EFFECT OF INOCULUM DENSITY ON DIFFERENT FUNGI

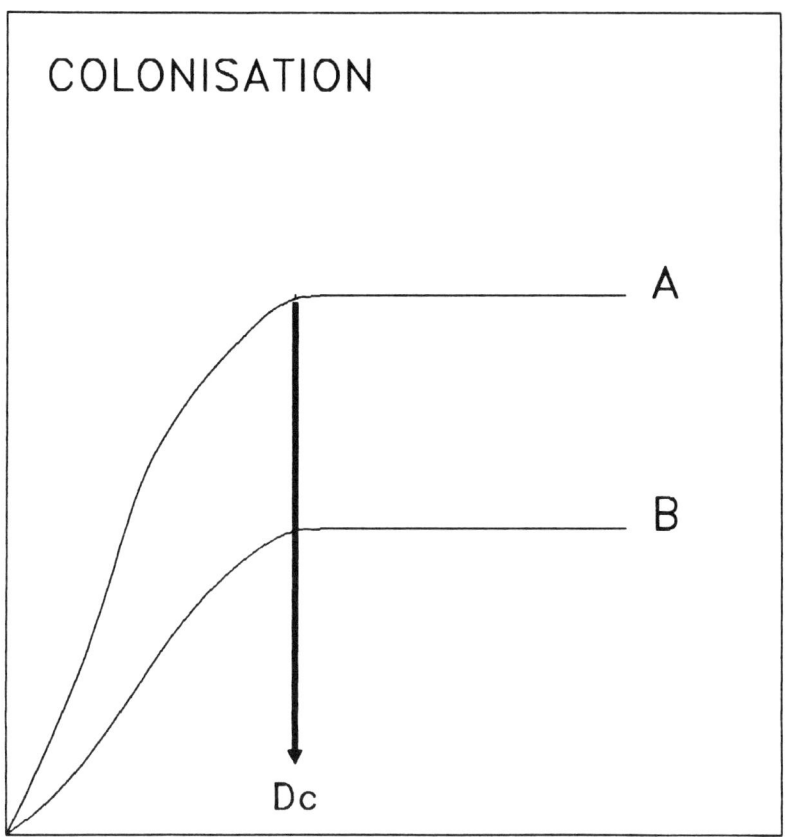

From Wilson, 1983

Figure 7.2. The relationship between inoculum density and infection for two fungi (A and B) at a particular harvest time. Both fungi have the same critical inoculum density (Dc), but A is more infective than B. From Wilson (1983).

1986, 1990; Miller et al., 1985, 1989; Lopez-Aguillon & Mosse, 1987; Hepper et al., 1988).

On standardization of the propagule density, infectivity becomes a measure of the ability of each fungus to spread within a root (Wilson, 1984a). For example, she showed that *Gl. fasciculatum, Gl. tenue,* and *Gig. decipiens* produced similar amounts of infection from single propagules, but they were differently infective in that they produced quite different amounts of infection as the inoculum density increased. *Gl. fasciculatum* became the most infective and *Gig. decipiens* the least. The reasons they differ in their infectivities may have been related, in part, to differences in intraspecific competition. In some fungi (e.g., *Gig. decipiens*), existing colonies limited the ability of the expanding hyphae to form secondary entry points. Differences in the abilities of the fungi to spread inside (rather than outside) the root probably also contributed to their infectivities.

Infection Development

Because infectivity is a significant factor in determining effectiveness, the infection characteristics of different fungi need to be compared. The infection process can be divided into a series of steps as follows: propagule activation, hyphal growth in the soil, interception of the hyphae by a root, formation of a primary entry point, ramification of mycelium within the root to form an infection unit, formation of secondary entry points and infection units, formation of discrete areas of infection (or infection segments) composed of several infection units, formation of extraradical hyphae, formation of propagules such as spores, and so on (Wilson, 1984a). The process has been described in detail by Cox and Sanders (1974), and various mathematical representations have been proposed (Smith & Walker, 1981; Buwalda et al., 1982; Sanders & Sheikh, 1983; Walker & Smith, 1984).

Infection development by VAM fungi is essentially polycyclic (Pfender, 1982; Fig. 7.1), and both primary infection (penetration) and secondary spread are important factors in the colonisation of a root system. To be able to quantify and model the process of infection it is important to differentiate between these two stages (Bassett & Gilligan, 1988). Such differentiation is also significant if VAM fungi are to be compared as they can possess characteristics that will give them a different competitive advantage at each of the two stages. The basic processes contributing to infection development can thus be broadly grouped into two parts: those occurring outside the root prior to infection, which influence primary penetration, and those occurring between root segments, which are related to secondary spread. In addition, external factors such as inoculum density, physiological, soil, and microclimatic effects will influence fungal expansion. Each of these matters will be discussed in turn.

Initiation of Infection

A primary entry point is formed the first time a hypha from any propagule, including one from another mycorrhiza, enters a particular root. (For theoretical purposes it is assumed that this is the first infection on the plant, but in practice infection by another VAM fungus may already have occurred elsewhere on the root system; this will be discussed later with respect to the effect of prior colonisation on the competitive ability of an endophyte.) The number of primary entry points formed on a plant by a particular fungus is equivalent to its inoculum potential sensu Garrett (1956), with the practical proviso of Bouhot (1979). (The latter stated that "the number of successful infections . . . is the only valid measure of inoculum potential.") Thus the inoculum potential does not necessarily equal and may be less than or even greater than the number of fungal propagules present in the soil. The best way of measuring the number of primary entry points is by using the Most Probable Number method (Wilson & Trinick, 1982). As this is an end-point dilution method, it excludes the possibility of counting secondary entry points.

The number of primary infections formed at any particular instance in time is a reflection of interactions between inoculum density, propagule size and activation, and the rate of root interception. From an equivalent number of propagules, the main difference between fungi in relation to the initiation of infection appears to be the speed with which primary entry points are formed. Different fungi can produce similar amounts of infection from single infection points, but large propagules initiate first (Wilson, 1984a). For fungi with small propagules such as *Gl. tenue,* or to a lesser extent *Gl. fasciculatum,* the lag phase prior to entry-point formation may be considerable (Wilson, 1984a; Abbott & Robson, 1982b, 1985b). Because root interception and the initiation of infection are influenced by propagule activation, root geometry, and the rate of root growth, the number of primary entry points will also be affected by external factors such as soil temperature, phosphorus, and pH. These are discussed later.

With respect to characterizing primary entry points, the whole root system is, in theory, a single entity. However, in practice, a root system may be comprised of a number of parts which are far enough apart to act interdependently. Interactions between different parts of the root may be dependent on factors such as the ability of the fungus to form secondary infections (Wilson, 1984a), its particular demand for resources from the root (Thompson et al., 1986, 1990), changes in root respiration rate (Bass & Lambers, 1988), and the efficiency of translocation of carbohydrates within the root system. Thus, the formation of primary entry points a sufficient distance from one another to avoid secondary interference is an important factor in obtaining the maximum possible infection development. This can be achieved by distributing the inoculum more widely, in a larger volume of soil (Smith & Smith, 1981; Abbott & Robson, 1984b), and is probably of particular relevance for species with low infectivity.

Progress of Colonisation: Internal and External Spread

Once infection has been initiated, colonisation proceeds both within a root to form discrete infection units, and by the growth of external hyphae that "run" along the root and repeatedly form new ("secondary") entry points. Secondary entry points can arise in several ways (Wilson, 1984a, Fig. 7.3), including hyphal spread from an existing infection to a new part of the root. In a mature infection, it is often impossible to distinguish between the individual parts of an infection because infection units and segments progressively overlap one another. Thus, it is possible to measure them accurately only in very young infections. Nevertheless, measurements of the dimensions of infection units and segments have proved useful, and have been incorporated into the mathematical modeling and quantitative analysis of the development of infection.

Differences in infectivity among fungi could be caused by differential rates of hyphal growth on or in roots. Wilson (1983) showed that single infection units of *Gig. decipiens, Gl. tenue,* and *Gl. fasciculatum* in subterranean clover spread longitudinally 1.3, 1.5, and 2.5 mm per day (in one direction), respectively. Using a mixed inoculum, Smith and Walker (1981) and Walker and Smith (1984) reported expansion rates of 1.2 mm/day. Sanders and Sheikh (1983) calculated rates for longitudinal spread of *Gl. caledonium* in red clover of 0.4 mm/day and of *Gl. geosporum* of 1.0 mm/day (F.E. Sanders, personal communication). Because of the technical difficulties inherent in such experiments, it is impossible to know whether these differences are real, but the values are sufficiently similar to suggest that hyphal growth rates per se could account for only part of the differences in infectivity among fungi. However, a fungus that can spread rapidly (either internally or externally) or keep pace with root growth for a longer time, may be able to achieve greater infectivity than a competitor by postponing the inevitable decline in the rate of growth of the infection units (Walker & Smith, 1984).

A factor of major importance in comparing the infectivities of VAM fungi is that they can exhibit different modes of infection spread. Two strategies are significant: the ability to spread within the root cortex and form large infection units, or the ability to form large numbers of infection units and segments, usually from "runner hyphae," traversing the length of the root. In a few comparative studies published, *Gl. fasciculatum* formed fewer entry points, but large infection segments, whereas species of *Gigaspora* and *Scutellospora* were dependent on the formation of many small infection segments (Wilson, 1984a; Thomson et al., 1986). *Gl. tenue* appeared to behave in an intermediate fashion (Wilson, 1984a), although it produced 10 times as many entry points per unit of infected root than *Gl. monosporum* in another study (Wilson & Trinick, 1983).

The formation of a plateau in infection at relatively low inoculum densities, and differences between fungi in the position of the plateau (Fig. 7.2), may be due to intraspecific (or intra-isolate) competition. *Gl. fasciculatum* appears to be less sensitive than other fungi in this respect. This may be due to its ability to spread inside the root, thereby reducing its reliance on the formation of secondary entry points. Different levels of intraspecific competition occur in interactions between plants (Donald, 1963; Harper, 1977), and a similar phenomenon could

INFECTION DEVELOPMENT

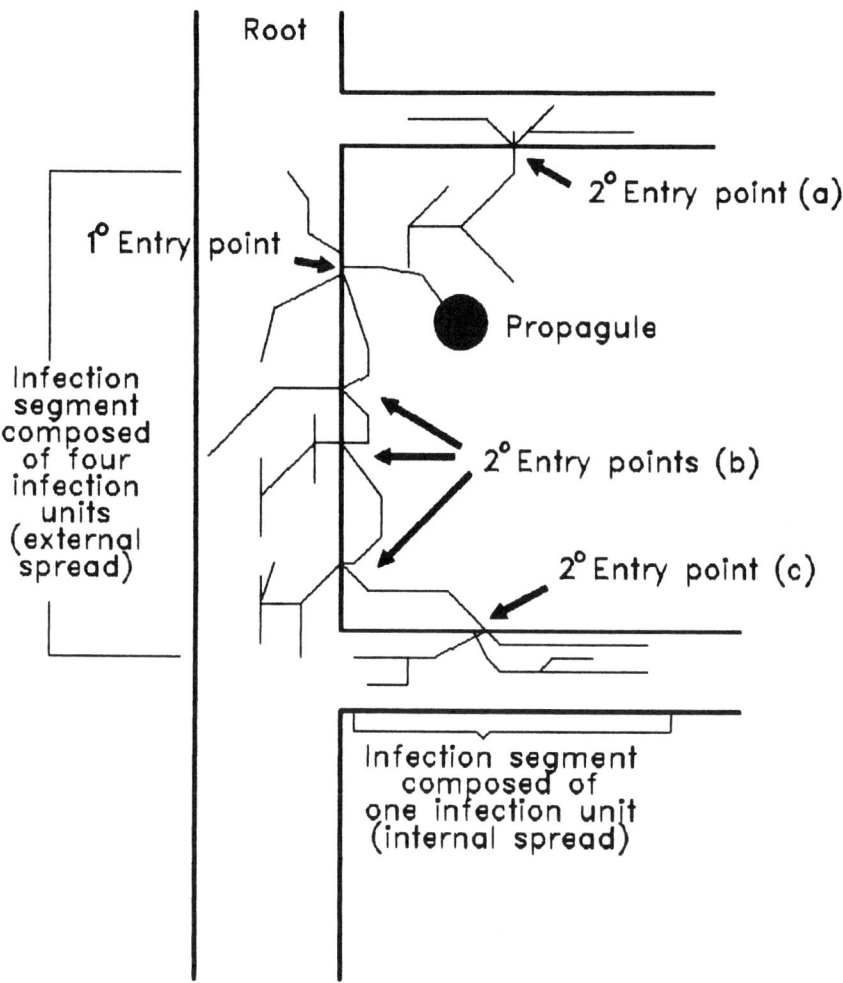

Figure 7.3. Diagrammatic representation of the terms used to describe infection development: infection unit, infection segment, primary (1°) and secondary (2°) entry points, and internal and external spread. Secondary entry points can arise in several ways, for example (a) directly from a propagule, (b) by longitudinal spread of an infection segment, or (c) by hyphal spread from existing infection to a new part of the root. From Wilson (1984a).

occur with fungi. Plants can sometimes be protected against fungal pathogens by prior inoculation with the same fungus (Kuc et al., 1975), and prior occupation by one VAM fungus can reduce infection by a second coloniser (Wilson & Trinick, 1983; Wilson, 1984b). A similar mechanism may inhibit the formation of secondary entry points close to previously colonised tissue in a single fungal isolate. New infective propagules may also be prevented from forming primary entry points. The concept of "controlled hyphal growth" as formulated by Amijee et al. (1986) is probably relevant here. We suggest that the phenomenon is likely to be at least partly related to the carbohydrate status of the root, which will decline with increasing colonisation. In theory, the fungi most affected would be those such as *Gigaspora* and *Scutellospora* spp., which rely for infection spread on the formation of secondary entry points, and which are sensitive to reduced root carbohydrates. This hypothesis is supported by Thomson et al. (1986, 1990), in which *S. calospora* was less infective than *Gl. fasciculatum*.

Factors Determining the Success of Colonisation

Root Interception

To successfully form a primary entry point, a fungus must first arrive at the root surface. This may be achieved by either the growth (or movement) of the fungal propagules or by the growth of the root. Because of the three-dimensional structure of the soil, both the propagule expansion limits and the root density, distribution, and growth rates are important (Bowen, 1979; Huisman, 1982). In developing models of root infection, three characteristics appear to determine the success of primary infection by a root pathogen: the size of the pathozone, the inoculum efficiency, and the decline in infection efficiency with the distance of the propagule from the root (Gilligan & Simons, 1987; Gilligan, 1989a,b). The theory is directly applicable to infection by VAM fungi (e.g. Sanders & Sheikh, 1983; Walker & Smith, 1984). However, because it is difficult to differentiate between primary penetration and secondary spread, in some quantitative studies of the development of infection only secondary spread may have been measured (e.g., Buwdala et al., 1982; Abbott and Robson, 1984b). This may underestimate the importance of the initiation of primary infection.

Because prior colonization is a determinant of competitive success (Wilson & Trinick, 1983), any factor that promotes the initiation of primary entry points by one fungus rather than another will have a significant effect on the outcome of interactions between fungi. Root interception will be influenced by many factors, including the rate of root growth, the density of roots available for infection, the activation of propagules, or the rate of hyphal extension from them. For instance, temperature differentially affects spore germination and hyphal growth of different fungi (Tommerup, 1983b) and root growth. Thus, increasing the soil temperature can speed up primary entry point formation. In a study by Wilson and Trinick

(1982), soil temperature was shown to have a greater effect on *Gl. tenue* than on *Gl. monosporum*. Similarly, moist incubation of the inoculum can speed up infection by various fungi, including *A. laevis* and *Gl. tenue* (Abbott & Robson, 1984a; Tommerup, 1984a; Wilson, 1984b). Dormancy will also be important by altering the inoculum efficiency. VAM fungi differ in their dormancy requirements. For example, *A. laevis* has a very long dormant phase (Tommerup, 1983a). Root density will affect interception in a similar way to propagule density (Warner & Mosse, 1982; Abbott & Robson, 1984b), and will be more important for fungi with small propagules or those able to grow only short distances. Soil phosphorus can also have differential effects. Thomson et al. (1986) showed that *S. calospora* produced more entry points that *Gl. fasciculatum* at seven different phosphorus levels, but was also much more sensitive to increasing concentrations of phosphorus. Amijee et al. (1989) showed that *Gl. mosseae* was similarly affected by phosphorus. However, it is probable that a large proportion of these entry points was secondary rather than primary ones, and hence the effect was related to infection spread rather than to the initiation of infection per se.

Inoculum Density and Propagule Distribution

Infection development by a single fungus is influenced by inoculum density in several ways. The most obvious is that increasing the inoculum density decreases the lag phase before infection commences and thereby increases the rate at which infection develops (Fig. 7.4; Wilson & Trinick, 1983). Very low inoculum densities may not result in a plateau of infection for a long time.

At any particular time, there is also a plateau in infection at a certain critical density (Fig. 7.2). However, this plateau will vary among fungi. For example, in Wilson's studies *Gl. fasciculatum* was the most infective, then *Gl. tenue, Gig. decipiens,* and *Gl. monosporum* in decreasing order (Wilson & Trinick, 1983; Wilson, 1984a). The critical density (Dc) may be different for different fungi and may change with time. However, the value of Dc in young subterranean clover roots was similar for four different fungi, and lay between 2 and 5 propagules per gram (Wilson & Trinick, 1983; Wilson, 1984b). As discussed by Wilson (1984a), other workers have observed maximum infection at similarly low levels of inoculum density (Wilson, 1984a).

The level of the plateau depends on the volume of inoculum as well as its density, as the spatial distribution of propagules with respect to the growing root system will affect the amount of infection spread possible. Thus, if the inoculum density is greater than the critical density, infectivity could be increased by distributing the same level of propagules through a larger volume of soil (Fig. 7.5). This both decreases the inoculum density and changes the spatial distribution of propagules in relation to the growing root system. Whereas delayed root interception will slow the rate of infection development, a wider distribution of propagules will reduce the amount of interference between infection units and

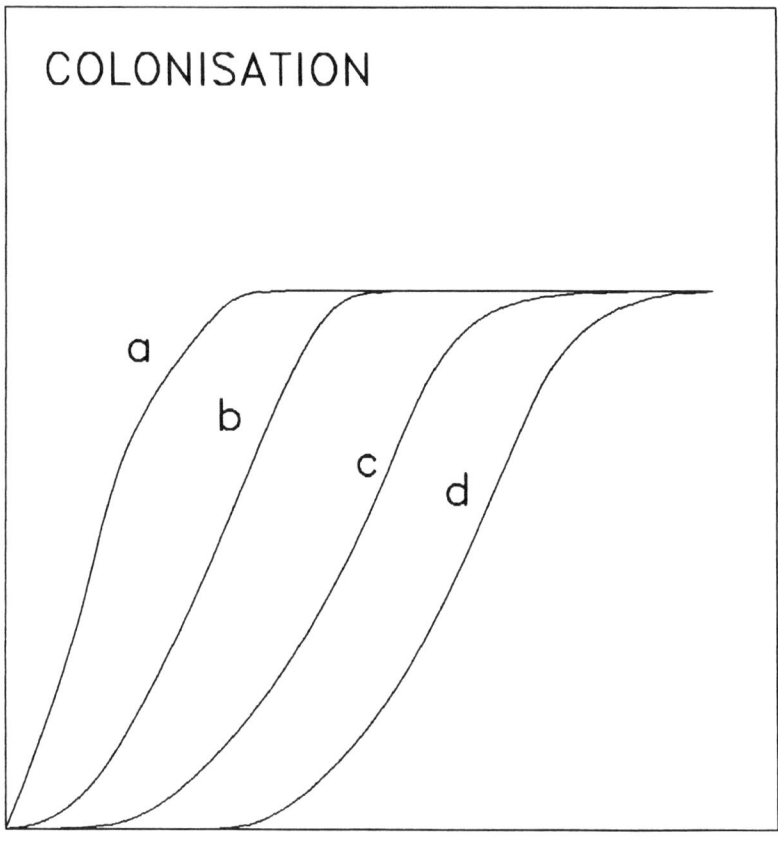

Figure 7.4. The effect of inoculum density on the development of VAM infection by any one fungus. (a)–(d) Different inoculum densities that vary from high (a) to low (d) in decreasing order. From Wilson (1983).

EFFECT OF PROPAGULE DISTRIBUTION OVER TIME

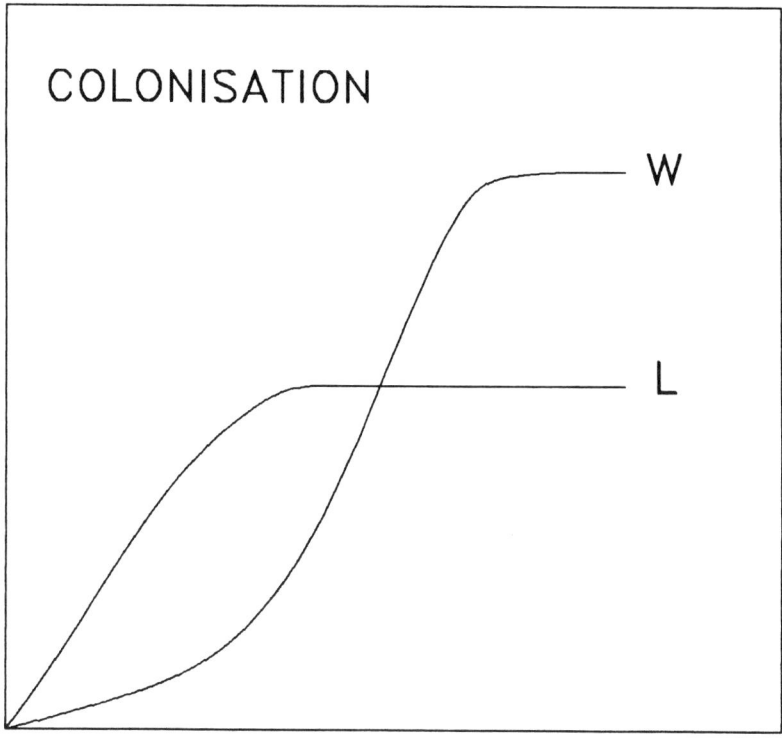

Figure 7.5. The development of infection by the same fungus, from inocula containing the same total number of propagules, but with different propagule distributions. W represents a wide distribution of propagules and L a localized distribution. From Wilson (1983).

allow greater spread of infection from each propagule (Smith & Smith, 1981; Abbott & Robson, 1984b, 1985b; Wilson, 1984b).

We predict that the differences between curves W and L in Figure 7.5 will be greatest for fungi that have low infectivity from localized inoculum. A fungus that is infective in such a system is less affected by inoculum density. Although dispersing the inoculum will delay infection, it may have little effect on the plateau (compare A_w and A_L in Fig. 7.6). By contrast, for a fungus that has

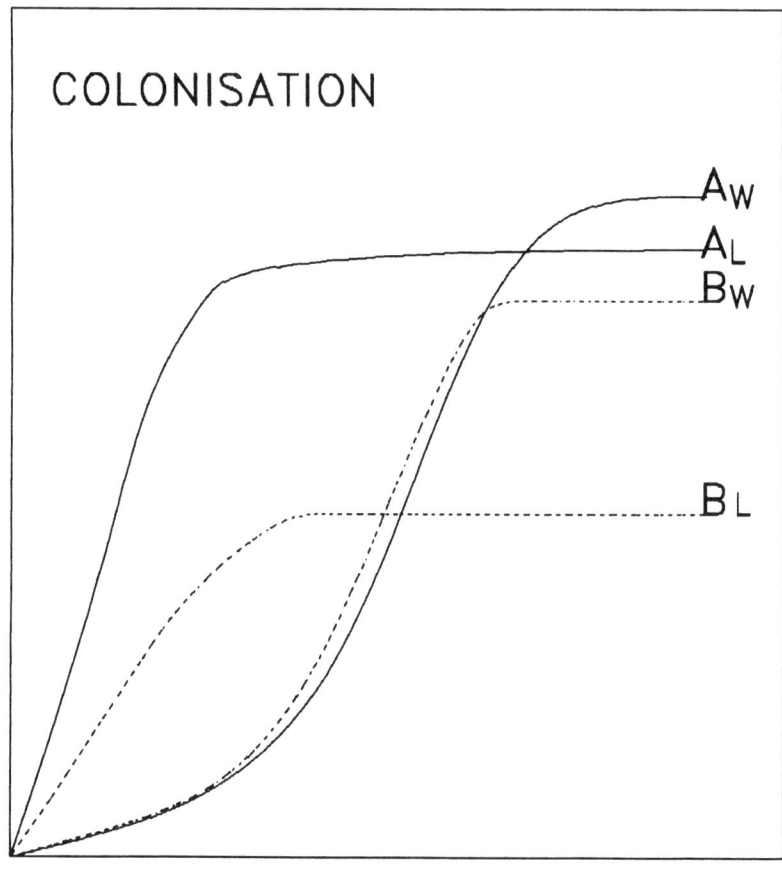

Figure 7.6. The predicted effect of propagule distribution on the development of infection by two fungi that exhibit either high (A,———) or low (B, – – –) infectivity from localized inocula. The inoculum levels are the same, but the propagules are either locally distributed (A_L, B_L) or dispersed (A_w, B_w). From Wilson (1983).

low infectivity from localized inoculum (fungus B, Fig. 7.6), dispersing the propagules may increase the infection plateau (the "infectivity") considerably. If this is so, then factors such as the density and distribution of different fungi will have considerable effects on the outcome of any interactions between them.

Propagule Size, Dormancy, and Growth

An interesting analysis by Smith et al. (1986) has shown that the apparent width of the "rhizosphere" (a term synonymous with Gilligan's "pathozone") at any particular time is variable and dependent on factors that alter the activation or growth of a propagule. Because of the importance of root interception, any factor that affects the speed of formation of primary entry points is important. Although asexual spores are a common propagule of VAM fungi (Wilson & Trinick, 1982; Tommerup, 1984a), infection can be initiated by several other structures, including severed hyphal fragments, fresh and dried roots, and sexual spores (Hall, 1976; Powell, 1976; Tommerup & Abbott, 1981; Tommerup, 1988a). However, because of the difficulty of equalizing inoculum potential there is little information available on the comparative effectiveness of different inoculum forms. Nevertheless, the usefulness of various forms of inoculum may vary among fungi. For instance, Tommerup and Abbott (1981) showed that dried roots of subterranean clover were an effective form of inoculum for *Gl. fasciculatum, Gl. monosporum* and *S. calospora,* but not for *Gl. caldeonium* or *A. laevis*.

Some data suggest that the size of propagules may also influence the rate of infection. The data of Wilson et al. (1983) and Wilson (1984a), for example, show that *Gig. decipiens,* which has very large spores, and produces a large mass of germ tubes, infected much more rapidly than either *Gl. fasciculatum* or *Gl. tenue*. Daniels et al. (1981) showed that *Gl. mosseae* had both the largest spore size and the most rapid infection compared with five other fungal isolates. While infection efficiency is theoretically not directly proportional to propagule size (Gilligan, 1989a), the spores of VAM fungi are largely (though not exclusively) dependent on the use of endogenous rather than exogenous nutrients (Beilby, 1980; Hepper, 1983a). Thus their size may well be important. Smith et al. (1986) showed that the effective "rhizosphere" for *Gl. mosseae* was larger than for some unspecified indigenous fungi whose small-spored propagules that could not be isolated by wet sieving (Smith & Bowen, 1979).

The rate and extent of hyphal growth also differ among fungi. The germ tube of *Gl. caledonium* grew at a much slower rate than that of either *Gig. calospora* or *A. laevis* (Tommerup, 1983b). However, temperature may differentially affect these processes. For example, several species of *Glomus* and *A. laevis* have similar optima for spore germination of 20–25°C decreasing above 25°C. Alternatively, *Gigaspora* spp. tend to have the capacity to germinate at higher temperatures (Tommerup, 1983b). A complicating factor is that spores of many species of VAM fungi are unable to germinate when they are first formed. Tommerup (1983a) found that the dormancy period in wet soil was 6 weeks for *Gl. caledonium* and *Gl. monosporum* and 12 weeks for *S. calospora*. This time period was reduced by at least half in dry soil. However, the dormancy period of *A. laevis* was very long (6 months), and was not affected by soil moisture.

Although some research suggested that propagule activation and growth are not influenced by the presence of a host root (Daniels & Trappe, 1980; Tommerup, 1983a,b), root exudates may affect the subsequent formation and development of infection (Ratnayake et al., 1978). Hosts may differ in this respect. Thompson et al. (1986, 1990) argued that in subterranean clover, the conditions inside the root are more critical than root exudates. No studies have so far shown that fungal isolates or types of propagules are differentially affected by root exudates per se.

Production of External Hyphae

The term "external hyphae" refers to hyphae which are formed by and maintain links with an infected root. Their putative significance is in the uptake of nutrients. They act by extending the absorption surface available, and in the uptake of an immobile nutrient such as phosphorus, their wide distribution is probably as important as their absolute length (Abbott & Robson, 1985b). A detailed discussion of the role of external hyphae in determining the effectiveness of the symbiosis is presented in Read (this volume, chapter 4). However, they also act to spread secondary infection along and between roots.

Although fungi differ in their ability to produce secondary entry points, it is unclear whether different isolates produce differing hyphal lengths. Sanders et al. (1977) found a high correlation between infected root length and external hyphae in onions, and noted that the quantity of external mycelium was similar for *Gl. mosseae, Gl. macrocarpum, Gl. microcarpum,* and *S. calospora*. On the other hand, others reported differences in external hyphal mass and root infection frequency between genera, species, and isolates of fungi (Graham et al., 1982; Abbott & Robson, 1985b). Graham et al. (1982) suggested that four isolates of *Gl. fasciculatum* produced more external hyphae than *Gl. etunicatum, Gl. macrocarpum,* and two other isolates of *Gl. fasciculatum* in citrus. In contrast, Abbott and Robson (1985b) found that in subterranean clover *Gl. fasciculatum* produced less external hyphae than *Gl. tenue, A. laevis,* and *Gig. calospora*. Interpretation of these results is complicated by the difficulties of measuring external hyphae and the different methods used. It is difficult to comment further on the significance of any differences, except to note that the effects may be complicated by other factors, including soil pH and host phosphorus concentrations (Hayman & Tavares, 1985; Abbott et al., 1984; Thomson et al., 1990). It is worth noting that despite the apparent paucity of external mycelium produced by *Gl. fasciculatum,* this fungus still managed to increase plant growth (Abbott & Robson, 1985b).

With respect to infection development, the expansion of the fungus by external hyphae will depend on the rate of internal growth by that fungus. For example, both Abbott and Robson (1985b) and Wilson (1984a) reported that *Gl. fasciculatum* produced less external hyphae but spread internally more rapidly than other fungi.

Propagule Production

Growth and reproduction of a VAM fungus are determined by the life cycle of that fungus. Much of the life cycle occurs in association with host plants, especially the production of new propagules. Critical phases in the survival of fungal populations include persistence of innately dormant or quiescent propagules and subsequent germination or regrowth of those propagules, as they contact a susceptible root, colony survival, and production of new propagules before root death. Progeny of VA fungi include colonized roots, extraradicle hyphae, and other propagules (Tommerup, 1985a, 1988a). The other propagules are structures which persist separately from living roots and spores such as detached extraradicle hyphae and hyphae associated with dormant, senesced, or detached roots (Jasper et al., 1989; Tommerup, 1985a, 1988a,b; Tommerup & Abbott, 1981; Tommerup & Bett, 1985; Tommerup & Sivasithamparam, 1990). Major roles of propagules in interactions between fungi are perennation, display of genetic variation, and dispersal.

Asexual spores are a dominant propagule for many fungi in agricultural soils, particularly those having a fallow period. Zygospores have been found only for *Gigaspora decipiens* and they are produced from extraradicle mycelium or germ tubes of asexual spores (Tommerup, 1988a; Tommerup & Sivasithamparam, 1990). Dead root fragments containing viable mycelium can be classified as asexual spores (Tommerup, 1988b,c). The ability to produce these mycelial propagules is species specific for the genus *Glomus* but may be genus specific for *Acaulospora* (Tommerup, 1982, 1988c; Tommerup & Abbott, 1981). Mother cells of *A. laevis* can be propagules but those of *A. trappei* have not behaved similarly (Tommerup, 1988c). Inoculum consisting of young detached roots is effective inoculum for many species but has not been critically examined in comparative studies. Likewise, comparative studies of detached extraradicle hyphae have not been made.

Survival of propagules is influenced by their physiological state, storage conditions, and time (Tommerup, 1985b, 1988c; Tommerup & Bett, 1985). Spore chemistry is greatly influenced by the condition under which spores form. When spores of *A. laevis, Gl. caledonium, Gl. monosporum,* and *S. calospora* were produced under similar conditions, then stored at a series of different temperatures and water potentials, survival was affected by the storage conditions, but all species behaved similarly. Additionally, for all these fungi, when spores of an isolate were produced under a series of different conditions and stored in one environment they had a corresponding series of survival levels that were correlated with the production conditions and not the isolate. For many soil physicochemical factors, the effects on quiescent spore survival may be the same for all fungi.

Dormancy is a physiological state enabling the spores to persist when conditions are unfavorable for growth and development. The physiological capacity to germinate immediately after formation in soil depends on whether spores are heritably dormant (dormant) or non-innately dormant (quiescent). Innate dormancy of asexual spores varies with the species. For some (such as *A. laevis*), it lasts for several months, but for others, it is a few weeks (e.g. *S. calospora*) or a few days (e.g., *Gl. caledonium*). Dormancy is broken in asexual spores of some species by desiccation, but in others, such as *A. laevis* and *Gig. decipiens*, time is the only known factor. For most species of VA fungi there is no information about spore dormancy. Biological factors can prevent germination of asexual spores in the soil in which they formed and in other soils (Wilson, 1984c; Tommerup, 1985c). In examining interactions among fungi, it is important that dormancy and soil factors inducing quiescence are considered.

Provided there are no biological factors preventing germination of quiescent spores, the time lapsing before germination differs among the species when compared at the optimum temperature and water potential for each species (Tommerup, 1983a,b, 1984b, 1988b). For fungal populations produced and stored under a standardized condition and behaving near synchronously, the time between the initiation of the first biochemical events of germination and germ tube emergence for half-maximum germination varied from 5½ days for *Gl. monosporum* and *Gl. caledonium* to 7 days for *Gl. monosporum* and 9 days for *A. laevis*. The lag before any germination was 7 days for *A. laevis* and 3 days for the other species. In near-synchronous populations the lag phase differed but the germination rate was similar, whereas in asynchronous populations both were significant characteristics and may affect interactions. Differences in the lag phase are inherent for isolates and possibly for species (Tommerup, 1988b). The differences are due to variable lengths of the four major processes that precede germ-tube emergence (Tommerup, 1988a). The lag phase can be reduced to 1 day for *S. calospora, Gl. monosporum,* and *Gl. caledonium* and 2 days for *A. laevis* by priming spores using repeated short periods of hydration and controlled desiccation (Tommerup, 1984b; Tommerup, 1988c). In investigations of interaction between fungi, processes occurring prior to penetration can have a significant effect. Minimization of the lag phase increases uniformly in germination time and helps to identify differences between fungi in colony initiation (Tommerup, 1985a).

Prepenetration phases, colony initiation, colonization of the host, and spread of the fungus to other roots are all developmental phases that directly affect propagule production both quantitatively and qualitatively. Some soil factors, particularly very high or very low host phosphorus or nitrogen status, reduce the development of all stages, and, in effect, are equivalent to reduction in the host susceptibility. Although spore germination and hyphal growth to the host may not be affected by a host of low susceptibility, further development after penetration is greater reduced (Tommerup, 1984c). The size of propagule populations in field soils is depressed by previous crops of low to negligible susceptibility (Iqbal & Quereshi, 1976; Black & Tinker, 1979). Growth of VAM fungi in some plants of highly resistant hosts such as lupin and oil seed rape may be a feature of an isolate rather than a species. At present, no series of differential hosts for distinguishing among VAM fungi have been elucidated and such factors are unavailable for investigating interactions among the fungi.

Soil Factors

Many factors will alter the ability of a fungus to colonise roots, with the most important being the phosphorus status of the soil relative to that for the maximum growth of the host. Others include climatic and environmental factors which interact with the soil textural and physicochemical characteristics such as soil pH, soil water potential, temperature, microbiology, and interactions among all the factors. Superimposed on them are the previous cropping history and the current crop, particularly in terms of root architecture and host susceptibility and their influence on propagule numbers and distribution.

Formation of penetration points, colonisation, length of external hyphae, and spore numbers are highest at phosphorus levels which are moderate for the particular host but are reduced at very low or very high phosphorus host levels (Abbott et al., 1984; Thompson et al., 1986, 1990). Germination and hyphal growth of various fungi is differentially affected by high phosphorus levels in agar, however, in soils they were not changed by levels up to 20% greater those adequate for maximum growth of several crop and pasture species (Hepper, 1983b; Pons & Gianinizzi-Pearson, 1984). Fungi can differ in their ability to colonise plants at the same high levels of phosphate supply (Thompson et al., 1990). Decline in colonisation is correlated with decreasing levels of soluble carbohydrates rather than internal plant phosphorus concentrations (Thompson et al., 1990). Soluble carbohydrate is likely to have a direct effect on sporulation but experimentation is required to determine whether it affects species differentially.

Soil pH may prove to be as important as phosphorus in determining both the success of colonisation and the outcome of interactions between fungi. Fungi clearly differ in their pH optima, with some being active at pH 4, but most preferring lower acidity levels (Abbott & Robson, 1985a; Hayman & Tavares, 1985). Penetration and colonisation are also decreased by increasing nitrogen fertilizers (Chambers et al., 1980), by high soil moisture levels (Sondergaard & Laegaard, 1977), and by inadequate light duration or intensity (Thompson et al., 1990).

Extraradicle mycelium, measured as hyphal walls by several techniques, varies with phosphorus supply, pH, and other factors (Abbott et al., 1984; Abbott &

Robson, 1985a; Bethlenfalvay et al., 1982; Graham et al., 1982; Schubert el al., 1987). However, whether the proportion of extraradicle mycelium that is living varies independently under changing conditions has not yet been measured and its significance to interactions is difficult to assess.

The Physiology and Anatomy of the Symbiosis

One limitation to the concept of infectivity is that most experimental comparisons have been based on measurements of root length infected (or percentage infection) and few have seriously taken into account differences in the volume of root infected (infection "intensity"), or in the frequency of formation of arbuscules or vesicles within the root. Measurements of *Glomus mosseae* infection in leek have shown that a uniform level of hyphal density was produced in all parts of the root (Amijee et al., 1986, 1989). There have been observations of differences in infection intensity among fungi: Abbott and Robson (1985b) found that *Gl. fasciculatum* produced more intense infections in subterranean clover than *Gig. calospora*. However, Wilson (1984a) suggested that in young roots these effects may be related to inoculum density. She found no differences in infection intensities between *Gig. decipiens, Gl. fasciculatum,* and *Gl. tenue,* but in all cases, intensity increased with increasing inoculum density.

Although there are characteristic morphological differences between fungi in the presence or abundance of vesicles (Abbott, 1982), these are primarily storage organs (Rhodes & Gerdemann, 1980), and probably have only a minor effect on the effectiveness of the symbiosis. It is possible that the development of large numbers of vesicles denotes an aging infection (Abbott & Robson, 1984b). An infection containing many arbuscules but few vesicles is usually in its most active phase of colonisation (Douds & Chaney, 1982), and as arbuscules are the primary site of nutrient exchange (Cox & Sanders, 1974), their number and activity are very important. As arbuscules remain active for only a few days (Bevege & Bowen, 1975; Cox & Tinker, 1976), the extensive spread of infection and formation of new arbuscules are probably essential if the symbiosis is to remain operating at maximum efficiency. There is little comparative data available on arbuscules. Thomson et al. (1986) showed that in subterranean clover, *S. calospora* produced similar numbers of arbuscules to *Gl. fasciculatum* relative to root length infected, but fewer in terms of total fungal biomass. Miller et al. (1989) observed that colonisation of apple by *Gl. mosseae* was more "vigorous" than that of *Gl. macrocarpum* in that the former produced arbuscules for a longer time, and infection by the latter was characterized by vesicles after a relatively short period of time. However, for any particular fungal isolate, its anatomy in a host appears to be consistent. Amijee et al. (1989) showed that although the density of hyphae and arbuscules, and the numbers of entry points of *Gl. mosseae* were reduced by phosphorus, the ratios of the various parameters were unchanged. Nevertheless, there are clearly qualitative differences between fungi in the anatomy (and probably also the physiology) of infections. It is advisable that future work take this into account when measuring infectivity.

Another problem encountered when attempting to compare fungi, and discuss interactions, is that the host species will affect the success of the symbiosis. Although a few hosts, such as lupins and various brassicas, have low to negligible susceptibility, with the majority of plants there is an almost complete lack of host/fungus specificity in VA mycorrhizal relationships. Nevertheless, it appears that the infectivity of fungi may be affected by the host. For instance, Plenchette et al. (1982) showed that infection by several different fungi was much less in strawberry than in apple and leek, and the relative infectivities of the fungi differed according to the host plant. Warner and Mosse (1982) showed that the spread of colonisation by *Gl. fasciculatum* was greater in white clover than in fescue, and that the effect of the plant species was greater than that of root density. Lopez-Aguillon and Mosse (1987) found that the relative rates of colonisation by three different fungi were similar in white clover and sorghum, but the differences were more pronounced in the latter host. Daft and Hogarth (1983) suggested that the relative infection produced by four different fungi was different in maize and onion. Maize also supported greater sporulation than onion and the differences were attributed to the differing carbon metabolisms of the two hosts. However, whether such differences are due to differences in the phosphorus status, carbohydrate status, anatomy or root geometry of the hosts, physiology of the fungi, or to some other factor such as the genetics of the interactions, is unknown. Some workers would argue that differences between hosts are largely the result of differences in their phosphorus status, which in turn affects the concentration of soluble carbohydrate in their roots (Thomson et al., 1990), and that an understanding of the phosphorus response curve of a host species, and the relative position of the experimental plants on the curve is important (Abbott & Robson, 1984c). Unfortunately, most mycorrhizal workers have used only one host, usually because they approach the topic from an agricultural perspective that is focused on improving the growth of that particular plant species. This is quite valid, but it confuses the issue when we try to generalize about the relative advantages of one particular fungus over another, or try to predict the outcome of any interaction between fungi.

Competition between VAM Fungi for Root Colonisation

The Concept of "Aggression"

In analyzing interactions between fungi, one of the most difficult problems is to identify which of the many interacting factors are the most important. A fungus that appears to be more successful than its neighbor in colonising a root may do so for any of a number of reasons. For instance, it may grow faster through the soil to the root and penetrate more rapidly, it may have physiological attributes which allow it to exclude another coloniser, it may colonise so rapidly and extensively that it can exclude another coloniser, or it may gain a competitive advantage merely because its inoculum potential is very high or its propagules are distributed in a more favorable way in relation to the root geometry than its neighbor.

Irrespective of whether a particular fungus is "competitive" when in the presence of other fungi, it may be considered to be "effective" because it is "functionally superior" for host growth. For instance, it may form more external hyphae than a competitor or its external hyphae may be more widely distributed. To date there is no evidence that any fungi are intrinsically better at taking up phosphorus, but we cannot exclude the possibility. In addition, the balance between the ability to take up phosphorus and transfer it to the host, and the use by the fungus of the plant's carbohydrate reserves, may differ among fungi; that is, some fungi may be more "efficient" than others (Baas & Lambers, 1988; Thomson et al., 1990). Another complication when comparing the abilities of specific fungi is that interactions between them will have different outcomes depending on the physical and climatic conditions. As mentioned previously, the relative infectivities of different fungi are affected by factors such as soil pH, soil phosphorus level, soil temperature, soil water potential, and the plant host species.

Unfortunately, most work on interactions between fungi has been confounded by using unequal and usually unknown inoculum potentials for the fungi. We believe that the best experimental approach to this topic is to establish the relative competitiveness of different fungi in a system in which the inoculum potentials (i.e., the numbers of primary entry-points) and propagule distributions of each fungus are standardized. Subsequently, various factors in the system can be altered to assess how these changes affect the outcome of the interaction.

In this respect, the concept of the "replacement series," as devised by de Wit (1960) in his study of plant competition, is relevant. It has been used successfully in a study of interactions between mycorrhizal fungi by Wilson (1983), in which she manipulated the inoculum potentials by using the Most Probable Number technique (Wilson & Trinick, 1982). In a replacement series, the proportions of two interacting species (their "frequencies") are varied while the overall density remains constant. The results of such experiments are usually expressed in "replacement diagrams," which can take many different forms depending on the type of interaction occurring. The range of possible outcomes is outlined by Hall (1978), and a very clear explanation of the most common situations is given by Harper (1977). Examples of such diagrams for VA mycorrhizal fungi are given in Wilson and Trinick (1983) and Wilson (1984b), and a typical result is shown in Figure 7.7.

In experiments involving replacement series, it is usual to investigate simultaneously the behavior in monoculture of each species using the same series of frequencies (see Fig. 7.7). If this is done, the relative competitiveness of each species can be assessed. In this case, the concept of "aggression" (Trenbath, 1978)

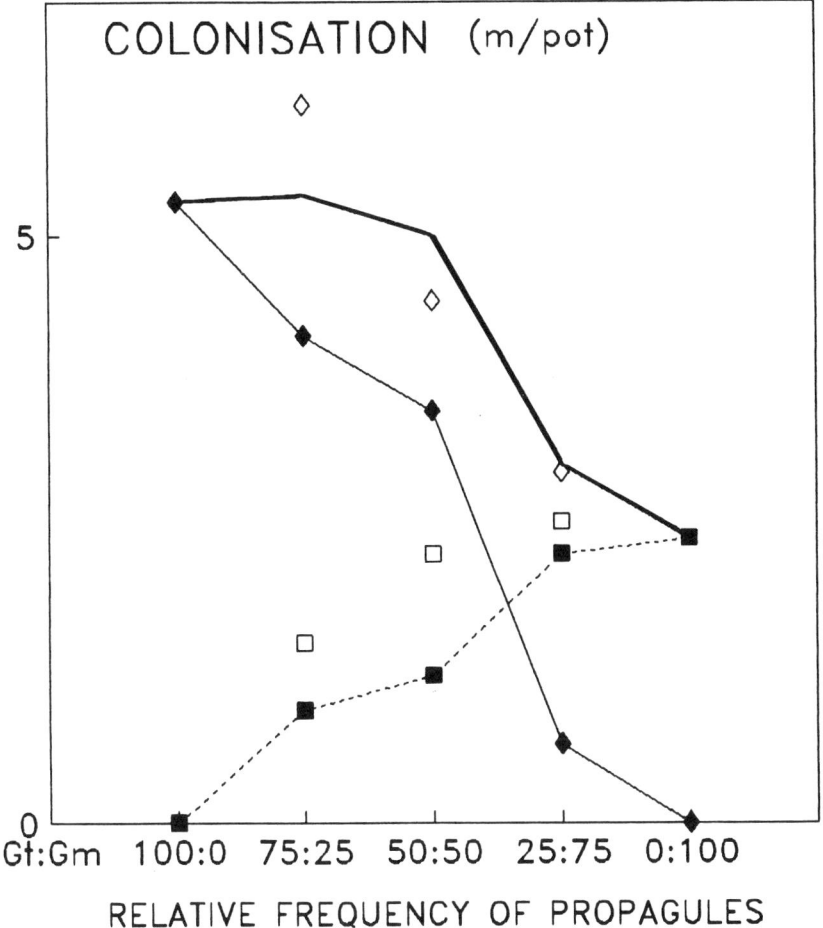

Figure 7.7. Comparison of colonisation of the roots of subterranean clover after 36 days. Replacement diagram for the interaction between *Gl. tenue* (Gt, ——) and *Gl. monosporum* (Gm———). Total infection is also shown (━━━). Open symbols show corresponding infection from single inocula. From Wilson (1983).

is useful. The aggressiveness of a particular species is its relative productivity in mixture compared with its performance alone. One of the simplest measures of it is the "relative crowding coefficient," which can be easily determined from the results of a replacement series experiment. The relative crowding coefficient (k_{ij}) of species i in competition with species j is defined as $k_{ij} = [O_iO_j^{-1}][M_iM_j^{-1}]^{-1}$ where O_i and O_j are the yields of the species in mixtures and M_i and M_j are their yields in monoculture. An even simpler concept is that of "relative yield," where the relative yield (r_i) of species i is $O_iM_i^{-1}$. Table 7.1 shows the relative yields and crowding coefficients for the data depicted in Figure 7.7. *Gl. monosporum* is clearly more aggressive than *Gl. tenue* and can thus maintain its level of colonisation in the competitive situation. Its aggression is greatest at the earlier harvest and when it is present at high frequencies. These data also serve to underline the differences between competitiveness and infectivity. Despite its superior aggression, *Gl. monosporum* is less infective than *Gl. tenue*.

The concept of aggression may be important when selecting a desirable VAM isolate. A fungus chosen because it has proved to be functionally superior (e.g., highly infective, or effective at increasing plant growth) in monoculture experiments will be most useful if it can maintain its infectivity in the face of competition. Wilson (1984b) showed that VA mycorrhizal fungi can vary in their aggressiveness. In her studies in subterranean clover, *Gig. decipiens* was more aggressive than both *Gl. fasciculatum* and *Gl. tenue,* and *Gl. tenue* was usually more aggressive than *Gl. fasciculatum*. In the study discussed earlier (Fig. 7.7), she showed that *Gl. monosporum* was more aggressive than *Gl. tenue* (Wilson & Trinick, 1983). Unfortunately, the trend in aggressiveness for these fungi was

Table 7.1. *Interactions between* Gl. tenue *(Gt) and* Gl. monosporum *(Gm) at a Series of Inoculum Frequencies at Two Harvest Times*[a]

Propagule frequencies Gt:Gm	Values for Gt		Values for Gm	
	At 24 days	At 36 days	At 24 days	At 36 days
Results expressed as % relative yield of fungal infection[b]				
75:25	47	72	88	79
50:50	27	84	107	65
25:75	26	25	187	83
The aggressiveness of the fungi toward one another as measured by the "crowding coefficient"[b]				
75:25	0.5	0.9	1.9	1.1
50:50	0.3	1.3	4.0	0.8
25:75	0.1	0.3	7.2	3.7

[a]Data from Wilson (1983) and calculated from data expressed as root length infected.

[b]Relative yield is the ratio of the yield in mixture to the yield when present alone (see text). The crowding coefficient is the ratio of the relative yields of the two interacting species (see text). For any particular set of data the values for the two fungi are reciprocals of one another.

in the opposite direction to the trend in their infectivities; the fungi that were the least "infective" were the most "aggressive." Similar inferences can be drawn from Daft and Hogarth (1983), who showed that sporulation ability when inoculated alone was inversely related to the ability to sporulate in a competitive situation.

Other Experimental Evidence

Only the studies of Wilson (1984b) and Daft and Hogarth (1983) attempted to examine interactions between fungi using equivalent inoculum levels and distributions. Most studies on interactions between VA mycorrhizal fungi have simply added specific fungi to the seed or planting furrow of soils containing indigenous fungi. Subsequent observations simply recorded plant growth (e.g. Mosse, 1977; Owusu-Bennoah & Mosse, 1979; Powell, 1979b, 1981; Azcon-Aguilar & Barea, 1981; Hall, 1984). Although such experiments can demonstrate the agricultural usefulness of VAM fungi, few have examined the relative infectivities, or any other specific attribute, of each of the interacting fungi. Thus, it is impossible to generalize from such studies about the competitive abilities of the fungi concerned. In addition, the placement of the inoculant fungus (near the seed) compared with the indigenous fungus (dispersed through the soil) complicates any conclusions. Such experiments were successful only where the inoculum of the indigenous population was low, such as in eroded soils or mine spoils (Hall, 1980; Lambert & Cole, 1980).

Nevertheless, a few studies, particularly by Abbott and her co-workers (Abbott & Robson, 1981a; Abbott, et al., 1983), have been useful because they have intensively and sequentially measured the relative infectivities of interacting fungi. These studies have been completed both in pots and in the field. A major finding of this work is that in soils with high levels of indigenous fungi, inoculation of subterranean clover did not alter the total amount of root colonised, but inoculant fungi displaced some of the indigenous fungal infections. In contrast, in soils with low levels of indigenous fungi, inoculation significantly increased the total infection. Of several different inoculant fungi, *Gl. fasciculatum* established readily but colonisation in the field was successful only where the natural endophytes exhibited both low infectivity and a large lag in the development of infection. When *Gl. monosporum* was introduced into field sites, it colonised poorly but persisted for at least 2 years. This latter observation corresponds with Wilson's (1984b) conclusions that *Gl. monosporum* has low infectivity in subterranean clover but is aggressive and can maintain its position in the face of competition.

Interesting studies on competition between VAM endophytes have also recently been published by Lopez-Aguillon and Mosse (1987) and Hepper et al. (1988). In both investigations the spatial dispersions of the interacting fungi were different; they found that of three fungi studied in two by two interactions, *Gl.*

fasciculatum was the most invasive, followed by *Gig. margarita* and *Gl. tenue*. Their term "invasiveness" referred to the ability of an endophyte to infect roots already colonised by another fungus. The poor performance of *Gl. tenue* in these studies may have been due to low inoculum density. Hepper et al. (1988) found *Gl. caledonium* to be slightly more competitive than *Gl. mosseae*, but *Gl. fasciculatum* was a poor competitor.

Some workers have tested the effectiveness of VAM fungi by using preinoculated, transplanted seedlings (e.g., Mosse & Hayman, 1971; Powell, 1977; Abbott & Robson, 1984a). Such a procedure gives the advantage of prior occupancy to the inoculant fungus over its indigenous competitors, and in most cases has produced significant improvements in plant growth. However, the procedure is not practical in field agriculture and would be most beneficial in systems where transplanted seedlings are the norm, such as the orchard industry (Menge et al., 1977), or in the rehabilitation of mined land (Khan, 1981). Nevertheless, the use of transplanted seedlings has added to our understanding of the competitive process. By eliminating any phases of competition that occur prior to root penetration, transplantation can help to establish the importance of spread by the prior occupant compared with primary penetration by a second fungus. In a study by Abbott and Robson (1984a), in which *Gl. fasciculatum* in transplanted seedlings competed with *A. laevis* and *Gl. tenue* in the surrounding soil, considerable colonisation by *Gl. tenue* occurred in the absence of *Gl. fasciculatum*, but infection by *Gl. tenue* was reduced when *Gl. fasciculatum* was already present in the roots. *A. laevis* germinates very slowly (Tommerup, 1983a,b, 1984a,b) and colonisation by this fungus was negligible. However, moist incubation of the soil for 3 weeks prior to transplanting greatly increased the infectivity of *A. laevis* and reduced colonisation by both *Gl. fasciculatum* and *Gl. tenue*. The effects were more pronounced at 20° than at 15°C, and were probably due to decreasing the lag time involved in root interception by *A. laevis*.

Another type of investigation that has provided information on interactions between VA mycorrhizal fungi is the study of the spread of inoculated fungi through field soils (e.g., Mosse et al., 1982). In a particularly interesting experiment, Powell (1979a) studied the spread of *Gl. tenue* from primary seedlings to companion seedlings previously infected with *Gl. fasciculatum, Gl. mosseae*, or *Gig. margarita*. He showed that the ease with which *Gl. tenue* spread through and inhibited the original endophyte in companion seedlings was inversely related to the spreading ability of the latter fungus. *Gig. margarita* was the slowest spreader, and suffered the greatest encroachment from *Gl. tenue*, whereas *Gl. fasciculatum* spread fast (second only to *Gl. tenue*) and suffered the least encroachment from *Gl. tenue*. The presence of mycorrhizal fungi in the soil as well as in the companion plants greatly reduced the spread of *Gl. tenue*. The observation by Powell that *Gl. fasciculatum* and *Gl. tenue* had high spreading abilities corresponds with the reports that these fungi had relatively high infectivities (Wilson, 1984a).

Factors Affecting an Interaction

Inoculum Frequency

Increasing the density of propagules decreases the lag phase before penetration. Thus, altering the relative densities of the interacting fungi (hereafter referred to as inoculum frequency) may change the order in which they gain entry to the root. Because of the importance of prior occupation, this may be a critical alteration. Empirical evidence suggests that the importance of inoculum frequency may be related to the particular interaction involved. Changing the inoculum frequencies had only a minor influence on the interaction between *Gl. fasciculatum* and *Gl. tenue* (Wilson, 1984b). Even a 10-fold increase in *Gl. tenue* in one experiment caused little increase in its aggressiveness. This may occur because most inoculum levels chosen produced maximum colonisation. In contrast, changing the frequencies had a marked effect on the interaction between *Gl. monosporum* and *Gl. tenue* (Table 7.1, Wilson & Trinick, 1983). For both these fungi, infection from mixed inoculum was directly proportional to frequency in the inoculum, as the replacement diagram shows (Fig. 7.7). Both fungi produced less infection in mixture than when alone, but appeared to be equivalent in their ability to interfere with one another. Thus the success of each fungus was directly related to its frequency in the inoculum. These observations highlight one of the problems in interaction experiments, the need to characterize the density–infection response curve for each fungus and to choose inoculum frequencies accordingly.

Spatial Relationships

Twenty years ago, Mosse and Hayman (1971) recognized the importance of the relationship between VAM propagules and root distribution. They predicted that "mycorrhizal fungi can be introduced against competition . . . providing that the new strains are already established in the roots (as inoculant transplants) or are favourably placed near emerging roots (as in seed-inoculated plants)." However, this advantage may not remain if the inoculum potentials of the indigenous strains are high. The dispersed propagules are in an ideal position to infect parts of the root distant from the point of inoculation, whereas the inoculated fungus has not spread sufficiently through the root. Propagules dispersed throughout the soil markedly reduced the ability of a fungus to spread within an existing infected root system and to expand through the soil between plants (Powell, 1979a). Lopez-Aguillon and Mosse (1987) and Hepper et al. (1988) also showed that the background inoculum will dominate colonisation unless the introduced endophyte is exceptionally infective.

Several investigations have examined the significance of inoculum placement in competitive studies involving *Gl. fasciculatum*. Because of its high infectivity, this fungus often performs well in competition. For instance, Lopez-Aguillon and

Mosse (1987) showed that *Gl. fasciculatum* was able to replace both *Gig. margarita* and *Gl. tenue*, irrespective of its relative position in the inoculum. Abbott and Robson (1984a) showed that *Gl. fasciculatum* already present in transplanted seedlings could maintain its level of colonisation when in competition with fungi distributed throughout the soil, although it was less competitive if the potting soil containing the indigenous endophytes was moistened before the time of planting. In a study of Wilson (1984b) colonisation from a central inoculum of *Gl. fasciculatum* of 10 propagules per gram was inhibited when *Gl. tenue* was present in the surrounding soil at 3.4 propagules per gram, but not when it was present at only at 0.34 propagules per gram. She concluded that a fungus more aggressive than *Gl. tenue* when distributed throughout the soil would probably have had a greater inhibitory effect. This was demonstrated by Abbott and Robson (1984a) with *Gl. fasciculatum* and *A. laevis*, although *A. laevis* required prior activation with moist incubation. In a study by Hepper et al. (1988), *Gl. fasciculatum* did not compete well against *Gl. caledonium* and *Gl. mosseae* but this may have been due to its low inoculum density.

Relative Propagule Size and State of Activation

Any factor that reduces the time lag prior to penetration is of competitive significance. Hence, a race between fungi with equal inoculum densities is likely to be won by the one with the biggest propagule or that is activated quickly. For instance, the differences in aggressiveness of *Gl. monosporum* and *Gl. tenue* at different harvest times, as shown in Table 7.1, were probably due to different times of entry into the root. *Gl. monosporum*, which has relatively large spores and infects early, was very aggressive at the 24 day harvest but less so at 36 days. In contrast, the aggressiveness of *Gl. tenue*, which has very small propagules, increased with time.

Wilson (1984b) and Abbott and Robson (1984a) also illustrated the importance of activation factors. With equal numbers of propagules, moist incubation of an inoculum containing *Gl. fasciculatum* and *Gl. tenue* increased the aggressiveness of the latter (Wilson, 1984b). This presumably occurred because moist incubation stimulated germination of *Gl. tenue* more than *Gl. fasciculatum*. In bulk soil containing both *Gl. tenue* and *A. laevis*, these two fungi were differentially competitive depending on the temperature, the time of harvest, and moist incubation of the inoculum prior to planting (Abbott & Robson, 1984a). The relative densities of the interacting propagules were not stated, so the low infectivity of *A. laevis* from dry soil may have been due either to low propagule numbers or slow germination, particularly at low temperatures (Tommerup, 1983b). In contrast, moist incubation prior to planting and incubation at 20 rather than 15°C increased infectivity of *A. laevis* and decreased infection by *Gl. tenue*. Such an effect was probably due to the fact that under these conditions, *A. laevis* reached the root before *Gl. tenue*.

Prior Occupation of the Root

A fungus already present in a particular part of a root has an advantage over another one attempting to gain entry at that point. The competitive outcome was probably due to decreases in the rate or extent of formation of entry points (Wilson & Trinick, 1983). Thus any factor that promotes the formation of entry points by one fungus, such as increasing inoculum density or activating propagules will give it an advantage. Why one endophyte should reduce penetration by another is uncertain, but it is clear that fungi vary in their abilities to cohabit. Those fungi which were most aggressive and most likely to reduce infection by another coloniser (e.g., *Gl. monosporum* nd *Gig. decipiens*) were also those least likely to be found as dual occupants (Wilson & Trinick, 1983; Wilson, 1984b). In contrast, *Gl. fasciculatum* and *Gl. tenue* were commonly found together in the root. The effect also may be due to changes in the physiological status of the host. For instance, an aggressive fungus may cause a localised depletion of the host's carbohydrates, to the extent that a previously infected part of the root becomes unattractive to penetration by a second coloniser. Recent data of Thompson and his co-workers (Thompson et al., 1986, 1990) showed that colonisation of subterranean clover by *S. calospora* was more sensitive than *Gl. fasciculatum* to reductions in the concentration of soluble root carbohydrates. Thus it follows that one fungus may reduce the concentration of carbohydrates in a host root more than another, and this supports the hypothesis that competition for nutrient-rich infection sites accounts for exclusion due to prior occupation. An alternative explanation for the competitive effects is that infection by one fungus induces a defence response in the host, which makes the root less susceptible to infection by another fungus (Wilson, 1983).

Prior occupation of a root does not appear to reduce penetration by another coloniser at a point distant from an existing infection. Studies on infected transplanted seedlings have provided an ideal situation in which to examine this aspect. They have usually shown that considerable infection from soil fungi can occur in distal parts of the root even when there is an existing infection at the time of transplanting. Such distal parts are usually areas of new growth which would be high in carbohydrate and very susceptible to invasion. If the soil inhabitant has a high inoculum potential or is very infective, it may be able to limit the spread of the primary coloniser (e.g., see Abbott & Robson, 1984a). Thus the competitive effect of "prior occupation" may be spatially limited. It is impossible to know what distance this limit may be, and it is difficult to envisage experimental techniques which could adequately measure it.

The Time Frame

One of the greatest difficulties in examining interactions between fungi is caused by the time frame of an experiment. Wilson and Trinick (1982) and Smith et al. (1986) demonstrated that both the inoculum potential and the size of the rhizosphere are time dependent. The relationship is complex, because all factors can be altered by changing the activation state of the propagules, for example by varying temperature, or by moist incubation of the inoculum.

Time can affect the processes in several ways. Fungi differ in the speed with which they reach the root and penetrate. The first coloniser will have a greater competitive advantage at an early harvest (e.g., see Table 7.1). At later harvests the first coloniser's advantage may progressively diminish, but this will depend on its inherent aggression and infectivity. The rate of root growth and the availability and dispersion of susceptible root tips also varies in time. If root growth slows down in response to space, nutrients, or seasonal changes, then the relative levels of colonisation by different fungi may change. Such factors, together with differences in sporulation, dormancy, and longevity of hyphae in old roots, will affect the competitive outcome and will influence the persistence of specified endophytes in a mixed community.

Other Factors

In the section above, we described several factors which will affect the progress of colonisation. Some of these are temperature, soil pH, soil water, soil phosphorus level, and the plant species involved. Such factors can also alter the balance between one fungus and another in mixed infections. Some of the parameters, e.g., temperature (e.g., Abbott & Robson, 1984a) have already been mentioned, but unfortunately there is insufficient information available on this topic to discuss it in any detail. At this stage in our understanding of interactions, the most important point to note is that any interaction is very dependent on the environment in which it occurs.

Likely Outcomes of Interactions between Different Fungi

Changes in the Effectiveness of the Symbiosis

One of our basic premises in this review has been that the main difference between fungi in terms of their effectiveness at increasing plant growth is their ability to infect rapidly and extensively. If this is so, then the only way in which inoculation could change the effectiveness of the symbiosis would be to increase or decrease in the total amount of colonisation. We have conceded that there may also be differences in the amount of external hyphae produced (per unit of infection) by different endophytes, although the data on this are conflicting. To clarify this problem, further comparisons need to be made of the abilities of different fungi to produce external hyphae, and of the lengths of time different hyphae are active (Schubert et al., 1987). These factors are crucial to an understanding of the effects of fungal interactions.

Changes in Total Colonisation

If fungi have similar abilities to take up phosphorus, then an increase in total colonisation (within a particular time) would be the primary aim of any inoculation program. The relative proportion of any particular fungus within the root system would be of no consequence, at least in the short term. Several studies have shown that when interacting fungi are placed in similar spatial positions relative to the root system, then the total amount of colonisation is similar to that obtained by single inoculation (Wilson & Trinick, 1983; Wilson, 1984b; Miller et al., 1989). The data of Daft and Hogarth (1983) suggest a similar phenomenon, although they measured sporulation rather than colonisation.

In experimental systems where the interacting fungi were placed in different positions, such as central inoculation of a soil containing indigenous endophytes, two different outcomes have been observed. First, and most commonly, the inoculant fungi displaced some of the infection associated with the indigenous endophytes, but total infection remained unchanged. In a second situation, the introduction of an inoculant fungus increased infection, but this occurred only where the indigenous fungi were of low infectivity or were slow to infect (Abbott & Robson, 1981a; Abbott et al., 1983; Lopez-Aguillon & Mosse, 1987; Hepper et al., 1988).

Significant reductions in total infection have been a rare occurrence in interaction studies. Such reductions probably only occur when an aggressive fungus (i.e., one able to exclude others) gains entry to the root first. Wilson (1984b) measured reduced total colonisation in an interaction between *Gig. decipiens* and *Gl. tenue* and a similar phenomenon may have occurred in Daft and Hogarth's study (1983), where the poorest growth responses always involved *Gl. mosseae*.

The observation that total colonisation usually remains unchanged has important implications for the success of inoculation programs in field soils. As Wilson (1984b) pointed out, where the dominant indigenous endophyte is slow to initiate infection, or has low infectivity, inoculation with a highly infective fungus may be beneficial. However, if the indigenous strains were both aggressive and widely dispersed throughout the soil, it would probably be difficult to introduce another fungus.

Alteration in Fungal Reproductive Ability

Although changes in the relative proportions of different fungi within roots may have little effect on the short-term effectiveness of the symbiosis, they may have a considerable influence on the production of propagules by the various fungi, and thus on the relative persistence of a fungus in the soil. Any alterations in root colonisation are likely to change the ability of a fungus to reproduce. If root pieces are a significant propagule for a particular fungus (Tommerup & Abbott, 1981), then this is clearly important. Sporulation is closely related to the development of infection (Furlan & Fortin, 1973), and fungi produce quite different amounts of spores per unit of infection. They also differ in their ability to sporulate when present as mixed inocula. Daft and Hogarth (1983) found that *Gl. mosseae* produced few spores but was "competitive" in that it maintained its sporulation in mixtures. *Gl. caledonium,* on the other hand, produced many spores when a single inoculum, but sporulation in a mixed inoculum was very poor. *Glomus clarum* and *Gl. geosporum* were intermediate. There was little effect of mixed inoculation on the total infection obtained, but because fungi exhibited different relationships between infection and sporulation, mixed inoculation affected the total number of spores produced. In any interaction involving *Gl. mosseae* total spore numbers were markedly reduced. Thus although *Gl. mosseae* acted "competitively," its presence did not increase the total amount of mycorrhizal inoculum in the soil.

Ecological and Agricultural Implications of Interactions

The Significance of Infectivity and Aggressiveness

Aggressiveness has been used in this review as a measure of the ability of different fungi to compete for infection. An aggressive fungus is one that maintains its infectivity from a mixed inoculum (relative to its infectivity from a single inoculum) better than the fungus with which it is competing. As aggressiveness does not take into account the absolute level of colonisation attained, it is only one of the factors contributing to the competitive ability of a fungus. Our interpretation of the relationships between infectivity and aggressiveness, and some of the other key factors such as inoculum density, are described in Figure 7.8.

In the sense in which the term infectivity has been used here, it is a measure of the ability of each fungus to produce secondary spread. Differences in infectivity between fungi may be partly due to differences in their aggressiveness. There is some evidence of an inverse relationship between infectivity and aggressiveness, but the reason for this is unknown. Differences in aggression are associated with the ability of a fungus to spread by secondary infection points. Differences between fungi in their abilities to spread inside (rather than outside) the root thus contributes to differences in their infectivity. The expression of aggressiveness is probably a localised reaction and may be related to differences in the carbohydrate requirements of the fungi, or to differences in the responses elicited in the host.

The relative importance of infectivity and aggressiveness in determining the outcome of a particular interaction is difficult to assess. A fungus with high infectivity but low aggression will perform relatively better than one with the opposite attributes when it is challenged by few competitors. Its ability to colonise roots rapidly and extensively is an important characteristic in the short term. However, if the plants are to benefit from mycorrhizas for a long period of time, then fungi with persistence are also important. Aggressive fungi are likely to be the most persistent, but as they may also have low infectivity, they may never

INTER-RELATIONSHIPS BETWEEN FACTORS

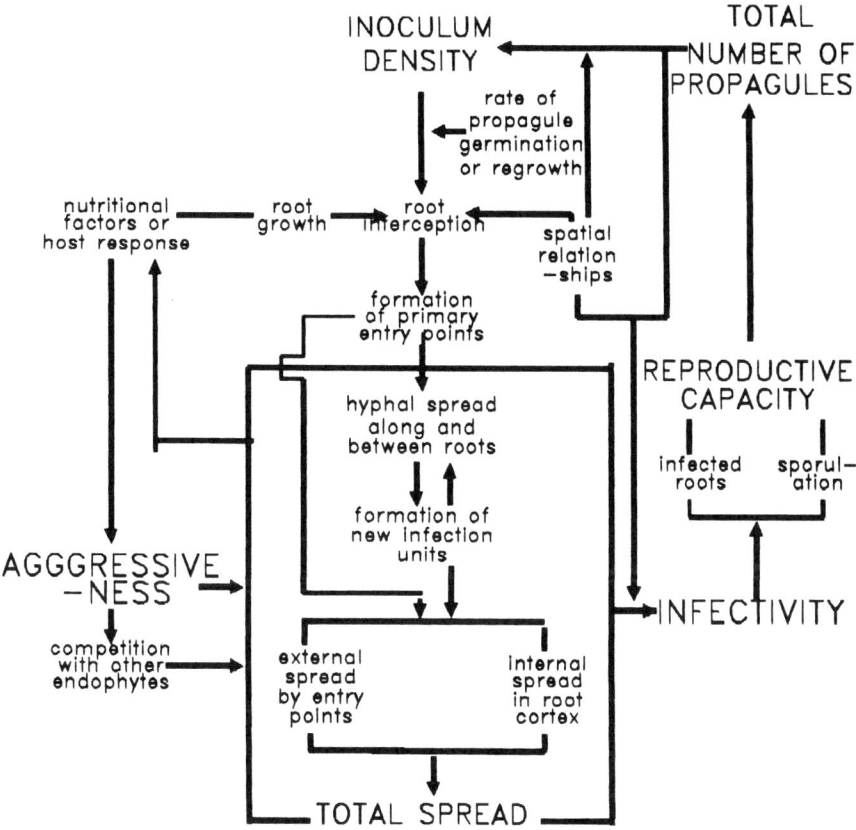

Figure 7.8. The interrelationships between some of the factors that affect infection by a VAM fungus. Other factors, such as environmental influences, propagule dormancy, and survival, and the effect of the fungus on plant nutrient uptake, could also be included.

have a large density. Fungi with high infectivities will be important endophytes to introduce into eroded soils, mine soils, sterilised horticultural soils, or any situation in which the indigenous population is low. However, in field soils with high indigenous populations, infectivity alone will not guarantee the success of an inoculant endophyte. Here, aggression will be an equally important attribute.

Our discussions on the concept of aggression have indicated that it is always a positive attribute as it is an important component of competitiveness; but as aggressiveness may be associated with higher demands for carbohydrates, it may also lead to a less efficient symbiosis from the point of view of plant growth. To determine whether there are intrinsic physiological differences in efficiency between mycorrhizas, it would be necessary to compare mycorrhizas that have the same levels of colonisation. This would be difficult. When more information on this topic is available, it may be necessary to modify the conclusions about the value of aggressiveness.

Parallels with Plant Biology: "r–k" Strategies

In other areas of biology, species are broadly categorized into two types according to whether their ecological strategies are based on fecundity or on their ability to withstand interference from their neighbors. The two types are named "r-species" and "k-species."[1] These concepts have been applied to the survival and reproduction of insect, animal, and plant populations (Cody & Diamond, 1975). In plant biology, plants which are "r-species" are widely dispersed, have large numbers of small seeds, and are pioneer colonisers, which usually give way to others in natural succession. "K-species" are specialists in resource-limited environments where there is intense competition from neighbors, and will tend to have larger seeds and maximise survival rather than fecundity (Harper, 1977).

Grime (1979) noted that fungi have strategies corresponding with those in plants. There are clearly parallels between the r–k strategies proposed for plants and those observed with VAM fungi. "Aggressiveness" corresponds to the "K-strategy" and "infectivity" is a major factor in the reproductive capacity (the "r-strategy"). In the studies of Wilson (Wilson, 1983; Wilson & Trinick, 1983; Wilson, 1984b), not only was aggressiveness inversely related to infectivity, but those fungi which were most aggressive were also those which produced fewer but larger propagules (e.g., *Gl. monosporum* and *Gig. decipiens*). Conversely, the most infective fungi (*Gl. fasciculatum* and *Gl. tenue*) produced many small propagules. The widespread occurrence of *Gl. fasciculatum* and *Gl. tenue,* particularly in pioneer localities (Crush, 1973, 1975; Molina et al., 1978; Forster & Nicolson, 1981; Abbott & Robson, 1982b), supports the idea that they are "r-type" organisms. It should be noted that the relationship between root infection and the resulting propagule numbers was the same for all the fungi examined (Wilson & Trinick, 1982), but as it was exponential small differences in infection resulted in large differences in propagule numbers.

However, speculation about ecological strategies need to be made with caution, as the r- and K-strategies form a continuum, and the majority of organisms fall

---

1 These terms are derived from the parameters in the idealised logistic equation for population growth:

$$dN/dt = rN\,[(K-N)/K]$$

where N is the original size of the population, t is time, r is the potential rate of increase of the population, and K is the maximum attainable population size in the system.

between the extremes. Genetic variation may cause populations of the same species to occupy different positions along the continuum (Tommerup, 1988a). More fungi need to be studied to determine whether the concepts are broadly applicable. Nevertheless, Daft and Hogarth (1983) have also suggested that VA mycorrhizal fungi differ in their ecological strategies. From their studies of interactions they concluded that in terms of persistence and reliability, field inoculation would be most successful where mixed inocula containing endophytes of various strategies were used.

Differences between Mixed Plant Communities and Monocultures

Although there is little specificity in VA mycorrhizal relationships, the host species can alter the relative infectivities of endophytes. Unfortunately, there are no data available on how this would affect competition between endophytes. If the plant system of interest is an agricultural monoculture, then it is relatively easy to devise experiments to measure the competitiveness and effectiveness of different fungi. However, if a mixed plant population is involved, such as in a pasture sward, the situation becomes complicated by competition among the plant species as well as among the fungal endophytes. The fact that hyphal links can occur between the mycorrhiza of different plant species, and that phosphorus can be transported between such plants (Heap & Newman, 1980a,b) is an interesting complication. This has implications for nutrient recycling, and it also emphasizes the complexity of the system when taken as a whole.

If fungi differ in their abilities to colonise different hosts, then the number and proportion of the various hosts present will have a direct effect on the balance of the fungal endophytes found in a particular community. A complex mixture of hosts is likely to support a diverse range of fungi. In contrast, in a plant monoculture (such as a wheat crop), we would expect a more limited range of fungi to be present, particularly in a situation where crop rotation is not practised. Ecological interactions will of course be influenced by many factors, such as whether the dominant hosts are annuals or perennials, and whether the environment supports plant growth year round or only seasonally.

Such ecological factors will have a profound effect on the outcome of interactions between VAM endophytes, but the data are too limited to be adequately discussed in this review. Our understanding of interactions has been gleaned from oversimplified experimental systems, and our knowledge of the natural population movements of VAM fungi is rather scanty, so any extrapolations to the field situation would probably lead to incorrect conclusions. A detailed discussion of the ecology of mycorrhizae and mycorrhizal fungi has previously been prepared by Mosse et al. (1981), but there is much still to be learned.

Success of Inoculation

There are four situations in which it may be desirable to inoculate plants or soils with VAM fungi to improve agricultural productivity: (1) transplanted seedings and (2) sterilised beds in the nursery and horticultural industries, (3) in field soils in eroded, mining, or pioneer locations, and in other soils with low populations of endophytes, and (4) in field soils with high populations of VAM endophytes.

To successfully introduce a new endophyte into the system and to obtain higher levels of colonisation than before, the inoculant fungi must compete successfully with any indigenous fungi. We believe that (1), (2), and (3) represent situations in which it will be relatively simple to achieve increases in colonisation with inoculation, providing the soil phosphorus levels are appropriate and a fungus with high infectivity is used. There are already many examples of successful inoculation in these situations (see earlier discussions). However, the plant responses may be short lived, which may pose problems in some situations. For example, situation (3), where long-term effects would be desirable, may develop a higher population level of mycorrhizal fungi over time. This location then would resemble situation (4).

Situation (4) presents many more difficulties for artificial inoculation. Success will depend on a thorough understanding of the ecology of the location and a careful choice of both the inoculant fungi and the inoculation procedure. An outline of the factors to consider in this circumstance has been discussed by Abbott and Robson (1982). It is possible that a combination of strategies will be best if more than one fungus is present in the inoculum. If possible, inoculant fungi should be made more aggressive by manipulating the environment in their favor.

Fungi should infect rapidly and extensively, but to persist, aggressiveness will also be critical. The ability to produce large numbers of propagules is an important component of persistence. Because of the importance of early infection the inoculant fungi should not exhibit any lag before penetration occurs. Hence they should be in a state of activation such that germination and growth from the propagules is very rapid. It may be desirable to use pregerminated spores (Tommerup, 1984a,b), although our knowledge of the physiology of propagules needs to be vastly improved (Tommerup, 1988a,b,c). Another consideration is the volume of inoculum applied. It is clear that the wider the dispersion of propagules, given an appropriate propagule density, the greater the resulting infection. Thus, methods of inoculation should be devised that distribute inoculum in a way to maximise contact with growing root tips. For example, in broad acre crops or pastures that are sown with tined instruments the inoculum should be distributed in a wide band below the seed. Appropriate machinery and methods need to be devised that will suit each particular circumstance.

Methodological Problems

Defining the System

In this review we have made several generalisations about interactions, and about the characteristics of particular endophytes. We are well aware of the dangers of comparing one experimental system with another, or of extrapolating from the pot to the field. Many factors are important, and these must be kept in mind when attempting to predict the outcome of any particular interaction or to use such information to devise an inoculation procedure. The relevant factors have been discussed earlier, but they include the host and fungal species, the relative propagule densities, the spatial interrelationships of the various fungi and the plant roots, the growth rate and physiological state of the roots, and the climatic and edaphic environment. It is essential to view each agricultural system with a fresh approach and to carefully define its components and identify any problem areas. In the past, inoculation experiments have often been unsuccessful because the inoculum potential, physiological state and ecological behavior of the inoculated fungi were unknown. Moreover, the inoculation method and even the inoculum quality may have been inadequate. Even less was known about the indigenous fungi; very often even their identity was a mystery. Until the system is defined and understood, it is unlikely that an attempt to introduce new fungi into a competitive situation will be successful.

Identifying Individual Fungi

One of the greatest difficulties in studying interactions between VAM fungi is to reliably identify and quantify them. This has traditionally been done using spore morphology, and this technique has been used effectively in interaction studies by Daft and Hogarth (1983). Since root colonisation is crucial to the mycorrhizal process, it is very important to be able to identify fungi within roots. VAM fungi produce characteristic infection patterns in roots, and some workers have found Abbott's (1982) systematic key to infection in subterranean clover to be a very useful identification tool (Abbott, 1982; Wilson & Trinick, 1983). Nevertheless, such keys have limited value when studying closely related species or isolates of the same species. A further consideration with VAM fungi is the identification of external hyphae. Using conventional microscopy it is difficult to distinguish the hyphae of endogonaceous fungi from those of other fungal classes, let alone other Zygomycetes. A method is thus required that distinguishes between the infection and hyphal structures of VAM fungi and non-VAM fungi, and if possible also between different genera, species, and isolates of VAM fungi. An immunological method would be useful, but although some progress has been made in this direction (Aldwell et al., 1983; Wilson et al., 1983), further refinements are necessary before such techniques can be used widely in competition studies. A promising technique is the identification of diagnostic enzymes using gel electrophoresis (Rosendahl et al., 1989), which has already been used successfully in competition studies (Hepper et al., 1988), and which has the advantage over other methods of detecting only metabolically active hyphae.

Measuring the Putative Benefits

The approach taken to interactions in this review has been based on a belief in the importance of both extensive colonisation and the formation of external hyphae as determinants of effectiveness. Although colonisation has been studied in considerable detail by many workers, the relative significance of the various colonisation parameters, such as absolute and percentage root length and root volume infected and infection intensity, is still not well understood. Even less is known about differences in the physiology of various host/fungus combinations, and there is conflicting evidence on the relationships between external hyphae, colonisation, and effectiveness. As more information is accumulated on such factors, and as the relative significance of the various measurements is established, our understanding of the relevance of manipulating interactions among fungi will improve.

Modeling the System

Mathematical models of biological systems rarely address all the interacting processes at a sufficiently complex level to be realistic. Nevertheless, they force the investigator to dissect and rebuild the components in a logical and thorough manner. In doing so, they can provide valuable insights into the nature of the processes themselves. Although models of VAM infection were originally oversimplistic (Walker & Smith, 1984) and they still require further development (Bassett & Gilligan, 1988), they have been extremely useful for defining a framework in which to assess the significance of interactions among endophytes in simplified, short-term pot systems. One of the important limitations so far has been the difficulty of applying time as a parameter, and of introducing the concept of the cyclical nature of the process, particularly in the field. It is to be hoped that someone will take up this challenge so that our understanding can progress further.

Conclusions

This chapter has reviewed the processes and factors that are important in determining the outcome of an interaction between VAM fungi. Of primary significance are the abilities of the individual fungi to penetrate and colonise roots. Infectivity is a measure of this ability, and it is governed by many factors including the speed at which infection is initiated, the progress of spread both inside and outside roots, and the extent of formation of external hyphae. Fungi differ markedly in their infectivities, but the extent of colonisation can be altered by changing the density, distribution, or state of activation of propagules. Provided a certain critical inoculum density is maintained, colonisation is greatest when the inoculum is widely dispersed.

Competition between VAM fungi for root colonisation is influenced by the ability to penetrate the root first and gain prior occupation. In addition, fungi differ in their aggressiveness or their ability to exclude another coloniser. Aggressiveness appears to be inversely related to infectivity. The outcome of competition is affected by the inoculum frequencies of the interacting fungi, their relative spatial relationships with the root, and the time frame under consideration. However, total colonisation by more than one fungus is only rarely different from that obtained by single inoculation with the most infective of the interacting fungi.

The outcome of any interaction will depend on whether the system under consideration consists of single plants in pots, monocultures in the field, or a mixed plant community. If inoculation is under consideration, then the choice of fungus and the inoculation process will depend on the infectivities and aggressiveness of the indigenous fungi. Inoculation will be difficult where there is an already well-established population of mycorrhizal fungi. It is imperative that the system is defined before consideration is given to inoculation. Unfortunately there are several methodological problems to overcome before further progress can be made in assessing the significance of interactions among VAM fungi. These include the critical assessment of the level of activity of hyphae both within and outside the root, the refinement of methods of identification of individual fungi, and further definition of the processes contributing to the effectiveness of the symbiosis.

References

Abbott, L. K. (1982). Comparative anatomy of vesicular arbuscular mycorrhizas formed on subterranean clover. *Australian Journal of Botany* **30**, 485–490.

Abbott, L. K., & Robson, A. D. (1977). Growth stimulation of subterranean clover with vesicular arbuscular mycorrhizas. *Australian Journal of Agricultural Research* **28**, 639–649.

Abbott, L. K., & Robson, A. D. (1978). Growth of subterranean clover in relation to the formation of endomycorhizas by introduced and indigenous fungi in a field soil. *New Phytologist* **81**, 575–585.

Abbott, L. K., & Robson, A. D. (1981a). Infectivity and effectiveness of five mycorrhizal fungi: Competition with indigenous fungi in field soils. *Australian Journal of Agricultural Research* **32**, 621–630.

Abbott, L. K., & Robson, A. D. (1981b). Infectivity and effectiveness of vesicular arbuscular mycorrhizal fungi: Effect of inoculum type. *Australian Journal of Agricultural Research* **32**, 631–639.

Abbott, L. K., & Robson, A. D. (1982a). The role of vesicular arbuscular mycorrhizal fungi in agriculture and the selection of fungi for inoculation. *Australian Journal of Agricultural Research* **33**, 389–408.

Abbott, L. K., & Robson, A. D. (1982b). Infectivity of indigenous vesicular arbuscular mycorrhizal fungi in agricultural soils. *Australian Journal of Agricultural Research* **33**, 1049–1059.

Abbott, L. K., & Robson, A. D. (1984a). Colonisation of the root system of subterranean clover by three species of vesicular-arbuscular mycorrhizal fungi. *New Phytologist* **96**, 275–281.

Abbott, L. K., & Robson, A. D. (1984b). The effect of root density and infectivity of inoculum on the development of VA mycorrhizas. *New Phytologist* **97**, 285–299.

Abbott, L. K., & Robson, A. D. (1984c). The effect of VA mycorrhiza on plant growth. In: *VA Mycorrhiza* (Ed. by C. L. Powell, & D. L. Bagyaraj), pp. 113–130. CRC Press, Boca Raton, FL.

Abbott, L. K., & Robson, A. D. (1985a). The effect of pH on the formation of VA mycorrhizas by two species of *Glomus*. *Australian Journal of Soil Research* **23**, 253–262.

Abbott, L. K., & Robson, A. D. (1985b). Formation of external hyphae in soil by four species of vesicular arbuscular mycorrhizal fungi. *New Phytologist* **99**, 245–255.

Abbott, L. K., Robson, A. D., & Hall, I. R. (1983). Introduction of vesicular arbuscular mycorrhizal fungi into agricultural soils. *Australian Journal of Agricultural Research* **34**, 741–749.

Abbott, L. K., Robson, A. D., & De Boer G. (1984). The effect of phosphorus on the formation of hyphae in soil by the vesicular-arbuscular mycorrhizal fungus *Glomus fasciculatum*. *New Phytologist* **97**, 437–446.

Aldwell, F. E. B., Hall, I. R., & Smith, J. M. B. (1983). Use of an enzyme-linked immunosorbent assay for the identification of endomycorrhizal fungi. *Soil Biology and Biochemistry* **15**, 377–378.

Allen, M. F., Moore, T. S., & Christensen, M. (1980). Phytohormone changes in *Bouteloua gracilis* infected by vesicular-arbuscular mycorrhizae: I. Cytokinin increases in the host plant. *Canadian Journal of Botany* **58**, 371–374.

Allen, M. F., Smith, W. K., Moore, T. S., & Christensen, M. (1981). Comparative water relations and photosynthesis of mycorrhizal and non-mycorrhizal *Bouteloua gracilis* H. B. K. Lag ex Steud. *New Phytologist* **88**, 683–693.

Amijee, F., Stribley, D. P., & Tinker, P. B. (1986). The development of endomycorrhizal root systems. VI. The relationship between development of infection, and intensity of infection in young leek roots. *New Phytologist* **102**, 293–301.

Amijee, F., Tinker, P. B., & Stribley, D. P. (1989). The development of endomycorrhizal root systems. *New Phytologist* **111**, 435–446.

Azcon-Aguilar, C., & Barea, J. M. (1981). Field inoculation of *Medicago* with V-A mycorrhiza and *Rhizobium* in phosphate-fixing agricultural soil. *Soil Biology and Biochemistry* **13**, 19–22.

Baas, R., & Lambers, H. (1988). Effects of vesicular-arbuscular mycorrhizal infection and phosphate on *Plantago major* ssp. *Pleiosperma* in relation to the internal phosphate concentration. *Physiologia Plantarum* **74**, 701–707.

Bassett, P. R., & Gilligan, C. A. (1988). A model for primary and secondary infection in botanical epidemics. *Zeitschrift für Pflanzenkrankheiten und Pfanzenschutz (Journal of Plant Diseases and Protection)* **95,** 352–360.

Beilby, J. (1980). Early metabolism in the germinating vesicular-arbuscular endophyte spore of *Glomus caledonius*. Ph.D. Thesis, University of Western Australia.

Bethenfalvay, G. J., Brown, M. S., & Pacovsky, R. S. (1982). Relationships between host and endophyte development in mycorrhizal soy beans. *New Phytologist* **90,** 537–543.

Bevege, D. T., & Bowen, G. D. (1975). Endogone strain and host plant differences in development of vesicular arbuscular mycorrhizas. In: *Endomycorrhizas* (Ed. by F. E. Sanders, B. Mosse, & P. B. Tinker), pp. 77–86. Academic Press, London.

Black, R., & Tinker, P. B. (1979). The development of endomycorrhizal root systems. Effect of agronomic factors and soil conditions on the development of vesicular arbuscular mycorrhizal infection in barley and on the endophyte spore density. *New Phytologist* **83,** 401–413.

Bouhot, D. (1979). Estimation of inoculum density and inoculum potential: techniques and their value for disease prediction. In: *Soil-Borne Plant Pathogens* (Ed. by B. Schippers & W. Gams), pp. 21–33. Academic Press, London.

Bowen, G. D. (1979). Integrated and experimental approaches to the growth of organisms around roots. In: *Soil-Borne Plant Pathogens* (Ed. by B. Schippers & W. Gams), pp. 209–227. Academic Press, London.

Buwalda, J. G., Ross, G. J. S., Stribley, D. P., & Tinker, P. B. (1982). The development of endomycorrhizal root systems. III. The mathematical representation of the spread of vesicular arbuscular mycorrhizal infection in root systems. *New Phytologist* **91,** 669–682.

Carling, D. E., & Brown, M. F. (1980). Relative effect of vesicular arbuscular mycorrhizal fungi on the growth and yield of soybeans. *Journal of the Soil Science Society of America* **44,** 528–532.

Carling, D. E., Brown, M. F., & Brown, R. A. (1979). Colonisation rates and growth responses of soybean plants infected by vesicular arbuscular mycorrhizal fungi. *Canadian Journal of Botany* **57,** 1769–1772.

Chambers, C. A., Smith, S. E., & Smith, F. A. (1980). Effects of ammonium and nitrate ions on mycorrhizal infection, nodulation and growth of *Trifolium subterraneum*. *New Phytologist* **85,** 47–62.

Clark, F. E. (1965). The concept of competition in microbial ecology. In: *Ecology of Soil-Borne Plant Pathogens* (Ed. by K. F. Baker & W. C. Snyder), pp. 339–345. University of California Press.

Cody, M. L., & Diamond, J. M. (Eds) (1975). *Ecology and Evolution of Communities*. The Belknap Press of Harvard University Press, Cambridge, MA.

Cooper, K. M., & Tinker, P. B. (1978). Translocation and transfer of nutrients in vesicular arbuscular mycorrhizas. II. Uptake and translocation of phosphorus, zinc and sulphur. *New Phytologist* **81,** 43–52.

Cox, G., & Sanders, F. E. (1974). Ultrastructure of the host-fungus interface in a vesicular arbuscular mycorrhiza. *New Phytologist* **73,** 901–912.

Cox, G., & Tinker, P. B. (1976). Translocation and transfer of nutrients in vesicular arbuscular mycorrhizas. I. The arbuscule and phosphorus transfer: a quantitative ultrastructural study. *New Phytologist* **77**, 371–378.

Crush, J. R. (1973). Significance of endomycorrhizas in tussock grasslands in Otago, New Zealand. *New Zealand Journal of Botany* **11**, 645–660.

Crush, J. R. (1975). Occurrence of endomycorrhizas in soils of the Mackenzie Basin, Canterbury, New Zealand. *New Zealand Journal of Agricultural Research* **18**, 361–364.

Daft, M. J., & Hogarth B. (1983). Competitive interactions amongst four species of *Glomus* on maize and onion. *Transactions of the British Mycological Society* **80**, 945–952.

Daniels, B. A., & Menge, J. A. (1981). Evaluation of the commercial potential of the vesicular arbuscular mycorrhizal fungus, *Glomus epigaeus*. *New Phytologist* **87**, 345–354.

Daniels, B. A., & Trappe, J. M. (1980). Factors affecting spore germination of the vesicular-arbuscular mycorrhizal fungus, *Glomus epigaeus*. *Mycologia* **72**, 457–471.

Daniels, B. A., McCool, P. M., & Menge, J. A. (1981). Comparative inoculum potential of spores of six vesicular arbuscular mycorrhizal fungi. *New Phytologist* **89**, 352–392.

Dart, P. (1977). Infection development of leguminous nodules. In: *A Treatise on Dinitrogen Fixation, Section III: Biology* (Ed. by R. W. F. Hardy & W. S. Silver), pp. 367–472. Wiley, New York.

deWit, C. T. (1960). On Competition. *Verslagen van Landbouwkundige Onderzoekingen* **668**, 1–82.

Donald, C. M. (1963). Competition among crop and pasture plants. *Advances in Agronomy* **15**, 1–118.

Douds, D. D., & Chaney, W. R. (1982). Correlation of fungal morphology and development to host growth in green ash mycorrhiza. *New Phytologist* **92**, 519–526.

Forster, S. M., & Nicolson, T. H. (1981). Microbial aggregation of sand in a maritime dune succession. *Soil Biology and Biochemistry* **13**, 205–208.

Furlan, V., & Fortin, J-1. (1973). Formation of endomycorrhizae by *Endogone calospora* on *Allium cepa* under three temperature regimes. *Naturaliste Canadian* **100**, 467–477.

Garrett, S. D. (1956). *Biology of Root-Infecting Fungi*. Cambridge University Press, Cambridge.

Gilligan, C. A. (1989a). Mathematical modelling of soilborne plant pathogens. In: *Epidemics of Plant Disease*, 2nd ed. (Ed. by J. Kranz). Springer-Verlag, Berlin pp. 96–137.

Gilligan, C. A. (1989b). Mathematical models of infection. In: *The Rhizosphere* (Ed. by J. M. Lynch). Wiley, Chichester pp. 207–232.

Gilligan, C. A., & Simons, S. A. (1987). Inoculum efficiency and pathozone with for two host-parasite systems. *New Phytologist* **107**, 549–566.

Graham, J. H., Linderman, R. G., & Menge, J. A. (1982). Development of external hyphae by different isolates of mycorrhizal *Glomus* spp. in relation to root colonisation and growth of troyer citrange. *New Phytologist* **91**, 183–189.

Grime, J. P. (1979). *Plant Strategies and Vegetation Processes*. Wiley, Chichester.

Hall, I. R. (1976). Response of *Coprosma robusta* to different forms of endomycorrhizal inoculum. *Transactions of the British Mycological Society* **67**, 409–411.

Hall, R. L. (1978). The analysis and significance of competitive and non-competitive interference between species. In: *Plant Relations in Pastures* (Ed. by J. R. Wilson), pp. 163–174. CSIRO, Melbourne.

Hall, I. R. (1980). Growth of *Lotus pedunculatus* Cav. in an eroded soil containing soil pellets infested with endomycorrhizal fungi. *N.Z. Journal of Agricultural Research* **23**, 103–105.

Hall, I. R. (1984). Field trials assessing the effect of inoculating agricultural soils with endomycorrhizal fungi. *Journal of Agricultural Science, Cambridge* **102**, 725–731.

Harper, J. L. (1977). *Population Biology of Plants*. Academic Press, London.

Hayman, D. S., & Tavares, M. (1985). Plant growth responses to vesicular arbuscular mycorrhiza. XV. Influence of soil pH on the symbiotic efficiency of different endophytes. *New Phytologist* **100**, 367–377.

Heap, A. J., & Newman, E. I. (1980a). Links between roots by hyphae of vesicular-arbuscular mycorrhizas. *New Phytologist* **85**, 169–171.

Heap, A. J., & Newman, E. I. (1980b). The influence of vesicular-arbuscular mycorrhizas on phosphorus transfer between plants. *New Phytologist* **85**, 173–179.

Hepper, C. M. (1983a). Limited independent growth of a vesicular-arbuscular mycorrhizal fungus *in vitro*. *New Phytologist* **93**, 537–542.

Hepper, C. M. (1983b). Effect of phosphate on germination and growth of vesicular-arbuscular mycorrhizal fungi. *Transactions of the British Mycological Society* **80**, 487–490.

Hepper, C. M. (1985). Influence of age of roots on the pattern of vesicular-arbuscular mycorrhizal infection in leek and clover. *New Phytologist* **101**, 685–693.

Hepper, C. M., Azcon-Aguilar, C., Rosendahl, S., & Sen, R. (1988). Competition between three species of *Glomus* used as spatially separated introduction and indigenous mycorrhizal inocula for leek (*Allium porrum* L.). *New Phytologist* **110**, 207–215.

Huisman, O. C. (1982). Interrelations of root growth dynamics to epidemiology of root-invading fungi. *Annual Review of Phytopathology* **20**, 303–327.

Iqbal, S. H., & Qureshi, K. S. (1976). The influence of mixed sowing (cereals and crucifers) and crop rotation on the development of mycorrhiza and subsequent growth of crops under field conditions. *Biologia* **22**, 287–298.

Jasper, D. A., Robson, A. D., & Abbott, L. K. (1979). Phosphorus and the formation of vesicular arbuscular mycorrhizas. *Soil Biology and Biochemistry* **11**, 501–505.

Jasper, D. A., Abbott, L. K., & Robson, A. D. (1989). Hyphae of a vesicular arbuscular mycorrhizal fungus maintain infectivity in dry soil, except when soil is disturbed. *New Phytologist* **112**, 101–107.

Khan, A. G. (1981). Growth responses of endomycorrhizal onions in unsterilised coal waste. *New Phytologist* **87**, 363–370.

Khan, A. G. (1988). Inoculum density of *Glomus mosseae* and growth of onion plants in unsterilised bituminous coal spoil. *Soil Biology and Biochemistry* **20**, 749–753.

Kuc, J., Shockley, G., & Kearney, K. (1975). Protection of cucumber against *Colletotrichum lagenarium* by *Colletotrichum lagenarium*. *Physiological Plant Pathology* **7**, 195–199.

Lambert, D. H., & Cole, H. (1980). Effects of mycorrhizae on establishment and performance of forage species in mine spoil. *Agronomy Journal* **72**, 257–260.

Lopez-Aguillon, R. & Mosse, B. (1987). Experiments on competitiveness of three endomycorrhizal fungi. *Plant and Soil* **97**, 155–170.

Menge, J. A., Lembright, H., & Johnson, E. L. V. (1977). Utilisation of mycorrhizal fungi in citrus nurseries. *Proceedings of the International Society of Citriculture* **1**, 129–132.

Miller, D. D., Domoto, P. A., & Walker, C. (1985). Colonisation and efficacy of different endomycorrhizal fungi with apple seedlings at two phosphorus levels. *New Phytologist* **100**, 393–402.

Miller, D. D., Bodmer, M., & Schuepp, H. (1989). Spread of endomycorrhizal colonisation and effect on growth of apple seedlings. *New Phytologist* **111**, 51–59.

Molina, R. J., Trappe, J. M., & Strickler, G. S. (1978). Mycorrhizal fungi associated with *Festuca* in the western United States and Canada. *Canadian Journal of Botany* **56**, 1691–1695.

Mosse, B. (1972). The influence of soil type and *Endogone* strain on the growth of mycorrhizal plants in phosphorus deficient soils. *Revue d'Ecologie et de Biologie du Sol* **9**, 529–537.

Mosse, B. (1977). Plant growth responses to vesicular arbuscular mycorrhiza. X. Responses to *Stylosanthes* and maize to inoculation in unsterile soils. *New Phytologist* **78**, 277–288.

Mosse, B., & Hayman, D. S. (1971). Plant growth responses to vesicular arbuscular mycorrhiza. II. In unsterilised field soils. *New Phytologist* **70**, 29–34.

Mosse, B., Stribley, D. P., & LeTacon, F. (1981). Ecology of mycorrhizae and mycorrhizal fungi. In: *Advances in Microbial Ecology* (Ed. by M. Alexander), pp. 137–210. Plenum Press, New York.

Mosse, B., Warner, A., & Clarke, C. A. (1982). Plant growth responses to vesicular arbuscular mycorrhiza. XIII. Spread of an introduced VA endophyte in the field and residual growth effects of inoculation in the second year. *New Phytologist* **90**, 521–528.

Odum, E. P. (1971). *Fundamentals of Ecology*. W. B. Saunders, Philadelphia.

Owusu-Bennoah, E., & Mosse, B. (1979). Plant growth responses to vesicular-arbuscular mycorrhiza. *New Phytologist* **83**, 671–679.

Pfender, W. F. (1982). Monocyclic and polycyclic root diseases: distinguishing between the nature of the disease cycle and the shape of the disease progress curve. *Phytopathology* **72**, 31–32.

Plenchette, C., Furlan, V., & Fortin, J. A. (1982). Effects of different endomycorrhizal fungi on five host plants grown on calcined montmorillonite clay. *Journal of the American Society of Horticultural Science* **107**, 535–538.

Plenchette, C., Perrin, R., & Duvert, P. (1987). The concept of soil infectivity and a method for its determination as applied to Endomycorrhizas. *Canadian Journal of Botany* **67**, 112–115.

Pons, F., & Gianinizzi-Pearson, V. (1984). Influence du phosphore, du potassium, de l'azote et du pH sur le comportement *in vitro* de champignons de endomycorhizogenes a vesicules et arbuscules. *Cryptogame Mycologie* **5**, 87–100.

Powell, C. Ll. (1976). Development of mycorrhizal infections from *Endogone* spores and infected root segments. *Transactions of the British Mycological Society* **66**, 439–445.

Powell, C. Ll. (1977). Mycorrhizas in hill country soils. III. Effect of inoculation on clover growth in unsterile soils. *N. Z. Journal of Agricultural Research* **20**, 343–348.

Powell, C. Ll. (1979a). Spread of mycorrhizal fungi through soil. *New Zealand Journal of Agricultural Research* **22**, 335–339.

Powell, C. Ll. (1979b). Inoculation of white clover and ryegrass seed with mycorrhizal fungi. *New Phytologist* **83**, 81–85.

Powell, C. Ll. (1981). Inoculation of barley with efficient mycorrhizal fungi stimulates seed yield. *Plant and Soil* **59**, 487–489.

Ratnayake, M., Leonard, R. T., & Menge, J. A. (1978). Root exudation in relation to supply of phosphorus and its possible relevance to mycorrhizal formation. *New Phytologist* **81**, 543–552.

Rhodes, L. H., & Gerdemann, J. W. (1980). Nutrient translocation in vesicular arbuscular mycorrhizae. In: *Cellular Interactions in Symbiosis and Parasitism* (Ed. by C. B. Cook, P. W. Pappas, & E. D. Rudolph), pp. 173–195. Ohio State University Press, Columbus.

Rosendahl, S., Sen, R., Hepper, C. M., & Azcon-Aguilar, C. (1989). Quantification of three vesicular-arbuscular mycorrhizal fungi (*Glomus* spp.) in roots of leek (*Allium porrum*) on the basis of activity of diagnostic enzymes after polyacrylamide gel electrophoresis. *Soil Biology and Biochemistry* **21**, 519–522.

Safir, G. R., Boyer, J. S., & Gerdemann, J. W. (1972). Nutrient status and mycorrhizal enhancement of water transport in soybean. *Plant Physiology* **49**, 700–703.

Sanders, F. E., & Sheikh, N. A. (1983). The development of vesicular arbuscular mycorrhizal infection in plant root systems. *Plant and Soil* **71**, 223–246.

Sanders, F. E., Tinker, P. B., Black, R. L. B., & Palmerley, S. M. (1977). The development of endomycorhizal root systems: I. Spread of infection and growth-promoting effects with four species of vesicular arbuscular endophyte. *New Phytologist* **78**, 257–268.

Schenck, N. C., & Kellam, M. K. (1978). The influence of vesicular arbuscular mycorrhizae on disease development. *Florida Agricultural Experiment Stations Technical Bulletin*, 198. University of Florida, Gainesville.

Schubert, A., Marzachi, C., Mazzitelli, M., Cravero, M. C., & Bonfante-Fasolo, P. (1987). Development of total and viable extraradical mycelium in the vesicular-arbuscular mycorrhizal fungus *Glomus clarum* Nicol. & Schenck. *New Phytologist* **107**, 183–190.

Smith, S. E., & Bowen, G. D. (1979). Soil temperature, mycorrhizal infection and nodulation. *Soil Biology and Biochemistry* **11**, 469–473.

Smith, F. A., & Smith, S. E. (1981). Mycorrhizal infection and growth of *Trifolium subterraneum:* Comparison of natural and artificial inocula. *New Phytologist* **88**, 311–325.

Smith, S. E., & Walker, N. A. (1981). A quantitative study of mycorrhizal infection in *Trifolium:* Separate determination of the rates of infection and mycelial growth. *New Phytologist* **89**, 225–240.

Smith, S. E., Walker, N. A., & Tester, M. (1986). The apparent width of the rhizosphere of *Trifolium subterraneum* L. for vesicular-arbuscular mycorrhizal infection: Effects of time and other factors. *New Phytologist* **104**, 547–558.

Sondergaard, M., & Laegaard, S. (1977). Vesicular-arbuscular mycorrhiza in some aquatic plants. *Nature (London)* **268**, 232–233.

Thomson, B. D., Robson, A. D., & Abbott, L. K. (1986). Effects of phosphorus on the formation of mycorrhizas by *Gigaspora calospora* and *Glomus fasciculatum* in relation to root carbohydrates. *New Phytologist* **103**, 751–765.

Thomson, B. D., Robson, A. D., & Abbott, L. K. (1990). Mycorrhizas formed by *Gigaspora calospora* and *Glomus fasciculatum* on subterranean clover in relation to soluble carbohydrate concentrations in roots. *New Phytologist* **114**, 217–225.

Tinker, P. B. H. (1975). Effects of vesicular arbuscular mycorrhizas on higher plants. In: *Symbiosis* (Ed. by D. H. Jennings & D. L. Lee), pp. 325–349. Cambridge University Press, London.

Tommerup, I. C. (1982). Airstream fractionation of VA mycorrhizal fungi; the concentration and enumeration of propagules. *Applied and Environmental Microbiology* **44**, 533–539.

Tommerup, I. C. (1983a). Spore dormancy in vesicular arbuscular mycorrhizal fungi. *Transactions of the British Mycological Society* **81**, 37–45.

Tommerup, I. C. (1983b). Temperature relations of spore germination and hyphal growth of vesicular arbuscular mycorrhizal fungi in soil. *Transactions of the British Mycological Society* **81**, 381–387.

Tommerup, I. C. (1984a). Persistence of infectivity by germinated spores of vesicular arbuscular mycorrhizal fungi in soil. *Transactions of the British Mycological Society* **82**, 275–282.

Tommerup, I. C. (1984b). Effect of soil water potential on spore germination by vesicular-arbuscular mycorrhizal fungi. *Transactions of the British Mycological Society* **83**, 193–202.

Tommerup, I. C. (1984c). Development of infection by a vesicular-arbuscular mycorrhizal fungus in *Brassica napus* L. and *Trifolium subterraneum* L. *New Phytologist* **98**, 487–495.

Tommerup, I. C. (1985a). Population biology of VA mycorrhizal fungi: Propagule behaviour. In: *Proceedings of the 6th North American Conference on Mycorrhizae* (Ed. by R. Molina), p. 331. USDA, Oregon State University, Corvallis.

Tommerup, I. C. (1985b). Strategies for longterm preservation of VA mycorrhizal fungi. In: *Proceedings of the 6th North American Conference of Mycorrhizae* (Ed. by R. Molina), pp. 87–89. USDA, Oregon State University, Corvallis.

Tommerup, I. C. (1985c). Inhibition of spore germination of vesicular-arbuscular mycorrhizal fungi. *Transactions of the British Mycological Society* **85**, 267–278.

Tommerup, I. C. (1988a). Genetics of vesicular-arbuscular mycorrhizal fungi. In: *Genetics of Pathogenic Fungi* (Ed. by G. S. Sidhu), pp. 81–91. Academic Press, London.

Tommerup, I. C. (1988b). Physiology and ecology of VAM spore germination and dormancy in soil. In: *Mycorrhiza and the Next Decade, Practical Applications and Research Priorities* (Ed. by D. M. Sylvia, L. L. Hung, & J. H. Graham), pp. 175–177. University of Florida, Gainesville.

Tommerup, I. C. (1988c). Long-term preservation by L-drying and storage of vesicular arbuscular mycorrhizal fungi. *Transactions of the British Mycological Society* **90**, 585–591.

Tommerup, I. C., & Abott, L. K. (1981). Prolonged survival and viability of VA mycorrhizal hyphae after root death. *Soil Biology and Biochemistry* **13**, 431–434.

Tommerup, I. C., & Bett, K. B. (1985). Cryopreservation of genotypes of VA mycorrhizal fungi. In: *Proceedings of the 6th North American Conference on Mycorrhizae* (Ed. by R. Molina), p. 235. USDA, Oregon State University, Corvallis.

Tommerup, I. C., & Sivasithamparam, K. (1990). Zygospores and asexual spores of *Gigaspora decipiens*, an arbuscular mycorrhizal fungus. *Mycological Research* **94**, 897–900.

Trenbath, B. R. (1978). Models and the interpretation of mixture experiments. In: *Nitrogen Fixation in Legumes* (Ed. by J. M. Vincent), pp. 229–236. Academic Press, Sydney.

Trinick, M. J. (1982). Competition between *Rhizobium* strains for nodulation. In: *Nitrogen Fixation by Legumes* (Ed. by J. M. Vincent), pp. 229–236. Academic Press, Sydney.

Walker, N. A., & Smith, S. E. (1984). The quantitative study of mycorrhizal infection: II. The relation of rate of infection and speed of fungal growth to propagule density, the mean length of the infection unit and the limiting value of the fraction of the root infected. *New Phytologist* **96**, 55–69.

Warner, A., & Mosse, B. (1982). Factors affecting the spread of vesicular arbuscular mycorrhizal fungi in soil. I. Root density. *New Phytologist* **90**, 529–536.

Wilson, J. M. (1983). *Interactions between vesicular arbuscular mycorrhizal fungi and their identification using immunofluorescence*. Ph.D. Thesis, University of Western Australia.

Wilson, J. M. (1984a). Comparative development of infection by three vesicular-arbuscular mycorrhizal fungi. *New Phytologist* **97**, 413–426.

Wilson, J. M. (1984b). Competition for infection between vesicular arbuscular mycorrhizal fungi. *New Phytologist* **97**, 427–435.

Wilson, J. M. (1984c). Inhibition of germination of spores of a *Gigaspora* species in sterilised soils. *Soil Biology and Biochemistry* **16**, 133–435.

Wilson, J. M., & Trinick, M. J. (1982). Factors affecting the estimation of numbers of infective propagules of vesicular arbuscular mycorrhizal fungi by the most probable number method. *Australian Journal of Soil Research* **21,** 73–81.

Wilson, J. M., & Trinick, M. J. (1983). Infection development and interactions between vesicular arbuscular mycorrhizal fungi. *New Phytologist* **93,** 543–553.

Wilson, J. M., Trinick, M. J., & Parker, C. A. (1983). Identification of vesicular arbuscular mycorrhizal fungi using immunofluorescence. *Soil Biology and Biochemistry* **15,** 439–445.

8

Interactions of Ectomycorrhizal Fungi

J. W. Deacon and L. V. Fleming

SUMMARY. Interactions of ectomycorrhizal fungi are discussed in relation to the sequences of mycorrhizal development on root systems of birch (*Betula* spp.) and other trees. In the absence of "parent" trees, seedlings are infected by "early stage" mycorrhizal fungi which behave as ruderals. As the plants age, early stage fungi remain dominant at the periphery of the root system but are progressively replaced by "late stage" fungi (*K*-selected) in older root regions. Experiments suggest that the distinction between early and late stage fungi lies in the size of the food base needed to initiate infection or support spread from established infections. Thus, only early stage fungi readily infect seedlings from spores or fragmented mycelia in unsterile soils; only they proliferate and persist when seedlings that are infected in aseptic culture are transferred to unsterile soil, and early stage fungi require less sugar for mycelial extension and infection in gnotobiotic conditions than do late stage fungi. Ingress by late stage fungi might be favored by exudates from older root regions. These fungi ultimately dominate the community by mycelial spread from established infections. This spread seems dependent on photosynthate, because late stage fungi do not infect seedlings planted into the root zone of established trees if the soil is cored or trenched. But soil conditions may favor late stage fungi in old woodland sites, where coring or trenching does not preclude infection.
 Antagonistic interactions between ectomycorrhizal fungi have received little study, but most aspects of mycorrhizal successions can be explained by competition and host–fungus interaction. These and other points are discussed in relation to mycorrhizal community development in natural and man-managed woodlands, the potential use of mycorrhizal inoculants in commercial forestry, and ecological strategy theory.

Introduction

Ectomycorrhizal fungi seldom occur alone on tree root systems. Even seedlings in nurseries or a glasshouse often support several fungal symbionts, and Zak and

Marx (1964) recorded an extreme case in which three fungi were isolated from a single mycorrhiza. This diversity reflects the fact that most mycorrhizal fungi are not host specific (Trappe, 1962), even if some have host preferences (Duddridge, 1987), and many of them are widely disseminated as air-borne spores. Clearly, there is ample opportunity for interaction between the fungi. But such interactions are best viewed in the wider context of communities of mycorrhizal fungi. These develop in response to numerous factors, including influences of genotype, age, and vigor of the tree, soil and environmental conditions, availability of inoculum and activities of the general soil microflora and microfauna, in addition to interactions between the fungi themselves. Analysis of the dynamics of mycorrhizal communities is important for the rational management of woodlands for economic or amenity value; it is also central to commercial inoculation programs (Marx & Cordell, 1989). We thus adopt a broad approach to interactions in this chapter, while also considering specific issues such as the following:

- How do individual fungi become established in soils containing mixtures of potential symbionts?
- How does the spectrum of mycorrhizal fungi change during the maturation of a tree or a stand of trees?
- How do introduced fungi persist in the face of competition from others or become replaced by them?
- How is mycorrhizal development on roots of regenerating seedlings or transplants influenced by site history?

A framework for analysis of such issues is provided by the recent concept of successions of ectomycorrhizal fungi (Deacon et al., 1983; Last et al., 1983; Mason et al., 1983a) shown schematically in Figure 8.1. Its strengths are those of all concepts. It enables us to see beyond the particular to the general, to formulate and test hypotheses, and to predict patterns of behavior from work in simple model systems. Its potential weakness, however, is that it is based on relatively few observations and, more important, on relatively few experiments. As explained below, the concept was developed mainly from studies on newly afforested sites, on a restricted range of soils (particularly brown earths), and with reference to birch (*Betula* spp.). It still requires rigorous testing outside of these conditions, and there is a dearth of information on the mechanisms involved. One of our principal aims, therefore, will be to identify these and other gaps in basic knowledge, while reviewing the progress that has been made to date.

Spatial and Temporal Sequences of Ectomycorrhizal Fungi

Based on observations of fruit bodies alone, there is a general trend for the number of mycorrhizal fungi associated with a stand of trees to increase with

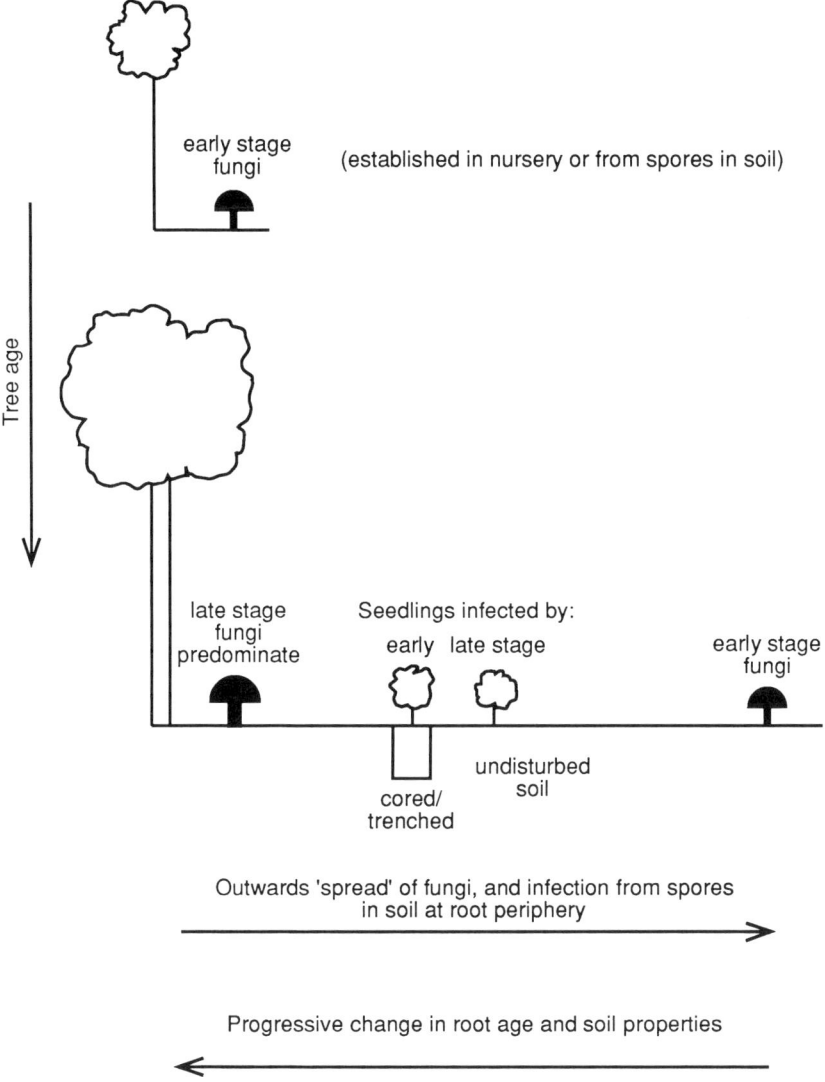

Figure 8.1. Schematic representation of sequences of ectomycorrhizal fungi as birch trees mature in a previously nonwooded site.

age. For example, Miller (1983) recorded fruit bodies of only five mycorrhizal fungi in a 15-year-old stand of western white pine, compared with up to 37 species in a 40-year-old stand and 78 species in 175- to 215-year-old stands. Similar evidence was presented by Richardson (1970), Miles (1985), and others, and was discussed by Mason et al. (1987). Such "floristic records," however,

Figure 8.2. Distribution of fruit bodies of *Hebeloma* spp. (H) and *Lactarius pubescens* (L) around a young birch tree south of Edinburgh.

have limited value for analysis of community structure. They are strongly biased toward the Basidiomycotina, especially those that readily form basidiomes; they are usually based on comparisons of different sites rather than sequential analyses of individual sites; they often lack replication or statistical support, and they do not relate to the microsites where these fungi interact with roots or with each other.

A more precise analysis of spatial and temporal sequences of ectomycorrhizal fungi was initiated by F. T. Last and his colleagues at the Institute of Terrestrial Ecology, near Edinburgh. It was done on a stand of 60 birch trees (*Betula pendula* and *B. pubescens*) of different genotypes that had been raised from surface-sterilized seed in a glasshouse and then outplanted to an agricultural brown earth. The site had not, for at least some decades, borne trees that are hosts of ectomycorrhizal fungi, although nearby trees and woodlands would have provided sources of air-borne spores for recruitment of inoculum. As the fruit bodies of mycorrhizal fungi appeared, so their distributions were mapped on orthogonal coordinates (Fig. 8.2). Recordings were made for 12 consecutive years in which every single fruit body was mapped. Then the detailed mapping was abandoned, but observations were continued to provide a record of the ingress of species to the site (P. M. Mason, personal communication). The results have been described in several papers (Ford et al., 1980; Mason et al., 1982, 1983a, 1987; Last et al., 1983) but the main findings can be summarized as follows.

Table 8.1. Sequence of the First Appearance of Fruit Bodies of Ectomycorrhizal Fungi in a Stand of Birch Trees South of Edinburgh

Age of stand (years)	Sequence of appearance
2	*Hebeloma crustuliniforme; Laccaria proxima*
3	*Laccaria tortilis; Thelephora terrestris*
4	*Hebeloma fragilipes; H. sacchariolens; H. mesophaeum; Inocybe lanuginella; Lactarius pubescens*
6	*Cortinarius* sp.; *Hebeloma leucosarx; Hymenogaster tener; Inocybe petiginosa; Leccinum roseofractum; L. scabrum; L. versipelle*
7	Other *Cortinarius* spp.; other *Hebeloma* spp.; *Lactarius glyciosmus; Leccinum subleucophaeum*
10	*Hebeloma vaccinum; Russula betularum; R. grisea; R. versicolor*
14	*Laccaria laccata; Lactarius spinulosus; Russula atropurpurea*

Based on Mason et al. (1987).

Fruit bodies of some mycorrhizal fungi appeared in early years after transplanting, perhaps because these fungi became established on the seedlings in the glasshouse. With time, fruit bodies of other fungi appeared, perhaps these fungi took longer to establish a "critical mass" on the root systems or because they were later arrivals (Table 8.1). The net result in the stand of 60 trees as a whole was that fungal diversity (reflected by types of fruit body in any one year) increased progressively up to at least the 14th year. Thus, there was evidence of a temporal sequence of mycorrhizal fungi. The second major finding was that the mean distance (from the tree base) of fruit bodies of any one fungus tended to increase in successive years. Ford et al. (1980) presented detailed information on this for one tree over a 3-year period; Mason et al. (1982) analyzed data over a 5-year period for four fungal species in the stand of 60 trees. In any one year, the earliest fungi in the "fruiting" sequence tended to fruit furthest from each tree, while the later fungi fruited closer to the tree bases. In other words, there was evidence of a spatial sequence in which the "early" fungi moved outward with the expanding root system, being replaced by "later" fungi nearer the tree base. The rate of outward movement differed between species, resulting in some overlap, but analysis of the spatial patterns revealed no obvious evidence of interaction between the fungi (Ford et al., 1980). In contrast to this, Murakami (1987) found a lack of spatial overlap between three *Russula* spp. in a Japanese forest site, which he attributed to antagonism between the fungi.

Even casual inspection of the stand of birch trees in Figure 8.2 revealed two points of interest. First, fungi with the larger and fleshier fruit bodies (e.g., *Lactarius, Cortinarius, Leccinum* spp.) were seen only in the later years—from the sixth year onward, except for *Lactarius pubescens* from the fourth year onward. Second, fungi with the smaller fruit bodies often produced them in abundance; in other words, to at least some degree numbers were inversely related

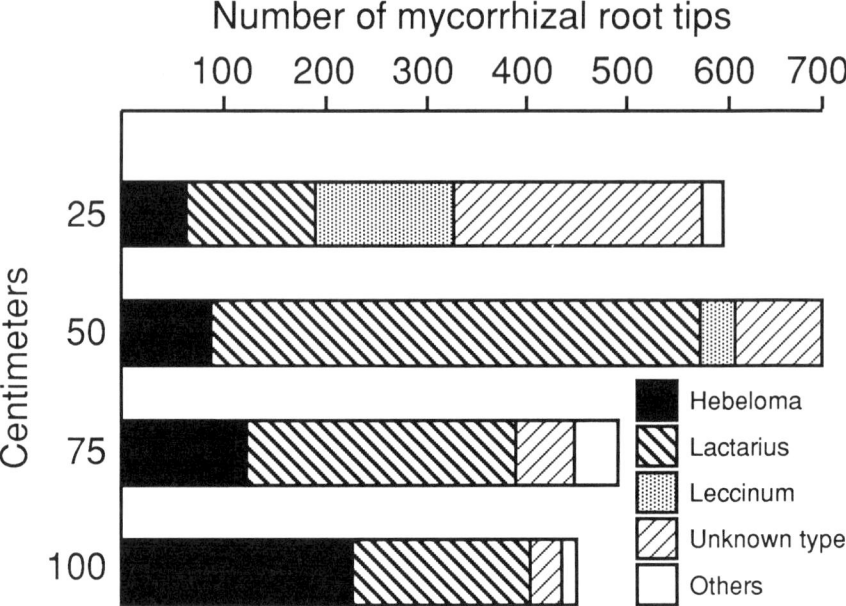

Figure 8.3. Numbers of mycorrhizas attributable to different fungi at 25, 50, 75, and 100 cm from the base of a birch tree on an experimental plot south of Edinburgh; data from 15 cores of soil taken at each distance along three radial transects. From Deacon et al. (1983).

to size. Such observations raise largely unexplored questions about the degree of "cooperation" within mycelia of the individual fungi, or about the abilities of different fungi to obtain host assimilates for fruiting.

There is much less information on the distributions of mycorrhizas than fruit bodies in this experimental plot. J. Warcup (cited in Mason et al., 1983a) examined the mycorrhizas in a number of random positions and confirmed the occurrence of different fungi (based on mycorrhizal morphology—Ingleby et al., 1990) where their fruit bodies had been recorded, although other mycorrhizal fungi also were present. Deacon et al. (1983) dissected soil cores taken at different distances along radial transects from the bases of three trees. The predominant mycorrhizal types changed with distance (Fig. 8.3), although not all of these types could be assigned to the species that formed the nearest fruitbodies. Fleming (1983) also found a change in predominant mycorrhizal types with distance from one of these trees (Table 8.2). Of the identifiable types, *Hebeloma* and *Lactarius* were most abundant at relatively large distances (125 and 100 cm, respectively) from the tree base; one unidentified type was dominant close to the tree base, and a second unidentified type was generally distributed. In the same year the position of every fungal fruit body was mapped around this tree (Fig. 8.4). Comparison with the data in Table 8.2 shows a general (although not absolute) correspondence

Table 8.2. Mean Percentage of Mycorrhizas of Different Types in Soil Cores Taken at Different Distances Along 10 Radial Transects from the Base of a 10-Year-Old Tree (Betula pubescens) in an Experimental Plot South of Edinburgh

Distance (cm) from base of tree	Mycorrhizal type			
	Hebeloma	Lactarius	Unidentified type 1	Unidentified type 2
25	15	5	9	72
50	31	3	11	55
75	14	21	9	40
100	31	32	13	25
125	59	12	9	20

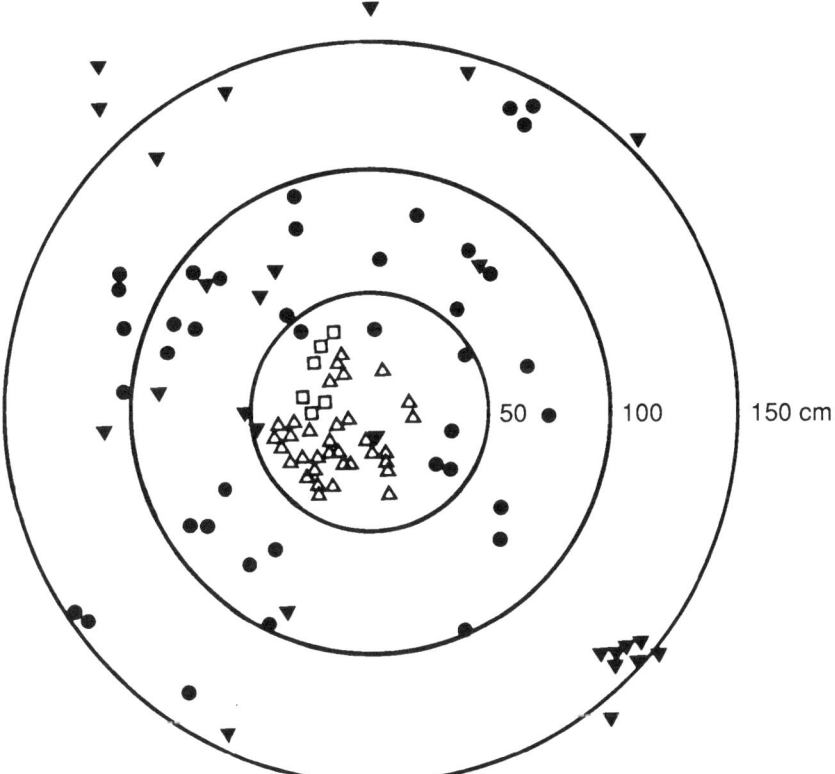

Figure 8.4. Map of the positions of all fruit bodies of ectomycorrhizal fungi that appeared in autumn 1981 beneath one birch tree on an experimental plot south of Edinburgh. Concentric circles represent distances 50, 100, and 150 cm from the tree base. (●) Hebeloma velutipes; (▼) Lactarius pubescens; (△) Cortinarius sp.; (□) Russula spp. (R. versicolor, R. betularum, and R. grisea).

between the zones of fruiting by *H. velutipes* or *L. pubescens* and the zones where *Hebeloma* or *Lactarius* mycorrhizas were found. However, the commonest mycorrhizal type close to the tree base could not be identified as belonging to either *Cortinarius* sp. or the three *Russula* spp. that fruited in this zone.

In work parallel to that above, Fleming et al. (1984) raised birch from seed in a glasshouse, in cores of soil taken from beneath one of the trees in the experimental plot ("mycorrhizal soil") or in cores from outside the plot (nonmycorrhizal soil). After 8 weeks, the seedlings in intact cores were transferred to pots of brown earth from just outside the plot and grown for 2 years in the glasshouse. Then the saplings were outplanted to this field site and grown for a further 2 years (4 years in all). Their mycorrhizal status was followed throughout, by destructive sampling of replicates. When the saplings were uprooted at the end of the experiment, there was a clear spatial sequence of mycorrhizas (Fig. 8.5). The periphery of each root system was dominated by *Inocybe* mycorrhizas, but nearer the tree bases there was a dominance of *Lactarius pubescens* on seedlings propagated initially in the nonmycorrhizal soil, and a fungus forming brown mycorrhizas with cream tips (now identified as *Tuber* sp.—Ingleby et al., 1990) on seedlings propagated in the "mycorrhizal" soil. These fungi (and others) had become established in the glasshouse, mainly in the second year, but became dominant only in the later years. Thus a spatial and a temporal sequence was found, similar to that from fruit body mapping in the larger plot, and based on mycorrhizas per se.

Gibson and Deacon (1988) extended these findings in a glasshouse study, using 4-year-old birch saplings uprooted from a coal spoil site. The roots were washed thoroughly to remove adhering soil and, inevitably, some of the mycorrhizas. Then the root systems were aligned in troughs filled with brown earth, which would have been inimical to the established mycorrhizal fungi, especially *Paxillus involutus*. The design of the troughs (Fig. 8.6) was such that they could be inverted periodically to examine the distributions of mycorrhizas in the bases of the troughs, and inoculations could be made in selected root regions without disturbance to the plants. In the 2 years of this study, there was no consistent pattern of change in distributions of mycorrhizal types on the root systems, although major changes occurred in individual troughs. However, it was shown by inoculation that *L. pubescens* could establish mycorrhizas only in older regions of the root systems (Fig. 8.7) whereas *Hebeloma* spp. established in both old and young root regions. The occurrence of fruit bodies in the troughs during the two seasons was reminiscent of "fruit body successions" insofar as the types of fruit body and their positions changed during this time. However, there was not a close correlation between the positions of fruit bodies and the intensity of mycorrhizal development by any one fungus (Fig. 8.8). This was confirmed for individual troughs by assessing the percentage occurrence of mycorrhizas of each fungus in zones where fruit bodies had or had not developed. For example, a mean 26.4% of root tips were mycorrhizal with *Thelephora* in zones where fruit bodies of *T.*

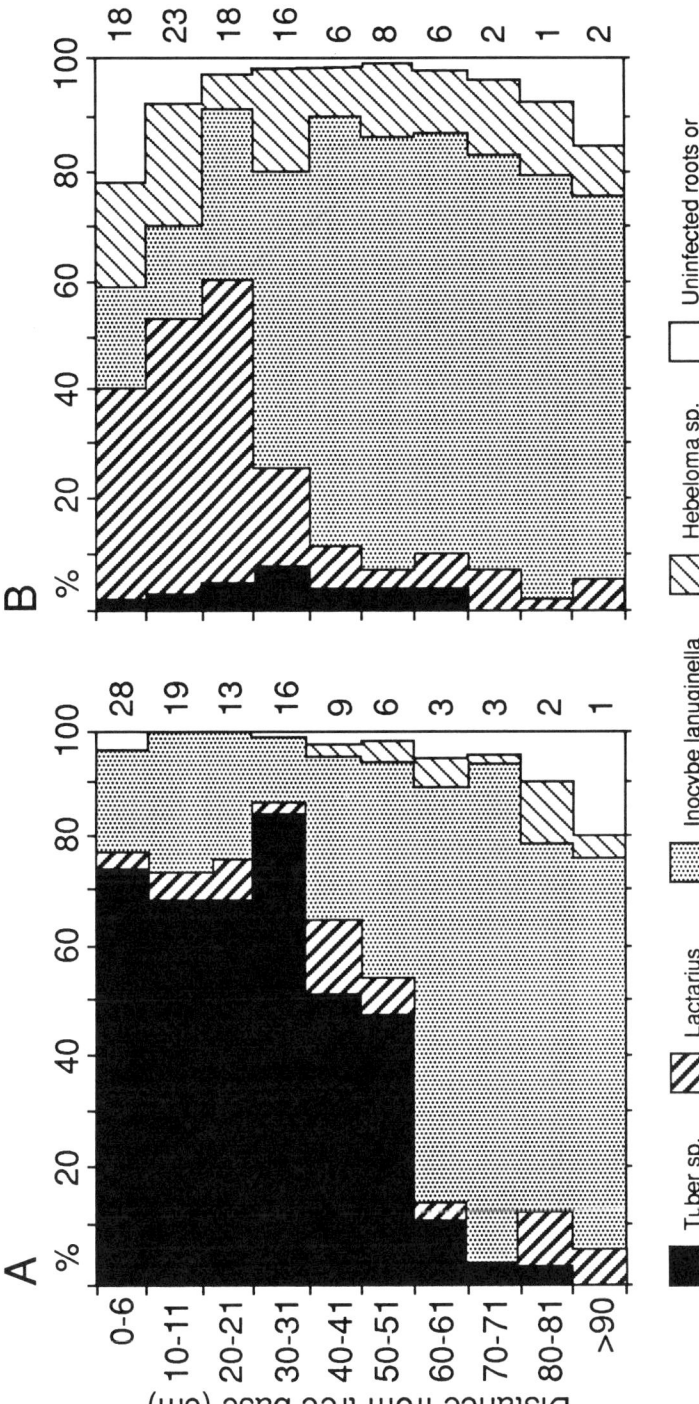

Figure 8.5. Frequency of occurrence (% root tips) of different mycorrhizas at various distances from the bases of birch saplings propagated initially in (A) "mycorrhizal" soil or (B) "nonmycorrhizal" soil and transplanted to a field site; combined data for 12 replicate trees when 35 or 45 months old. Numbers on right show percentage distribution of recorded root tips at different distances from the tree bases. From Fleming et al. (1984).

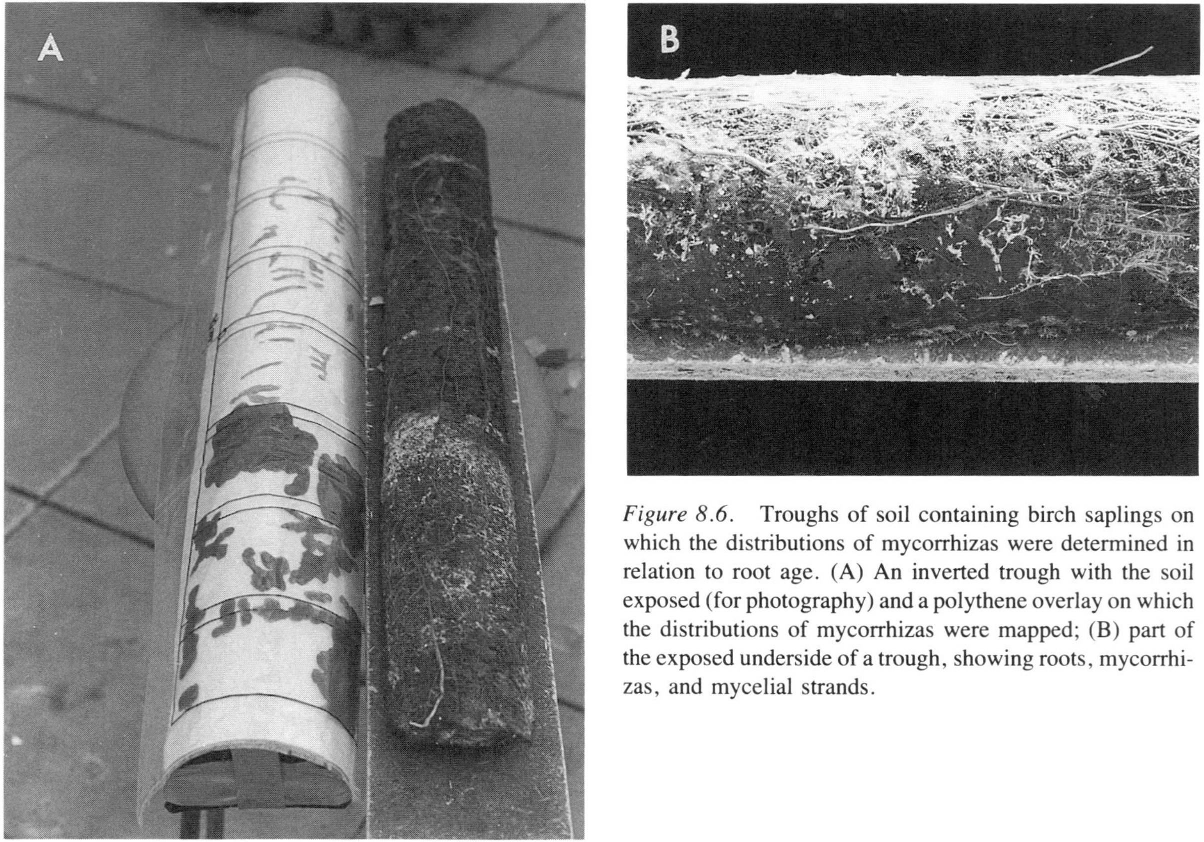

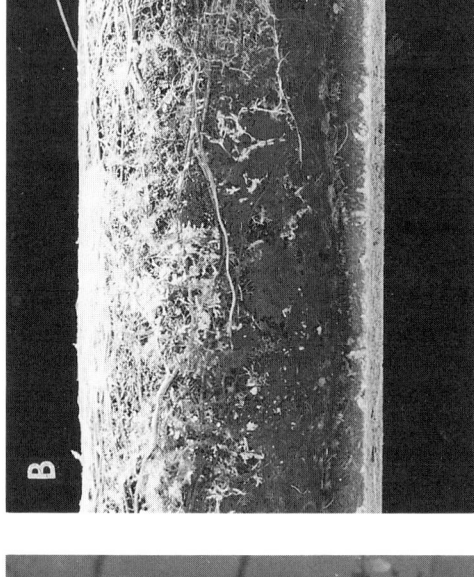

Figure 8.6. Troughs of soil containing birch saplings on which the distributions of mycorrhizas were determined in relation to root age. (A) An inverted trough with the soil exposed (for photography) and a polythene overlay on which the distributions of mycorrhizas were mapped; (B) part of the exposed underside of a trough, showing roots, mycorrhizas, and mycelial strands.

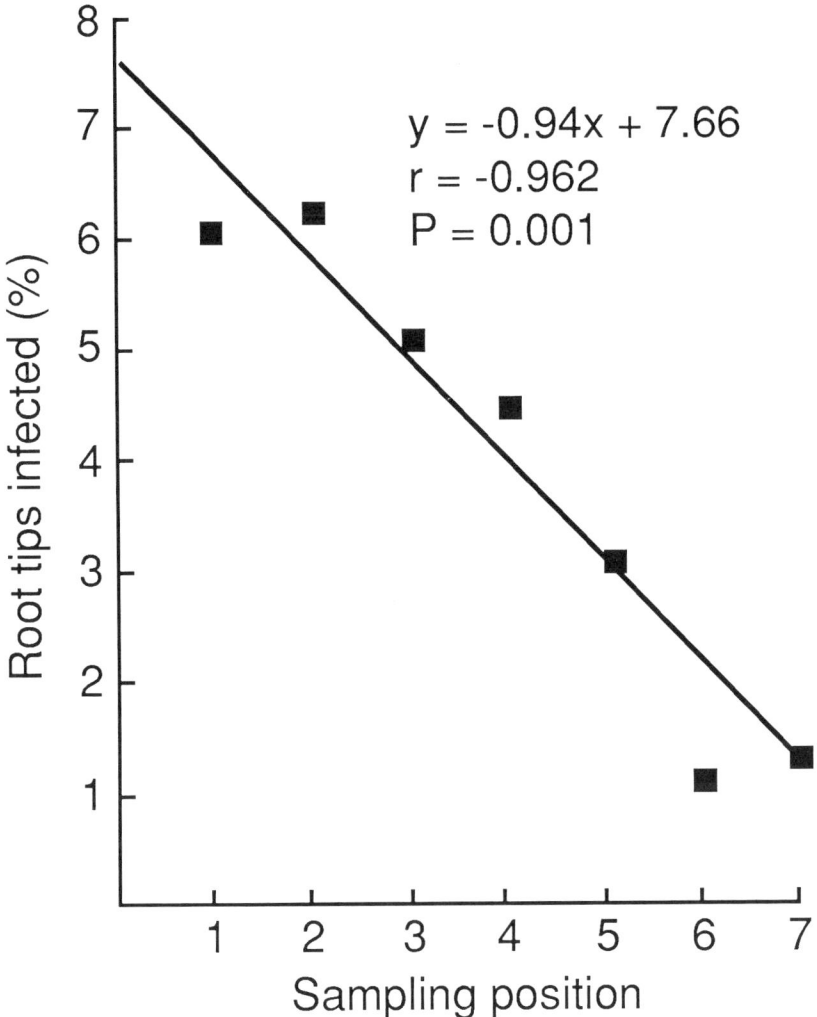

Figure 8.7. Relationship between root age (sampling position—1 is oldest) and infection of birch sapling roots from inoculum of *Lactarius pubescens* added to troughs of soil. From Gibson and Deacon (1988).

terrestris were seen, compared with a mean 22.4% *Thelephora* mycorrhizas where this fungus did not fruit. Corresponding data for *Inocybe* were 14.8 and 12.2%, and for *Hebeloma* 19.8 and 12.6%. The findings suggested that fruiting lagged behind current mycorrhizal status of any particular root region but nevertheless might give a general indication of changes of mycorrhizal status in space and time.

We have given these studies prominence because of our familiarity with them

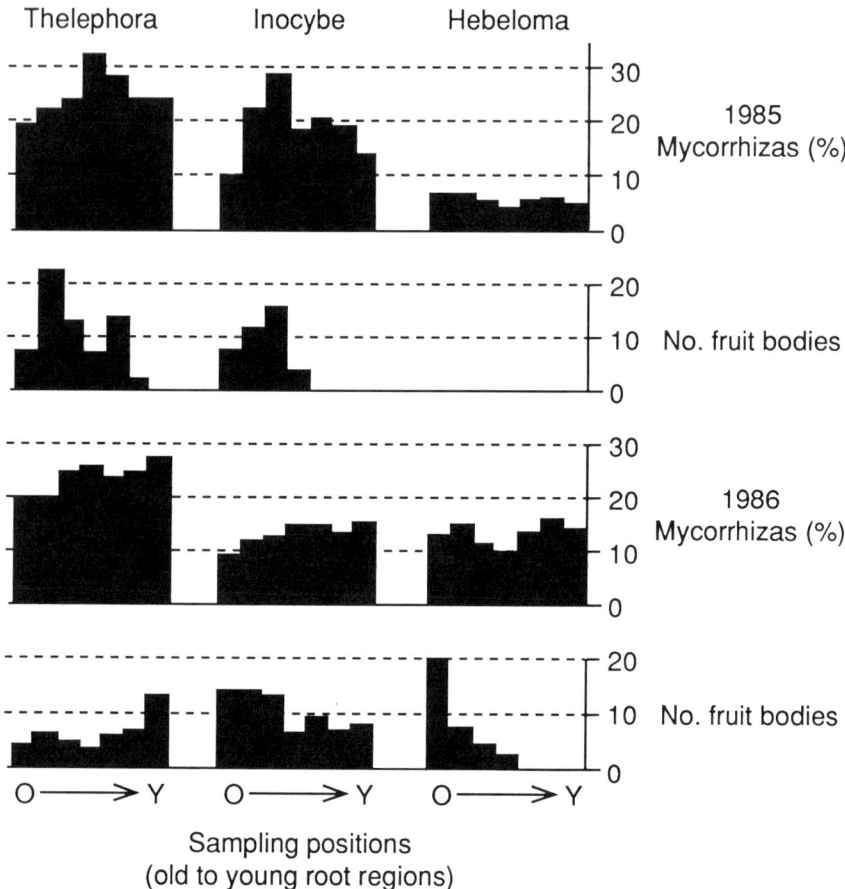

Figure 8.8. Distribution of mycorrhizas of *Thelephora, Inocybe,* and *Hebeloma* and of fruit bodies of *T. terrestris, I. lacera,* and *H. crustuliniforme* in seven assessment positions along the lengths of troughs of soil containing birch saplings in two successive years. Fruit bodies of each fungus first appeared (1985) in positions corresponding to the older (O) parts of root systems, but their distributions changed (*T. terrestris*) or extended (*I. lacera*) to younger parts (Y) in 1986. Based on data in Gibson and Deacon (1988).

and because they represent the most sustained, detailed investigation of ectomycorrhizal sequences. Similar evidence based on fruit bodies, mycorrhizas, or both has been recorded for *Pinus* spp., Douglas fir, *Eucalyptus* spp., and Sitka spruce (Becker, 1956; Tominaga, 1975; Chu-Chou, 1979; Chu-Chou & Grace, 1981, 1982, 1983; Thomas et al., 1983; Dighton et al., 1986). Evidently, mycorrhizal fungi occur in sequences in both space and time during the development of a tree root system—*at least in newly afforested sites*. We consider later how this relates to existing forest stands, to clear-felled sites, and to the regeneration of seedlings beneath established trees.

Early Stage and Late Stage Fungi

The studies on birch, above, suggested a broad distinction between "early stage" and "late stage" mycorrhizal fungi. These terms were coined to denote the times at which fruit bodies *first appeared* in age-related sequences but were not intended to imply that early stage fungi are necessarily restricted to young trees—a point that will be addressed later. This broad distinction was reinforced by experimental studies. For example, only early stage fungi were found to develop mycorrhizas on birch seedlings when mycelial or basidospore inocula were added to soil from treeless sites (Deacon et al., 1983; Fox, 1983, 1986a; Fleming, 1985). Similarly, only early stage fungi formed mycorrhizas on seedlings planted into cores of soil taken from beneath young (10–15 year) trees on the "birch plot," even if the soil itself contained a predominance of mycorrhizas of late stage fungi (Deacon et al., 1983; Fleming, 1983). Moreover, even if late stage fungi were established on seedlings in aseptic conditions—as can be done without difficulty (Mason, 1980; Mason et al., 1983b; Fleming, 1985)—they often did not persist and proliferate when the seedlings were transplanted to natural soils (Fleming, 1985; Last et al., 1985). These and other features, which are discussed later, are shown below.

Behavioral Features of Early Stage and Late Stage Mycorrhizal Fungi

Early stage (ruderal):

1. fruit bodies develop in early years beneath young trees
2. fruit bodies and mycorrhizas seen near periphery of expanding root system
3. infect readily from spores or mycelial inocula added to unsterile soil
4. persist when aseptically inoculated seedlings are transplanted to soil
5. have low sugar demand for extension growth and infection in culture
6. infection is not adversely affected by low phosphorus content of media
7. spores germinable in culture or in presence of plant roots
8. some are known to infect as monokaryons.

Late stage (K-selected):

1. fruit bodies develop in later years as trees mature
2. fruit bodies and mycorrhizas seen mainly in older root zones
3. do not readily infect from spores or mycelia added to unsterile soil
4. persist poorly after transplanting of aseptically inoculated seedlings
5. have high sugar demand for extension growth and infection in culture
6. infection is adversely affected by low phosphorus content of media
7. spores not readily germinable in culture
8. have not been shown to infect as monokaryons

The early stage–late stage distinction is unlikely to be absolute; it may serve only to identify the ends of a spectrum of behavior, which is also influenced by other factors. For example, in inoculation studies *Paxillus involutus* behaved as an early stage fungus on birch in colliery spoil, but did not infect birch seedlings from basidiospores in brown earth (Fox, 1986a); it is an early stage fungus that is markedly influenced by soil type. Also, *Lactarius pubescens* behaves as a late stage fungus by all of the above criteria but it often establishes on birch seedlings in their second season in glasshouse conditions, indicative of an intermediate position in age-related sequences (Fleming et al., 1984).

Despite such variations, during attempts to establish mycorrhizas on birch seedlings when basidiospores were added to soil, Fox (1986a) noted that the early stage–late stage distinction seems to be generic: all species of *Hebeloma, Inocybe*, and *Laccaria* that she tested behaved as early stage, whereas all of *Lactarius, Cortinarius, Russula,* and the Boletaceae behaved as late stage (Table 8.3). Support for this comes from other studies on other trees. Marx (1980) could not establish mycorrhizas on pines from basidiospores of three species of *Amanita* and three of *Lactarius*. Likewise, species such as *Laccaria laccata* and *Cenococcum geophilum* readily colonized jack pine or Sitka spruce from mycelial slurries or vermiculite-peat inocula added to soil, but species of *Hydnum, Amanita,*

Table 8.3. Groups of Early Stage and Late Stage Fungi[a]

Early stage fungi
Hebeloma crustuliniforme; H. leucosarx; H. sacchariolens; Inocybe geophila; I. lacera; I. lanuginella; Laccaria proxima; L. tortilis; Paxillus involutus

Late stage fungi
Amanita muscaria; Cortinarius bulbosus; C. debilitus; Elaphomyces muricatus[b]; Lactarius blennius; L. pubescens; L. rufus; L. spinulosus; L. turpis; L. vietus; Leccinum roseofractum; L. scabrum; Russula cyanoxantha; R. grisea; Scleroderma citrinum; Suillus luteus

[a]Groupings are based on ability to infect birch seedlings during 12–16 weeks from freshly harvested basidiospores added to unsterile brown earth, vermiculite-peat, or coal spoil. Based on data in Fox (1986a).

[b]Ascospore inoculum.

Amphimena, and *Lactarius* established poorly or not at all, whereas they could infect seedlings in pure culture (Shaw & Molina, 1980; Danielson et al., 1984). Marx and Cordell (1989) invoked the early stage–late stage distinction to explain why only some ectomycorrhizal fungi can be established in commercial inoculation programmes.

There is a parallel between some of these findings and those of Watling (1981, 1984) from field observations of fruit bodies. He found that *Laccaria laccata* and *Lactarius pubescens* are typically associated with pioneer birch woodland in Scotland, whereas *Lactarius torminosus* and species of Boletaceae and Russulaceae are found mainly in older birch woodlands. Similarly, Danielson (1984a) found that species of *Elaphomyces, Cortinarius, Cantharellus,* and *Hydnum* were restricted to mature jack pine stands whereas *Laccaria proxima, Scleroderma* sp., and *Rhizopogon* were typical of disturbed sites. In all these respects, however, we make the qualification that an isolate does not define a species in behavioral terms, and yet most of the experimental work has necessarily involved very few isolates. Fries et al. (1985) detected three intersterility groups within *Paxillus involutus,* one of the groups having an association with conifers and the other two with hardwood trees, notably birch. The delimitation of species is also difficult in genera such as *Hebeloma* and *Laccaria.* Mason et al. (1982) noted that what they had originally recorded as *Laccaria laccata* was later reidentified as *L. tortilis* and that fruit bodies of *L. laccata* did not appear until the fourteenth year in the experimental birch plot near Edinburgh. *L. laccata* has yet to be shown experimentally to be an early stage fungus, although its generic placement suggests that it is so.

In terms of modern strategy theory as applied to fungi (Pugh, 1980; Cooke & Rayner, 1984) the early stage mycorrhizal fungi can be regarded as ruderal or r-selected, and late stage fungi as perhaps exhibiting stress-tolerant (S-selected) or combative (C-selected) strategies. But whether this is entirely applicable to mycorrhizal fungi will be discussed in the final section of this chapter.

How are Mycorrhizal Sequences Determined?

In attempting to answer this question, we address three issues: (1) Why do early stage fungi start the sequence? (2) How do late stage fungi enter it? (3) How do late stage fungi then proliferate, ultimately to dominate the community of mycorrhizal fungi?

Early Stage Establishment: The Start of the Sequence

Ruderal species typically have a short vegetative phase followed by production of dispersal units, resulting in a widespread distribution; they also rapidly exploit an available resource. Early stage mycorrhizal fungi satisfy these criteria. They are widely distributed, as evidenced by, for example, the frequent invasion of forest nurseries by spores of *Thelephora terrestris*. They tend to be "generalists," with very wide host ranges, as again exemplified by *T. terrestris* (Mikola, 1970). They fruit early—even in nurseries, where the fruit bodies of *T. terrestris* can cause a problem by choking the seedlings. They fruit abundantly, as shown by the mapping studies of Ford et al. (1980) and Mason et al. (1982). Lastly, they readily establish mycorrhizas when dispersed inocula (either spores or mycelia) are added to normal, untreated soils (e.g., Fox, 1986a), so they exhibit a rapid response to a resource. In newly afforested sites they will need to establish infections from spores, so it is relevant to consider the evidence on spore germination by early and late stage fungi.

The Role of Basidiospores

Germination in Culture

The basidiospores of most ectomycorrhizal fungi do not germinate readily in culture but some degree of germination can be achieved in the presence of activated charcoal or microbial colonies (e.g., Fries, 1983, 1987; Ali & Jackson, 1989). The relevance of these factors to natural conditions is unknown, but the studies to date indicate some difference in the behavior of early compared with late stage fungi. For example, Fries (1978) reported up to 20% germination of basidiospores of *Laccaria laccata,* and 1% for *Paxillus involutus,* in appropriate experimental conditions. In similar conditions, *Thelephora terrestris* also germinated but this was described as "slow, sparse and erratic" (Birraux & Fries, 1981). Yet the spores of presumed late stage fungi (*Amanita muscaria, Leccinum scabrum, Lactarius helvus*) have shown even poorer germination, never exceeding 0.1% (Fries, 1978). This distinction was reinforced by Ali and Jackson (1989) who studied the effects of several bacteria and fungi on basidiospore germination in culture. *Hebeloma crustuliniforme* showed a maximum 21% germination, in the presence of a bacterium isolated from fruit bodies of this fungus. Germination by *P. involutus* and *L. laccata* was stimulated by other bacteria, but it was to maxima of <0.1% and <0.01%, respectively. Yet the late stage fungi *Amanita fulva, A. rubescens, Lactarius turpis,* and *Russula nitida* failed to germinate, even at a low level, in response to any treatment.

Fries (1979, 1981) found that spores of *Leccinum* were triggered to germinate by hyphae of *Leccinum,* the effect being species-group specific. Germination by *Laccaria laccata* also was triggered by fungal mycelia (Fries, 1983) but the effect was nonspecific, as also seems to be true for *Paxillus involutus* (Fries, 1978). Again, such findings provide preliminary evidence of a difference between early and late stage fungi. But caution is necessary because Straatsma et al. (1985) tried several methods to stimulate germination by *Cantharellus cibarius* (presumably late stage), and the only *unsuccessful* treatment was with mycelial cultures of this fungus. Ali and Jackson (1989) obtained virtually no germination by *H. crustuliniforme* (early stage) in the presence of nonspecific mycelium, and the presence of mycelium neither increased nor decreased the response to an activating bacterium.

The triggers for germination are unknown for any of these fungi. Hyphae of *Paxillus involutus* produce a volatile compound that stimulates germination by *Paxillus,* but the germination stimulant from *Leccinum* hyphae is nonvolatile (Bjurman & Fries, 1984). In related studies, germination of basidiospores of *Agaricus bisporus* was triggered by the volatile metabolite, isovaleric acid, from mycelium of *Agaricus* (Losel, 1964; Rast & Strobel, 1970). But isovaleric acid had no effect on germination of basidiospores of ectomycorrhizal fungi (Fries, 1978).

Germination on Roots

Several workers have found that plant roots stimulate germination by basidiospores. For example, Theodorou and Bowen (1987) obtained 46–69% germination for *Rhizopogon luteolus,* 34% for *Suillus luteus,* and 31% for *S. granulatus* on the surface of pine (*P. radiata*) seedling roots in aseptic conditions, and Birraux and Fries (1981) obtained some stimulation of germination by *Thelephora terrestris* near seedling roots of a range of trees. In experiments with *T. terrestris* (Birraux & Fries, 1981) and *Hebeloma mesophaeum* (Fries & Swedjemark, 1986), the effect was largely restricted to tree roots; in the case of *R. luteolus* (Theodorou & Bowen, 1987) it seems even more restricted because it occurred on pine but not *Eucalyptus globulus* (a nonhost tree). In other cases, however, it is more general; tomato roots can stimulate germination by a range of fungi, including *Russula* spp. (Melin, 1962) and *Cantharellus cibarius* (Straatsma et al., 1985).

Ali and Jackson (1988) compared germination by several ectomycorrhizal fungi of birch in the presence of a range of tree and nontree roots in aseptic conditions. Birch (*Betula pubescens*) was the most stimulatory of the plants tested, promoting 9% (*Paxillus involutus*), 13% (*Laccaria laccata*), and 30% (*Hebeloma crustuliniforme*) germination after 2 weeks. Four late stage fungi (*A. rubescens, A. fulva, L. turpis,* and *R. nitida*) showed little or no germination even after 8 weeks. Fries and Swedjemark (1986) also consistently failed to induce germination by species of *Leccinum, Boletus, Cortinarius,* and *Lactarius* in the presence of tree seedling roots. Ali and Jackson (1988) were the first to suggest specifically that such differences in germinability might be related to the distinction between early and late stage fungi. However, in their work germination (except by *P. involutus*) was generally poor on roots in soil compared with on roots in culture, and this anomaly must be resolved before the findings can be applied to natural situations.

Monokaryons versus Dikaryons

It is often assumed that the formation of mycorrhizas requires a dikaryotic mycelium. Consistent with this view, Ducamp et al. (1986) failed to establish mycorrhizas on maritime pine from single basidiospore isolates of *Suillus granulatus*, whereas mixtures of single-spore isolates, or dikaryotic mycelia from fruit bodies, did establish mycorrhizas. Similarly, monokaryotic mycelia of *Tuber melanosporum* fail to form mycorrhizas whereas dikaryotic mycelia do so (Fasolo-Bonfante & Brunel, 1972; Grente et al., 1972; Rouguerol & Payre, 1974). The specific effect of mycelia of *Leccinum* in inducing germination by *Leccinum* spores might also be seen in this light because it can lead to plasmogamy (Fries, 1981). However, recent work has shown that cultures of *Laccaria bicolor* derived from single basidiospores can form mycorrhizas on *Pinus banksiana* (Kroop et al., 1987), as can monokaryons of *Hebeloma cylindrosporum* on *P. pinaster* (Debaud et al., 1988). In the latter case, monokaryons of all four mating types formed mycorrhizas that were structurally and functionally similar to those of dikaryons. It is notable that both *L. bicolor* and *H. cylindrosporum* are in genera representing early stage fungi (Fox 1986a). At present there seems to be no information on mycorrhiza formation by monokaryons in field conditions. If it occurs, it should not be difficult to detect in fumigated nursery beds and the findings would add considerably to our understanding of mycorrhizal development in field conditions.

Infection from Chlamydospores, Ascospores, and Conidia

"Pseudomycorrhizas" or ectendomycorrhizas are widely distributed in forest nurseries, particularly on pines (see Harley & Smith, 1983), but sometimes on spruce (Thomas & Jackson, 1979) and other trees. For nursery-raised seedlings, they may represent the earliest stage of age-related sequences of mycorrhizas. The fungi concerned are taxonomically ill defined, but some have been characterized, including the hyphomycetes, *Phialophora finlandia, Phialocephala fortinii,* and *Chloridium paucisporum,* which are most often encountered as gray, sterile mycelia (Wang & Wilcox, 1985). The role of their conidia is unknown. In laboratory culture these were produced only after extended incubation—for 6 to 12 months (Wang & Wilcox, 1985)—but such a delay might not disadvantage a mycorrhizal fungus in nature. Equally interesting was the discovery of a new species of operculate discomycete, *Tricharina mikolae* (Yang & Wilcox, 1984). In its production of chlamydospores, its hyphal characteristics and its behavior on host roots it was typical of E-strain fungi frequently reported to form ectendomycorrhizas on pines. It is almost certainly the teleomorph (perfect stage) of one of the E-strain fungi, which accords with an earlier suggestion by Danielson (1982) that the likely teleomorphs of E-strain fungi would be operculate discomycetes. The ascospores of *T. mikolae* were found to germinate readily after discharge, and monospore isolates could form ectendomycorrhizas (Yang & Wilcox, 1984). Also, the large chlamydospores of this fungus could initiate mycorrhiza formation when placed near seedling roots. Lamb and Richards (1974) had earlier established mycorrhizas on *Pinus radiata* from chlamydospores of three unidentified symbionts of pines, especially in previously fumigated soils.

Another operculate discomycete, *Sphaerosporella brunnea*, was shown by Danielson (1984b) to form ectomycorrhizas on pine, spruce, larch, and poplar in axenic conditions or "open" systems in a glasshouse; the mantles were usually discontinuous or thin. As with *T. mikolae*, the ascospores germinated readily after discharge, so they might play an important role in infection. *S. brunnea* is interesting in another respect: it is found commonly as a carbonicolous discomycete, on burnt substrates, and some of the isolates shown by Danielson to form mycorrhizas had been obtained from burnt materials. Moreover, the fungus was found to degrade cellulose and to grow rapidly, consistent with a saprophytic (saprotrophic) life style. So it might infect seedlings (perhaps following burning) from a saprophytic food base.

Recently, Warcup and Talbot (1989) described three *Muciturbo* spp. that form ectomycorrhizas with thin sheaths on eucalypts in Australia. *M. reticulatus*, the most-studied species, produced both hypogeous ascocarps and conidia in pots of heat-treated soil containing seedlings. The conidia could not be germinated on agar, and the ascospores also would not germinate when newly formed. But the ascospores could survive heat treatment and even autoclaving of soil. These fungi may be equivalent to *Rhizina undulata*, the pathogen that causes the group dying disease of conifers, in their adaptation to fire-managed sites (Jalaluddin, 1967).

Summary of Infection from Spores

The limited number of studies to date reveals that spores of some ascomycetes that form ectendomycorrhizas, or ectomycorrhizas with a thin mantle, can germinate readily. This is consistent with the widespread occurrence of such fungi on seedlings in nurseries. Other ascomycetes, such as *Muciturbo* spp., may require heat activation of their ascospores. Chlamydospores of E-strain and other fungi also can initiate infections, but there is little information on the roles of conidia of the hyphomycetes or ascomycetes. The picture is also unclear for basidiospores, which merit further study on roots in unsterile soils. Nevertheless, if we take the current very incomplete evidence at face value we might conclude: (1) that basidiospores of early stage fungi germinate better than do those of late stage fungi in equivalent conditions in culture or on roots (at least 10 times better), and (2) that monokaryons of early stage fungi can form mycorrhizas whereas this has not yet been demonstrated for late stage fungi, which may need to be dikaryotic. All species of *Hebeloma* and most of *Laccaria* studied to date show bifactorial heterothallism—there are four mating types (Fries & Mueller, 1984; Bruchet et al., 1986; Debaud et al., 1988). Assuming this to be true for mycorrhizal fungi in general, as for wood-destroying basidiomycetes (Fries, 1987), then from (1) and (2) above we can make a simple calculation: from a low number of basidiospores, there would be at least a 400 times greater chance of infection by an early stage than by a late stage fungus. This comment is offered merely by way of illustration or provocation. If even remotely relevant to field conditions, it could largely account for the observed difference in establishment by early and late stage fungi in nature. It illustrates the importance of further studies on infection from spores.

We add a rider at this point: the ability of early stage fungi to infect seedlings in soil, where late stage fungi cannot do so, is seen in experiments involving *mycelial inocula* as well as spore inocula (Deacon et al., 1983; Fox, 1983, 1986a; Fleming, 1985). As the mycelia were derived from fruit bodies and thus were dikaryotic, there must be additional factors (apart from spore germinability and nuclear status) that govern the early–late stage distinction. Such factors are considered later.

Ingress by Late Stage Fungi: The Sequence Develops

The ingress of late stage fungi in previously nonwooded sites must occur from spores, but evidently more slowly than in the case of early stage fungi. It is difficult to study experimentally because of the long periods involved and the unpredictability of the timings and locations of events. This situation, however, is little different from that in root pathology, where much could be learned by studying the earliest events in the establishment of infections from natural, critically low inoculum levels in soil. The methods are available. Tree seedlings and saplings can be grown in soil containers that provide periodic and nondestructive access to root regions of different ages (Gibson & Deacon, 1988) and inocula can be applied to roots in agar (Theodorou & Bowen, 1987) or alginate gels (Deacon & Fox, 1988) that can later be retrieved for examination. By using an agar carrier in this way, Theodorou and Bowen (1987) found 42% germination by spores of *Rhizopogon luteolus* (early stage) on roots of pine seedlings in soil, similar to that on pine roots in aseptic conditions. A more natural experimental system would involve adding spores alone to roots and then retrieving the spores in films of varnish, gelatine, etc. applied to soil or to the roots at sampling (Lingappa & Lockwood, 1963). Such "direct" methods are underused.

Although the late ingress by some fungi in field conditions could be explained by a paucity of available inoculum, this is clearly not the case in experimental conditions where inoculum is supplied. So an explanation must be based on features associated with ageing of trees. There would seem to be five possibilities. (1) Tree age per se determines "receptivity" (equivalent to susceptibility) to different fungi; (2) the age of a root region determines such receptivity; (3) soil or environmental conditions change with age of a tree or root region and differentially influence the behavior of fungi; (4) age-related changes in root-associated microorganisms are involved; and (5) previously established mycorrhizal fungi influence those that follow. We review below the limited information in these respects.

Effects of Host and Edaphic Factors

To our knowledge, the only *direct* experimental evidence that changes in host or edaphic factors might influence the ingress by late stage fungi is for *Lactarius pubescens* on birch in troughs of soil (Gibson & Deacon, 1988). In this work, bare-rooted 4-year-old saplings were transferred to a brown earth from a treeless site so that soil conditions would be identical along the lengths of the root systems. Mycelial inoculum of *L. pubescens* was applied to various regions of the roots, and the fungus was found to establish progressively better with age of the root regions (Fig. 8.7). No such age-related effect was found with *Hebeloma subsaponaceum* (early stage). As the infectible root tips would almost certainly have represented current season's growth, the implication is that root tips have different receptivity to *L. pubescens* according to the age of the subtending root region.

The mechanisms involved in such an effect remain unknown. There are reports that the intensity of infection is correlated with root sugar levels (Bjorkman, 1942, 1970; Marx et al., 1977b). Similar findings have been made with VAM fungi on herbaceous plants (e.g., Azcon & Ocampo, 1981; Graham et al., 1981; Same et al., 1983). As Bowen (1987) emphasized, the correlation in these cases is between sugar levels and *length of infected root* rather than the number of individual infections as assessed by entry points of VAM fungi. In other words, sugar levels may influence mainly the postinfection stages. Perhaps also relevant is the finding by Smith (1970), that levels of carbohydrates in root exudates of sugar maple are higher for seedlings than for mature trees. In relation to ectomycorrhizas, it would be useful to have comparative data for trees of different ages as well as for root regions at different distances from the union with the stem. In terms of the establishment of infection, however, it might be unwise to concentrate solely on sugars and other water-diffusible compounds. Gemma and Koske (1988) found that unidentified volatile compounds from tomato roots induce both germination and germ-tube tropism by the VAM fungus *Gigaspora gigantea;* Fries et al. (1985) found that lipids in pine root exudate stimulate growth by *Laccaria* spp., and Fries (1987) records that a lipophilic material (possibly a glycolipid) from pine roots stimulates germination of the spores of *Hebeloma mesophaeum*.

In the work of Gibson and Deacon (1988) above, the possible influence of changing soil factors could not be excluded because *Lactarius* did not form many mycorrhizas until the second year of the experiment. But this was addressed in further work (Gibson & Deacon, 1990) where growth (on agar) and mycorrhiza formation (on aseptic birch seedlings on agar) were examined for five fungi when

Table 8.4. Numbers of Mycorrhizas and Percentage of Roots Mycorrhizal (in Parentheses) on Birch Seedlings after 12 Weeks on Agar Supplemented with Different Levels of Mineral Nutrients[a]

	Mineral nutrient medium			
	Complete	One-tenth P	One-tenth N	5% LSD
Early stage fungi	8.3 (23)	5.7 (21)	1.3 (8)	3.4 (6.5)
Late stage fungi	12.8 (19)	2.9 (6)	1.7 (4)	4.8 (4.8)

[a] Data are means of 4 fungi (early stage) or 5 fungi (late stage), 10 seedlings for each fungus; the "complete" medium was Melin–Norkrans agar. Based on data in Gibson and Deacon (1990).

glucose and mineral nutrients were supplied at different concentrations. Four isolates of early stage fungi (two each of *Laccaria proxima* and *Hebeloma crustuliniforme*) were found to grow continuously across agar plates in the absence of glucose, whereas five isolates of late stage fungi (*L. pubescens, Tricholoma fulvum,* and *Amanita muscaria*) required quite high concentrations of glucose (0.1% or more) for continued hyphal extension. A similar difference was found for infection of aseptic birch seedlings: the early stage fungi could infect maximally in the absence or virtual absence of exogenous glucose, whereas four of the five isolates of late stage fungi infected maximally only at high glucose levels. The exceptional isolate (of *L. pubescens*) infected independently of glucose, but it was unusual among the late stage fungi in making initial (but not sustained) growth on agar in the absence of glucose. Despite the artificial conditions, these experiments indicate that different fungi have different degrees of dependence on sugars, which in turn influence their abilities to infect seedlings.

More directly relevant to natural conditions were equivalent experiments with altered mineral nutrition (Gibson & Deacon, 1990; Table 8.4). The development of mycorrhizas by the late stage fungi was severely reduced by a reduction of phosphorus to one-tenth of the "standard" concentration in modified Melin and Norkrans medium, whereas mycorrhiza formation by early stage fungi was much less affected. The development of mycorrhizas by all fungi was severely affected by reduction of nitrogen to one-tenth of standard. There is no basis here for thinking that reductions in the availability of mineral nutrients with age of a tree (or root region) would facilitate the ingress of the late stage fungi; quite the opposite may be true. This does not, of course, rule out the possible influences of other soil factors that change with age of root regions.

Effects of Microorganisms

Soil microorganisms are known to influence mycorrhizal development (e.g., Bowen & Theodorou, 1979; Garbaye & Bowen, 1987; McAfee & Fortin, 1988; Malajczuk, 1988) and germination of basidiospores in culture (Ali & Jackson, 1989). However, there are few data relevant to mycorrhizal sequences. It was noted earlier that late stage mycorrhizal fungi often fail to infect from mycelial inocula added to soil, whereas the same inocula are effective in sterile conditions (e.g., Shaw & Molina, 1980; Deacon et al., 1983; Danielson et al., 1984). This suggests that soil microorganisms not only influence the establishment of ectomycorrhizal fungi but also differentially affect the early and late stage types. The explanation may be simple: if late stage fungi have the higher sugar requirement to maintain their growth and to support the establishment of mycorrhizas (Gibson & Deacon, 1990), then nonspecific competition for nutrients from soil microorganisms would have the greater effect on late stage fungi.

Treatment of small samples of soil in a domestic microwave (MW) oven for as little as 30 sec can raise the soil temperature to 60–70°C and markedly reduce the soil microbial population. Yet the soil cools rapidly after treatment and may be exposed to temperatures >50°C for only a few minutes (Fig. 8.9). By adding mycelial inocula and aseptic birch seedlings to variously treated tubes of soil, Gibson et al. (1988) found that MW treatment could greatly enhance the establishment of mycorrhizas by *L. pubescens* (late stage), but had no effect on that by *Hebeloma subsaponaceum* (early stage), which established well even in untreated soils. However, MW treatment did not enable *A. muscaria* (late stage) to infect from added inoculum, and MW treatment was not always effective for *L. pubescens*. From several experiments of this type, it was concluded that a temporary reduction or simplification of the soil microflora could significantly benefit late stage mycorrhizal fungi, but *only* if these fungi could establish mycorrhizas to at least some degree in the corresponding untreated soil. In other words, the microflora had a modifying effect rather than a primary effect on mycorrhizal establishment. Again, this might relate to the need for a food base.

The most spectacular response to MW treatment of soil was an enhancement of birch seedling growth (Fig. 8.10). The larger responses (8.0- and 11.4-fold) occurred in two "old" birchwood soils that contained large amounts of organic matter, whereas two young (11–16 year) birchwood soils gave only 1.9- and 2.9-fold growth responses (Gibson et al., 1988). These responses were not related either to mycorrhizal status of the seedlings or to treatment times per se, but rather to exposure times (20–60 sec) that raised the temperature of particular soil samples to ca. 60°C or more and that markedly reduced the soil microbial populations. Yet, the soils were maintained in open containers and would rapidly have been recolonized by microorganisms. Such findings are reminiscent of the effects of aerated steam and soil fumigants on tree seedling growth (e.g., Benzian, 1970). Their significance lies in the fact that woodland soils had markedly different *potentials* to support seedling growth, but these different potentials were neither reflected in nor predicted by the actual growth observed in the untreated soil samples (Gibson et al., 1988). For agricultural crops, there is evidence that "deleterious rhizosphere bacteria" and fungal "minor pathogens" (Salt, 1979) are the cause of yield declines in monoculture or other intensive cropping systems (e.g., Schippers et al., 1985). These yield declines can be overcome experimen-

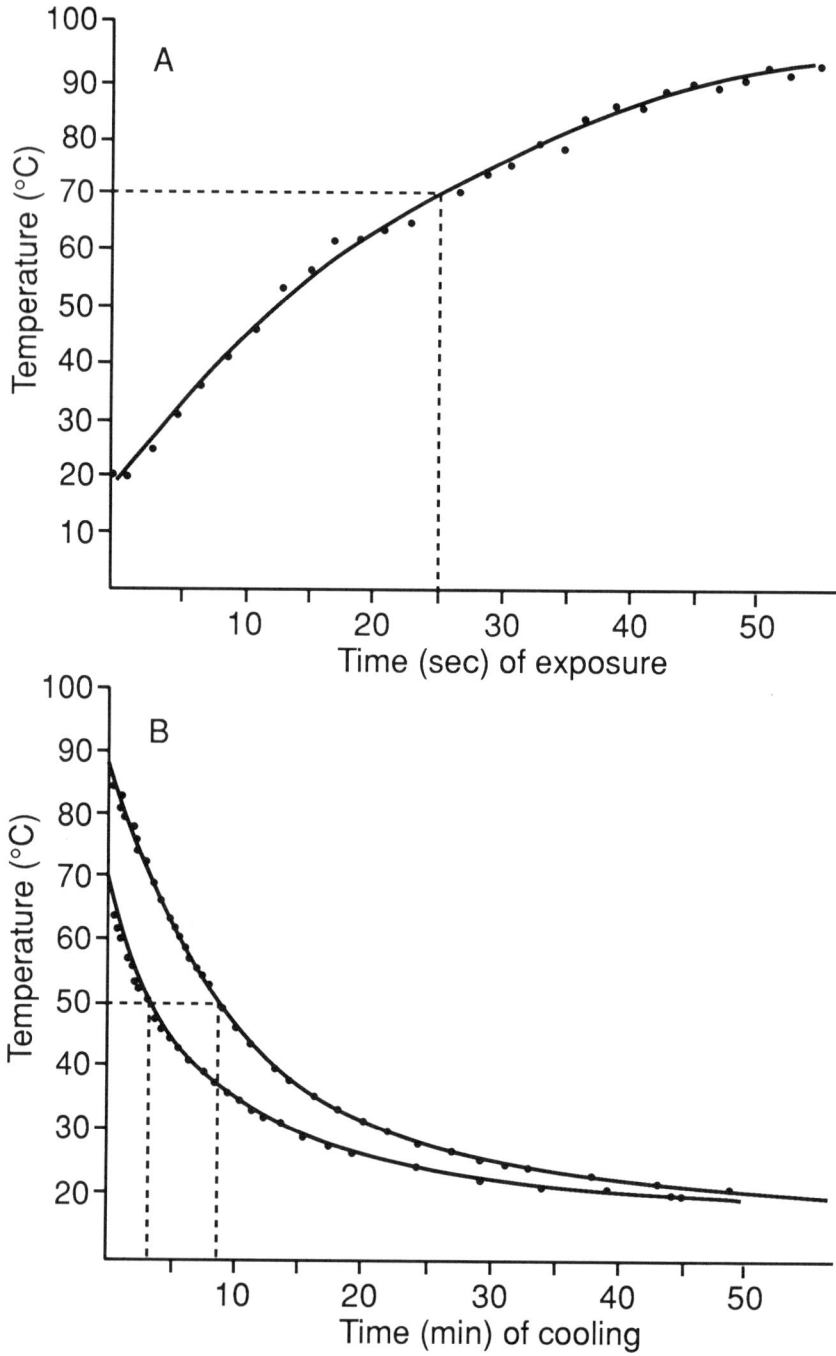

Figure 8.9. (A) Heating curve of 20 ml samples of woodland soil (53% w/w water content) exposed for different times in a 1 kW microwave oven. (B) Cooling curves of samples held at room temperature after reaching 70° and 90°C. Growth of birch seedlings in similar soil (Struan) is shown in Figure 8.10.

		Exposure to MW (sec)			
	Soil	0	20	40	80
Mean shoot wt.	Hayes	21	211	208	269
(mg)	Struan	6	15	35	51
	Newtongrange	13	8	<1	6
Maximum temp.	Hayes	26	70	86	>95
(°C)	Struan	26	58	93	>95
	Newtongrange	26	68	91	>95

Figure 8.10. Growth of birch seedlings after 12 weeks in tubes of soil exposed to microwave (MW) irradiation for different times. Back row, soil from beneath birch (>30 years old) (Hayes); middle row, soil from an ancient birch woodland (Struan); front row, coal spoil (Newtongrange). Left to right, exposure to MW for 0, 20, 40, and 80 sec. Mean shoot weights (mg) and maximum temperatures during MW exposure are shown below the figure. Data from Gibson et al. (1988).

tally by relatively mild treatments with aerated steam, partial soil fumigants, or solarization, and also by inoculation with "plant growth-promoting" fungi and bacteria—organisms that are not beneficial per se, only when applied in soils that exhibit the yield-decline phenomenon. It seems reasonable to suppose that, with time, a stand of trees becomes equivalent to a monoculture of an annual crop in terms of its accumulating load of potentially deleterious microbes. The implications for natural regeneration in established woodlands may be profound. The implications for mycorrhizal development may be equally great but are unexplored. When used as inoculants, some but by no means all mycorrhizal fungi have a pronounced stimulatory effect on early seedling growth—an effect not necessarily related to the "success" of fungal establishment. Indeed, Frankland and Harrison (1985) found no relationship between growth parameters of birch seedlings and the occurrence of "mature" (i.e., macroscopically recognizable) ectomycorrhizas on them in a multiple regression analysis involving seedlings grown in 25 different soil samples. But "immature" mycorrhizas, with incipient or well-developed sheaths detectable only by microscopy, did account for a small (ca. 10%) but highly significant proportion of the variation in seedling heights. The growth–response phenomenon in general would seem to merit reinvestigation, as would the microbiology of the rhizosphere (or mycorrhizosphere) of trees in this respect.

Effects of Previously Established Mycorrhizal Fungi

The possibility that previously established mycorrhizal fungi might facilitate or otherwise influence the ingress of late stage mycorrhizal fungi was raised by the work of Fleming et al. (1984). Birch seedlings initially propagated (8 weeks) in cores of soil from beneath a tree developed mycorrhizas of *Inocybe* in the first year, followed by *Tuber* sp. in the second year, whereas seedlings propagated in "nonmycorrhizal" soil developed *Hebeloma* mycorrhizas in the first year and *L. pubescens* in the second year. Yet, from 8 weeks onward, all seedlings had been treated identically.

To investigate whether these were causal relationships in which particular fungi follow others, Fleming (1985) established the early stage fungi *Hebeloma sacchariolens* and *Laccaria proxima* on birch seedlings in aseptic culture (*Inocybe* spp. could not be used because they could not be cultured); then the plants, and uninoculated controls, were transferred to pots of soil supplemented with mycelial inocula of late stage fungi (*Lactarius pubescens, Leccinum scabrum,* or *Amanita muscaria*). None of the late stage types infected the seedlings, so there was no evidence that early stage fungi directly facilitate their establishment. However, in the second year (following dormancy) *Lactarius* mycorrhizas developed from naturally occurring inoculum (presumably spores) on seedlings that had initially been nonmycorrhizal (but were soon colonized by *Hebeloma* sp. and *Thelephora* in the soil) and also on seedlings inoculated with *H. sacchariolens*. *Lactarius* mycorrhizas did not develop on seedlings initially mycorrhizal with *Laccaria proxima;* instead, these became infected by *Cenococcum geophilum,* which did not occur in the other treatments (L. V. Fleming, unpublished). All pots had received the same soil and had been fully randomised, so this was experimental confirmation of the tendency for some fungi to follow others, and in both experiments it occurred following plant dormancy. We defer discussion of this to the section on "Role of Root Dormancy."

Spread of Late Stage Fungi

With time, late stage fungi become dominant in the older parts of root systems (Fig. 8.5), although they may not entirely replace early stage fungi, such as *Laccaria* spp. A likely explanation of their ultimate dominance was found in studies on coring and trenching of soils beneath trees in field sites (Fleming, 1983, 1984; Fig. 8.11). In each of three experiments on birch trees approximately 10 years old, aseptically raised seedlings planted beneath the trees in undisturbed positions developed mycorrhizas of late stage fungi (*Lactarius pubescens, L. tabidus, Leccinum roseofractum, L. scabrum*). But seedlings planted into identical positions where the soil had been cored or trenched, to sever connections with the older tree, developed few or no late stage mycorrhizas but only early stage types (Table 8.5). A similar "coring" experiment in an ancient birchwood (Fleming et al., 1986) showed that *Amanita muscaria* infected seedlings only when the soil was undisturbed, but *Leccinum roseofractum* and *Lactarius* (probably *L. tabidus*) infected in both cored and undisturbed positions. From this it seems that there is a difference in behavior in young and old birchwood soils; but, for at least young birchwood soils, several late stage fungi may infect seedlings only from inocula that remain organically connected to an older tree. By implication, they may also increase their dominance on the roots of single trees by this means.

The inoculum in these cases is probably a hyphal network or a system of mycelial strands that derive nutrients from mycorrhizas on the older tree (Fleming, 1984). Experiments with transparent chambers have shown such a role of mycelial strands, or hyphal wefts associated with them, in initiating infections of pine seedlings (Brownlee et al., 1983). Mycelial strands have been found to extend at rates of 2.4 mm (*Pisolithus tinctorius*) and 4.3 mm/day (*Suillus bovinus*) when growing across the surface of peat from established mycorrhizas on seedlings of *Pinus contorta* (Finlay & Read, 1986). A similar rate was reported by Skinner and Bowen (1974) for strands of *R. luteolus* on *Pinus radiata*. It thus seems reasonable to assume, as Chilvers and Gust (1982) suggest, that the development of intense localized clusters of mycorrhizas, followed by spread of mycelial strands, becomes more important than primary infections arising from airborne spores or other dispersed forms of inoculum in the maturing root zone.

Figure 8.11. Design of experiments involving trenching and coring of soil around birch trees in an experimental plot. (A) Trenches during preparation. (B) Trenches "sealed" with corrugated plastic; positions of birch seedlings planted inside and outside the trenched areas are marked with labels.

Table 8.5. Mean Percentage of Mycorrhizas Attributable to Late Stage Fungi on Birch Seedlings Grown in Nonisolated (Undisturbed) and Isolated (Cored or Trenched) Positions Around Mature Birch Trees and on Roots of the Mature Trees in Equivalent Positions.

	Seedlings			
Fungus	Nonisolated	Isolated	Mature tree	Reference
Lactarius pubescens	38.5	2.1	12.0	Fleming (1983)
Lactarius pubescens	62.0	5.9	40.9	Fleming (1984)
Lactarius glyciosmus	17.6	0.9	38.9	Fleming (1984)
Leccinum spp.	59.5	4.2	38.9	Fleming (1984)

The Importance of a Food Base

The discussion of mycelial strands, above, recalls work on the rhizomorphs of *Armillaria mellea* (Redfern, 1970) and the mycelial strands of various pathogens and saprophytes, such as *Serpula lacrymans*. In all cases, the ability of the mycelial aggregate to grow and to initiate infection or colonization of a new substrate depends on provision of nutrients from a food base. The question arises: does this requirement for a food base apply generally to infection by late stage fungi? Could it, for example, explain why *dispersed* mycelia of these fungi cannot infect when added to soil, whereas mycelia of early stage fungi can infect? Could it also explain why basidiospores of late stage fungi have low infectivity in soil (Fox, 1986a)? And could it explain why late stage fungi can be established on seedlings in aseptic culture, which often entails the use of sugars (e.g., Trappe, 1967; Mason et al., 1983b; Palm & Stewart, 1984) and yet the same fungi often do not persist and proliferate when the seedlings are transplanted to soil? Then the mycorrhizas would need to be supported by assimilates from the seedlings, which might be inadequate to maintain the nutrient-demanding late stage fungi (Gibson & Deacon, 1990), while the root exudates that had been available to the mycorrhizal fungi alone in aseptic conditions can now be utilized by microbial competitors. Two possibilities stem from this, and both might be rewarding to investigate. First, the strand-forming fungi might channel a disproportionate amount of plant assimilates into their mycelial strands, with consequences for host productivity and even for the allocation of assimilates to the less "aggressively" spreading mycorrhizal fungi on other parts of the root system. This could be investigated by radiolabeling studies in experiments with split root systems. Second, the grazing activities of the soil microfauna, which have been little studied with regard to ectomycorrhizal fungi (Cromack et al., 1988), could have important consequences for the integrity of mycelial strand systems and their roles in the spread of infection.

Infection of Seedlings in Old Woodland Soils

Naturally, regenerating seedlings in an existing forest may often be infected from mycelial strands as described above. Carbon transfer from larger (donor) to smaller (receiver) seedlings has been demonstrated to occur via hyphae and hyphal aggregates of mycorrhizal fungi (Read et al., 1985; Finlay & Read, 1986). The ecological consequences of this were discussed by Newman (1988) and are considered further in Chapter 9 by Miller and Allen. If seedling roots will develop mycorrhizas in any case (e.g., by early stage fungi), and if mycorrhizas represent a potential drain on host resources (Harley & Smith, 1983), then seedlings would benefit by being infected from mycelial strands even if these channel only enough carbon from an older tree to support the mycorrhizas on the seedling.

Other studies suggest that the sequence in previous nonwooded sites can be shortened in old woodland soils, even when there is no direct involvement of a "parent" tree. For example, mycorrhizas of *Lactarius* (*L. tabidus?*), *Amanita* (presumed to be *A. muscaria*), *Cortinarius,* and *Leccinum* developed in a glasshouse, when aseptically raised birch seedlings were grown in pots of soil from an ancient birchwood site (Fleming et al., 1986). Birch seedlings planted in this field site also developed *Lactarius* and *Leccinum* mycorrhizas in both cored (isolated) and undisturbed positions. In a similar study (F. M. Fox & J. W. Deacon, unpublished) birch seedlings developed mycorrhizas of *Lactarius* (probably *L. rufus*) in pots of soil taken from a long-established (40–50 year) pine plantation. Pine seedlings similarly developed *Lactarius* mycorrhizas when planted in both cored and undisturbed positions in the field site, around the base of a living 50-year-old pine tree or around the stumps of two recently felled trees.

Such results contrast with those of Fleming (1983, 1984) for seedlings planted into a younger birch stand. It is not known whether the difference relates to features of the resident inoculum in old compared with young woodland soils or whether the soil properties in old woodlands are conducive to establishment by late stage fungi. F. M. Fox and J. W. Deacon (unpublished) could not establish mycorrhizas of *Lactarius glyciosmus* on birch seedlings when basidiospores of this fungus were added to pots of old birchwood soil or soil from an old pine plantation. But *Lactarius* and other late stage mycorrhizas developed from natural inocula in these soils, and mycorrhizas of *Laccaria* (early stage) developed on the seedlings from added basidiospores of *Laccaria proxima* or *Laccaria amethystina*. Such findings, although preliminary, indicate that the differences might be sought in terms of the inoculum of late stage fungi. Perhaps related to this is the finding of Fleming et al. (1986) that mycorrhizas of *Amanita* and *Leccinum* did not develop on seedlings planted in soil samples collected directly beneath the fruit bodies of these fungi, even though appropriate freshly excised mycorrhizas were present in the soil samples and there would have been an abundance of spores immediately beneath the fruit bodies. Instead, *Amanita* and *Leccinum* developed mycorrhizas on seedlings planted in nearby soil samples from beneath the fruit bodies of *other* fungi, where there were fewer excised mycorrhizas of *Amanita* and *Leccinum* in the soil. It was suggested that the infections developed from excised mycorrhizas, as Robertson (1954) and Fleming (1985) have demonstrated. If so, then the failure of *Amanita* and *Leccinum* to infect in soils from beneath their own fruit bodies might indicate a depletion of nutrient reserves in the mycorrhizas following fruiting—a phenomenon well characterized in mycelia of the saprophyte, *Schizophyllum commune* (Wessels & Sietsma, 1979).

At the opposite end of the spectrum of behavior, some fungi that occur early in a sequence evidently have the ability to persist in woodland soils. One of the clearest demonstrations of this was by Shaw and Sidle (1982) who found that sclerotia of *Cenococcum geophilum* survive for several years in clear-cut forestry sites in Western Ontario. Even low numbers of surviving sclerotia were sufficient to provide good establishment of mycorrhizas on seedlings of *Tsuga heterophylla* planted into these sites. Sclerotia are also produced by the early stage fungi *Pisolithus tinctorius* and *Paxillus involutus* (Laiho, 1970; Dennis, 1980; Grenville et al., 1985a; Grenville et al., 1985b). But their significance in long-term survival may be questioned, at least for *Paxillus*, because Fox (1986b) found that its sclerotia decayed rapidly in moist, unsterile soil; after 16 weeks burial they had lost much of their initial infectivity and after 40 weeks they were degenerate. F. M. Fox (personal communication) also examined sclerotia of *Cenococcum geophilum* that were retrieved from a coal spoil heap, and found evidence of penetration of their cells by both fungi and actinomycetes (Fig. 8.12). Whether such invasion significantly affects their survival remains to be determined.

In the papers cited above, the sclerotia of *C. geophilum*, *P. involutus*, and *P. tinctorius* were shown to be structurally similar to sclerotia of other fungi (Townsend & Willetts, 1954). But *Hebeloma sacchariolens* produced unique sclerotium-like bodies (Fox, 1986c) that are extremely common on roots infected by this fungus (Fig. 8.13). They are intimately embedded in the mycorrhizal sheath, and consist of an outer zone of thin-walled hyaline cells and an inner zone of very thick-walled cells virtually filled with lipid-like material. When detached from roots of glasshouse-grown birch and buried in moist soil, they survived for at least 40 weeks and remained fully infective to birch seedlings. Previously air-dried sclerotium-like bodies also survived in moist soil, but less well and with less subsequent infectivity. A potential role in survival was thus demonstrated for these structures. Supporting this, *H. sacchariolens* was found to infect birch seedlings (apparently from natural soil-borne inoculum) following prolonged microwave treatment of a woodland soil (to 100°C) that eliminated other mycorrhizal fungi (Gibson et al., 1988). This extreme persistence was probably due to sclerotium-like bodies in the soil, because microwaves act by transferring energy between water molecules, and the high lipid content of the sclerotial cells (and correspondingly low water content) could have conferred a high degree of protection from the effects of microwaves.

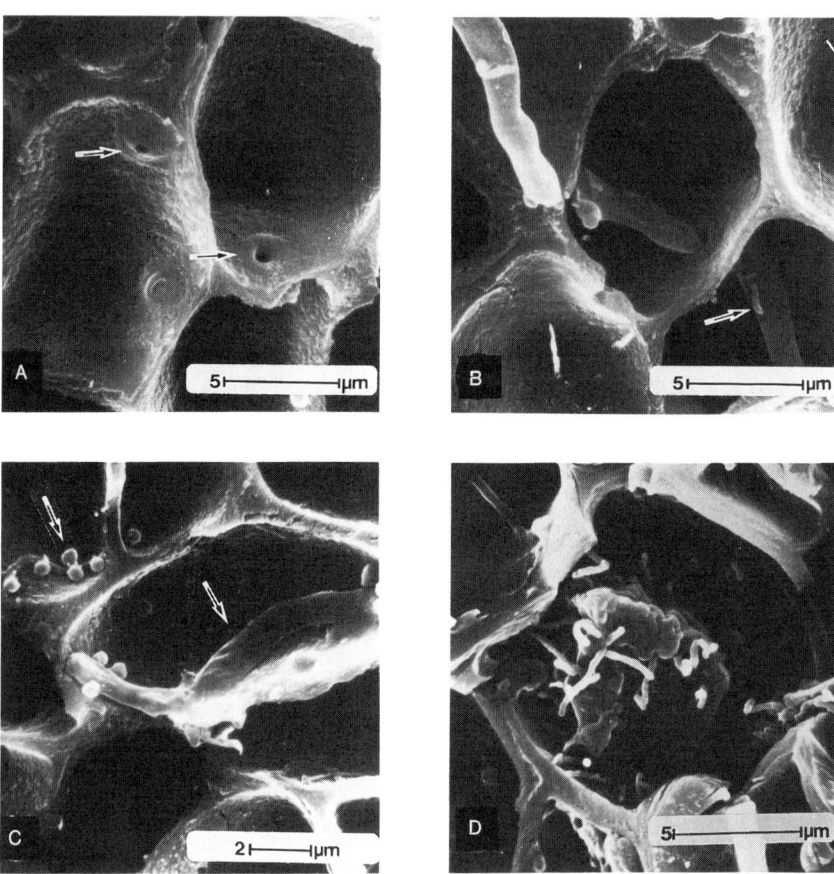

Figure 8.12. Scanning electron micrographs showing microbial invasion of degenerate sclerotia. (A–C) *Cenococcum geophilum* sclerotia retrieved from a coal spoil site, showing erosion pits in the walls of internal sclerotial cells (A) associated with fungal hyphae (B, C); bacteria (B, arrow) and actinomycete hyphae (C, top left) are also present. (D) Sclerotia of *Hebeloma sacchariolens* that were air-dried for 24 hr then buried in moist soil for 16 weeks, showing actinomycete hyphae in a sclerotial cell. Photographs courtesy of F. M. Fox.

Interactions during Establishment and Persistence of Inoculant Fungi

Seedlings are infected from natural sources of inoculum in nursery beds, the common indigenous fungi including *Thelephora terrestris, Laccaria* spp., *Hebeloma* spp., *Inocybe* spp., and "E-strain" or similar fungi (Trappe & Strand, 1969; Chu-Chou & Grace, 1983; Thomas et al., 1983; Croghan, 1984). Nevertheless, inoculation may be advantageous if the trees are to be outplanted in sites with

Interactions of Ectomycorrhizal Fungi / 281

Figure 8.13. Sclerotium-like bodies of *Hebeloma sacchariolens* associated with mycorrhizal roots. Photograph courtesy of F. M. Fox.

special conditions that only some fungi can tolerate. *Pisolithus tinctorius,* for example, is well suited to adverse sites such as mine spoils, where it significantly improves the establishment and growth of trees (Marx & Cordell, 1989). Inoculation also can be beneficial in the nursery itself if inoculant strains are selected for their growth-promoting effects. Attention has focused, therefore, on factors that affect the establishment of inoculants, and on how these fare when the seedlings are outplanted to their final sites. These issues are directly linked to the discussion of sequences above. In effect, one is asking, to what degree can the normal patterns be modified by inoculation?

Effects of Partial Soil Sterilants on Establishment

The establishment of inoculant fungi is often facilitated by prior fumigation of nursery beds. For example, Le Tacon et al. (1983) added inoculum of *Hebeloma cylindrosporum* in various formulations to beds that were untreated or fumigated with methyl bromide. Fumigation destroyed the indigenous inoculum, and the inoculant fungus established well and dominated the seedling root systems. In nonfumigated beds, however, the seedlings became mycorrhizal with indigenous fungi (mainly *Thelephora terrestris* and *Laccaria laccata*), and the inoculant fungus established poorly. An interesting point revealed by this study was that the total "mycorrhizal index," an estimate of the proportion of all root tips that became infected, was markedly lower in the nonfumigated than fumigated beds. In other words, root tips that should have been available for infection by the inoculant fungus or indigenous fungi did not become infected [unless they were occupied by inconspicuous mycorrhizas—Frankland and Harrison (1985)]. So it seems that the failure of the inoculant fungus in nonfumigated soil was not caused solely by competition for sites on the root systems. A similar conclusion can be drawn from the glasshouse study by Garbaye (1983) involving infection of *Fagus sylvatica* by *Hebeloma crustuliniforme* in soils untreated or fumigated with methyl bromide, and from earlier studies such as that of Lamb and Richards (1974) involving spore inocula on pine.

We lack a full explanation of the effects of heat, fumigants, or other partial soil sterilants on the establishment of inoculant fungi. The selective killing of soil microorganisms is often advanced as an explanation. However, McAfee and Fortin (1988) found, in experiments with sterilized vermiculite-peat potting mix, that the addition of soil washings containing microbial propagules could *enhance* mycorrhiza formation from mycelial inocula of *Laccaria bicolor* or *Pisolithus tinctorius* on jack pine, although the washings had no effect on establishment by *Hebeloma cylindrosporum* on jack pine, and no effect on any of these fungi on larch seedlings. If the microorganisms as such were involved in these effects then clearly they did not operate in a simple, general way. In the same study, mycelial inocula of the three fungi were mixed in equal proportions, with or without soil washings, in sterilized potting mix. *P. tinctorius* did not infect from the mixed inocula, *Laccaria* was usually dominant (but not always), and the degree of establishment by *Hebeloma* was reduced, while that by *Laccaria* was increased, by the addition of soil washings. Unfortunately, the mixed inocula contained proportionately less of each fungus than did the single inocula, but McAfee and Fortin (1988) explained the results in terms of the competitive abilities of the fungi, *Laccaria* being a fast grower and *Pisolithus* a slow grower. In addition, it is clear that components of the soil washings (perhaps the microorganisms or other factors) could differentially influence the inoculant fungi and their interactions with one another.

Positional Effects and Establishment

An alternative to soil fumigation was described by Deacon and Fox (1988) whereby Sitka spruce seedlings were allowed to develop mycorrhizas from indigenous inoculum in nursery seedbeds, then the bare-rooted plants were inoculated during lining-out at the start of their second year. The bare root systems were immersed in sodium alginate solution containing vermiculite-peat inoculum of *Hebeloma subsaponaceum*, then into a solution of a calcium salt to gel the inoculum on the roots before lining-out in nonfumigated soil. One million 1-year-old seedlings were treated in this way, during normal commercial operations. The inoculant fungus established well and soon dominated the mycorrhizal community, even though the seedlings already bore mycorrhizas of other fungi and the untreated controls developed further "indigenous" mycorrhizas from the nonfumigated lining-out beds. The method relied on the fact that the inoculant was placed (and held) where it could infect all new root tips as they developed. Also, the inoculum did not have to survive for long in soil because a large number of infectible roots are produced shortly after seedlings are lined out. A further contributory factor may be that the inoculum was protected by the gel. Alginate gel beads have been applied successfully to seedbeds (Le Tacon et al., 1983, 1985) but the inoculant fungus then must survive at least until the seed has germinated and produced infectible roots [4 to 6 weeks according to Marx & Cordell (1989)]. Even then, there are relatively few roots on which it can establish and it must colonise subsequent roots from this tenuous base. These factors perhaps explain why soil fumigation is beneficial in seedbed inoculations.

Persistence and Spread of Inoculants

There are many reports that mycorrhizal fungi established on seedlings in a nursery are replaced by indigenous fungi after the seedlings are transplanted to field sites (e.g., Benecke & Gobl, 1974; Lamb, 1979; Thomas et al., 1983). This is undesirable if the aim is to produce a lasting symbiosis with a particular fungus, although in practice it may be unimportant if the inoculant fungus stimulates early seedling or sapling growth, leading to the development of further mycorrhizas of any type, and if tree growth is correlated with numbers rather than types of mycorrhizas (Cline & Reid, 1982).

Bledsoe et al. (1982) found that all new roots of inoculated seedlings of Douglas fir were colonized by indigenous mycorrhizal fungi within 5 months of transplanting. Marx et al. (1977a) found that the inoculant *Pisolithus tinctorius* persisted for 2 years after outplanting of pine seedlings, but during this time increasing proportions of the root systems were colonized by other fungi. Grossnickle and Reid (1982) found that the inoculants *Pisolithus* and *Cenococcum* were no longer present 4 years after outplanting of seedlings, but the inoculant *Suillus granulatus* still persisted. McAfee and Fortin (1986) also found that *Pisolithus* could not compete with indigenous mycorrhizal fungi after outplanting. They suggested (McAfee & Fortin, 1988) that the well-known success of this fungus on acidic mine spoils and in other inhospitable sites is due to the low level of competition, presumably from other mycorrhizal fungi.

Last et al. (1985) studied the persistence and spread of *Hebeloma sacchariolens, Paxillus involutus,* and *Amanita muscaria* when these had been established on birch in aseptic culture. The inoculated seedlings were transferred to pots of four unsterile soils—two peats and two mineral soils from treeless sites. *Amanita* did not persist or spread on the roots in any soil, even though it was initially well established on the seedlings. *Hebeloma* persisted and remained dominant for 2 years in three of the soils; in the fourth (an acid peat) it persisted during the first season but had disappeared after 2 years. *Paxillus* persisted and spread to some degree in all soils, but less well in one mineral soil than in the others. Thus two separate phenomena were revealed: the late stage fungus (*Amanita*) did not persist, and persistence or spread of the early stage fungi was soil dependent. This was confirmed in similar work with *P. involutus* and *Laccaria proxima* on Sitka spruce (Mason et al., 1986).

In an extension of these studies, Fleming (1985) found that *A. muscaria* and *Leccinum scabrum,* when established on birch seedlings in aseptic culture, did not persist or spread after the seedlings were transplanted to pots of a brown earth. The decline of these late stage fungi were evident before other mycorrhizal fungi became established on the roots from natural soil-borne inocula; it could not be arrested by supplementing the soil with vermiculite-peat inoculum of *Amanita* or *Leccinum,* and it occurred even if the soil had been steam sterilized (but subsequently exposed in the pots). In short, the seedlings could not support the late stage fungi in anything but aseptic conditions. This might be explained if small transplanted seedlings lack the photosynthetic capacity to support the assimilate demands for spread by late stage fungi (Gibson & Deacon, 1990). This hypothesis could be tested by using older and larger transplants, perhaps coupled with shading or defoliation treatments. An alternative possibility is that nonwoodland soils are inimical to these fungi and thus cause their demise. It is perhaps an oversight in the experimental work to date that there have been no such transplanting studies in older woodland soils.

In the same experiments as above (Fleming, 1985), the early stage fungi, *Hebeloma sacchariolens* and *Laccaria proxima,* persisted on roots of transplants and spread with the growing root systems, remaining the dominant mycorrhizal types even after 441 days in pots of soil. Intermediate behavior was shown by *Lactarius pubescens.* It persisted after transplanting but colonized the new roots very slowly, representing only 15% of all mycorrhizas after 79 days. But it then started to spread and represented 33% of all mycorrhizas by 147 days. These findings suggest that *L. pubescens* may experience an initial check in its ability to spread after transplanting (perhaps related to assimilate supply) but can subsequently recover, consistent with its intermediate position in age-related sequences.

Persistence of early stage fungi was studied by Fleming (1985) who established *P. involutus, H. sacchariolens,* and *T. terrestris* on aseptic birch seedlings (with uninoculated controls for comparison) then transplanted the seedlings to either a brown earth (pH 5.8) or a Sphagnum peat (pH 3.9). These soils were either untreated or supplemented factorially with vermiculite-peat inocula of the same three fungi. The experiment was sampled 77 days after transplanting, and it enabled the behavior of each fungus to be studied both as a primary (seedling) and a secondary (soil) inoculant, in the presence or absence of competing mycorrhizal fungi, in two soils.

Disregarding, for the present, the effects of soil-applied inocula, *T. terrestris* could persist on the transplants in both soils, *H. sacchariolens* persisted in brown earth but not in peat (consistent with the results of Last et al., 1985), and *P. involutus* persisted poorly, if at all, in both soils. From soil-borne inocula (disregarding the effects of previously established fungi), *T. terrestris* established on newly formed roots in both soils, as did *P. involutus,* but *H. sacchariolens* established only in the brown earth. Thus, *P. involutus* showed contrasting behavior as a primary or secondary inoculant, which is difficult to explain. But the other two fungi showed consistent patterns indicating that *Thelephora* was tolerant of both soils whereas *H. sacchariolens* was intolerant of acid peat. Reddell and Malajczuk (1984) also found evidence of this sort, within the different horizons of a single soil, because the mycorrhizal types that developed on eucalypt roots in the organic-rich horizons were different from those in the underlying mineral layers. In the transplanting experiment above, there was interaction between the fungi used as primary and secondary inoculants. *Hebeloma,* when established as a primary inoculant, colonized the new roots and remained virtually the only mycorrhizal type in untreated brown earth. It also remained dominant in soil supplemented with *Thelephora* inoculum, but it spread poorly in soil supplemented with *Paxillus,* which became dominant on the new root system. *Thelephora* (as a primary inoculant) similarly spread and dominated the root system in untreated soil, but was supplanted to a greater or lesser degree by *Hebeloma* or *Paxillus* in soil supplemented with these fungi.

Mechanisms of Persistence and Replacement

Based on various assessments, mycorrhizal rootlets are thought to have a longevity in the order of months to 1 year (Harley & Smith, 1983). So, to persist, a mycorrhizal fungus must spread from its initial sites of establishment. There are two components to this: (1) spread with the existing mycorrhizal root tips and onto branches that arise from them and (2) infection of new root tips in uninfected parts of the root system. Usually the root tip grows slowly within the mantle of a mycorrhiza, and branches that develop from the mycorrhiza become colonized by the fungus as they emerge. The many detailed studies of this were discussed by Harley and Smith (1983). If site conditions are suitable for a fungus then it should be able to maintain itself locally in this way. Spread to other root tips typically occurs by external growth of hyphae or mycelial strands along roots or through soil, but in pine and spruce the fungi can grow internally along the root (Robertson, 1954; Ashton, 1976; Theodorou, 1980). In these conditions, much will depend on the relative growth rates of fungal hyphae and of roots, and on the distribution of inocula of other fungi in soil.

Vigorously growing roots can outgrow the hyphae of fungi (Wilcox, 1968). For example, Bowen and Theodorou (1973) recorded growth rates of 3–4 mm/day for roots of *Pinus radiata* when the hyphae of its mycorrhizal symbiont *Rhizopogon luteolus* were recorded as extending at only 1.5 mm/day. An inoculant fungus in such cases will increasingly be confined to the older part of the root zone, and the expanding periphery of the root system will be infected from inocula of other fungi in the soil. The results of Bledsoe et al. (1982) and Fleming et al. (1984) show a clear pattern of restriction of preestablished mycorrhizal fungi to older parts of the root systems of transplanted trees. But such experiments must be done in field plots, where root systems are free to expand; experiments in containers can give potentially misleading results because the periphery of a root system remains physically close to previously colonized root regions, which may serve as an inoculum source (Fleming, 1985). The finding that mycorrhiza formation can occur very quickly, for example, within 4 days in optimal conditions (Horan et al., 1988), may help explain why the persistence and spread of inoculant fungi are often facilitated by prior fumigation of soil, because competition from the indigenous soil inoculum would be correspondingly reduced. For example, Ruehle (1983) found that *Pisolithus tinctorius,* once established on roots, spread more quickly by mycelial strands and formed more mycorrhizas on outgrowing roots when seedlings were transferred to a fumigated compared with a nonfumigated soil. In the latter, more roots were infected by *T. terrestris*.

Role of Root Dormancy

The comments above cannot easily explain how one mycorrhizal fungus can replace another in the older parts of root systems; this is where late stage fungi first become established, ultimately to dominate the community. The problem is that the sheath of an existing mycorrhiza must present a barrier to colonization by other mycorrhizal fungi, just as it does to some plant pathogens (Marx, 1969, 1970). A practical consequence of this was noted by Chilvers et al. (1987), in that ectomycorrhizal fungi can replace the initially established vesicular-arbuscular mycorrhizal fungi on roots of *Eucalyptus* spp., but not vice versa. Similarly, Thomas and Jackson (1979) noted that fungi that formed ectendomycorrhizas on spruce seedlings in forest nurseries could be replaced by the ectomycorrhizal *T. terrestris* on the same roots in the early years of seedling growth. Even so, dual associations of ectomycorrhizal fungi have been observed on single roots (Zak & Marx, 1964). Marks and Foster (1967) recorded a frequency of 1–4% for this phenomenon on roots of *Pinus radiata*. In our experience with birch roots the phenomenon is much rarer than this; one of the few occasions when it was seen involved replacement of *Hebeloma sacchariolens* by *Lactarius pubescens*.

Marks and Foster (1967) suggested that the replacement of one fungus by another on single root members occurs only when roots resume growth after dormancy. This is consistent with the observation by Wilcox (1968) that root regrowth can precede regrowth of a mycorrhizal fungus after dormancy, enabling colonization of the new root from inoculum sources in soil. In a number of our experiments the incidence of mycorrhizas of *Lactarius pubescens* has, indeed, increased substantially following a natural or imposed dormancy of birch seedlings (Fleming et al., 1984; Fleming, 1985; Gibson & Deacon, 1988). It was mentioned earlier that *Lactarius* colonized seedlings that initially bore mycorrhizas of *Hebeloma* and not those bearing mycorrhizas of *Inocybe* or *Laccaria* (Fleming et al., 1984; Fleming, 1985). A possible explanation of this might be sought by studying the speed of regrowth of established mycorrhizal fungi following root dormancy, or, indeed, following transplant shock.

Types of Interaction between Ectomycorrhizal Fungi

In this section we try to relate the sequences of ectomycorrhizal fungi to current concepts of microbial interactions, based mainly on work with saprophytic (saprotrophic) fungi. For these, Cooke and Rayner (1984) recognized two major types of interaction for substrates, termed primary resource capture and combat. These are broadly equivalent to exploitation and interference competition (Culver, 1981; Lockwood, 1981). Primary resource capture is achieved by attributes such as rapid response to the presence of the resource and rapid exploitation of it (competitive saprophytic ability in the terms of Garrett, 1970). Combat (or secondary resource capture) involves the wresting of resources of one fungus by another, brought about by antibiosis, hyphal interference, or some other antagonistic attribute.

If ectomycorrhizal fungi are to be accommodated in such a scheme, then the "resource" must be defined. Some ectomycorrhizal fungi have been shown to degrade polymers such as pectin, cellulose, and lignin (Troyanowski et al., 1984; Linderberg, 1986), and work by Danielson (1984a) suggests that the ascomycete *Sphaerosporella brunnea* may be particularly active as a saprophyte. Also, several ectomycorrhizal fungi have been shown to make at least limited growth in the rhizosphere of nonhost plants (Theodorou & Bowen, 1971; Bowen & Theodorou, 1983; Duddridge, 1987). But it is generally held that such behavior—facultative mycotrophy—is uncommon or unimportant, and that most ectomycorrhizal fungi depend primarily on assimilates of their plant hosts for their carbon and energy supplies (Harley & Smith, 1983). If so, then the resources for which mycorrhizal fungi compete is, strictly speaking, the individual infectible root tip. This typically becomes sealed off by development of a mantle when it is infected. So most interactions between ectomycorrhizal fungi probably involve only primary resource capture, or successive "rounds" of primary resource capture as more root tips are produced. Success of one fungus over another could then be achieved by only two mechanisms, which we term *competitive exclusion* and *antagonistic (interference) exclusion*. Competitive exclusion would occur when one fungus infects before another by virtue of the density and position of its inoculum, its speed of response, and the suitability of the host, the soil, and other conditions. Most studies on infection of seedlings from mixed inocula are probably explicable in this way (Garbaye, 1983; McAfee & Fortin, 1988). Antagonistic exclusion has, to our knowledge, never been demonstrated, but would involve the ability of one fungus to infect from a mixture by virtue of its direct antagonistic effects on other fungi.

Broadening this picture, we might define the resource as the population of infectible root tips on a developing root system. Again competitive exclusion and antagonistic exclusion could operate, but three other types of interaction—or variants of the above—would be possible: competitive replacement, antagonistic (interference) replacement, and facilitated replacement.

Competitive replacement could occur, for example, when a late stage fungus such as *Lactarius pubescens* is established in one part of the root system and uses host-derived nutrients to support the development of mycelial strands that serve as inoculum (Fleming, 1983, 1984). It is conceivable that mycorrhizal fungi differ in ability to create a sink for host assimilates; those best able to do so could gain a competitive advantage over others if the assimilates were channeled into production of more inoculum. To our knowledge this has not been investigated for plants bearing mixtures of mycorrhizas, but it should be achievable using radiotracers and split-root systems as suggested earlier. The information would have wider relevance to the carbohydrate drain caused by different fungal symbionts (Harley, 1989; Harley & Smith, 1983). Competitive replacement would also occur if changes in soil conditions associated with progressive ageing of a tree are more favorable to one fungus than another. In this respect the growth of some mycorrhizal fungi is stimulated by polymers such as pectin, cellulose, and lignin in aseptic culture (Giltrap, 1982; Linderberg, 1986). Extracts of soil, litter, and bark have also been shown to influence mycorrhiza formation by different fungi (Chu-Chou, 1978; Schoenberger & Perry, 1982; Rose et al., 1983). A third way in which competitive replacement could occur is on individual root members during regrowth following dormancy, if the established mycorrhizal fungus does not resume growth fast enough to exclude potential competitors.

Antagonistic replacement could occur by direct effects of one fungus on another but as yet we know little about the combative roles of ectomycorrhizal fungi. Antibiosis has been documented, but against pathogenic fungi such as *Phytophthora cinnamomi* and *Fusarium oxysporum* (Marx, 1969, 1970; Sylvia & Sinclair, 1983; Duchesne et al., 1988) rather than against other mycorrhizal fungi. A well-documented form of contact (or near-contact) inhibition termed hyphal interference has been described for several nonmycorrhizal basidiomycetes (Ikediugwu & Webster, 1970a,b; Ikediugwu, 1976). In unpublished studies (S. J. Donaldson & J. W. Deacon), factorial pairings were made between *Hebeloma crustuliniforme* (two isolates), *H. sacchariolens, Lactarius pubescens, Leccinum versipelli,* and *Amanita muscaria* on agar containing Ingestad's (1971) medium and 0.1% glucose. Only *A. muscaria* caused hyphal interference, and then only against *H. sacchariolens, L. versipelli,* and one isolate of *H. crustuliniforme*. The phenomenon was evidenced by uptake of neutral red dye by the susceptible hyphae where they contacted those of *A. muscaria*. If such interactions occur in nature, they might be especially significant between the mycelial wefts that are common in the litter layer of mature forests. This merits investigation.

The possibility of *facilitated replacement* was investigated by Fleming (1985), as discussed in the section on "Effects of Previously Established Mycorrhizal Fungi." There was no evidence that early stage fungi directly facilitate the establishment or development of late stage fungi, although delayed effects were observed following root dormancy.

Ectomycorrhizal Fungi and Ecological Theory

There is always a danger in extrapolating from experimental model systems (even in field plots) to situations in mature forests and natural woodlands. Nevertheless, the distinction between early stage and late stage fungi seems to be valid in the experimental systems in which it has been examined to date, and it also has practical consequences (Marx & Cordell, 1989). For instance, it is fruitless to try to establish late stage fungi on seedlings in nursery conditions: even if they can be established, they are unlikely to persist and spread after outplanting.

Early stage fungi have most, if not all, of the features of ruderals (r-selected), whereas late stage fungi have many of the features of K-selected organisms (Harper & Ogden, 1970; Gadgil & Solbrig, 1972). However, in current strategy theory it is proposed that three primary strategies (ruderal, stress-tolerant, and competitive or combative, designed as R, S, and C) and combinations of these (secondary strategies, designated, $S–C$, $C–R$, and $S–R$) exist within the $r–K$ continuum (Grime, 1977, 1979; Pugh, 1980; Cooke & Rayner, 1984). The question arises, can ectomycorrhizal fungi be accommodated in this scheme?

As was argued above, the resource unit of an ectomycorrhizal fungus is the individual root tip. Once occupied, this is adequately defended by the fungal sheath and by the restriction of new root growth within this. Also, there is little evidence for antagonistic activities among ectomycorrhizal fungi—and no evidence of this from soil-based studies. According to Cooke and Rayner (1984) such activities are "the essence of combative strategies." These authors continue: "In the absence of such mechanisms it is possible to envisage competition between organisms amounting only to an initial struggle for primary resource capture." This is, indeed, what we suggest. It follows that, on present knowledge, most if not all ectomycorrhizal fungi are excluded from the C, $S–C$, and $C–R$ strategies; only ruderal and stress-tolerant criteria apply. The early stage fungi would be R or $S–R$ strategists, but all late stage fungi would have to be viewed as S strategists.

Examples of $S–R$ strategies among early stage fungi include *Pisolithus tinctorius* and *Paxillus involutus*, which often are associated with adverse sites (mine spoils, etc.). In contrast, at least some species of *Hebeloma* are intolerant of such sites or acid peats (see the section on "Persistence and Spread of Inoculants") and qualify as R strategists. *Scleroderma citrinum* may be an example of an S strategist, based on evidence in Fox (1986a) that it could not infect seedlings from spores, and the findings of Ingleby et al. (1985) that it occurred at relatively deep levels in a coal spoil site, where the temperature was apparently too high for

P. involutus. Thus at least some ectomycorrhizal fungi can be accommodated in the strategy groupings, especially where the nature of a stress is obvious. The problem, however, is to identify the nature of the stress for *all* late stage fungi. It might be a critical shortage of available mineral nutrients in soil, or some other soil property associated with ageing of a tree. Newman (1988) discussed the role of bridging hyphae of mycorrhizal fungi in "cycling" of mineral nutrients such that they do not enter the soil pool, and this would certainly contribute to nutrient stress for fungi that are not part of an existing mycorrhizal community. Yet to designate all late stage fungi as stress-tolerant on this basis, especially without direct supporting evidence, would be premature. Indeed, the limited evidence from aseptic culture (Gibson & Deacon, 1990) suggests quite the opposite: late stage fungi were less infective at low phosphorus levels than were early stage fungi (see the section on "Effects of Host and Edaphic Factors").

In short, we still cannot place most late stage fungi in the framework of strategy theory. We have still to resolve a more fundamental issue: to what degree is the behavior of these fungi determined by soil-related factors as opposed to host-related factors, insofar as these are separable?

It seems likely from field and glasshouse observations that sequences of ectomycorrhizal fungi are the rule rather than the exception—at least in commercial forestry where seedlings are transferred from nurseries to outplanting sites. In parallel with plant ecology (fungi are not plants!, Whittaker, 1969), these sequences or successions can be described as seral rather than substrate successions (see Park, 1968; Frankland, 1981). Perhaps a further parallel can be found with the seral successions of plants on virgin sites such as sand dunes, where the primary colonizers serve a role in facilitating the development of later colonists (by stabilization of the sand, increase in soil organic matter, etc.). In an earlier section on facilitated replacement, we drew attention to the effects of established mycorrhizal fungi on those that follow. One can envisage this as a general (but indirect) phenomenon if early stage mycorrhizal fungi, as has often been shown, enhance plant growth and thereby facilitate the survival of seedlings. A better growing seedling will produce more infectible roots, while the progressive ageing of the root system and progressive changes in soil conditions brought about by tree growth might create opportunities for ingress by further mycorrhizal species. Thereby, the initial mycorrhizal fungi may facilitate their own replacement, and mycorrhizal successions will be, in part, self-generating.

References

Ali, N. A., & Jackson, R. M. (1988). Effects of plant roots and their exudates on germination of spores of ectomycorrhizal fungi. *Transactions of the British Mycological Society* **91**, 253–260.

Ali, N. A., & Jackson, R. M. (1989). Stimulation of germination of spores of some ectomycorrhizal fungi by other micro-organisms. *Mycological Research* **93**, 182–186.

Ashton, D. H. (1976). Studies on the mycorrhizae of *Eucalyptus regnans* F. Muell. *Australian Journal of Botany* **24,** 723–741.

Azcon, R., & Ocampo, J. A. (1981). Factors affecting the vesicular-arbuscular infection and mycorrhizal dependency of thirteen wheat cultivars. *New Phytologist* **87,** 677–685.

Becker, G. (1956). Observations sur l'ecologie des champignons superieurs. *Annales Scientifiques de l'Universitaire de Besancon* **7,** 15–128.

Benecke, U., & Gobl, F. (1974). Influence of different mycorrhizae on growth nutrition and gas exchange of *Pinus mugo* seedlings. *Plant and Soil* **40,** 21–32.

Benzian, B. (1970). Nutrition of young conifers and soil fumigation. In: *Root Diseases and Soil-Borne Pathogens* (Ed. by T. A. Toussoun, R. V. Bega, & P. E. Nelson), pp. 234–239. University of California Press, Berkeley.

Birraux, D., & Fries, N. (1981). Germination of *Thelephora terrestris* basidiospores. *Canadian Journal of Botany* **59,** 2062–2064.

Bjorkman, E. (1942). Uber die Bedingungen der Mykorrhizabildung bei Keifer und Fichte. *Symbolae Botanicae Upsaliensis* **6,** 1–190.

Bjorkman, E. (1970). Mycorrhiza and tree nutrition in poor forest soils. *Studia Forestia Suecica* **83,** 1–24.

Bjurman, J., & Fries, N. (1984). Purification and properties of the germination inducing factor in the ectomycorrhizal fungus *Leccinum aurantiacum* (Boletaceae). *Physiologia Plantarum* **62,** 465–471.

Bledsoe, C. S., Tennyson, K., & Lopushinsky, W. (1982). Survival and growth of outplanted Douglas-fir seedlings inoculated with mycorrhizal fungi. *Canadian Journal of Forest Research* **12,** 720–723.

Bowen, G. D. (1987). The biology and physiology of infection and its development. In: *Ecophysiology of VA Mycorrhizal Plants* (Ed. by G. R. Safir), pp. 27–57. CRC Press, Boca Raton, FL.

Bowen, G. D., & Theodorou, C. (1973). Fungal growth around seeds and roots. In: *Ectomycorrhizae—Their Ecology and Physiology* (Ed. by C. G. Marks & T. T. Kozlowski), pp. 107–150. Academic Press, New York.

Bowen, G. D., & Theodorou, C. (1979). Interactions between bacteria and ectomycorrhizal fungi. *Soil Biology and Biochemistry* **11,** 119–126.

Brownlee, C., Duddridge, J. A., Malibari, A., & Read, D. J. (1983). The structure and function of mycelial systems of ectomycorrhizal roots with special reference to their role in forming inter-plant connections providing pathways for assimilate and water transport. *Plant and Soil* **71,** 433–443.

Bruchet, G., Debaud, J. C., & Gay, G. (1986). Genetic variations in the physiology of *Hebeloma*. In: *Physiological and Genetical Aspects of Mycorrhizae* (Ed. by V. Gianinazzi-Pearson & S. Gianinazzi), pp. 121–131. INRA, Paris.

Chilvers, G. A., & Gust, L. W. (1982). The development of mycorrhizal populations on pot-grown seedlings of *Eucalyptus st-johnii*. *New Phytologist* **90,** 677–699.

Chilvers, G. A., Lapeyrie, F. F., & Horan, D. P. (1987). Ectomycorrhizal vs. endomycorrhizal fungi within the same root system. *New Phytologist* **107,** 441–448.

Chu-Chou, M. (1978). Effects of root residues on growth of *Pinus radiata* seedlings and a mycorrhizal fungus. *Annals of Applied Biology* **90**, 407–416.

Chu-Chou, M. (1979). Mycorrhizal fungi of *Pinus radiata* in New Zealand. *Soil Biology and Biochemistry* **11**, 557–562.

Chu-Chou, M., & Grace, L. J. (1981). Mycorrhizal fungi of *Pseudotsuga menziesii* in the north island of New Zealand. *Soil Biology and Biochemistry* **13**, 247–249.

Chu-Chou, M., & Grace, L. J. (1982). Mycorrhizal fungi of *Eucalyptus* in the north island of New Zealand. *Soil Biology and Biochemistry* **14**, 133–137.

Chu-Chou, M., & Grace, L. J. (1983). Characterization and identification of mycorrhizas of Douglas-fir in New Zealand. *European Journal of Forest Pathology* **13**, 251–260.

Cline, M. L., & Reid, C. P. P. (1982). Seed source and mycorrhizal fungus effects on growth of containerized *Pinus contorta* and *Pinus ponderosa* seedlings. *Forest Science* **28**, 237–250.

Cooke, R. C., & Rayner, A. D. M. (1984). *Ecology of Saprotrophic Fungi*. Cambridge University Press, Cambridge.

Croghan, C. F. (1984). Survey of mycorrhizal fungi in lake states tree nurseries. *Mycologia* **76**, 951–953.

Cromack, K., Fichter, B. L., Moldenke, A. M., Entry, J. A., & Ingham, E. R. (1988). Interactions between soil animals and ectomycorrhizal fungal mats. *Agriculture, Ecosystems and Environment* **24**, 161–168.

Culver, D. C. (1981). Introduction to the theory of species interactions. In: *The Fungal Community: Its Organization and Role in the Ecosystem* (Ed. by D. T. Wicklow & G. C. Carroll), pp. 281–294. Marcel Dekker, New York.

Danielson, R. M. (1982). Taxonomic affinities and criteria for identification of the common ectendomycorrhizal symbiont of pines. *Canadian Journal of Botany* **60**, 7–18.

Danielson, R. M. (1984a). Ectomycorrhiza formation by the operculate discomycete *Sphaerosporella brunnea* (Pezizales). *Mycologia* **76**, 454–461.

Danielson, R. M. (1984b). Ectomycorrhizal associations in jack pine stands in northeastern Alberta. *Canadian Journal of Botany* **62**, 932–939.

Danielson, R. M., Visser, S., & Parkinson, D. (1984). The effectiveness of mycelial slurries of mycorrhizal fungi for the inoculation of container-grown Jack pine seedlings. *Canadian Journal of Forest Research* **14**, 140–142.

Deacon, J. W., & Fox, F. M. (1988). Delivery of microbial inoculants into the root zone of transplant crops. *Proceedings of the Brighton Crop Protection Conference—Pests and Diseases* **1988**, 645–653.

Deacon, J. W., Donaldson, S. J., & Last, F. T. (1983). Sequences and interactions of mycorrhizal fungi on birch. *Plant and Soil* **71**, 257–262.

Debaud, J. C., Gay, G., Prevost, A., Lei, J., & Dexheimer, J. (1988). Ectomycorrhizal ability of genetically different homokaryotic and dikaryotic mycelia of *Hebeloma cylindrosporum*. *New Phytologist* **108**, 323–328.

Dennis, J. J. (1980). Sclerotia of the gasteromycete *Pisolithus tinctorius*. *Canadian Journal of Microbiology* **26**, 1505–1507.

Dighton, J., Poskitt, J. M., & Howard, D. M. (1986). Changes in occurrence of basidiomycete fruitbodies during forest stand development: With specific reference to mycorrhizal species. *Transactions of the British Mycological Society* **87**, 163–171.

Ducamp, M., Poitou, N., & Olivier, J. M. (1986). Comparison cytologique et biochimique entre cultures monospores et boutures de carpophores chez *Suillus granulatus* (Fr. ex L.)/Kuntze. In: *Physiological and Genetical Aspects of Mycorrhizae* (Ed. by V. Gianinazzi-Pearson & S. Gianinazzi), pp. 575–579. INRA, Paris.

Duchesne, L. C., Peterson, R. L., & Ellis, B. E. (1988). Pine root exudate stimulates the synthesis of antifungal compounds by the ectomycorrhizal fungus *Paxillus involutus*. *New Phytologist* **108**, 471–476.

Duddridge, J. A. (1987). Specificity and recognition in ectomycorrhizal associations. In: *Fungal Infection of Plants* (Ed. by G. F. Pegg & P. J. Ayres), pp. 25–44. Cambridge University Press, Cambridge.

Fasalo-Bonfante, P., & Brunel, A. (1972). Caryological features in the mycorrhizal fungus: *Tuber melanosporum* Vitt. *Allionia* **18**, 5–11.

Finlay, R. D., & Read, D. J. (1986). The structure and function of the vegetative mycelia of ectomycorrhizal plants. I. Translocation of ^{14}C-labelled carbon between plants interconnected by a common mycelium. *New Phytologist* **103**, 143–156.

Fleming, L. V. (1983). Succession of mycorrhizal fungi on birch: Infection of seedlings planted around mature trees. *Plant and Soil* **71**, 263–267.

Fleming, L. V. (1984). Effects of soil trenching and coring on formation of ectomycorrhizas on birch seedlings grown around mature trees. *New Phytologist* **98**, 143–153.

Fleming, L. V. (1985). Experimental study of sequences of ectomycorrhizal fungi on birch (*Betula* sp.) seedling root systems. *Soil Biology and Biochemistry* **17**, 591–600.

Fleming, L. V., Deacon, J. W., Last, F. T., & Donaldson, S. J. (1984). Influence of propagating soil on the mycorrhizal succession of birch seedlings transplanted to a field site. *Transactions of the British Mycological Society* **82**, 707–711.

Fleming, L. V., Deacon, J. W., & Last, F. T. (1986). Ectomycorrhizal succession in a Scottish birch wood. In: *Physiological and Genetical Aspects of Mycorrhizae* (Ed. by V. Gianinazzi-Pearson & S. Gianinazzi), pp. 259–264. INRA, Paris.

Ford, E. D., Mason, P. A., & Pelham, J. (1980). Spatial patterns of sporophore distribution around a young birch tree in three successive years. *Transactions of the British Mycological Society* **75**, 287–296.

Fox, F. M. (1983). Role of basidiospores as inocula of mycorrhizal fungi of birch. *Plant and Soil* **71**, 269–273.

Fox, F. M. (1986a). Groupings of ectomycorrhizal fungi of birch and pine, based on establishment of mycorrhizas on seedlings from spores in unsterile soils. *Transactions of the British Mycological Society* **87**, 371–380.

Fox, F. M. (1986b). Ultrastructure and infectivity of sclerotia of the ectomycorrhizal fungus *Paxillus involutus* on birch (*Betula* spp.). *Transactions of the British Mycological Society* **87**, 627–631.

Fox, F. M. (1986c). Ultrastructure and infectivity of sclerotium-like bodies of the ectomycorrhizal fungus *Hebeloma sacchariolens,* on birch (*Betula* spp.). *Transactions of the British Mycological Society* **87**, 359–369.

Frankland, J. C. (1981). Mechanisms in fungal successions. In: *The Fungal Community: Its Organization and Role in the Ecosystem* (Ed. by D. T. Wicklow & G. C. Carroll), pp. 403–426. Marcel Dekker, New York.

Frankland, J. C., & Harrison, A. F. (1985). Mycorrhizal infection of *Betula pendula* and *Acer pseudoplatanus:* Relationships with seedling growth and soil factors. *New Phytologist* **101**, 133–151.

Fries, N. (1978). Basidiospore germination in some mycorrhiza-forming Hymenomycetes. *Transactions of the British Mycological Society* **70**, 319–324.

Fries, N. (1979). The taxon-specific spore germination reaction in *Leccinum*. *Transactions of the British Mycological Society* **73**, 337–341.

Fries, N. (1981). Recognition reactions between basidiospores and hyphae in *Leccinum*. *Transactions of the British Mycological Society* **77**, 9–14.

Fries, N. (1983). Spore germination, homing reaction, and intersterility groups in *Laccaria laccata* (Agaricales). *Mycologia* **75**, 221–227.

Fries, N. (1987). Ecological and evolutionary aspects of spore germination in the higher Basidiomycetes. *Transactions of the British Mycological Society* **88**, 1–7.

Fries, N., & Mueller, G. M. (1984). Incompatibility systems, cultural features and special circumscriptions in the ectomycorrhizal genus *Laccaria* (Agaricales). *Mycologia* **76**, 633–642.

Fries, N., & Swedjemark, G. (1986). Specific effects of tree roots on spore germination in the ectomycorrhizal fungus, *Hebeloma mesophaeum* (Agaricales). In: *Physiological and Genetical Aspects of Mycorrhizae* (Ed. by V. Gianinazzi-Pearson & S. Gianinazzi), pp. 725–730. INRA, Paris.

Fries, N., Bardet, M., & Serck-Hanssen, K. (1985). Growth of ectomycorrhizal fungi stimulated by lipids from a pine root exudate. *Plant and Soil* **86**, 287–290.

Gadgil, M., & Solbrig, O. T. (1972). The concept of r- and K-selection: evidence from wild flowers and some theoretical considerations. *American Naturalist* **106**, 14–31.

Garbaye, J. (1983). Premiers resultats de recherches sur la competitivite des champignons ectomycorrhiziens. *Plant and Soil* **71**, 303–308.

Garbaye, J., & Bowen, G. D. (1987). Effect of different microflora on the success of mycorrhizal inoculum on *Pinus radiata*. *Canadian Journal of Forest Research* **17**, 941–943.

Garrett, S. D. (1970). *Pathogenic Root-Infecting Fungi*. Cambridge University Press, Cambridge.

Gemma, J. N., & Koske, R. E. (1988). Pre-infection interactions between roots and the mycorrhizal fungus *Gigaspora gigantea:* Chemotropism of germ-tubes and root growth response. *Transactions of the British Mycological Society* **91**, 123–132.

Gibson, F., & Deacon, J. W. (1988). Experimental study of establishment of ectomycorrhizas in different regions of birch root systems. *Transactions of the British Mycological Society* **91**, 239–251.

Gibson, F., & Deacon, J. W. (1990). Establishment of ectomycorrhizas in aseptic culture: Effects of glucose, nitrogen and phosphorus in relation to successions. *Mycological Research* **94**, 166–172.

Gibson, F., Fox, F. M., & Deacon, J. W. (1988). Effects of microwave treatment of soil on growth of birch (*Betula pendula*) seedlings and infection of them by ectomycorrhizal fungi. *New Phytologist* **108**, 189–204.

Giltrap, N. J. (1982). Production of polyphenol oxidases by ectomycorrhizal fungi with special reference to *Lactarius* spp. *Transactions of the British Mycological Society* **78**, 75–81.

Graham, J. H., Leonard, R. T., & Menge, J. A. (1981). Membrane-mediated decrease in root exudation responsible for phosphorus inhibition of vesicular-arbuscular mycorrhiza formation. *Plant Physiology* **68**, 548–552.

Grente, J., Chevaliers, G., & Pollacsek, A. (1972). La germination de l'ascospore de *Tuber melanosporum* et la synthese sporale des mycorrhizes. *Compte Rendu Hebdomadaire des Seances de l'Academie des Sciences, Series D* **275**, 743–746.

Grenville, D. J., Peterson, R. L., & Piche, Y. (1985a). The development, structure and histochemistry of sclerotia of ectomycorrhizal fungi. II. *Paxillus involutus*. *Canadian Journal of Botany* **63**, 1412–1417.

Grenville, D. J., Piche, Y., & Peterson, R. L. (1985b). Sclerotia as viable sources of mycelia for the establishment of ectomycorrhizae. *Canadian Journal of Microbiology* **31**, 1085–1088.

Grime, J. P. (1977). Evidence for the existence of three primary strategies in plants and its relevance to ecological and evolutionary theory. *American Naturalist* **111**, 1169–1194.

Grime, J. P. (1979). *Plant Strategies and Vegetation Processes*. Wiley, Chichester.

Grossnickle, S. C., & Reid, C. P. P. (1982). The use of ectomycorrhizal conifer seedlings in the revegetation of a high-elevation mine site. *Canadian Journal of Forest Research* **12**, 354–361.

Harley, J. L. (1989). The significance of mycorrhiza. *Mycological Research* **92**, 129–139.

Harley, J. L., & Smith, S. E. (1983). *Mycorrhizal Symbiosis*. Academic Press, New York.

Harper, J. L., & Ogden, J. (1970). The reproductive strategy of higher plants. I. The concept of strategy with special reference to *Senecio vulgaris* L. *Journal of Ecology* **58**, 681–689.

Horan, P., Chilvers, G. A., & Lapeyrie, F. F. (1988). Time sequence of the infection process in eucalypt mycorrhizas. *New Phytologist* **109**, 451–458.

Ikediugwu, F. E. O. (1976). The interface in hyphal interference by *Peniophora gigantea* against *Heterobasidion annosum*. *Transactions of the British Mycological Society* **66**, 291–296.

Ikediugwu, F. E. O., & Webster, J. (1970a). Antagonism between *Coprinus heptemerus* and other coprophilous fungi. *Transactions of the British Mycological Society* **54**, 181–204.

Ikediugwu, F. E. O., & Webster, J. (1970b). Hyphal interference in a range of coprophilous fungi. *Transactions of the British Mycological Society* **54**, 205–210.

Ingestad, T. (1971). A definition of optimum nutrient requirements in birch seedlings. *Physiologia Plantarum* **24**, 118–125.

Ingleby, K., Last, F. T., & Mason, P. A. (1985). Vertical distribution and temperature relations of sheathing mycorrhizas of *Betula* spp. growing on coal spoil. *Forest Ecology and Management* **12**, 279–285.

Ingleby, K., Mason, P. A., Last, F. T., & Fleming, L. V. (1990). *Identification of Ectomycorrhizas*. ITE Research Publication No. 5. HMSO, London.

Jalaluddin, M. (1967). Studies on *Rhizina undulata*. I. Mycelial growth and ascospore germination. *Transactions of the British Mycological Society* **50**, 449–459.

Kropp, B. R., McAfee, B. J., & Fortin, J. A. (1987). Variable loss of ectomycorrhizal ability in monokaryotic and dikaryotic cultures of *Laccaria bicolor*. *Canadian Journal of Botany* **65**, 500–504.

Laiho, O. (1970). *Paxillus involutus* as a mycorrhizal symbiont of forest trees. *Acta Forestalia Fennica* **106**, 1–65.

Lamb, R. J. (1979). Factors responsible for the distribution of mycorrhizas of *Pinus* in eastern Australia. *Australian Forestry Research* **9**, 25–34.

Lamb, R. J., & Richards, B. N. (1974). Inoculation of pines with mycorrhizal fungi in natural soils. I. Effect of density and time of application of inoculum and phosphorus amendment on mycorrhizal infection. *Soil Biology and Biochemistry* **6**, 167–171.

Last, F. T., Mason, P. A., Wilson, J., & Deacon, J. W. (1983). Fine roots and sheathing mycorrhizas: Their formation, function and dynamics. *Plant and Soil* **71**, 9–21.

Last, F. T., Mason, P. A., Wilson, J., Ingleby, K., Munro, R. C., Fleming, L. V., & Deacon, J. W. (1985). "Epidemiology" of sheathing (ecto-) mycorrhizas in unsterile soils: A case study of *Betula pendula*. *Proceedings of the Royal Society of Edinburgh* **83B**, 299–315.

Le Tacon, F., Jung, G., Michelot, P., & Mugnier, M. (1983). Efficacite en pepiniere forestiere d'un inoculum de champignon ectomycorrhizien produit en fermenteur et inclus dans une matrice de polymeres. *Annales des Sciences Forestieres* **40**, 165–176.

Le Tacon, F., Jung, G., Mugnier, J., Michelot, P., & Mauperin, C. (1985). Efficiency in a forest nursery of an ectomycorrhizal fungus inoculum produced in a fermentor and entrapped in polymeric beads. *Canadian Journal of Botany* **63**, 1664–1668.

Linderberg, G. (1986). Physiology of interactions between mycorrhizae and rhizosphere microorganisms. In: *Physiological and Genetical Aspects of Mycorrhizae* (Ed. by V. Gianinazzi-Pearson & S. Gianinazzi), pp. 197–206. INRA, Paris.

Lingappa, B. T., & Lockwood, J. L. (1963). Direct assay of soils for fungistasis. *Phytopathology* **53**, 529–531.

Lockwood, J. L. (1981). Exploitation competition. In: *The Fungal Community: Its Organization and Role in the Ecosystem* (Ed. by D. T. Wicklow & G. C. Carroll), pp. 319–350. Marcel Dekker, New York.

Losel, D. M. (1964). The stimulation of spore germination in *Agaricus bisporus* by living mycelium. *Annals of Botany N. S.* **28,** 541–554.

Malajczuk, N. (1988). Interaction between *Phytophthora cinnamoni* zoospores and microorganisms on non-mycorrhizal and ectomycorrhizal roots of *Eucalyptus marginata*. *Transactions of the British Mycological Society* **90,** 375–383.

Marks, G. C., & Foster, R. C. (1967). Succession of mycorrhizal associations on individual roots of radiata pine. *Australian Forestry* **31,** 194–201.

Marx, D. H. (1969). The influence of ectotrophic mycorrhizal fungi on the resistance of pine roots to pathogenic infections. I. Antagonism of mycorrhizal fungi to root pathogenic fungi and bacteria. *Phytopathology* **59,** 153–163.

Marx, D. H. (1970). The influence of ectotrophic mycorrhizal fungi on the resistance of pine roots to pathogenic infections. II. Resistance of mycorrhizae to infection by vegetative mycelium of *Phytophthora cinnamoni*. *Phytopathology* **60,** 1472–1473.

Marx, D. H. (1980). Ectomycorrhizal fungus inoculations: A tool for improving forestation practices. In: *Tropical Mycorrhiza Research* (Ed. by P. Mikola). Clarendon Press, Oxford.

Marx, D. H., & Cordell, C. E. (1989). The use of specific ectomycorrhizas to improve artificial forestation practices. In: *Biotechnology of Fungi to Improve Plant Growth* (Ed. by J. M. Whipps & R. D. Lumsden), pp. 1–25. Cambridge University Press, Cambridge.

Marx, D. H., Bryan, W. C., & Cordell, C. E. (1977a). Survival and growth of pine seedlings with *Pisolithus* ectomycorrhizae after two years on reforestation sites in North Carolina and Florida. *Forest Science* **23,** 263–373.

Marx, D., Hatch, A. B., & Mendicino, J. F. (1977b). High fertility decreases sucrose content and susceptibility of loblolly pine roots to ectomycorrhizal infection by *Pisolithus tinctorius*. *Canadian Journal of Botany* **55,** 1569–1574.

Mason, P. A. (1980). Aseptic synthesis of sheathing (ecto-) mycorrhizas. In: *Tissue Culture Methods for Plant Pathologists* (Ed. by D. S. Ingram & J. P. Helgeson), pp. 173–178. Blackwell Scientific Publications, Oxford.

Mason, P. A., Last, F. T., Pelham, J., & Ingleby, K. (1982). Ecology of some fungi associated with an ageing stand of birches (*Betula pendula* and *B. pubescens*). *Forest Ecology and Management* **4,** 19–39.

Mason, P. A., Wilson, J., & Last, F. T. (1983a). The concept of succession in relation to the spread of sheathing mycorrhizal fungi on inoculated tree seedlings growing in unsterile soils. *Plant and Soil* **71,** 247–256.

Mason, P. A., Dighton, J., Last, F. T., & Wilson, J. (1983b). Procedure for establishing sheathing mycorrhizas on tree seedlings. *Forest Ecology and Management* **5,** 47–53.

Mason, P. A., Last, F. T., & Wilson, J. (1986). Effects of different soils on the establishment and influence of sheathing mycorrhizas. In: *Physiological and Genetical Aspects of Mycorrhizae* (Ed. by V. Gianinazzi-Pearson & S. Gianinazzi), pp. 767–772. INRA, Paris.

Mason, P. A., Last, F. T., Wilson, J., Deacon, J. W., Fleming, L. V., & Fox, F. M. (1987). Fruiting and successions of ectomycorrhizal fungi. In: *Fungal Infection of*

Plants (Ed. by G. F. Pegg & P. J. Ayres), pp. 253–268. Cambridge University Press, Cambridge.

McAfee, B. J., & Fortin, J. A. (1986). Competitive interactions of ectomycorrhizal mycobionts under field conditions. *Canadian Journal of Botany* **64**, 848–852.

McAfee, B. J., & Fortin, J. A. (1988). Comparative effects of the soil microflora on ectomycorrhizal inoculation of conifer seedlings. *New Phytologist* **108**, 443–449.

Melin, E. (1962). Physiological aspects of mycorrhizae of forest trees. In: *Tree Growth* (Ed. by T. T. Kozlowski), pp. 247–263. Ronald Press, New York.

Mikola, P. (1970). Mycorrhizal inoculation in afforestation. *International Review of Forestry Research* **3**, 123–196.

Miles, J. (1985). Soil in the ecosystem. In: *Ecological Interactions in Soil: Plants, Microbes and Animals* (Ed. by A. H. Fitter, D. Atkinson, D. J. Read, & M. B. Usher), pp. 407–427. Blackwell Scientific Publications, Oxford.

Miller, O. K. (1983). Ectomycorrhizae in the Agaricales and Gasteromycetes. *Canadian Journal of Botany* **61**, 909–916.

Murakami, Y. (1987). Spatial distribution of *Russula* species in *Castanopsis cuspidata* forest. *Transactions of the British Mycological Society* **89**, 187–193.

Newman, E. I. (1988). Mycorrhizal links between plants: Their functioning and ecological significance. *Advances in Ecological Research* **18**, 243–270.

Palm, M. E., & Stewart, E. L. (1984). *In vitro* synthesis of mycorrhizae between presumed specific and nonspecific *Pinus* + *Suillus* combinations. *Mycologia* **76**, 579–600.

Park, D. (1968). The ecology of terrestrial fungi. In: *The Fungi: An Advanced Treatise, Vol. III. The Fungal Population* (Ed. by G. C. Ainsworth & A. S. Sussman), pp. 5–39. Academic Press, New York.

Pugh, G. J. F. (1980). Strategies in fungal ecology. *Transactions of the British Mycological Society* **75**, 1–14.

Rast, D., & Stauble, E. J. (1970). On the mode of action of isovaleric acid in stimulating the germination of *Agaricus bisporus*. *New Phytologist* **60**, 557–566.

Read, D., Francis, R., & Finlay, R. D. (1985). Mycorrhizal mycelia and nutrient cycling in plant communities. In: *Ecological Interactions in Soil: Plants, Microbes and Animals* (Ed. by A. H. Fitter, D. Atkinson, D. J. Read, & M. B. Usher), pp. 193–217. Blackwell Scientific Publications, Oxford.

Reddell, P., & Malajczuk, N. (1984). Formation of mycorrhizae by Jarrah (*Eucalyptus marginata* Donn ex Smith) in litter and soil. *Australian Journal of Botany* **32**, 511–520.

Redfern, D. B. (1970). The ecology of *Armillaria mellea*: rhizomorph growth through soil. In: *Root Diseases and Soil-Borne Pathogens* (Ed. by T. A. Toussoun, R. V. Bega, & P. E. Nelson), pp. 147–149. University of California Press, Berkeley.

Richardson, M. J. (1970). Studies on *Russula emetica* and other agarics in a Scots pine plantation. *Transactions of the British Mycological Society* **55**, 217–229.

Robertson, N. F. (1954). Studies on the mycorrhiza of *Pinus sylvestris*. I. The pattern of development of mycorrhizal roots and its significance for experimental studies. *New Phytologist* **53**, 253–283.

Rose, S. L., Perry, D. A., Pilz, D., & Schoenberger, M. M. (1983). Allelopathic effects of litter on the growth and colonization of mycorrhizal fungi. *Journal of Chemical Ecology* **9**, 1153–1162.

Rouguerol, T., & Payre, H. (1974). Observations sur la comportement de *Tuber melanosporum* dans un site naturel. *Revue de Mycologie* 39, 107–117.

Ruehle, J. L. (1983). The relationship between lateral-root development and spread of *Pisolithus tinctorius* ectomycorrhizae after planting of container-grown loblolly pine seedlings. *Forest Science* **29**, 519–526.

Salt, G. A. (1979). The increasing interest in 'minor pathogens'. In: *Soil-Borne Plant Pathogens* (Ed. by B. Schippers & W. Gams), pp. 289–312. Academic Press, London.

Same, B. I., Robson, A. D., & Abbott, L. K. (1983). Phosphorus, soluble carbohydrates and endomycorrhizal infection. *Soil Biology and Biochemistry* **15**, 593–597.

Schippers, B., Geels, F. P., Hoekstra, O., Lamers, J. G., Maehout, C. A. A. A., & Scholte, K. (1985). Yield depressions in narrow rotations caused by unknown microbial factors and their suppression by selected pseudomonads. In: *Ecology and Management of Soilborne Plant Pathogens* (Ed. by C. A. Parker et al.), pp. 127–130. American Phytopathological Society, St. Paul, MN.

Schoenberger, M. M., & Perry, D. A. (1982). The effect of soil disturbance on growth and ectomycorrhizae of Douglas fir and western hemlock seedlings: A greenhouse bioassay. *Canadian Journal of Forest Research* **12**, 343–353.

Shaw, C. G., & Molina, R. (1980). Formation of ectomycorrhizae following inoculation of containerized Sitka spruce seedlings. *U.S.D.A. Forestry Service Research* Note PNW-351.

Shaw, C. G., & Sidle, R. C. (1982). Evaluation of planting sites common to a southeast Alaska clear-cut. II. Available inoculum of the ectomycorrhizal fungus *Cenococcum geophilum*. *Canadian Journal of Forest Research* **13**, 9–11.

Skinner, M. F., & Bowen, G. D. (1974). The penetration of soil by mycelial strands of ectomycorrhizal fungi. *Soil Biology and Biochemistry* **6**, 57–61.

Smith, W. H. (1970). Root exudates of seedling and mature sugar maple. *Phytopathology* **60**, 701–703.

Straatsma, G., Konings, R. N. H., & Van Griensven, L. J. L. D. (1985). A strain collection of the mycorrhizal mushroom *Cantharellus cibarius* obtained by germination of spores and culture of fruit body tissue. *Transactions of the British Mycological Society* **85**, 689–697.

Sylvia, D. M., & Sinclair, W. A. (1983). Suppressive influence of *Laccaria laccata* on *Fusarium oxysporum* on douglas fir seedlings. *Phytopathology* **73**, 384–389.

Theodorou, C. (1980). The sequence of mycorrhizal infection of *Pinus radiata* D. Don following inoculation with *Rhizopogon luteolus* Fr. and Nordh. *Australian Forestry Research* **10**, 381–387.

Theodorou, C., & Bowen, G. D. (1971). Effects of non-host plants on growth of mycorrhizal fungi of radiata pine. *Australian Forestry* **35**, 17–22.

Theodorou, C., & Bowen, G. D. (1987). Germination of basidiospores of mycorrhizal fungi in the rhizosphere of *Pinus radiata* D. Don. *New Phytologist* **106**, 217–223.

Thomas, G. W., & Jackson, R. M. (1979). Sheathing mycorrhizae of nursery-grown *Picea sitchensis*. *Transactions of the British Mycological Society* **73**, 117–125.

Thomas, G. W., Rogers, D., & Jackson, R. M. (1983). Changes in the mycorrhizal status of Sitka spruce following outplanting. *Plant and Soil* **71**, 319–323.

Tominaga, Y. (1975). Studies on the mycorrhiza of "fairy ring" of *Tricholoma matsutake* (S. Ito et Imai) Sing. VI. On the mycorrhiza of *Pinus densiflora* Sieb. et Zucc. *Bulletin of the Hiroshima Agricultural College* **5**, 159–163.

Townsend, B. B., & Willetts, H. J. (1954). The development of the sclerotia of certain fungi. *Transactions of the British Mycological Society* **37**, 213–221.

Trappe, J. M. (1962). Fungus associates of ectotrophic mycorrhizae. *Botanical Review* **28**, 538–606.

Trappe, J. M. (1967). Pure culture synthesis of Douglas-fir mycorrhizae with species of *Hebeloma, Suillus, Rhizopogon,* and *Astraeus*. *Forest Science* **13**, 121–130.

Trappe, J. M., & Strand, R. F. (1969). Mycorrhizal deficiency in a Douglas-fir region nursery. *Forest Science* **15**, 381–389.

Trojanowski, J., Haider, K., & Huttermann, A. (1984). Decomposition of ^{14}C-labelled lignin, holocellulose and lignocellulose by mycorrhizal fungi. *Archives of Microbiology* **139**, 202–206.

Wang, C. J. K., & Wilcox, H. E. (1985). New species of ectendomycorrhizal and pseudomycorrhizal fungi: *Phialophora finlandia, Chloridium paucisporum,* and *Phialocephala fortinii*. *Mycologia* **77**, 951–958.

Warcup, J. H., & Talbot, P. H. B. (1989). *Muciturbo:* A new genus of hypogeous ectomycorrhizal ascomycetes. *Mycological Research* **92**, 95–100.

Watling, R. (1981). Relationships between macromycetes and the development of higher plant communities. In: *The Fungal Community: Its Organization and Role in the Ecosystem* (Ed. by D. T. Wicklow & G. C. Carroll), pp. 427–458. Marcel Dekker, New York.

Watling, R. (1984). Macrofungi of birch woods. *Proceedings of the Royal Society of Edinburgh* **85B**, 129–140.

Wessels, J. G. H., & Sietsma, J. H. (1979). Wall structure and growth in *Schizophyllum commune*. In: *Fungal Walls and Hyphal Growth* (Ed. by J. H. Burnett & A. P. J. Trinci), pp. 27–48. Cambridge University Press, Cambridge.

Whittaker, R. H. (1969). New concepts of kingdoms of organisms. *Science* **163**, 150–160.

Wilcox, H. E. (1968). Morphological studies of the roots of red pine, *Pinus resinosa*. II. Fungal colonization of roots and the development of mycorrhizae. *American Journal of Botany* **55**, 686–700.

Yang, C. S., & Wilcox, H. E. (1984). An E-strain ectendomycorrhiza formed by a new species, *Tricharina mikolae*. *Mycologia* **76**, 675–684.

Zak, B., & Marx, D. H. (1964). Isolation of mycorrhizal fungi from roots of individual slash pines. *Forest Science* **10**, 214–222.

9

Mycorrhizae, Nutrient Translocation, and Interactions Between Plants

Steven L. Miller and Edith B. Allen

SUMMARY. Mycorrhizae and mycorrhizal fungi may be involved in both competitive and facilitative interactions between plants. Mediation of competition by mycorrhizae can be attributed to changes in inoculum density, to the degree of infection of neighboring hosts, or to differential physiological responses of the hosts to the mycobiont. Mycorrhizae may also induce facilitation of one plant by another. Facilitation is direct if there are hyphal linkages between plants, or indirect if, for instance, the fungus affects the growth of an unlinked "nurse" plant. The subject of direct facilitation via hyphal linkages has received much attention. Transfer of nutrients and carbohydrates has been shown to occur between fungus and host, host and fungus, and individual host plants connected by hyphal linkages, yet the degree to which hyphal transport occurs and can influence growth and fitness of one plant by another is highly controversial. If net nutrient transport between plants is common and provides a substantial mechanism by which facilitation can occur, our concept of plant interaction at the community level may need to be reevaluated. If mycorrhizal-mediated facilitation is not important, simple energetics or competition theory can serve to explain observed plant interactions. This chapter is a review of physiological evidence used to support hyphal transport between plants and competition mediated by mycorrhizae.

Introduction

Interactions between individual plants may range from competition to facilitation, and there is evidence that mycorrhizae are involved in both processes. Competition is defined as depletion of resources by an individual plant to the detriment of a neighbor, while facilitation is an increase in plant growth or fitness caused by the presence of a neighbor. Documented cases of facilitation are most often caused by amelioration of the microenvironment of one plant by another (e.g., Allen & Allen, 1988, Franco & Nobel, 1988), but this is an indirect facilitation in that the plant is responding to the changed microenvironment, not the neighboring plant. Mycorrhizae may also cause direct facilitation when they create hyphal linkages between plants, where one plant provides nutrients or carbohydrates to its linked neighbor.

Competitive interactions between plants may in some cases be explained by changes in inoculum density, by degree of mycorrhizal infection, or by differential physiological response to mycorrhizae by neighboring plants (Allen & Allen, 1990). There is mounting evidence that mycorrhizae do affect plant competition, although we are not yet able to predict the outcome of competition with assurance, even when we know the physiological response of individuals to mycorrhizae. By contrast, direct facilitation via mycorrhizal hyphal connections is controversial. Researchers question whether hyphal transport of carbon and nutrients between plants occurs on a significant scale, and thus whether facilitation can be mediated by mycorrhizae. If nutrient transport between plants is significant or if other functions can be ascribed to plant–plant mycorrhizal linkages, several researchers have suggested that we must rethink our notions of how plants interact in communities (Chiariello et al., 1982). For instance, the existence of a common hyphal network may make plants more similar in their niches than they were thought to be (Janos, 1980). The hyphal network in fact may create a "superorganism" (sensu Clements, 1916) that interconnects the plants (Björkman, 1960; Perry et al., 1989a). However, these suggestions, taken to their extreme, violate competition theory, which holds that organisms will compete when resources are limiting. When resources are limiting, the plant with the greater capacity to take up nutrients, or to deplete the nutrient pool to a greater extent, will have greater growth than a neighbor with lesser capacities (Tilman, 1982, 1988). When plants are connected via hyphae, competition theory suggests that facilitation via hyphae cannot occur because the weaker of the two neighbors will not have the sink strength to withdraw nutrients from a rapidly assimilating neighbor. When resources are not limiting, then hyphal nutrient transport might occur from a plant that has already absorbed sufficient nutrients to a slower growing neighbor.

The classical competition theory argument holds when the primary focus is on the host plants and if the mycorrhizal fungal network is considered to be a static, nonmetabolizing conduit functioning in nutrient transport between plants. However, if the fungus is considered to be a physiologically active entity capable of taking up nutrients from the soil and carbohydrates from the host, yet able to segregate both from the hosts and pool them within its own separate metabolic framework, a different hypothesis emerges. Rather than the differential sink strength of the hosts controlling movement of nutrients via the fungus, the fungus may regulate the kinetics, direction of translocation, and amount of nutrients and carbohydrates alloted each host. In this scenario, the fungus may interact individually with each host connected by the hyphal network. Therefore competition or facilitation may not exist directly between individual host plants because of the intervening fungal network or between each host and its associated mycorrhizal fungus since the relationship between them is one of mutualistic symbiosis. Energetics of the system as a whole including mycorrhizal fungi, mycorrhizae, and host plants would determine nutrient uptake and movement.

Hyphal Transport between Plants and Fungi

The primary function of mycorrhizal hyphae is considered to be the absorption of resources from the soil. The relative efficiency of mycorrhizae in taking up and translocating water and nutrients compared to that of nonmycorrhizal roots is caused by active, rather than passive, uptake and transport of nutrients by mycorrhizae. The effect that an active system of uptake and translocation has over a passive one on interactions between individual plants is yet to be determined.

Various types of mycorrhizae including ectomycorrhizae, vesicular-arbuscular (VA), ericoid, and monotropoid mycorrhizae have been shown to take up and transport cations, nutrients, organic materials, and water efficiently in both laboratory experiments and in natural systems (Alexander, 1983; Harley & Smith, 1983; Read et al., 1985). Most physiological and biochemical research has been accomplished using labeled forms of nitrogen, phosphorus and carbon.

Ectomycorrhizal fungi have the ability to utilize nitrogen sources from soil pools as either organic nitrogen, including protein and free amino acids (Mikola, 1948; Melin & Norkrans, 1948; Alexander, 1983; Spinner & Hasselwandter, 1985; Read et al., 1985; Botton et al., 1986), or inorganic nitrogen including ammonium and nitrate (Lundeberg, 1970; Le Tacon, 1972; Lee & Stewart, 1978; Salsac et al., 1982; Alexander, 1983). Ectomycorrhizal fungi also contribute to nitrogen nutrition of the host by conversion of absorbed nitrogen into forms that are more readily metabolized by the host and by long distance transport of absorbed nitrogen to the host plant (Melin & Nilsson, 1953). Refer to Chapter 12 by Bledsoe in this volume for detailed information on forms of N absorbed and used by the fungus.

Phosphorus is another nutrient actively taken up by mycorrhizal fungi and transported to the host plant, and has been central in showing effectivity in endomycorrhizal fungi. Phosphorus uptake by nonmycorrhizal plants is limited by immobility of the anion in the soil. Hattingh et al. (1973) showed that the ability of plants to retrieve phosphate is greatly enhanced by the additional absorbing surface provided by the external hyphae of endomycorrhizal fungi. Similar functions have been observed for the extramatrical hyphae of ectomycorrhizal fungi (Melin & Nilsson, 1950; Finlay & Read, 1986c). Additional information on P transport by hyphae is reviewed in Chapter 4 by Read in this volume.

The ability of the host plant to supply carbohydrates to the mycobiont in symbiotic, putatively mutualistic, mycorrhizal relationships is considered basic to the nature of the symbiosis. Björkman (1942) determined that many factors that negatively influence photosynthesis including shading, decreasing day length, and defoliation also negatively affect the growth of fungal tissues in ectomycorrhizal rootlets. Early attempts to quantify carbon movement in ectomycorrhizal *Pinus sylvestris* seedlings showed that labeled carbon compounds produced during photosynthesis were transported to the mantle and extramatrical hyphae within 1 day (Melin & Nilsson, 1957). More recent ^{14}C-labeling studies have confirmed that carbon originating as host photosynthate is rapidly translocated to the mycobiont in both endo- and ectomycorrhizae (Bevege et al., 1975; Snellgrove et al., 1982; Reid et al., 1983; Miller et al., 1989a).

Only crude estimates of the magnitude of the drain on photosynthates by mycorrhizal fungi are available. Vogt et al. (1982) estimated that 15% of net primary production in Pacific silver fir forest was consumed by the ectomycorrhizal component. Bevege et al. (1975) found that after 24 hr of exposure to $^{14}CO_2$ ectomycorrhizal root tips had 15 times more label than nonmycorrhizal roots. Reid et al. (1983) reported 22–65% of carbon assimilated by the host was allocated to roots, while Söderström and Read (1987) estimated that 30% of belowground respiration was attributed to the extramatrical mycelium alone. There is little question that mycorrhizal fungi can impose a sizable drain on their host. Even the same species of mycobiont under different physiological conditions appears to induce different degrees of drain on the host's carbon (Miller et al., 1989a), and the same species of host plant experiences different degrees of drain with different species of mycobionts (Dosskey et al., 1990). How this drain might affect interactions between plants can only be hypothesized. For instance, a fungus that causes a large drain on its host might make its host less competitive than a neighbor that experiences less drain.

Another subject concerning carbon movement that may be relevant to plant interactions is the ability of the fungi to act as saprophytes. Although there are no data to show that VA mycorrhizal fungi have any saprophytic abilities, there is mounting evidence that some ectomycorrhizal fungi can decompose lignin, holocellulose, lignocellulose, and protein to some extent (Lundeberg, 1970; Theodorou, 1971; Ho & Zak, 1979; Giltrap, 1982; Dighton, 1983; Trojanowski et al., 1984; Abuzinadah & Read, 1986a,b, 1989a,b; Abuzinadah et al., 1986; Dighton et al., 1987). In fact, the ectomycorrhizal fungus *Suillus luteus* was shown by Dighton et al. (1987) to be highly active in the decomposition of hide powder and cotton, and there was little difference in the rates of decomposition activity between mycorrhizal and saprophytic fungi in the absence of the tree host. Whether plants can benefit from saprophytically derived carbon is unknown, although the amount of C contributed to the plant would probably be small.

The form of carbon transported by the fungal hypha is important in determining whether it can be used by the plant. There is evidence that fungi convert plant carbon sources, such as glucose or fructose, to trehalose, arabitol, and mannitol (Söderström et al., 1986). Since there is no evidence that plants can use these compounds, this would suggest that carbon flow goes one way from plant to fungus. However, some carbon is transported between plants (see below), so conversion of forms by the fungus apparently does not eliminate carbon movement between plants. In addition, Martin et al. (1985) indicated that mannitol and trehalose fed an endogenous pool of glucose, with a high pool turnover. Since glucose is readily absorbed by host root cells, they speculated that this glucose pool may be a valuable source of carbon for translocation when the plant is deprived of carbon. Additional, more detailed information on this subject is reported in Chapter 5 by Finlay and Söderström in this volume. The next section deals with the problem of hyphal transport between plants, and considers carbon as well as nutrients.

Hyphal Transport between Plants

Chlorophyllous Plants

There is a great deal of evidence for transport of nutrients and carbohydrates between plant and fungus, as reviewed above. The extramatrical phase of mycorrhizae, or the mycelium perfusing through the soil is the exploratory organ that functions mainly in the uptake of water and nutrients. Brownlee et al. (1983) pointed out that mycelial strands or hyphae growing from mycorrhizal roots provide a functional pathway through which assimilates can be transported. However, because the mycelium of a single fungal individual is known to spread for great distances, often several meters, through the soil (Butler, 1957; Thompson & Rayner, 1983; Brownlee & Jennings, 1982; Skinner & Bowen, 1974a; Dahlberg & Stenlid, 1990) it is likely that many plants and their associated mycorrhizae and extramatrical hyphae are in close contact in the soil. In addition host specificity is low for many host–fungal interactions and a single fungal species may infect the roots of several different plant species in a given area. These observations have led to the hypothesis that a belowground community of mycorrhizal fungi can physically and physiologically connect individual host plants through a network of extramatrical hyphae (Björkman, 1960; Woods & Brock, 1964).

Using labeling techniques, anastomosing systems of mycorrhizal fungi have been shown in the laboratory to link the root systems of different individual plants (Finlay & Read, 1986a,b; Skinner & Bowen, 1974a,b; Brownlee et al., 1983; Foster, 1981), and to facilitate the translocation of water (Duddridge et al., 1980; Read & Malibari, 1978), carbon (Reid & Woods, 1969; Finlay & Read, 1986a,b; Stribley & Read, 1974), phosphorus (Skinner & Bowen, 1974b), and cations (Melin et al., 1958) between individual host plants. While most studies on plant–plant nutrient transport via VA mycorrhizae have emphasized P movement (Francis et al., 1986; Read et al., 1985), a recent study showed elevated ^{15}N in a neighboring plant but no significant change in ^{32}P, possibly due to the greater mobility of N than P (Eissenstat, 1990). Englander and Hull (1980), Pearson and Read (1973), and Stribley and Read (1974) have shown that reciprocal transfer

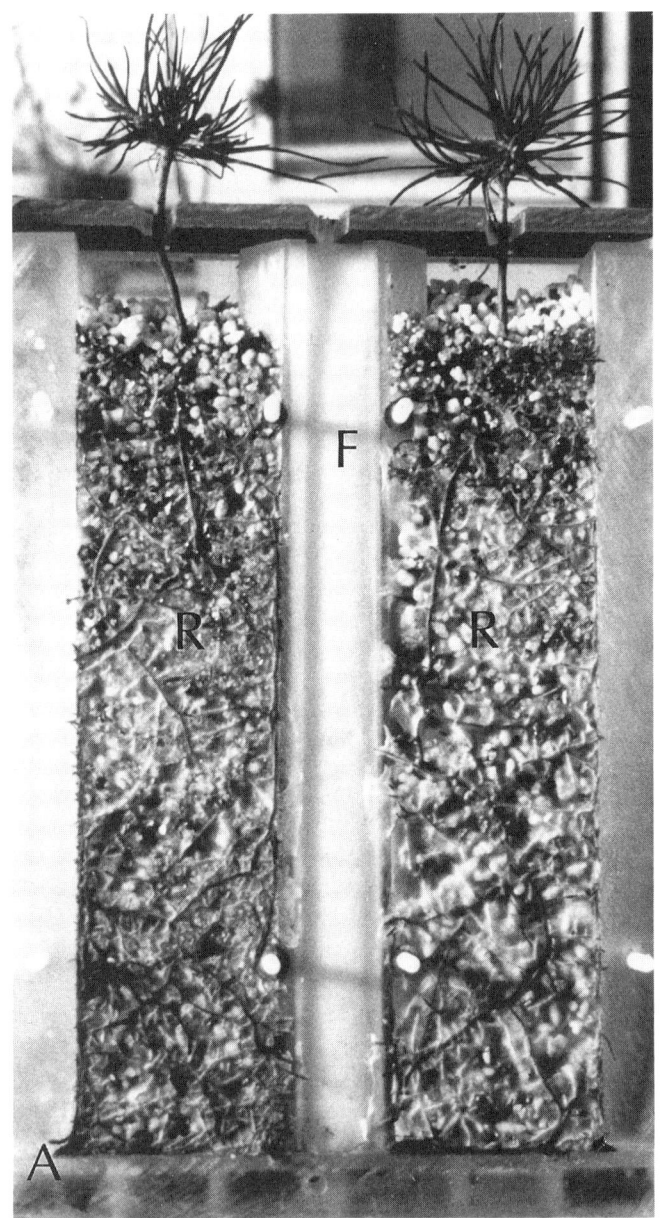

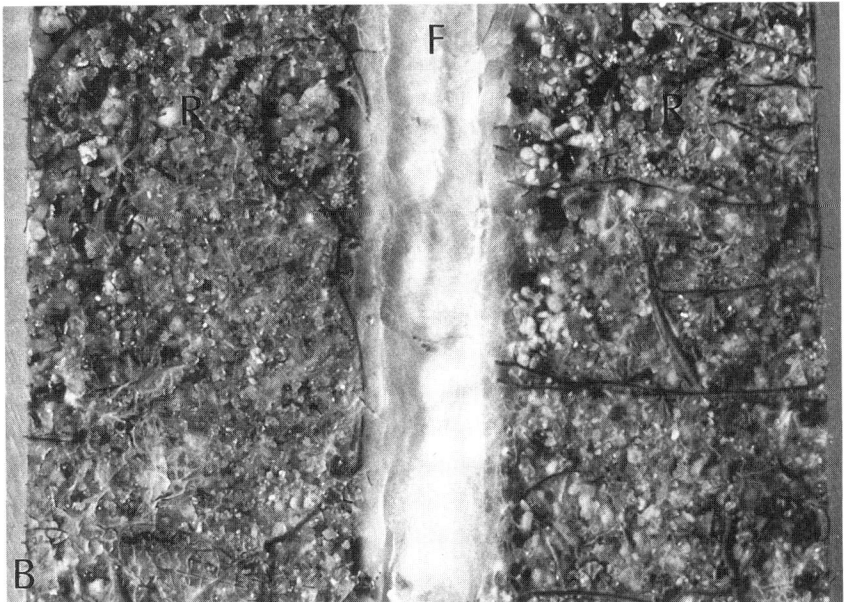

Figure 9.1. Experimental "root-mycocosms" used in linkage experiments at the University of Wyoming. (A) A root-mycocosm, shown 50% of actual size, with two *Pinus contorta* seedlings growing in the root chambers (R) separated by a single fungal chamber (F). Mycelium emanating from the ectomycorrhizae in the root chambers are allowed to anastomose in the fungal chamber, thereby linking the two seedlings. (B) Root and fungal chambers from one root-mycocosm shown full size. Note the profuse mycelial growth in the central fungal chamber.

of carbon and phosphorus is also possible between host and mycobiont in an ericoid type of mycorrhiza.

Although connections between plants via ectomycorrhizal and endomycorrhizal fungi have been produced in the laboratory, such as the root-mycocosms in use by the first author (Fig. 9.1), quantification of naturally occurring connections is yet to be determined (Newman, 1988). Naturally occurring ectomycorrhizal fungal linkages between one to several individual lodgepole pine trees were recently examined by tracing the origin of roots in the top 5 cm of soil to aggregates of ectomycorrhizae and extramatrical hyphae, and appear to be a common phenomenon (Miller et al., 1989b and unpublished). The laboratory work that has been accomplished with interconnected plants has largely been reflective of the "unit mycelium" concept of Buller (1931) where mycorrhizal hosts are interconnected belowground by an extensive mycelial network derived from a single individual fungal species. According to this theory the same or different species of host plants can become interconnected by a single mycorrhizal fungus, possibly providing a direct pathway for interplant nutrient transfer and physiological communication.

It has been established that host plants can be interconnected by mycorrhizal fungi, however, direct net transfer of a nutrient or metabolite from one plant to another has yet to be demonstrated (Newman, 1988). An alternative hypothesis to direct transfer is that a substance provided by a donor plant is transported to the fungus connecting the plants and is released, on death of a portion of the fungus, into the soil pool. The substance is then taken up by the same or another fungal species or individual and transported to the recipient plant. Newman and Ritz (1986) in fact found that the timing of phosphorus movement between plants connected with an endomycorrhizal fungus fits a time model approximating the soil pool hypothesis and suggested that P movement through direct hyphal links was not important in their study plants.

In addition, Newman (1988) strongly emphasized that, although much of the relevant evidence supporting transport of substances between plants is from movement of isotopes, it is important to realize that movement of an isotope does not itself prove net movement of the substance. Using Newman's example, carbon in amino acids that is moved from fungus to host may be exchanged by carbon in other organic compounds passing from host to fungus. Therefore, carbon movement would not demonstrate net carbon loss by the host or net carbon gain by the fungus. This is crucial to our discussion of mycorrhizal interaction between individual host plants.

Yet a few studies show that hyphal transport between plants may be an important process for regulating plant interactions. For instance, in one pot study the acceptor plant was a nonmycotrophic species (*Arabis hirsuta*), which took up none of the ^{32}P used to label the donor plant, while a mycotrophic acceptor plant took up the ^{32}P (Francis et al., 1986). However, the amount of ^{32}P taken up by the acceptor plant is almost always small, several orders of magnitude lower than the amount applied, and labeling experiments give instantaneous rather than long-term results. A better approach to this type of experiment was taken by Francis et al. (1986), who showed that fertilization of a donor plant with phosphate resulted in increased growth and P content of the acceptor after several months in a pot experiment. The evidence for mycelial transport of significant quantities of C is better than for P transport. A seedling that is connected to a mature plant via hyphae may have one to two orders of magnitude greater ^{14}C uptake in the sun, but may have three orders of magnitude greater uptake if it is shaded (Read et al., 1985). The implications of this to plant competition must be carefully considered, as a mature plant could potentially nurture both its own and its neighbor's offspring. If carbohydrates are not limiting in this system, then the nurturing of neighboring genotypes is not contradictory to competition theory. This could occur in any system where mature plants shade seedlings. If carbohydrates are limiting, there is no known mechanism by which a parent plant can exclude neighboring offspring. It is more likely that connections between mature plants and all seedlings would be severed in this situation. Read et al. (1985) concluded that the entire mycelial network must be examined, not just the connections between two experimental plants, to determine the importance of nutrient transport between plants.

One way to demonstrate net movement between plants is to examine the fate of a nutrient following death of a root or whole plant. Since the root is already inhabited by fungal hyphae, either surrounding cortical and epidermal cells and the whole rootlet as in ectomycorrhizae, or by intracellular penetration as in endo- and other types of mycorrhizae, the fungus is in prime position to capture nutrients and other substances released as the root dies before they are lost through leaching or mineralization by other rhizosphere organisms. This is especially true if the fungus, while being connected to still living roots, remains alive and functioning in the dying and dead roots. Tommerup and Abbott (1981) found that some VA mycorrhizal fungi such as *Glomus fasciculatus, G. monosporus,* and *Gigaspora calospora* were capable of germination and regrowth from dried mycorrhizal roots. MacLeod et al. (1986) showed that, although inoculation with VA mycorrhizal fungi did not markedly affect plant growth, it delayed root cortex death in phosphorus-deficient wheat and rape plants and accelerated root cortex death in phosphorus-adequate plants. Park (1970) found that the color of *Cenococcum* ectomycorrhizae changed markedly on aging, possibly suggesting a response to changing host root physiology. Al Abras et al. (1988) have suggested that hartig nets in 1- and 2-year-old ectomycorrhizae lacking mantles could still be contributing significantly to the metabolism and physiology of *Picea excelsa,* and therefore could function in nutrient extraction from dying roots.

A question can be raised regarding the fate of portions of mycelia after death of the immediate host plant. Harvey et al. (1980, 1986) have shown for forests in the Pacific Northwest that roots and ectomycorrhizae are able to remain alive for several months after clearcutting, but that mycorrhizal viability drops off significantly after that. Seedlings naturally recruited or planted during the window of mycorrhizal system viability are rapidly infected by the network of extramatrical hyphae present in the soil. Work in southwestern Oregon suggests that if seedlings are not planted into clearcuts within the spring following cutting, the mycorrhizal network is lost and transplanted seedlings are difficult to establish (Perry, personal communications). Likewise, Marshall and Perry (1987) observed that after decapitation, Douglas-fir seedlings inoculated with different ectomycorrhizal fungi exhibit rates of maintenance respiration that slowly drop off with time, until no respiration can be detected, suggesting that the roots and fungi remain alive for periods of time following decapitation. In a system where several host plants are interconnected by anastomosing mycelia of one fungal species, the death of a single host plant may be relatively unimportant; the fungus is connected with other host plants and possibly can contribute significantly to the nutrient status of the components of the interconnected community by capturing nutrients from the dying and dead root system while maintaining its energy base through connections with the other host plants. An unpublished experiment by Miller et al. using pairs of linked lodgepole pine seedlings growing in root-mycocosms (Miller et al., 1989a; Fig. 9.1) showed that starch stored in fine roots of lodgepole pine seedlings was solubilized and possibly translocated via hyphal linkages to the mycorrhizae of decapitated seedlings over a 4-week period. Current photosynthate labeled with ^{14}C was only translocated from the intact seedlings to the fine roots and mycorrhizae of the decapitated seedlings at the end of the fourth week following decapitation, after starch content in the intact seedlings had nearly been depleted.

Other work in progress by Miller et al. at the University of Wyoming suggests that mycorrhizae are conservators of nutrients in undisturbed pine forests. We inoculated replicate pairs of lodgepole pine trees with *Hebeloma crustuliniforme*, and allowed extramatrical hyphae from both trees to anastomose in central fungal chambers of the root-mycocosms, thereby linking them. Unlinked pairs of saplings were used as controls. Nitrogen in soil solution was flushed from the mycocosms to background levels and either 0 or 1 tree in each linked and unlinked set of trees was decapitated at soil level. Twice weekly for 8 weeks NO_3^- and NH_4^+ concentrations of leachates were determined. Patterns of N loss differed significantly between intact vs. cut and linked vs. unlinked treatments. In both uncut-linked and unlinked trees, NO_3^- and NH_4^+ remained at background levels throughout the experiment. In cut-unlinked trees, both NO_3^- and NH_4^+ concentration increased by 1–2 orders of magnitude over controls and NO_3^- exhibited two peaks, one at 14 and one at 46 days. A similar pulse in NH_4^+ concentration was observed at 14 days in the cut-unlinked treatment, but was only 60% of the highest NO_3^- concentration. In cut-linked trees NO_3^- losses began at 50 days and the highest concentration was only about 20% of the cut-unlinked trees. The magnitude of NH_4^+ losses in the cut-linked treatment was similar to NO_3^- levels in the cut-unlinked treatments, but began at fairly low levels about 14 days and continued to increase gradually. The dominant ion liberated into solution in the linked system was the relatively immobile NH_4^+, while the dominant form of N lost from the unlinked system was the highly mobile NO_3^-. Similar results were obtained from a larger scale in the field where we cut 0, 1, 5, 15, or 30 trees to create small to large canopy gaps (Parsons et al., 1990). Most nitrogen was lost into the soil solution in areas where 30 trees were cut contiguously.

Both studies suggest that available N may be conserved in disturbed systems that retain intact ectomycorrhizal fungal linkages between surviving trees. On disturbance of the aboveground forest to such a degree that a threshold number of trees are killed, a belowground gap in the roots and extramatrical hyphae is created. Hyphae and roots from living trees near the edge of the gap are no longer able to emanate into the gap, the nutrient-conserving ability of the mycorrhizal fungi is lost or reduced, and large amounts of nutrients such as nitrate are mineralized and leached into the groundwater. Haines and Best (1976) found a similar conservation ability in endomycorrhizal fungi (*Glomus mosseae*) associated with *Liquidambar styraciflua* L. The ability to conserve and possibly store nutrients has implications for hyphal transport between plants in that fungi may play a greater role in regulating nutrient movement, and would reduce competition between plants by assuring a greater supply of nutrients.

Achlorophyllous Plants

Early work on plants interconnected by mycorrhizal fungal hyphae involved conifers and an achlorophyllous ericaceous angiosperm, *Monotropa hypopitys*. Björkman (1960) was able to trace ^{14}C from an injection point in the bark of spruce and pine trees to nearby *Monotropa* plants. Since Björkman was able to synthesize mycorrhizae on pine in the laboratory with an ectomycorrhizal fungus isolated from *Monotropa*, he hypothesized that *Monotropa* obtains its carbon from the trees through hyphal connections by a common ectomycorrhizal fungus. Although the relationship has been variously termed parasitic (Lees, 1841, Neumann, 1844), epiparasitic (Björkman, 1960), symbiotic (Furman, 1966), and mycorrhizal (MacDougal & Floyd, 1990; Lutz & Sjolund, 1973), ultrastructural work by Duddridge and Read (1982) showed that monotropoid mycorrhizae are connected to pine and spruce ectomycorrhizae, and that carbohydrate and nutrient demand is probably greatest in early summer. Whether there is net carbon gain by *Monotropa* over the course of the season is yet to be determined.

As previously mentioned, the rate and amount of net transfer of carbohydrates or nutrients from one plant to another through the common hyphal network may be dependent on a source-sink relationship between the two plants. In the case of *Monotropa*, evolutionary loss of chlorophyll and subsequent loss of energy production may now signal a sink response to the trees, resulting in carbon translocation through the linking fungus to the *Monotropa* plants. However, it is also possible that the tree, which is already providing energy to the mycorrhizal fungus, is physiologically "unaware" of the additional loss of carbon to the *Monotropa* plants. The sink response may be detected by the mycorrhizal fungus and since carbon from the tree is sequestered by the fungus into forms of carbon no longer available to the host (Martin et al., 1987), the fungus may control the fate of it, and may exchange it for different carbon sources from the *Monotropa* plant.

The association between *Monotropa* and the conifers may also provide an additional function to the system. Although mycorrhizal hyphae are known to extend several meters into the soil from individual hosts, the distance of possible exploration by extramatrical hyphae may be limited by edaphic features, competition, physiological integrity, or other factors discussed below. In this case, *Monotropa* plants scattered among the conifers may act as an intermediate staging point for fungal metabolic processing, and may extend the effective distance of hyphal exploration and nutrient uptake. If competition is severe between inter- and intraspecific conifer individuals or similarly sized individuals, the small *Monotropa* plants may be a wise economic investment to be able to capitalize on additional resources without increasing competitive ability by using more expensive mechanisms. Alternatively, this system may show that carbon is not limiting, and the small amount utilized by *Monotropa* is not significant.

Properties of Fungal Communities That Affect Plant Interactions

Spatial Interactions

Although the unit mycelium concept is intriguing and can serve as a simple model for understanding processes belowground, likely the true behavior of mycorrhizal fungal communities is much more complex than this simple approach. Several ectomycorrhizal fungal species commonly occur on a single root tip, and many more can inhabit the root system of a single host plant. Several fungal species may occupy a given area of soil rather than only one as the unit mycelium approach would suggest. Thus interaction among and competition between different host and fungal species as well as different host–fungus combinations would be proportionately greater. Figure 9.2 shows that the extent to which the unit mycelium concept operates in natural conditions may be on a much smaller scale than originally conceived and may exist only in small areas. However, it would still be possible to have one or several hosts, or portions of their root systems interconnected with a single fungal species. The fungal community concept where several hosts are interconnected by fungal hyphae may still be possible, but in a much more complex manner.

In addition, the work by Rayner, Todd, and co-workers, summarized in Rayner and Todd (1982a,b), strongly suggests that a certain degree of vegetative incompatibility and "noncommunication" may exist among genetically dissimilar intraspecific isolates of woodrotting fungi. Rayner and Todd (1982a) used autoradiographic experiments to demonstrate that there was no exchange of ^{86}Rb between paired, antagonistic isolates of the wood-rotting fungus *Coriolus versicolor*, but showed that there was free exchange between genetically identical colonies of *C. versicolor*. A vegetative incompatibility factor prevented anastomoses and physiological communication between different isolates of the same fungal species. Interestingly, Fries (1987) and Dahlberg and Stenlid (1990) have observed this vegetative incompatibility phenomenon in the genus *Suillus*, an ectomycorrhizal fungus associated with conifers. Thus, even though two or more hosts or portions of their rooting systems may be interconnected by the same mycorrhizal fungal species, physical and physiological communication may still not be possible. Competition may occur not only between different fungal species, but also between genetically different individuals of the same species. How strong the communal nature of ectomycorrhizal fungi is and whether large scale interplant connections occur in spite of inter- and intraspecific incompatibility mechanisms are questions that must be addressed.

Several other properties of mycorrhizal fungi affect spatial and temporal inter-

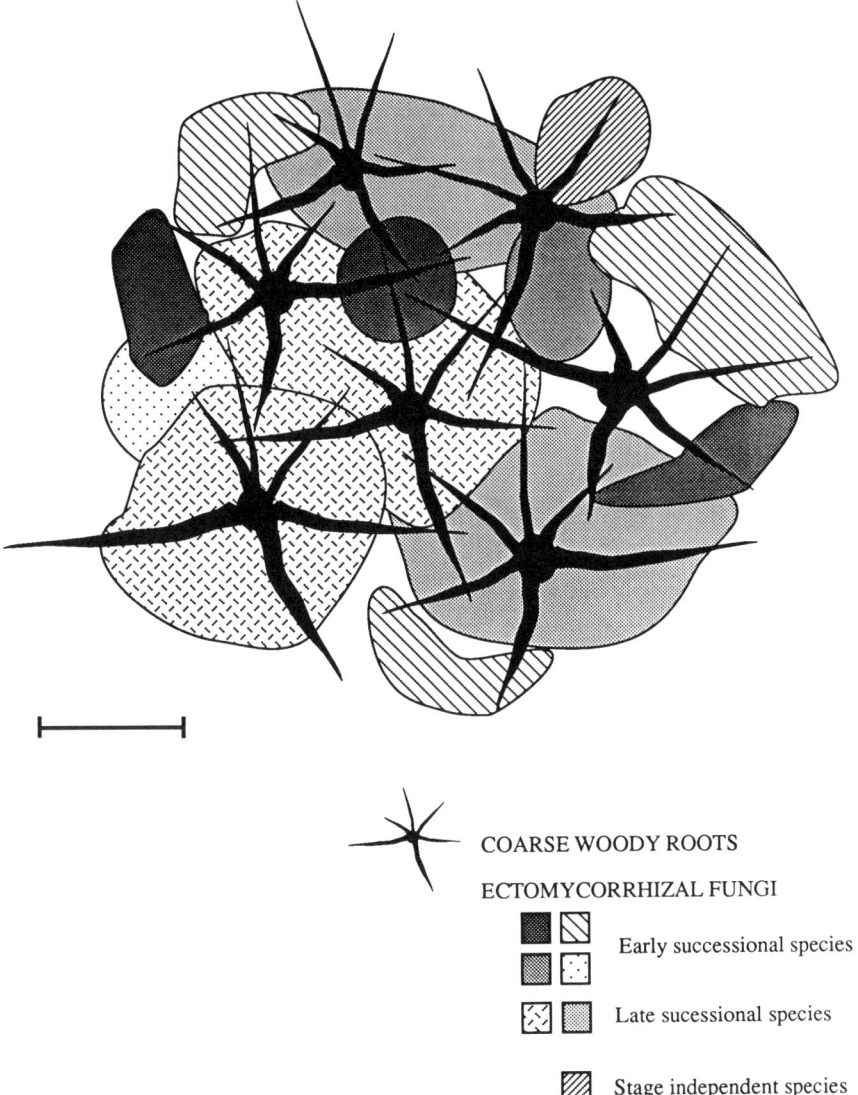

Figure 9.2. Graphic representation of the coarse root system of a forest and location of ectomycorrhizal fungal populations as viewed from directly above the forest. Fine roots are assumed to be distributed evenly throughout the position of the coarse roots. Note that a single tree may be associated with few or many ectomycorrhizal fungi, a single mycelium can link roots of one or more trees, and that most or all of the root systems shown can be effectively linked by complex spatial and temporal interactions of several fungal species and hosts. The scale is approximately 1 m.

actions between plants. These include fungal autonomy, host specificity, successional and phenological changes in fungi, and vegetative incompatibility.

Several fungal species and particular strains or isolates are repeatedly used in laboratory experiments, yet little is known about the basic biology and phenotypic plasticity of most mycorrhizal fungi. It appears, however, that each mycorrhizal fungal species exhibits its own degree of one or more superior attributes under certain circumstances or in association with certain hosts that make them desirable from an applied perspective. Such attributes include effectivity, uptake of a particular elemental nutrient, ability to withstand severe drought conditions, or low or high host specificity and may play an important role in determining competitive interactions between fungi and hosts. Other fungal species or fungus–host combinations show less of a response for a particular attribute. Antibus et al. (1981), for example, found that ectomycorrhizae of *Entoloma sericeum* + *Salix rotundifolia* exhibited surface acid phosphatase activity that was 10–40 times that demonstrated by mycorrhizae formed on the same host by other mycobionts. Parke et al. (1983) found that Douglas-fir seedlings inoculated with *Rhizopogon vinicolor* were less affected by drought than seedlings inoculated by any other mycobiont used in their study. Maintenance respiration rates of decapitated Douglas-fir seedlings inoculated with *Hebeloma crustuliniforme* were much lower than of seedlings inoculated with *Laccaria laccata,* suggesting a greater reliance of *L. laccata* on the host (Marshall & Perry, 1987). *Hebeloma crustuliniforme* was also more effective at taking up nitrogen from protein sources and providing it to *Betula pendula,* the host, than were *Amanita muscaria* or *Paxillus involutus* (Abuzinadah & Read, 1989a,b).

The expression of superior attributes may extend to particular strains or isolates of a species or even to the same strain under different physiological conditions. Stahl and Smith (1984) found that *Agropyron smithii* plants inoculated with an isolate of the VA mycorrhizal fungus *Glomus macrocarpum,* collected from the driest geographic area in their study, exhibited lower leaf resistance to water vapor loss than isolates obtained from more mesic geographic areas, thereby allowing increased stomatal opening under dry soil conditions. Marx (1981) tested 21 isolates of *Pisolithus tinctorius* for ectomycorrhizal development and *in vitro* growth response. Ten isolates formed few or no ectomycorrhizae, while the remaining eleven isolates formed abundant ectomycorrhizae. Individual isolates differed greatly in *in vitro* growth response at various temperatures. Likewise Miller et al. (1989a) found that differences in carbon allocation to the same isolate of *Hebeloma crustuliniforme* grown under different physiological conditions were probably sufficient to affect seedling establishment and survival.

From an applied point of view individual species of fungi may be desirable because of a particular attribute, such as increased plant productivity. From an ecological perspective observed differences in a particular attribute may indicate increased competition between superior and inferior fungi. For instance, mycorrhizal fungi able to increase drought tolerance of their host would be more likely to survive in an arid environment. Likewise fungi capable of increasing phosphorus mobilization in a P-limited area would also likely have the advantage. However, from an altruistic standpoint, fungal autonomy may result in less competition between host plants because of differing phenologies and differing responses to fungi resulting in finely divided niches. Each fungus and fungus–host combination could interact favorably because each contributes its own superior attribute to the fitness of the system.

Temporal Interactions

Mycorrhizae, mycorrhizal fungi, and plants interact not only spatially but also through time. Classical ecological theory speculates about succession of plants, but until recently few ecologists were concerned with the contributions that mycorrhizae and mycorrhizal fungi could have on the direction and trajectory of vascular plant succession. Allen and Allen (1984) and Yocum (1983) have speculated that the mycorrhizal flora of a particular area may regulate vascular plant succession and community composition.

Mycorrhizae and mycorrhizal fungi affect plant–plant interactions because of their host specificity and through succession of the fungi themselves. Succession of mycorrhizal fungi, or the progressive change in the composition of mycorrhizal fungal species over time, may occur throughout the lifetime of a single rootlet or a single host, or may occur in relation to transition from one host to another. Marks and Foster (1967) found that 1–4% of pine and fir ectomycorrhizae had dual associations with two or more mycobionts on the same root tip. They interpreted that the dual associations are formed when the portion of the root tip bearing one mycobiont goes dormant, and on regrowth, the mycobiont forming the mycorrhiza is a different strain. Likewise, Miller et al. (1991) found that most mycorrhizae of red alder were composed of more than one mycobiont. Several studies have shown that species of ectomycorrhizal fungi inoculated onto nursery stock before planting in the field are superceded within a short time by native fungi present in the soil at the planting site (Last et al., 1985). It is possible that such rapid changes in mycorrhizal fungal species on a single host are due to competition between introduced and native fungi. Changes in the host physiology over the life of the plants, from seedling to mature tree, for example, may also contribute to mycorrhizal fungal succession. Mason et al. (1983) and Flemming (1983, 1984) found that there was a succession of mycorrhizal fungi in even-aged stands of *Betula* spp., where a distinct flora of mycorrhizal fungi occurring early in succession gave way to another distinct flora that occurred late in succession. Birch seedlings planted in soil isolated from parent tree roots developed only mycorrhizae characteristic of early successional species of fungi, while seedlings raised in contact with root systems of mature parent birch trees developed ectomycorrhizae with fungal species characteristic of mature forests. Last et al. (1985) hypothesized that the ability to form ectomycorrhizae in a variety of unsterile soils, to the near exclusion of other fungi, is a characteristic of early successional fungi. In contrast, late stage fungi seem capable of propagating ectomycorrhizae in field situations only after early stage fungi have disappeared, or lost their effectiveness. However, Lodge and Wentworth (1990) showed that *Populus* and *Salix* may form both VA mycorrhizae and ectomycorrhizae, but on different root segments, and suggested that the ectomycorrhizae gradually replace the VA mycorrhizae.

Fungal Specificity

The specificity of mycorrhizal fungi may also play an important part in the interaction between plants, and in the establishment and succession of plants. Endomycorrhizae are well known to exhibit little host specificity, forming mycorrhizae with a variety of herbaceous and woody plant hosts. This may be conducive to large numbers of interspecific plant interconnections in the field and possibly minimalization of competition between interconnected plant individuals. However, it may also lead to greater competition between plants by facilitating expression and selection of attributes of superior fungal–plant combinations (Allen et al., 1984; Allen & Allen, 1990). Ectomycorrhizae on the other hand often exhibit a higher degree of host specificity, even though some species are capable of synthesis with a broad range of hosts. Molina (1979, 1981) and Miller et al. (unpublished) found that many species of fungi associated with red alder, an early successional tree species from the Pacific northwestern United States, appear to be highly host specific for species of alder, and that alder-specific ectomycorrhizal fungi induce incompatible, hypersensitive responses in many conifer roots (Molina & Trappe, 1982). Interestingly, most conifer host-specific fungi from the Pacific Northwest are also associated with pioneering tree species as well (Molina & Trappe, 1982; Kropp & Trappe, 1982).

Both the specialized and often host-specific nature of some ectomycorrhizal fungi, and the more cosmopolitan nature of some host-generalist fungi, may be essential to the successful establishment of early successional tree species on disturbed sites. Mikola (1970) suggested that given the strongly coevolved nature of many host–fungus associations, host-specific fungi are more specialized with respect to their host and so may be more effective at promoting survivability, competitive ability, and growth than cosmopolitan fungi that can associate with many hosts. It seems likely that at least some mycobionts in transitional seral stages would be broad-host ranging species that could accommodate both incoming and outgoing hosts. Harley and Smith (1983) argued that ectomycorrhizal associations have evolved toward a lack of host specificity to increase the likelihood of establishing a relationship rapidly in severely disturbed areas. In the Pacific Northwest, both red alder and Douglas-fir are early successional colonizers of disturbed sites, however, red alder often overtops and prevents establishment and growth of the Douglas-fir. Miller et al. (unpublished) found that the greatest diversity of ectomycorrhizal types on alder was in a conifer clearcut soil, while on Douglas-fir diversity was highest in a rotation-aged conifer soil. These results were obtained from a greenhouse bioassay of soil from a successional sequence, including a 1-year-old conifer clearcut, a young conifer plantation, rotation-aged and old aged conifer stands, and young and old alder stands. Ectomycorrhizal infection on alder was low in soils from sites where conifers were actively growing, yet was much higher in the same soil only a year after clearcutting. Initial ectomycorrhizal infection of red alder was much more rapid than for Douglas-fir, due primarily to infection by *Alpova diplophloeus,* an adept early successional, host-specific species. *Thelephora terrestris,* the only mycobiont that formed ectomycorrhizae on both red alder and Douglas-fir, is a notorious broad-host ranging species, and formed the first mycorrhizae on Douglas-fir in this study. Presumably, red alder may have a competitive advantage over Douglas-fir because of faster initial ectomycorrhizal infection.

The implications of high host specificity and succession are perhaps most dramatic for regions of old growth forest. If the fungi are specific to a particular species, and have evolved through the processes of succession controlled by host physiology and ecosystem maturation, sudden elimination of the host as in clearcut situations may result in a loss of the mycorrhizal fungal species necessary to reestablish the old growth ecosystem.

Plant Competition Mediated by Mycorrhizae

The above discussion reviewed the evidence for transport of nutrients between plants via hyphal connections, and indicated that this has occurred sufficiently in only a few instances to cause increased growth in a neighboring plant. The section that followed discussed the specificity of fungi that might infect a plant, and the changes in time and space that determine which plants and fungi might associate. Now we will turn to the effects of mycorrhizae on competition between plants.

Most of the work that has been published on the effects of mycorrhizae on competition has been done using VA mycorrhizae, although there is one recent study on competition between ectomycorrhizal plants (Perry et al., 1989b, reviewed below). Harper (1977) suggested that mycorrhizae may be important in changing the competitive balance between species when one is mycorrhizal and the other does not form a mycorrhizal association (i.e., is nonmycotrophic). Four studies bear out this suggestion. Crowell and Boerner (1988) showed that the old-field annual *Ambrosia artemisiifolia* had greatly increased growth with VA mycorrhizal inoculum in poor nutrient soil, and was competitively inferior to the cooccurring nonmycotrophic *Brassica nigra* in the absence of inoculum. In two other studies, inoculation in part overcame the negative growth effects of the nonmycotrophic *Salsola kali* on the grasses *Agropyron smithii* and *Bouteloua gracilis* (Allen & Allen, 1984) and *S. kali* and the nonmycotrophic *Atriplex rosea* on *Agropyron dasystachyum* (Benjamin & Allen, 1987). In a multispecies mixture of mycotrophic and nonmycotrophic garden flowers, there was an increase in biomass by the mycotrophic species when they were inoculated (Yocum, 1983).

Most plant species form mycorrhizae, so for mycorrhizae to change the competitive balance between plants, they must have different physiological effects on individual species. Plants may be nonmycotrophic, facultatively mycotrophic, or obligately mycotrophic, with a continuum in the extent of their response to infection. Plants that exhibit different degrees of mycotrophy co-occur in many biomes, especially in successional sites (Allen & Allen, 1990). Studies on competition between early and late seral species generally show both greater response to mycorrhizal infection and a greater competitive ability by the later seral species (Allen & Allen, 1984, 1990; Benjamin & Allen, 1987; Schwab & Loomis, 1987). There may also be differences between competitive abilities and mycorrhizal response of plants that are considered to be of the same seral stage. Three competition studies on the legume–grass combination *Trifolium repens–Lolium perenne* showed a similar pattern, with the legume more responsive to mycorrhizal infection in monoculture and more competitive with the grass when infected in a low P soil (Crush, 1974; Hall, 1978; Buwalda, 1980). In a multispecies mixture, there was a relative increase in herb compared to grass biomass after inoculation (Grime et al., 1987). Even two grass species that both have a relatively small response to mycorrhizal infection, *Holcus lanatus* and *Lolium perenne*, showed a skewed competitive response, with *H. lanatus* being the superior competitor (Fitter, 1977).

The above examples suggest that, to some degree, the competitive ability of plants with mycorrhizal infection and their response to mycorrhizae in monoculture may be related. To test this hypothesis, we calculated the biomass ratio of plants with and without infection in monoculture and mixed from the available literature (Table 9.1). The only studies included were those that maintained a constant density of plants in monocultures and mixtures. Of these studies, only two used species that had a very large response to mycorrhizal infection in either monoculture or mixture, *Andropogon gerardii* (Hetrick et al., 1989) and *Trifolium repens* (Hall, 1978). In both cases, the growth response was greatly reduced with phosphorus fertilization. In addition, one of the species used in both was a grass with a much smaller mycorrhizal response. A regression of the monoculture vs. the mixture mycorrhizal/nonmycorrhizal ratio shows that most plant pairs responded predictably, with response in monoculture related to response in mixture (Fig. 9.3). The two exceptions were the two plants with high response to infection. *Trifolium repens* had a greater than expected response in mixture, while *A. gerardii* had a response in mixture that was lower than expected. The reason for this discrepancy is not apparent, but the sample size for a regression of this type is quite small at this time. Newman (1992) also suggested that the response of individual plants to mycorrhizae does not predict their interaction in competition. Mycorrhizal fungi may affect the rate of plant growth and of nutrient uptake

Table 9.1. Biomass Ratios of Mycorrhizal (M) to Nonmycorrhizal (NM) Plants in Monocultures and Mixtures

Species	Monoculture M/NM	Mixture M/NM	Source
Andropogon gerardii	248.00	91.60	Hetrick et al. (1989)
Koelaria pyranidata	0.84	0.14	
A. gerardii +P[a]	2.48	3.10	
K. pyranidata +P	0.81	0.18	
Holcus lanata	1.11	1.39	Fitter (1977)
Lolium perenne	0.88	0.54	
Trifolium repens	9.00	20.00	Hall (1978)
Lolium perenne	1.19	0.97	
T. repens +P[b]	0.64	0.84	
L. perenne +P	0.89	1.00	
Agropyron smithii	0.97	1.39	Allen and Allen (1984)
Salsola kali	0.83	0.91	
Bouteloua gracilis	1.01	1.43	
S. kali	0.83	0.93	
Agropyron dasystachyum	1.37	3.29	Benjamin and Allen (1987)
S. kali	1.23	1.10	
A. dasystachyum	1.37	1.45	
Bromus tectorum	1.03	1.27	
A. dasystachyum	1.37	1.84	
Hordeum jubatum	1.64	1.40	
A. dasystachyum	1.37	1.22	
Atriplex rosea	1.02	1.10	

[a]50 ppm P added.
[b]108 kg P/ha added.

differently from the way a neighboring plant affects these parameters, so perhaps a high correlation should not be expected. More similar types of experiments, using more plant species with varying degrees of responsiveness to mycorrhizae, are needed.

Some still more complex interactions of mycorrhizae and plants were observed by Perry et al. (1989b) in the one competition study to date that utilized ectomycorrhizae. They assessed the competitive interactions between *Pinus ponderosa* and *Pseudotsuga menziesii* using four different species of ectomycorrhizal fungi and uninoculated controls. Nonmycorrhizal controls with ectomycorrhizal fungi are difficult to maintain, and in the study by Perry et al. (1989b), these plants were contaminated by the early successional fungus *Thelephora terrestris*. Interestingly, the two plant species had the greatest biomass in monoculture with *T. terrestris* compared to the inoculated mycorrhizal species. However, in mixture there was mutual inhibition, e.g., reduced biomass of the two tree species with the contaminant fungus, but not with the inoculated fungal species. The total amount of N and P taken up by both plant species was less with the contaminant

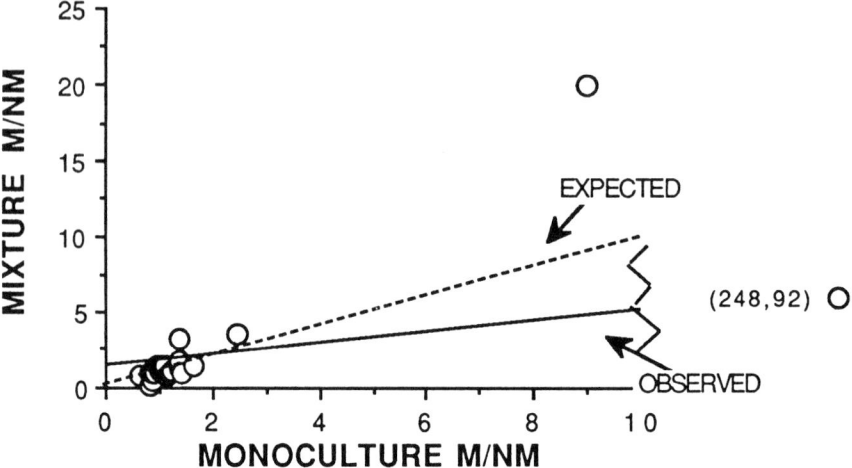

Figure 9.3. Regression of the ratio of shoot dry mass of plants grown with and without mycorrhizal inoculum in monoculture vs. mixture. Data are taken from Table 9.1. The two outlier points are from studies by Hetrick et al. (1989) and Hall (1978). The expected curve is assumed to have a slope = 1. The equation for the observed line is $y = 0.365x + 1.51$, $r^2 = 0.97$, $p = <0.001$.

than with the inoculated species. This study paints a complex picture of the interactions of plants with individual species of mycorrhizal fungi, which becomes compounded when neighboring plants interact.

The potentially interesting interactions between plants that form different kinds of mycorrhizae have not been experimentally studied, although some researchers are now considering these. Plant communities may consist of woody plants that form ecto-, ericoid/arbutoid, and VA mycorrhizae associated with herbaceous plants that form VA mycorrhizae. For instance, the pinyon-juniper woodlands of the southwestern United States have ectomycorrhizal *Pinus* spp. and VA mycorrhizal *Juniperus* spp., interspersed by VA mycorrhizal shrubs and herbs. Even more complex interactions may occur in chaparral vegetation, which is dominated by ectomycorrhizal oaks, VA mycorrhizal *Adenostema* spp. and *Ceanothus* spp., and arbutoid mycorrhizal *Arctostaphylos* spp. Such interactions may be especially important in seral communities (Allen & Allen, 1990), where the density and kind of inoculum as well as the plant species are changing. In some cases, two plant species will form a different type of mycorrhizal association with the same species of fungus, as in the case of the early seral *Arctostaphylos* spp. followed by *Pinus* spp. (Perry et al., 1989a). In other cases, the same plant species will form different types of mycorrhizae during different stages of its life cycle or under different environmental conditions, as for the genera *Populus, Salix* (Lodge & Wentworth, 1990), and *Eucalyptus* (Lapeyrie & Chilvers, 1985).

How changes in types of mycorrhizal fungi affect neighboring species still needs to be investigated.

The studies reviewed show that mycorrhizal fungi may shift the competitive balance between two species, whether they have high or low response to infection, or if one of the partners is nonmycotrophic. In addition to shifting the biomass composition of plant mixtures, mycorrhizae may cause an increase in the summed biomass of the mixture (Benjamin & Allen, 1987; Perry et al., 1989b). In this case, both neighbors benefit from the mycorrhizae, although one may benefit more than the other. What is not apparent from any of these studies is what the contribution of hyphal transport may be between neighbors that are competing, at least according to the standard measurements of shifts in biomass. In the previous sections we described how important hyphal connections may be for nutrient and carbon transfer, at least under some conditions. The final section addresses the situation in which neighboring plants that may be competing are also exchanging substances via hyphae.

Competition and Hyphal Transport

In a recent thoughtful review, Newman (1988) concluded that simply because there may be some degree of nutrient exchange via hyphae between plants, this does not imply that we need to alter our thinking on plant interactions. When two unequal competitors are neighbors, one will grow larger than the other in spite of hyphal connections. However, his conclusions overall were colored by the fact that very little of the labeled nutrient ever goes to the receiver plant in any published experiment. In fact, we know little about how much these small amounts of nutrient transfer might affect growth of the receiver. Newman criticizes perhaps the best experiment to date (Francis et al., 1986) because, although the receiver plant grew larger, there was no adequate control for competition between donor and receiver. The design of such experiments is problematical. To label or supply nutrients from donor to receiver, a high enough amount of nutrient could be added that might change the source-sink relationship of nutrient flow between the two plants, and might change the competitive relationship between them. Some other approaches are needed, such as to measure competition, plant growth, and nutrient transport simultaneously.

Such an approach was taken in two experiments. In a study on intraspecific competition, Eissenstat and Newman (1990) showed that seedlings of *Plantago lanceolata* did not have improved growth, whether the mature plant they were growing with was mycorrhizal or not, nor did they have increased ^{32}P uptake from the potential mature donor plant. However, this is a plant with very little response to mycorrhizae, so it may not be the best species to test the interactions of mycorrhizae and competition. In another experiment using several species with varied responses to mycorrhizal infection, the results varied by species (Benjamin & Allen, 1987; Allen, unpublished). The donor plant *Agropyron dasystachyum* was grown in the same pot with each of five recipients, including another *A. dasystachyum*, and its leaves were labeled with ^{32}P (Table 9.2). The label was detected in leaves of all recipient species, but at more than 1–2 orders of magnitude lower than the 700,000 cpm applied to the donor. Only two species showed a greater uptake of ^{32}P when the donor plant was inoculated, and one of them, *Atriplex rosea*, is nonmycotrophic. Considering the small amount transferred, and the variability (and lack of statistical significance) of some of the other values, the small amount of difference for *A. rosea* may not have biological significance. The response of *Hordeum jubatum* must be interpreted by examining the ratios of aboveground biomass of donor and recipient (Table 9.2). Both donor (*A. dasystachyum*) and recipient (*H. jubatum*) were larger when they were mycorrhizal than nonmycorrhizal, but *H. jubatum* was smaller relative to the donor in either the mycorrhizal or nonmycorrhizal treatment. In spite of being relatively smaller, it received a larger share of the label when inoculated. This indicates that the greater uptake of P by *H. jubatum* may not be related to the leaf "sink strength," but rather to the hyphal linkages between the plants.

Although instantaneous measurements of ^{32}P showed some differences in this experiment (Table 9.2), there were no effects of mycorrhizae on leaf or root P concentrations of any of the species (Benjamin & Allen, 1987 and unpublished). The soil P was moderate, about 21 μg P/g soil. This experiment showed that there was a potential for small amounts of hyphal nutrient transfer between plants, but this did not offset the competitive interactions between them, where nutrients other than P may have been more limiting. The possibility of upsetting the "source-sink" relationship of transport between donor and recipient is still a

Table 9.2. *Counts per minute (cpm) of ^{32}P Detected in Leaves of Recipient Plants 48 hr after Labeling Leaves of* Agropyron *Donor, Biomass per Plant of Each Species, and Donor/Recipient Biomass Ratio (D/R) for Mycorrhizal Inoculated (M) and Uninoculated (NM) Plants*[a]

| | | | Biomass/plant | | | | | |
| | cpm | | M | | | NM | | |
Recipient	M	NM	Donor	Recipient	D/R	Donor	Recipient	D/R
Agropyron	3215	3707 n.s.	0.26	0.26	1.00	0.19	0.19	1.00
Hordeum	3679	2392 *	0.57	0.31	1.80	0.31	0.22	1.40 *
Bromus	11596	39630 n.s.	0.16	0.42	0.38	0.11	0.33	0.33 n.s.
Atriplex	2882	2614 *	0.33	0.25	1.32	0.27	0.23	1.17 n.s.
Salsola	14370	19136 n.s.	0.25	1.18	0.21	0.08	1.07	0.07 *

[a] Significant differences (* is 0.95 probability, n.s. is nonsignificant) are shown for cpm and for D/R ratios. In each case the donor plant is *Agropyron*, and the recipient is the species listed. Specific names as in Table 9.1. Data from Benjamin and Allen (1987) and unpublished.

possibility in such a short-term experiment where a relatively high amount of label must be used to detect uptake. As mentioned previously, the addition of the label does not show directionality of movement of a nutrient. So we are still left with some evidence that transport occurs, but its importance is unclear.

Some new types of experiments may help to clarify the current dilemma. Consider the case of a phosphate ion traversing a hypha at the junction between two linked plants. The direction of its movement will depend on the sink strength of the two plants, and it will likely travel toward the plant that has the greater demand for P, which is in all likelihood the plant with greater growth rates and/or higher tissue P concentrations. This would suggest that a larger, more "competitive" plant would never give up nutrients to a smaller plant. However, this scenario assumes that fungi have no control over the rate at which nutrients are released, and that soil nutrients are homogeneous. Fungi have the ability to sequester a fraction of the nutrient pool, as reviewed above. The importance of the fungi in conserving nutrients for plant growth is accepted, but whether fungi provide nutrients to different plants at different rates is unexplored. Patches of high and low nutrients do occur, and roots (Jackson & Caldwell, 1989) and mycorrhizal fungi (M. Andrews, J. Richards, & M. Allen, unpublished data) proliferate in these patches. Thus there are nutrient gradients in the soil, and these gradients may affect hyphal nutrient transport gradients between plants. The next experiments to explain hyphal transport should incorporate the ability of fungi to sequester nutrients and the effects of soil nutrient gradients on movement of nutrients between plants.

The discussion on facilitation vs. competition has been presented largely from the point of view of the plant rather than of the fungus. However, a balanced discussion must include both views. From the plant's point of view, hyphal connections might enable nutrient transport, while the benefit to the fungus is to connect more plants into the network and obtain additional carbon. In a recent careful observation of the mycelial network of VA mycorrhizal plants, Friese and Allen (1991) showed specialized large diameter hyphae that they termed "runner hyphae" connecting individual roots, while narrower hyphae termed "absorbing hyphae" were found connecting roots to soil. The runner hyphae appear to be similar to the "arterial hyphae" described by Read (Chapter 4, this volume). Runner hyphae function to promote a mycelial network by connecting roots, but at the same time they probably provide the major, perhaps only, pathway for hyphal nutrient transport between plants. Ectomycorrhizal hyphae also have specializations, with rhizomorphs for nutrient versus water transport, although runner or arterial hyphae have not been described (Read, Chapter 4). These specialized functions of hyphae may be of different value to the plants than the fungi, and need to be understood in greater detail.

We suggest several future research directions that may help determine whether hyphal transport between plants is important. The first problem is the controversy of whether a labeled nutrient is actually transported directly via the hypha, or leaks into the soil and is then picked up by the neighboring plant (Newman, 1988). To understand this, the rate of movement of nutrients in a particular soil vs. via the hyphal network must both be calculated. The limiting factor here may be knowing the total number of hyphal connections in order to make this calculation, but the morphological study cited above (Friese & Allen, 1991) may make this possible. If runner hyphae of VA fungi are considered the connectors between roots of neighboring plants, they can be quantified microscopically, and rate of nutrient movement between plants can be calculated.

The second problem is the complexity of hyphal transport in the field where many species of plants and fungi may be involved. The fungi and plants may form patches that are dominated by certain species, as for the ectomycorrhizal fungi of Figure 9.2. A morphological approach may help to explain why some neighboring plant species experience relatively high rates of transport and others show none (Chiarello et al., 1982). Finally, the third problem that needs to be tackled is exactly how much of the nutrient that is transported gets to the neighboring plant, and how effective this is in increasing nutrient intake by the neighbor. Most of the studies reported here used instantaneous measurements of isotope movement, with very few attempts at examining the longer term effects of a plant that has excess nutrient on the nutrient uptake and growth of its neighbor (e.g., Francis et al., 1986). Although this nutrient transport is small, in a patchy soil, and over the lifetime of a plant, there may be times when this small increase in nutrient substantially increases the fitness of the host.

Acknowledgments

Research for this chapter was funded in part by grants from the National Science Foundation (BSR88-05983 to S.L.M., BSR88-18076 to E.B.A.), the USDA Competitive Grants Program (83-CRCR-1-1229 to E.B.A.), and the University of Wyoming/National Park Service Research Center to S.L.M.

References

Abuzinadah, R. A., & Read, D. J. (1986a). The role of proteins in the nitrogen nutrition of ectomycorrhizal plants. I. Utilization of peptides and proteins by ectomycorrhizal fungi. *New Phytologist* **103**, 481–493.

Abuzinadah, R. A., & Read, D. J. (1986b). The role of proteins in the nitrogen nutrition of ectomycorrhizal plants. III. Protein utilization by *Betula, Picea* and *Pinus* in mycorrhizal association with *Hebeloma crustuliniforme*. *New Phytologist* **103**, 507–514.

Abuzinadah, R. A., & Read, D. J. (1989a). The role of proteins in the nitrogen nutrition of ectomycorrhizal plants. IV. The utilization of peptides by birch (*Betula pendula* L.) infected with different mycorrhizal fungi. *New Phytologist* **112**, 55–60.

Abuzinadah, R. A., & Read, D. J. (1989b). The role of proteins in the nitrogen nutrition of ectomycorrhizal plants. V. Nitrogen transfer in birch (*Betula pendula*) grown in association with mycorrhizal and nonmycorrhizal fungi. *New Phytologist* **112,** 61–68.

Abuzinadah, R. A., Finlay, R. D., & Read, D. J. (1986). The role of proteins in the nitrogen nutrition of ectomycorrhizal plants. II. Utilization of protein by mycorrhizal plants of *Pinus contorta*. *New Phytologist* **103,** 495–506.

Al Abras, K., Bilger, I., Martin, F., Le Tacon, F., & Lapeyrie, F. (1988). Morphological and physiological changes in ectomycorrhizas of spruce (*Picea excelsa* (Lam.) Link) associated with ageing. *New Phytologist* **110,** 535–540.

Alexander, I. J. (1983). The significance of ectomycorrhizas in the nitrogen cycle. In: *Nitrogen as an Ecological Factor* (Ed. by J. A. Lee, S. McNeill, & I. H. Rorison), pp. 69–93. Blackwell, Oxford.

Allen, E. B., & Allen, M. F. (1984). Competition between plants of different successional stages: mycorrhizae as regulators. *Canadian Journal of Botany* **62,** 2625–2629.

Allen, E. B., & Allen, M. F. (1988). Facilitation of succession by the nonmycotrophic colonizer *Salsola kali* (Chenopodiaceae) on a harsh site: Effects of mycorrhizal fungi. *American Journal of Botany* **75,** 257–266.

Allen, E. B., & Allen, M. F. (1990). The mediation of competition by mycorrhizae in successional and patchy environments. In: *Perspectives on Plant Competition* (Ed. by J. B. Grace & G. D. Tilman), pp. 367–389. Academic Press, New York.

Allen, M. F., Allen, E. B., & Stahl, P. D. (1984). Differential niche response of *Bouteloua gracilis* and *Pascopyrum smithii* to VA mycorrhizae. *Bulletin of the Torrey Botanical Club* **3,** 361–365.

Antibus, R. K., Croxdale, J. G., Miller, O. K. Jr., & Linkins, A. E. (1981). Ectomycorrhizal fungi of *Salix rotundifolia*. III. Resynthesized mycorrhizal complexes and their surface phosphatase activities. *Canadian Journal of Botany* **59,** 2458–2465.

Benjamin, P. K., & Allen, E. B. (1987). The influence of VA mycorrhizal fungi on competition between plants of different successional stages in sagebrush-grassland. In: *Mycorrhizae in the Next Decade, Practical Applications and Research Priorities* (Ed. by D. M. Sylvia, L. L. Hung, & J. H. Graham), p. 144. IFAS, Gainesville, Florida.

Bevege, D. I., Bowen, G. D., & Skinner, M. F. (1975). Comparative carbohydrate physiology of ecto- and endo-mycorrhizas. In: *Endomycorrhizas* (Ed. by F. E. Sanders, B. Mosse, & P. B. Tinker), pp. 149–174. Academic Press, London.

Björkman, E. (1942). Uber die Bedingungen der Mykorrhizabildung bei Kiefer und Fichte. *Symbiolae Botanicae Upsaliensis* **6,** 2.

Björkman, E. (1960). *Monotropa hypopitys* L.-an epiparasite on tree roots. *Physiologia Plantarum* **13,** 308–327.

Botton, B., Khalid, A., Boukroute, A., & Martin, F. (1986). Purification and properties of the glutamate oxaloacetate transaminase of the ectomycorrhizal fungus *Cenococcum geophilum*. In: *Physiological and Genetical Aspects of Mycorrhizae* (Ed. by V. Gianinazzi-Pearson & S. Gianinazzi), pp. 407–411. Institut National de la Recherche Agronomique, Paris.

Brownlee, C., & Jennings, D. H. (1982). Long distance translocation in *Serpula lacri-*

mans: velocity estimates and the continuous monitoring of induced perturbations. *Transactions of the British Mycological Society* **79**, 143–148.

Brownlee, C., Duddridge, J. A., Malibari, A., & Read, D. J. (1983). The structure and function of mycelial systems of ectomycorrhizal roots with special reference to their role in forming inter-plant connections and providing pathways for assimilate and water transport. *Plant and Soil* **71**, 433–443.

Buller, A. H. R. (1931). *Researches on Fungi*, Vol. 4. Longmans, London.

Butler, G. M. (1957). The development and behavior of mycelial strands in *Merulius lacrymans* (Wulf.) Fr. I. Strand development during growth from a food-base through a non-nutrient medium. *Annals of Botany* **21**, 523–537.

Buwalda, J. G. (1980). Growth of a clover-ryegrass association with vesicular arbuscular mycorrhizas. *New Zealand Journal of Agricultural Research* **23**, 379–383.

Chiarello, N., Hickman, J. C., & Mooney, H. A. (1982). Endomycorrhizal role for interspecific transfer of phosphorus in a community of annual plants. *Science* **217**, 941–943.

Clements, F. E. (1916). *Plant Succession*. Carnegie Institute of Washington Publication 242.

Crowell, H. F., & Boerner, R. E. J. (1988). Influence of mycorrhizae and phosphorus on belowground competition between two old-field annuals. *Environmental and Experimental Botany* **28**, 381–392.

Crush, J. R. (1974). Plant growth responses to vesicular-arbuscular mycorrhizas. VII. Growth and nodulation in some herbage legumes. *New Phytologist* **73**, 743–749.

Dahlberg, A., & Stenlid, J. (1990). Population structure and dynamics in *Suillus bovinus* as indicated by spatial distribution of fungal clones. *New Phytologist* **115**, 487–493.

Dighton, J. (1983). Phosphate production by mycorrhizal fungi. *Plant and Soil* **71**, 455–462.

Dighton, J., Thomas, E. D., & Latter, P. M. (1987). Interactions between tree roots, mycorrhizas, a saprotrophic fungus and the decomposition of organic substrates in a microcosm. *Biology and Fertility of Soils* **4**, 145–150.

Dosskey, M. G., Linderman, R. G., & Boersma, L. (1990). Carbon-sink stimulation of photosynthesis in Douglas-fir seedlings by some ectomycorrhizas. *New Phytologist* **115**, 269–274.

Duddridge, J. A., Malibari, A., & Read, D. J. (1980). Structure and function of mycorrhizal rhizomorphs with special reference to their role in water transport. *Nature (London)* **287**, 834–836.

Duddridge, J. A., & Read, D. J. (1982). An ultrastructural analysis of the development of mycorrhizas in *Monotropa hypopitys* L. *New Phytologist* **92**, 203–214.

Eissenstat, D. M. (1990). A comparison of phosphorus and nitrogen transfer between plants of different phosphorus status. *Oecologia* **82**, 342–437.

Eissenstat, D. M., & Newman, E. I. (1990). Seedling establishment near large plants: Effects of vesicular-arbuscular mycorrhizae on the intensity of plant competition. *Functional Ecology* **4**, 95–99.

Englander, L., & Hull, R. J. (1980). Reciprocal transfer between ericaceous plants and a *Clavaria* sp. *New Phytologist* **84,** 661–667.

Finlay, R. D., & Read, D. J. (1986a). The structure and function of the vegetative mycelium of ectomycorrhizal plants. I. Translocation of ^{14}C-labelled carbon between plants interconnected by a common mycelium. *New Phytologist* **103,** 143–156.

Finlay, R. D., & Read, D. J. (1986b). The structure and function of the vegetative mycelium of ectomycorrhizal plants. II. The uptake and the distribution of phosphorus by mycelial strands interconnecting host plants. *New Phytologist* **103,** 157–165.

Finlay, R. D., & Read, D. J. (1986c). The uptake and distribution of phosphorus by ectomycorrhizal mycelium. In: *Physiological and Genetical Aspects of Mycorrhizae* (Ed. by V. Gianinazzi-Pearson & S. Gianinazzi), pp. 351–355. INRA, Paris.

Fitter, A. H. (1977). Influence of mycorrhizal infection on competition for phosphorus and potassium by two grasses. *New Phytologist* **79,** 19–125.

Flemming, L. V. (1983). Succession of mycorrhizal fungi on birch: Infection of seedlings planted around mature trees. *Plant and Soil* **71,** 263–267.

Flemming, L. V. (1984). Effects of soil trenching and coring on the formation of ectomycorrhiza on birch seedlings grown around mature trees. *New Phytologist* **98,** 143–153.

Foster, R. C. (1981). Mycelial strands of *Pinus radiata* D. Don: Ultrastructure and histochemistry. *New Phytologist* **88,** 705–712.

Francis, R., Finlay, R. D., & Read, D. J. (1986). Vesicular-arbuscular mycorrhiza in natural vegetation. IV. Transfer of nutrients in inter- and intra-specific combinations of host plants. *New Phytologist* **102,** 103–111.

Franco, A. C., & P. S. Nobel. (1988). Interactions between seedlings of *Agave deserti* and the nurse plant *Hilaria rigida*. *Ecology* **69,** 1731–1740.

Fries, N. (1987). Somatic incompatability and field distribution of the ectomycorrhizal fungus *Suillus luteus* (Boletaceae). *New Phytologist* **107,** 735–739.

Friese, C. F., & Allen, M. F. (1991). The spread of VA mycorrhizal fungal hyphae in the soil: Inoculum type and external hyphal archtecture. *Mycologia* **83,** 409–418.

Furman, T. E. (1966). Symbiotic relationships of *Monotropa*. *American Journal of Botany* **53,** 627.

Giltrap, N. J. (1982). Production of polyphenol oxidases by ectomycorrhizal fungi with special reference to *Lactarius* spp. *Transactions of the British Mycological Society* **78,** 75–81.

Grime, J. P., Mackey, J. M. L., Hillier, S. H., & Read, D. J. (1987). Floristic diversity in a model system using experimental microcosms. *Nature (London)* **328,** 420–422.

Haines, B. L., & Best, G. R. (1976). *Glomus mosseae* endomycorrhizal with *Liquidambar styraciflua* L. seedlings retards NO$_3$, NO$_2$, and NH$_4$ nitrogen loss from a temperate forest soil. *Plant and Soil* **45,** 257–261.

Hall, I. R. (1978). Effects of endomycorrhizas on the competitive ability of white clover. *New Zealand Journal of Agricultural Research* **21,** 509–515.

Harley, J. L., & Smith, S. E. (1983). *Mycorrhizal Symbiosis*. Academic Press, London.

Harper, J. L. (1977). *Population Biology of Plants*. Academic Press, London.

Harvey, A. E., Jurgensen, M. F., & Larsen, M. J. (1980). Clearcut harvesting and ectomycorrhizae: survival of activity on residual roots and influence on a bordering forest stand in western Montana. *Canadian Journal of Forest Research* **10**, 300–303.

Harvey, A. E., Jurgensen, M. F., Larsen, M. J., & Schleiter, J. A. (1986). Distribution of active ectomycorrhizal short roots in forest soils of the inland Northwest: Effects of site and disturbance. *U.S. Forest Service Research Paper* **INT-374**, 1–9.

Hattingh, M. J., Gray, L. E., & Gerdemann, J. W. (1973). Uptake and translocation of ^{32}P-labelled phosphate to onion roots by endomycorrhizal fungi. *Soil Science* **116**, 383–387.

Hetrick, B. A. D., Wilson, G. W. T., & Hartnett, D. C. (1989). Relationship between mycorrhizal dependence and competitive ability of two tallgrass prairie grasses. *Canadian Journal of Botany* **67**, 2608–2615.

Ho, I., & Zak, B. (1979). Acid phosphatase activity of six ectomycorrhizal fungi. *Canadian Journal of Botany* **57**, 1203–1205.

Jackson, R. B., & M. M. Caldwell. (1989). The timing and degree of root proliferation in fertile soil microsites for three cold desert microsites. *Oecologia* **81**, 149–153.

Janos, D. P. (1980). Mycorrhizae influence tropical succession. *Biotropica* **12**, 56–64.

Kropp, B. R., & Trappe, J. M. (1982). Ectomycorrhizal fungi of *Tsuga heterophylla*. *Mycologia* **74**, 479–488.

Lapeyrie, F. F., & Chilvers, G. A. (1985). An endomycorrhiza-ectomycorrhiza succession associated with enhanced growth of *Eucalyptus durmosa* seedlings planted in a calcareous soil. *New Phytologist* **100**, 93–104.

Last, F. T., Mason, P. A., Wilson, J., Ingleby, K., Munro, R. C., Flemming, L. V., & Deacon, J. W. (1985). 'Epidemiology' of sheathing (ecto-) mycorrhizas in unsterile soils: A case study of *Betula pendula*. *Proceedings of the Royal Society of Edinburgh* **85B**, 299–315.

Le Tacon, F. (1972). Disponibilite de l'azote nitrique et ammoniacal dans certains soles de l'Est de la France. Influence sur la nutrition et la croissance de l'Epicea commun (*Picea abies* Karst.). *Annales des Sciences Forestières* **30**, 183–205.

Lee, J. A., & Stewart, G. R. (1978). Ecological aspects of nitrogen assimilation. *Advances in Botanical Research*, Vol. 6, pp. 1–43. Academic Press, London.

Lees, E. (1841). On the parasitic growth of *Monotropa hypopitys*. *Phytologis* **1**, 97–101.

Lodge, D. J., & Wentworth, T. R. (1990). Negative associations among VA-mycorrhizal fungi and some ectomycorrhizal fungi inhabiting the same root system. *Oikos* **57**, 347–356.

Lundeberg, G. (1970). Utilisation of various nitrogen sources, in particular bound soil nitrogen, by mycorrhizal fungi. *Studia Forest Suecica* **79**, 5–95.

Lutz, R. W., & Sjolund, R. D. (1973). *Monotropa uniflora*: Ultrastructural details of its mycorrhizal habit. *American Journal of Botany* **60**, 339–345.

MacDougal, D. T., & Floyd, F. E. (1900). The roots and mycorrhizas of some of the Monotropaceae. *Bulletin of the New York Botanical Garden* **1**, 419–429.

MacLeod, W. J., Robson, A. D., & Abbott, L. K. (1986). Effects of phosphate supply and inoculation with a vesicular-arbuscular mycorrhizal fungus on the death of root cortex of wheat, rape and subterranean clover. *New Phytologist* **103**, 349–357.

Marks, G. C., & Foster, R. C. (1967). Succession of mycorrhizal associations on individual roots of radiata pine. *Australian Forestry* **31**, 94–201.

Marshall, J. D., & Perry, D. A. (1987). Basal and maintenance respiration of mycorrhizal and nonmycorrhizal root systems of conifers. *Canadian Journal of Forest Research* **17**, 872–877.

Martin, F., Canet, D., & Marchal, J.-P. (1985). ^{13}C nuclear magnetic resonance study of mannitol cycle and trehalose synthesis during glucose utilization by the ectomycorrhizal ascomycete *Cenococcum graniforme*. *Plant Physiology* **77**, 499–502.

Martin, F., Ramstedt, M., & Soderhall, K. (1987). Carbon and nitrogen metabolism in ectomycorrhizal fungi and ectomycorrhizas. *Biochemie* **69**, 569–581.

Marx, D. H. (1981). Variability in ectomycorrhizal development and growth among isolates of *Pisolithus tinctorius* as affected by source age and re-isolation. *Canadian Journal of Forest Research* **11**, 168–174.

Mason, P. A., Wilson, J., & Last, F. T. (1983). The concept of succession in relation to the spread of sheathing mycorrhizal fungi on inoculated tree seedlings growing in unsterile soils. In: *Tree Root Systems and Their Mycorrhizas* (Ed. by D. Atkinson, K. K. S. Bhat, M. P. Coutts, P. A. Mason, & D. L. Read), pp. 247–256. Martinus Nijhoff and Dr. W. Junk Publishers, The Hague.

Melin, E. & Norkrans, B. (1948). Amino acids and the growth of *Lactarius deliciosus*. *Physiologia Plantarum* **1**, 176–184.

Melin, E. & Nilsson, H. (1950). Transfer of radioactive phosphorus to pine seedlings by means of mycorrhizal hyphae. *Physiologia Plantarum* **3**, 88–92.

Melin, E. & Nilsson, H. (1953). Transfer of labelled nitrogen from glutamic acid to pine seedlings through the mycelia of *Boletus-variegatus*(Sw.) Fr. *Nature* (London) **171**, 134.

Melin, E., Nilsson, H., & Hacskaylo, E. (1958). Translocation of cations to seedlings of *Pinus virginiana* through mycorrhizal mycelium. *Botanical Gazette* **119**, 243–246.

Mikola, P. (1948). On the black mycorrhiza of birch. *Archivum Societatis Zoologicae-Botanicae Fennicae* **1**, 81–85.

Mikola, P. (1970). Mycorrhizal inoculations in afforestation. *International Review of Forestry Research* **3**, 123–196.

Miller, S. L., Durall, D. M., & Rygiewicz, P. T. (1989a). Temporal allocation of ^{14}C to extramatrical hyphae of ectomycorrhizal ponderosa pine seedlings. *Tree Physiology* **5**, 239–249.

Miller, S. L., Parsons, W. F. J., & Knight, D. H. (1989b). Small scale hydro-excavation of soil monoliths from a lodgepole pine forest. *Bulletin of the Ecological Society of America* **70**, 205–206.

Miller, S. L., Koo, C. D., & Molina, R. J. (1991). Characterization of red alder ectomycorrhizae—a preface to monitoring belowground ecological responses. *Canadian Journal of Botany* **69**, 515–531.

Molina, R. (1979). Pure culture synthesis and host specificity of red alder mycorrhizae. *Canadian Journal of Botany* **57**, 1223–1228.

Molina, R. (1981). Ectomycorrhizal specificity in the genus *Alnus*. *Canadian Journal of Botany* **59**, 325–334.

Molina, R. J., & Trappe, J. M. (1982). Patterns of ectomycorrhizal host specificity and potential among Pacific Northwest conifers and fungi. *Forest Science* **28**, 423–457.

Neuman, E. (1844). Notes on the supposed parasitism of *Monotropa hypopitys*. *New Phytologist* **1**, 297–299.

Newman, E. I., & Ritz, K. (1986). Evidence on the pathways of phosphorus transfer between vesicular-arbuscular mycorrhizal plants. *New Phytologist* **104**, 77–87.

Newman, E. I. (1988). Mycorrhizal links between plants: Their functioning and ecological significance. In: *Advances in Ecological Research 18* (Ed. by M. Begon, A. H. Fitter, E. D. Ford, & A. MacFadyn), pp. 243–270. Academic Press, London.

Newman, E. I. (1992). Interactions between plants: The role of mycorrhizae. *Mycorrhiza* **1**, in press.

Park, J. Y. (1970). A change in color of aging mycorrhizal roots of *Tilia americana* formed by *Cenococcum graniforme*. *Canadian Journal of Botany* **48**, 1339–1341.

Parke, J. L., Linderman, R. G., & Black, C. H. (1983). The role of ectomycorrhizas in drought tolerance of Douglas-fir seedlings. *New Phytologist* **95**, 83–95.

Parsons, W. F. J., Miller, S. L., & Knight, D. H. (1990). Nitrogen transformations following disturbances of different scales in Rocky Mountain lodgepole pine forests (abstract). *Bulletin of the Ecological Society of America, Supplement* **71**, 280.

Pearson, V., & Read, D. J. (1973). The biology of mycorrhiza in the Ericaceae. II. The transport of carbon and phosphorus by the endophyte and the mycorrhiza. *New Phytologist* **72**, 1325–1331.

Perry, D. A., Amaranthus, M. P., Borchers, J. G., Borchers, S. L., & Brainerd, R. E. (1989a). Bootstrapping in ecosystems. *BioScience* **39**, 230–237.

Perry, D. A., Margolis, H., Choquette, C., Molina, R., & Trappe, J. M. (1989b). Ectomycorrhizal mediation of competition between coniferous tree species. *New Phytologist* **112**, 501–511.

Rayner, A. D. M., & Todd, N. K. (1982a). Ecological genetics of basidiomycete populations in decaying wood. In: *Decomposer Basidiomycetes, 4th Symposium of the British Mycological Society* (Ed. by J. Frankland, B. Heger, & M. Swift), pp. 129–142. Cambridge University Press, Cambridge.

Rayner, A. D. M., & Todd, N. K. (1982b). Population structure in wood-decomposing basidiomycetes. In: *Decomposer Basidiomycetes, 4th Symposium of the British Mycological Society* (Ed. by J. Frankland, B. Heger, & M. Swift), pp. 109–128. Cambridge University Press, Cambridge.

Read, D. J., & Malibari, A. (1978). Water transport through mycelia and nutrient cycling in plant communities. In: *Proceedings of the International Union of Forestry Research Organizations Symposium* (Ed. by A. Riedacker), pp. 410–424. Nancy, France.

Read, D. J., Francis, R., & Finlay, R. D. (1985). Mycorrhizal mycelia and nutrient cycling in plant communities. In: *Ecological Interactions in Soil* (Ed. by A. H. Fitter, D. Atkinson, D. J. Read, & M. B. Usher), pp. 193–217. Blackwell Scientific Publications, Oxford.

Reid, C. P. P., & Woods, F. W. (1969). Translocation of C^{14}-labeled compounds in mycorrhizae and its implications in interplant nutrient cycling. *Ecology* **50,** 179–181.

Reid, C. P. P., Kidd, F. A., & Ekwebelam, S. A. (1983). Nitrogen nutrition, photosynthesis and carbon allocation in ectomycorrhizal pine. *Plant and Soil* **71,** 415–432.

Salsac, L., Mention, M., Plassard, C., & Mousain, D. (1982). Observations on the nitrogen nutrition of ectomycorrhizal fungi. In: *I.N.R.A. Publ. Colloq. 13* (Ed. by S. Gianinazzi, V. Gianinazzi-Pearson, & A. Trouvelot), pp. 129–140. INRA, Paris.

Schwab, S. M. & Loomis, P. A. (1987). VAM effects on growth of grasses in monocultures and mixtures. In: *Mycorrhizae in the Next Decade: Practical Applications and Research Priorities.* (Ed. by D. M. Sylvia, L. L. Hung, & J. H. Graham), p. 264. IFAS, Gainesville, FL.

Skinner, M. F., & Bowen, G. D. (1974a). The penetration of soil by mycelial strands of ectomycorrhizal fungi. *Soil Biology and Biochemistry* **6,** 57–61.

Skinner, M. F., & Bowen, G. D. (1974b). The uptake and translocation of phosphate by mycelial strands of pine mycorrhizas. *Soil Biology and Biochemistry* **6,** 53–56.

Snellgrove, R. C., Splittstoesser, W. E., Stribley, D. P., & Tinker, P. B. (1982). The distribution of carbon and the demand of the fungal symbiont in leek plants with vesicular-arbuscular mycorrhizas. *New Phytologist* **92,** 75–87.

Söderström, B., & Read, D. J. (1987). Respiratory activity of intact and excised ectomycorrhizal mycelial systems growing in unsterilized soil. *Soil Biology and Biochemistry* **19,** 231–236.

Söderström, B., Finlay, R. D., & Read, D. J. (1986). Qualitative analysis of carbohydrate contents of mycorrhizal mycelia after feeding of interconnected host plants with $^{14}CO_2$. In: *Mycorrhizae: Physiology and Genetics* (Ed. by V. Gianinazzi-Pearson & S. Gianinazzi), pp. 307–309. INRA, Paris.

Spinner, S., & Haselwandter, K. (1985). Protein as nitrogen sources for *Hymenoscyphus* (=*Pezizella*) *ericae*. In: *Proceedings of the 6th North American Conference on Mycorrhizae* (Ed. by R. Molina), p. 442. U.S.D.A. Forestry Service Research Laboratory Publication, Corvallis, Oregon.

Stahl, P. D., & Smith, W. K. (1984). Effects of different geographic isolates of *Glomus* on the water relations of *Agropyron smithii*. *Mycologia* **76,** 261–267.

Stribley, D. P., & Read, D. J. (1974). The biology of mycorrhiza in Ericaceae. III. Movement of carbon-14 from host to fungus. *New Phytologist* **73,** 731–741.

Theodorou, C. (1971). The phytase activity of the mycorrhizal fungus *Rhizopogon luteolus*. *Soil Biology and Biochemistry* **3,** 89–90.

Thompson, W., & Rayner, A. D. M. (1983). Extent, development and functions of mycelial cord systems in soil. *Transactions of the British Mycological Society* **81,** 333–345.

Tilman, D. (1982). *Resource Competition and Community Structure*. Princeton University Press, Princeton, N.J.

Tilman, D. (1988). *Dynamics and Structure of Plant Communities*. Princeton University Press, Princeton, N.J.

Tommerup, I. C., & Abbott, L. K. (1981). Prolonged survival and viability of VA mycorrhizal hyphae after root death. *Soil Biology and Biochemistry* **13,** 431–433.

Trojanowski, J., Haider, K., & Huettermann, A. (1984). Decomposition of carbon-14-labeled lignin, holocellulose and lignocellulose by mycorrhizal fungi. *Archives of Microbiology* **139,** 202–206.

Vogt, K. A., Grier, C. C., Meier, C. E., & Edmonds, R. L. (1982). Mycorrhizal role in net production and nutrient cycling in *Abies amabilis* ecosystems in western Washington. *Ecology* **63,** 370–380.

Woods, F. W., & Brock, K. (1964). Interspecific transfer of Ca^{45} and P^{32} by root systems. *Ecology* **45,** 886–889.

Yocum, D. H. (1983). The costs and benefits to plants forming mycorrhizal associations. Ph.D. Dissertation, State University of New York at Stony Brook.

10

Interactions with the Soil Fauna

A. H. Fitter and I. R. Sanders

SUMMARY. Soil animals are known to have important effects on the population dynamics of soil fungi, but their interaction with mycorrhizal fungi has not been extensively studied. Mycorrhizal fungi may comprise a large proportion of the soil fungal biomass, however, and it is to be expected that the animal–mycorrhizal interaction is of considerable significance. The spatial and temporal patterns exhibited by the fungi and soil animals mean that they come into direct contact frequently, and there are observations of grazing by both collembola and aphids. Grazing of external mycelium by soil animals might limit its development, simply disconnect it from the internal or attached mycelium, or even stimulate fungal growth. In the latter case, there might be benefits to the host plant, but in either of the other cases, the consequences for the host are likely to be deleterious. There is experimental evidence that such interactions do occur, and grazing has been shown to eliminate any benefits derived by the host plant from the symbiosis. Very little work has yet been performed under field conditions, however, and the ecological significance of mycorrhiza–fauna interactions remains uncertain.

Introduction

The soil is a region of great biological complexity and often also of intense activity. In particular, trophic interactions in soil appear to be more complex than in many aboveground habitats, with food-webs apparently having greater connectivity. Coleman (1985) displays a food-web model in which chains with up to 8 links appear possible, for example, roots–bacteria–flagellates–amoebae–omnivorous nematodes–predatory nematodes–nematophagous mites–predatory mites. Since the latter are presumably prey for larger organisms, it is possible to envisage as many as 10 links in a chain. It is unlikely that significant energy and material flows occur through such long chains, but their possible existence indicates that the structure and functioning of belowground communities may be controlled in importantly different ways than those above ground.

The principal source of energy to belowground communities is normally roots, and a significant proportion of the carbon sent to the root system by many plants is diverted to mycorrhizal fungi. These fungi therefore represent an important link in the belowground food-web that has hitherto received little attention. The soil fauna contains numerous fungivorous species, particularly in the nematodes, mites, and collembola, that are potentially important consumers of mycorrhizal fungi.

Nematodes are small (mostly in the range 0.1–1.0 mm), unsegmented worms that feed either by stylets inserted into the food or by ingesting it through an enlarged mouth. The latter is commonest among carnivorous species. Because of their size range, they can be important feeders on plant cells and on fungal hyphae, both of which tend to have diameters around 0.01 mm. Collembola are minute, wingless insects ranging from 0.1 up to nearly 10 mm in length; mites (Acarina) are arachnids with a similar size range, of which the most important fungivorous group is the Cryptostigmata. Both cryptostigmatid mites and collembola have biting mouthparts.

In this chapter we assess the evidence that these soil animals interact with mycorrhizal fungi sufficiently strongly to affect the interaction between fungus and host plant. Because most work on mycorrhizas has been performed in sterilized soil under laboratory conditions, where necessarily animals are excluded, few workers have been aware of the potential significance of the interaction. Since so little work has been performed on interactions between the soil fauna and mycorrhizal fungi, we shall examine first the effects of animals on soil fungi in general, seeking general patterns of response.

Population Patterns of Mycorrhizas and Animals

Predation and Dynamics in Nonmycorrhizal Fungi

The interactions between the major components of the soil—roots, soil fauna, and microorganisms—are complex (Coleman et al., 1978; Coleman, 1985), and the interaction between soil fauna and fungi must be considered in the context of the complete ecosystem. Hunt et al. (1987) have used a systems approach to this problem, identifying the belowground food web in a short-grass prairie. Other workers have concentrated in greater detail on the importance of specific interactions on the cycling of phosphorus (Stuart & McKercher, 1982), nitrogen (Parton et al., 1984), and carbon (Elliot et al., 1984) and also the consequences of these interactions on population dynamics (Anderson, 1978; Moore, 1988; Seastedt et al., 1988).

Predation on soil fungi by soil fauna can directly influence the dynamics of both of these components while indirectly influencing other components of the soil biota and processes within the soil, e.g., nutrient cycling (Anderson & Ineson, 1984). The mechanisms by which soil fauna affect the composition of fungal communities have been classified by Visser (1985). These are comminution, or the fragmentation and mastication of plant debris which includes microflora growing within plant residues, grazing of the microflora, and dispersal of microbial propagules. These direct interactions bring about changes to the composition of the soil microflora and fauna with respect to the size of populations present, the relative abundance of species, and their spatial distribution.

Direct grazing of fungi could either be deleterious by reducing fungal biomass or advantageous by the stimulation of greater productivity. Both of these situations could, in turn, lead to changes in respective grazer populations. The latter, termed direct nonsymbiotic mutualism (Moore, 1988), is thought to be caused by the removal of growth-inhibitory, secondary metabolites produced during hyphal senescence. This enhanced fungal growth and activity has been observed in arthropod–fungus (Hanlon, 1981) and nematode–fungus interactions. Whether grazing becomes mutualistic depends on grazing pressure, growth rate and nutritional quality of the grazed organism, and the success of its defence strategy (Visser, 1985).

Feeding preferences of soil fauna could also be significant in determining fungal species composition. Selectivity of some grazers to different fungal species could result in competitive advantages arising for less preferred species. The collembolan *Onychiurus subtenuis* appears to be important in the competitive interaction of two commonly occurring litter fungi, Basidiomycete 290 and Sterile dark 298 (Parkinson et al., 1979). In experimental microcosms Basidiomycete 290 was more successful in colonising leaf macerate than Sterile dark 298, in the absence of *O. subtenuis*. However, in the presence of *O. subtenuis* the colonising ability of Sterile dark 298 was significantly reduced; in other words, *O. subtenuis* reinforces an existing competitive relationship.

Grazing activity could also influence the spatial distribution of fungal populations. Newell (1984a) has shown that in the presence of two coexisting decomposer basidiomycetes *Mycena galopus* and *Marasmius androsaceus*, the collembola *Onychiurus latus* shows a grazing preference for the latter. In laboratory experiments, *M. androsaceus* was able to colonise litter from the L and F1 horizons more than twice as fast as *M. galopus*, although in the field *M. androsaceus* was restricted almost exclusively to the L horizon. Further experiments investigating the interactions of these three species in the field (Newell, 1984b) support the hypothesis that selective grazing by collembola alters competition between these two fungal species. This may be an important factor in determining their vertical distribution within leaf litter.

Fungal Biomass

There is clear evidence for the interaction between soil animals and fungi, but the evidence reviewed above refers exclusively to saprotrophic fungi. Such fungi are likely to be an important food source for many fungivorous soil animals; whether mycorrhizal fungi are a significant resource will depend in part on their contribution to the total soil fungal biomass. This has been estimated by direct measurement of fungal biomass occurring in the soil, and by calculations based on other variables such as whole plant and root biomass production and the allocation of carbon to roots.

Direct measurements of vesicular-arbuscular mycorrhizal (VAM) fungal spore biomass have been recorded from pot experiments. Sieverding et al. (1989) have extrapolated values recorded in pot experiments to field conditions. They estimate dry matter production in tropical soils to be 20–200 kg/ha/yr, although extreme values could be as high as 1000 kg/ha/yr. The uncertainty associated with such extrapolations is large. Dry weights of VAM fungal hyphae have also been measured in pots for different fungal species (Sanders et al., 1977; Bethlenfalvay et al., 1982). However, obtaining an accurate measure of VAM fungal biomass in natural ecosystems is difficult, since this involves measurements or estimations of both the internal and extramatrical hyphae. The proportion of the internal and external components of mycorrhizas will also be important in relation to the cost to the host plant in terms of carbon allocation.

It has been suggested that the total fungal biomass of VAM is between 5 and 20% of root weight (Smith & Gianinazzi-Pearson, 1988). If this figure is considered together with values for productivity and allocation of resources to plant roots then such a value would suggest that VAM comprise a large resource within the soil ecosystem. For example, in woodland ecosystems root biomass comprises 15–25% of total tree biomass, while the percentage biomass of roots in grassland and tundra ecosystems attains 75–98% of total biomass (Fogel, 1985). Other studies have considered the cost to the plant in terms of carbon allocation to mycorrhizas. It has been estimated that between 6 and 10% of total plant photosynthate is diverted to mycorrhizas (Snellgrove et al., 1982; Koch & Johnson, 1984). However, the total amount of carbon that is diverted cannot be considered as being an available resource. Roots that contain mycorrhizal fungi have a higher respiration rate than uninfected roots. Part of this increase can be attributed to the fungi and part is suggested to be a response in infected plant tissue (Smith & Gianinazzi-Pearson, 1988). Thus, it is likely that although figures for carbon allocation to mycorrhizas are high, only part of this large allocation of resources to mycorrhizal fungi is available to mycorrhizal grazer organisms.

The biomass of ectomycorrhizal (ECM) fungi has also been investigated in relation to the productivity of coniferous forests (Fogel & Hunt, 1979, 1983). In stands of Douglas fir *Pseudotsuga menziesii,* ECM have been shown to comprise 6% of total tree biomass and the mycorrhizal standing crop is estimated to range from 1 to 25 metric tons per ha (Fogel, 1985).

It appears that both VAM and ECM constitute a large, available resource for potential grazer organisms in the soil ecosystem. In grassland and forest ecosystems a substantial proportion of the net primary productivity is diverted to mycorrhizas. However, further investigation is required into how much of this carbon allocation results in fungal biomass, and how much of this is a suitable food source for fauna that may graze on it.

Palatability of Mycorrhizal Fungi

Although it is clear that an extensive amount of mycorrhizal fungal biomass occurs in the soil, it is not known how much of this resource is suitable for or palatable to consumers. The only study investigating the palatability of mycorrhizal fungi has been undertaken by Shaw (1985, 1988). Shaw (1988) carried out a series of choice experiments among pairs of 12 different fungal species (8 of which were ectomycorrhizal) using the collembola *Onychiurus armatus* as the grazer organism. By subtracting the number of fungal isolates that were significantly less preferred from those that were significantly more preferred, Shaw (1988) was able to establish the relative palatability of the 12 fungal species to the collembola. These were then ranked to give palatability rankings or a hierarchy of preference. Of the 12 fungal isolates, the ectomycorrhizal ones ranged from being the most palatable, i.e., *Lactarius rufus,* to those that are relatively unpalatable, i.e., *Pisolithus tinctorius* and *Hebeloma crustuliniforme.*

This investigation gives no indication of what factors affect palatability, although Shaw (1988) suggests that less preferred isolates may allocate a larger proportion of their metabolites to herbivore deterrence. However, the results of such choice experiments must be considered with caution. It is possible that a species of ECM-forming fungus could differ in its palatability when grown in culture and when growing in symbiotic association with plant roots. Additionally, in Shaw's choice experiment, *O. armatus* was only given the choice of ectomycorrhizal fungi and other fungal species occurring in that particular soil. Thus, although the results show that ECM are comparatively palatable to collembola, with respect to other fungi in the soil, it does not provide insight into whether this group of soil-dwelling animals would readily consume this resource in the natural environment.

Although the palatability of VAM fungi has not been investigated, hyphae of these fungi have been recorded in the guts of many animals suggesting that they are a worthwhile food source. However, this assumes that ingested hyphae are subsequently digested: although many soil animals also ingest VAM fungal spores, some of these pass through the gut and have been shown to be viable after extraction from the animal's faeces (Rabatin & Stinner, 1988).

Spatial and Temporal Patterns of Mycorrhizas and Animals

Since mycorrhizal fungi are a readily available and apparently palatable resource, they are likely to be grazed. The extent of grazing would depend on the spatial and temporal patterns of roots, mycorrhizas, and the relevant soil fauna in soil.

Traditionally, studies of the patterns of these organisms have been either from a zoological or botanical/mycological viewpoint; only a few more recent studies have attempted to consider the patterns of these organisms jointly. Much of the literature concentrates on soil animals and VAM and few studies have investigated the spatial and temporal patterns of soil fauna with ECM.

The spatial and temporal distribution of cryptostigmatid mites has been investigated in detail. Usher (1976) showed that these mites form distinct aggregations and that their spatial distribution was nonrandom. A recent study by Fogel and Lussenhop (unpublished data) investigated the distribution of soil invertebrates in relation to plant roots within the soil. They took a census of soil invertebrates along transects at three different times of the year. The observations were made through the windows of a soil biotron (a large root observation chamber) and record the density of individuals occurring in the soil and on plant roots in a mixed northern forest in Michigan. They found that the density of Collembola, Acarina (Tydeidae), and Enchytraeidae was consistently higher on roots than in the bulk soil during July, August, and October, to the extent that these animals were almost exclusively associated with roots. Other invertebrate groups such as the Nematoda, Acarina (Mesostigmata and Oribatida), and Protura also exhibited higher densities on roots than in the soil, although these were not consistent throughout the period of study.

Studies of spatial and temporal patterns of VAM have shown that they are abundant in a variety of natural ecosystems although levels of infection vary greatly throughout the year (Brundrett & Kendrick, 1988; McGonigle, 1987; Rosendahl et al., 1989) and between different plant species. However, at present, patterns of VAM occurrence are not well understood since none of the published data regarding these patterns covers more than 1 year.

A thorough investigation of both temporal patterns of VAM, roots, and soil microarthropods has been carried out by McGonigle (1987). He observed changes in coarse and fine endophyte infection in the roots of eight species of a seminatural grassland in North Yorkshire, England, for a period of 1 year. At the same field site, he also recorded the spatial arrangement of endophytes on the plant roots and the density of collembola and acari at different depths in the soil. Mycorrhizal infection was largely restricted to the top 6 cm of the soil, and the low density of roots below this level implied very low populations of VAM fungi. Collembola and acari also occurred at their greatest densities in the top 6 cm of the soil, supporting the supposition that the grazing of VAM is ecologically realistic.

However, McGonigle's survey shows only that the interaction can occur, since no gut content analyses were carried out to show that these animals were eating fungal hyphae from the soil. It is possible that the collembola and acari were not utilizing VAM fungi and that the aggregation of these animals in the top 6 cm of the soil occurred because they were exploiting another available resource, e.g., plant roots, other soil animals, or other organisms associated with the rhizosphere. The omnivorous nature of some soil fauna that have been considered as fungivorous has been demonstrated. For example, Walter et al. (1986) have shown that fungivorous mites from a short–mid-grass prairie ecosystem will readily consume nematodes.

Other evidence exists for the co-occurrence of VAM and possible grazing animals. Rich and Schenk (1981) have shown that seasonal changes in VAM fungal spore density in the soil are similar to changes in the density of plant-parasitic nematodes. However, VAM fungal spore density is not a reliable measure of VAM occurrence. The association of endoparasitic nematodes and VAM fungi appears to be complex. Evidence shows that these organisms do occur in close proximity and can have significant effects on population densities and VAM sporulation. However, they appear to be mutually inhibitory; mycorrhizal fungi do not colonise roots infected with nematodes or *vice versa* (Ingham, 1988).

Evidence for the co-occurrence of likely mycorrhizal grazers and ECM fungi is extremely scarce. Cromack et al. (1988) found that arthropods and nematodes were more abundant on rather than off ECM mycelial mats (*Hysterangium setchellii*), though with no apparent difference in biomass (this is probably a sampling problem rather than a real difference in response between numbers and biomass), and both Riffle (1967) and Sutherland and Fortin (1968) demonstrated nematode feeding in culture. There are several other, rather anecdotal, accounts of animals feeding on ECM fungi, but as yet no systematic studies have been attempted. To get a clearer understanding of the interactions between soil fauna and ECM more detailed investigations are required into the spatial and temporal patterns of these organisms and the possible consequences of their interactions.

Direct Evidence of Grazing on Mycorrhizal Fungi

Although it would seem that finding direct evidence of grazing on mycorrhizal fungi should be relatively easy, surprisingly few studies have been able to demonstrate this, perhaps because such observational science is no longer fashionable. A number of investigations have shown that the ingestion of VAM fungal spores occurs in a variety of different animals (e.g., Rabatin & Stinner, 1985, 1988), since they can easily be identified from gut contents or faeces. In most cases spore digestibility or palatability has not been investigated. To distinguish VAM fungal hyphal remains from other fungal hyphae in gut contents is more difficult. For example, the results of Warnock et al. (1982) imply that grazing of hyphae occurred. In this investigation greater numbers of the collembola *Folsomia candida* from +VAM pots were shown to contain fungal hyphae than those from −VAM pots. Although it appears likely that the hyphae were of mycorrhizal origin this could not be proved. With the development of techniques whereby mycorrhizal fungi can be identified by isoenzyme separation, following electrophoresis, the future identification of VAM hyphae in gut contents may be possible.

Consequently, evidence for hyphal grazing has relied on direct observation or experiments that have unavoidably created a highly artificial experimental environment. Of these, the observation of aphids (probably *Pemphigus piceae*) feeding on ECM of *Pseudotsuga menziesii* (Douglas fir) (Zak, 1965) is one of the only records of direct grazing on ECM. The investigation by Shaw (1988) also confirms that grazing of ECM hyphae occurs.

With respect to VAM fungi, direct evidence of grazing has been demonstrated only by Moore et al. (1985). In their investigation, four different microarthropods (*Folsomia candida, Onychiurus encarpatus, O. folsomi*, and *Proisotoma minuta*) were released onto agar with germinated spores of either *Gigaspora margarita, Gigaspora rosea, Glomus fasciculatum*, or *Glomus mosseae*. Feeding was observed only on *G. rosea* and *G. fasciculatum*. The hyphae were noted to have been severed rather than entirely ingested. Unfortunately, these observations were not supported by gut content analysis, which would have revealed how much of the hyphae was ingested. Although the investigation of Moore et al. (1985) demonstrates that grazing of VAM hyphae can occur, it is difficult to speculate from this evidence on whether the grazing of VAM hyphae occurs in the field and to the extent of this interaction.

Consequences of Fauna–Mycorrhiza Interactions

Types of Interaction

The soil fauna impinge on mycorrhizal fungi in three main ways: by affecting dispersal and the development of both the internal mycelium (colonisation) and the external mycelium. Of these, the effect on colonisation is a special case, since it can only be brought about by root-inhabiting animals, which in practice means pathogenic nematodes. These are known to influence the development of the internal mycelium in VAM-colonised plants (see the very thorough review by Ingham, 1988), but the mechanisms of the interaction and its significance are not yet clear. There seems to be little evidence that ECM fungi are adversely affected in this way, though some interactions certainly occur.

Effects on dispersal are better documented. VAM fungal spores have been found in the guts of a wide range of soil animals, including ants (Formicidae) and earthworms (Lumbricidae; McIlveen & Cole, 1976), woodlice (Isopoda), millipedes (Diplopoda), moth larvae (Lepidoptera) and beetles (Coleoptera; Rabatin & Stinner, 1988), and springtails (Collembola; Warnock et al., 1982; Finlay, 1985; Moore et al., 1985). It is unclear whether any of these organisms selectively feed on VAM fungal spores, but Rabatin and Stinner (1988) report that woodlouse and millipede faeces can act as inoculum to infect alfalfa (*Medicago sativa*) roots.

Some ECM fungi are certainly dispersed by animals. Fogel and Trappe (1978) demonstrated the importance of small mammals in this regard, especially for hypogeous fungi such as truffles. Epigeous fungi, in contrast, have such efficient wind-dispersal systems that it seems implausible that animals play more than a trivial role.

The ecological significance of transport by soil animals within the soil is unclear. Many VAM and ECM fungal species are widespread geographically, and it is often assumed that they are effectively ubiquitous. Sampling within defined areas, however, may reveal complex spatial patterns of inoculum density, and these might possibly arise partly through animal transport. Where fungal species interact differently with various plant species, such patterns might have important implications for community structure.

It seems likely, however, that the greatest influence exerted by the soil fauna on mycorrhizas is the consumption of the mycelium. Grazing is known to be a major factor in controlling the structure and function of the aboveground parts of ecosystems, and since it can both be observed and measured with ease, it has been intensively studied there. Grazing can, for example, cause the death of plants, reduce their competitive ability or in consequence increase that of neighboring plants, alter resource allocation patterns and especially reduce reproductive success, or even stimulate growth in isolated plants by the removal of apical dominance. Grazing on fungal mycelia can, as shown above, have a similar range of effects.

Consequences of Mycorrhizal Grazing

There are three main possible effects of grazing on a mycorrhizal fungus:

1. Grazing might limit the development of the external mycelium. This would occur if the animals were well distributed in soil or if the fungal tissue occurred in sufficiently dense patches to permit high animal population densities to persist.

2. Grazing might result in the disconnection of the external mycelium from the internal mycelium or sheath. This would occur if the animals predominantly fed in the rhizosphere, and is possible because of the unique structure of the mycorrhizal mycelium with a limited number of hyphal strands connecting the internal and external parts.

3. Grazing might stimulate fungal growth, for example, by recycling minerals locked up in senescing tissue, by importing minerals and depositing them in faeces, or by removal of growth inhibitors (see above).

The effects of such interactions on the plants are predictable. If the external mycelium is severely reduced or simply disconnected, then the internal mycelium or sheath will become parasitic, since it will still exert a demand for carbon and, presumably, other nutrients, but can offer nothing in exchange. Since this part of the mycelium may represent a significant proportion of root biomass (up to

20%), this cost may be not inconsiderable. This situation would continue until one or more of the following occurred:

1. The external mycelium might regenerate from the internal mycelium or sheath. There seems no doubt that this can occur, since root pieces can be used as inoculum, but it would certainly represent a very significant additional resource cost on the host plant.
2. Where disconnection has occurred, it is possible that some reconnection process might take place. This would involve anastomosis presumably.
3. There might be reinfection of the root system by the surviving or new external mycelium. This process requires the development of a new internal mycelium and would leave the question of the fate of the original internal mycelium unknown.

In all these cases, grazing could be expected to produce significant deleterious effects on plants, which would be specific to mycorrhizal individuals. In contrast, if grazing stimulated growth of the fungus, the predictions are more complex. It is possible that this might result in increased benefits to the host, on the grounds that if some mycelium can give a small benefit, more must give a greater one. This would imply that, as faunal population density and hence grazing pressure increased, there would be an initial benefit to the plant and then a phase of increasing disadvantage. It is equally possible, however, that an increased external mycelium, which would certainly represent a greater carbon drain on the host, would not bring any proportional benefits. This would be the case if, for instance, the growth of the plant was not limited by resources supplied by the fungus. Clearly, the net effect of grazing on the mycorrhizal association depends critically on the precise nature of the grazer–fungus interaction and the responses of the fungi to grazing. We need to examine the experimental evidence for these. Unfortunately, most of it is indirect and many critical experiments remain to be done.

Experimental Evidence

Virtually all the experimental work on mycorrhizal grazing is on VAM systems. This is odd, since ECM mycelia are very much easier to work with and their effects on plant performance are usually more dramatic. In addition, ECM mycelia represent a very large biomass in soil, are an obvious food source for the soil fungi, and are known to be eaten by soil animals (see above).

VAM fungal systems, on the other hand, have been relatively well studied, and these studies will be discussed in more detail here; it seems reasonable to suppose that similar principles will govern the interactions in both VAM and ECM mycorrhizas. Most work on VAM–faunal interactions has been in pot cultures, in which the typical, grossly simplified soil "ecosystem" of the pot experiment (sterilised soil, mycorrhizal inoculum with associated undetermined microbes, plant with different set of undetermined microbes) is made a little more realistic by the addition of a single animal species. Such experiments have been used to demonstrate pronounced effects of collembola and nematodes, though mites, another large group of soil fungivores, have not yet been used successfully.

The typical result of such an experiment (or at least of those reported in the literature) is a reduction in the benefit to the plant of being mycorrhizal. This was the case when Salawu and Estey (1979) added the nematode *Aphelenchus avenae* to soybean cultures inoculated with a *Glomus* sp. In the absence of nematodes, mycorrhizal plants grew better than nonmycorrhizal ones, but when nematodes were present, shoot yield was reduced by 40% and nodulation by nearly 90%. Similarly, Warnock et al. (1982) found that the collembola *Folsomia candida* reduced the yield of mycorrhizal leek plants (infected with *Glomus* strain E3) to levels similar to those of nonmycorrhizal plants (Fig. 10.1). P concentrations were actually lower in mycorrhizal plants with collembola than in nonmycorrhizal plants, while P inflows declined from 68 fmol/cm/sec in mycorrhizal plants without collembola to 17 when collembola were added. These figures compared to values of 23 and 10 for nonmycorrhizal plants; none of the three lower values was significantly different from each other.

These differences were not the result of different intensities of infection in the roots, since plants were inoculated 8 weeks before collembola were introduced, and between 74 and 84% of the root length was infected in all cases after 12 weeks. Nor could direct effects of collembola on the plants, such as root grazing, be implicated, since nonmycorrhizal plants were unaffected and root weight ratios unaltered. The guts of the collembola contained fungal material, but nothing recognisably of root origin, and the population density of collembola in mycorrhizal pots was three times that in nonmycorrhizal pots after 12 weeks. The two main treatments were combined factorially with a third: the addition of soil leachings, to restore soil bacteria and other fungi. This increased plant growth (by between 25 and 90%, depending on treatment) but had no effect on P inflows or collembola population growth.

Taken together, all this evidence strongly points to the likelihood that collembola and nematodes can influence plants by damaging the external mycelium of VAM fungi, so reducing P inflows and the benefit to the plant of infection. In some experiments, however, there may be no effect of adding animals: Hussey and Roncadori (1981) allowed the nematode *Aphelenchus avenae,* which Salawu and Estey (1979) had found seriously to inhibit VAM function in a soybean–*Glomus* association, to feed on cotton infected with *Gigaspora margarita* and *Glomus etunicatus*. They could only find inhibitory effects at unrealistically high nematode densities, which could not in any case be sustained.

Some of the most detailed studies are those of Finlay (1985), who performed a series of experiments in which he varied the species of plant, fungus, and collembola, and also investigated the role of collembola density. He was able to

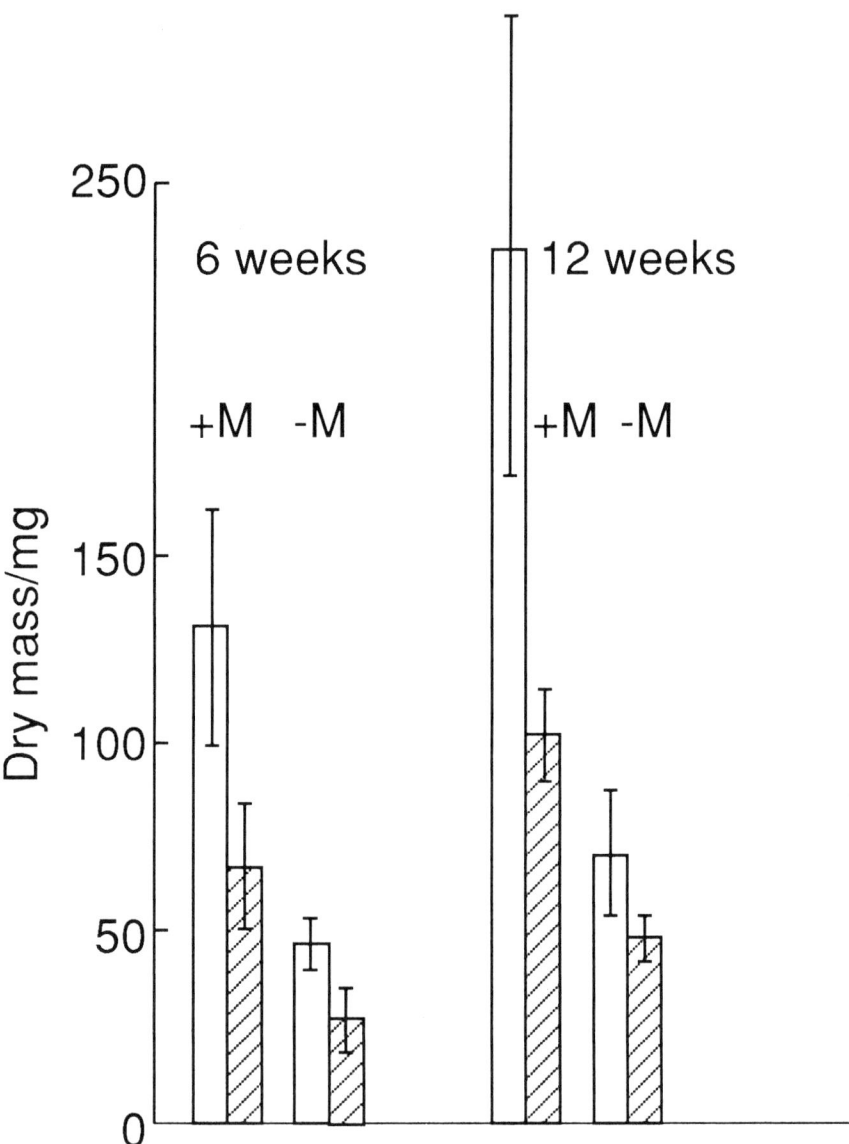

Figure 10.1. Influence of the collembola *Folsomia candida* on dry mass production in leek *Allium porrum*, infected (+M) or not (−M) with *Glomus* strain E3. Hatched columns represent pots containing *F. candida;* bars represent 95% confidence limits. Data from Warnock et al. (1982).

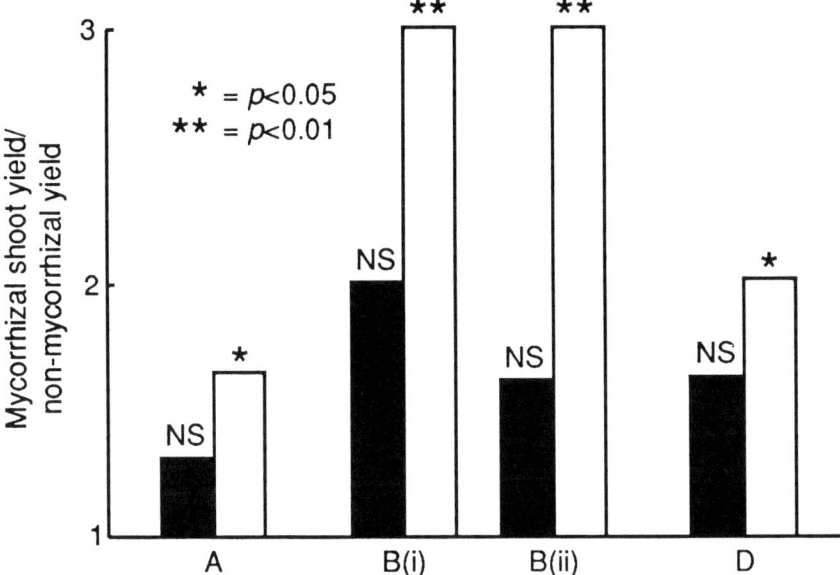

Figure 10.2. Mycorrhizal stimulation of shoot yield in a range of laboratory experiments (A–D) with (■) and without (□) collembola added to pots. Significance levels refer to t tests between the means of ln-transformed mycorrhizal and nonmycorrhizal treatments; the data are presented here as the quotient of yield in mycorrhizal pots to that in nonmycorrhizal pots. The experimental details were as follows (plant, fungus, collembola): (A) *Allium porrum, Glomus tenue, Onychiurus ambulans;* [B(i)] *A. porrum, G. fasciculatum, Folsomia candida;* [B(ii)] *A. porrum, G. fasciculatum, O. ambulans;* (D) *Trifolium pratense, G. caledonium, O. ambulans*. From Finlay (1985).

confirm the stimulation of collembola growth by mycorrhizal infection, and found that increased plant size brought about by phosphorus fertilisation was ineffective in promoting collembola numbers, whereas infection by a mycorrhizal strain (*Glomus tenue*) that did not promote plant growth, did stimulate animal numbers. When considered with the demonstration that *Onychiurus ambulans* guts contained fungal spores and hyphae, these results confirm that the effect of collembola on the association is initially on the fungus not the root.

In four of five experiments, Finlay (1985) showed that mycorrhizal infection stimulated plant growth only in the absence of collembola (Fig. 10.2). Since these experiments used two species of plant, two of collembola and six of *Glomus* (not all combinations are shown in Fig. 10.2), their generality seems good. Most striking, however, was the nonlinearity in the relationship between plant P uptake and collembola density (Fig. 10.3): low densities of collembola increased P uptake, whereas high densities decreased it. This is what would be expected if either mycorrhizal biomass was supraoptimal (from the plant's standpoint), so that grazing initially brought it to an optimum level and then reduced it below

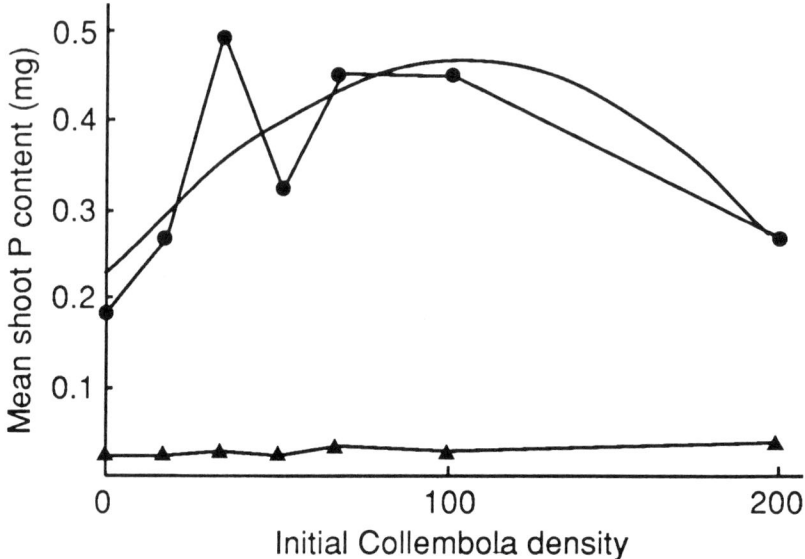

Figure 10.3. Relationship between mean shoot P content and initial collembola density for mycorrhizal (●) and nonmycorrhizal (▲) *Allium porrum* plants grown with seven densities of *Onychiurus ambulans* and with *Glomus fasciculatus* as the mycorrhizal symbiont. Data are from a pot experiment (Finlay, 1985). The line shown for mycorrhizal plants is a fitted quadratic relationship.

that, or grazing stimulated fungal growth and initially raised it from a suboptimal level. The former seems, however, to be more likely to give the observed effects (see Fig. 10.4).

Recently, Harris and Boerner (1990) completed experiments very similar to those of Warnock et al. (1982) and Finlay (1985), except that the plant was the woodland herb (and occasional weed) *Geranium robertianum*. They too found that collembola reduced VAM infection and that the effects of initial collembola density on plant growth were nonlinear. There also was some evidence for parallel changes in P uptake. Unfortunately they did not measure collembola densities at the end of their experiments and so it is not possible to conclude that the collembola were actively utilizing fungal resources.

In pot experiments, therefore, it has been demonstrated that soil animals, especially collembola and nematodes, feed on the external mycelium of VAM fungi and in so doing reduce the P inflow to plants relative to animal-free conditions. Plants do not normally grow in the absence of soil animals, so it is reasonable to suppose that these effects will be of particular importance under field conditions. Unfortunately, a conclusive demonstration of the effect in the field is technically difficult.

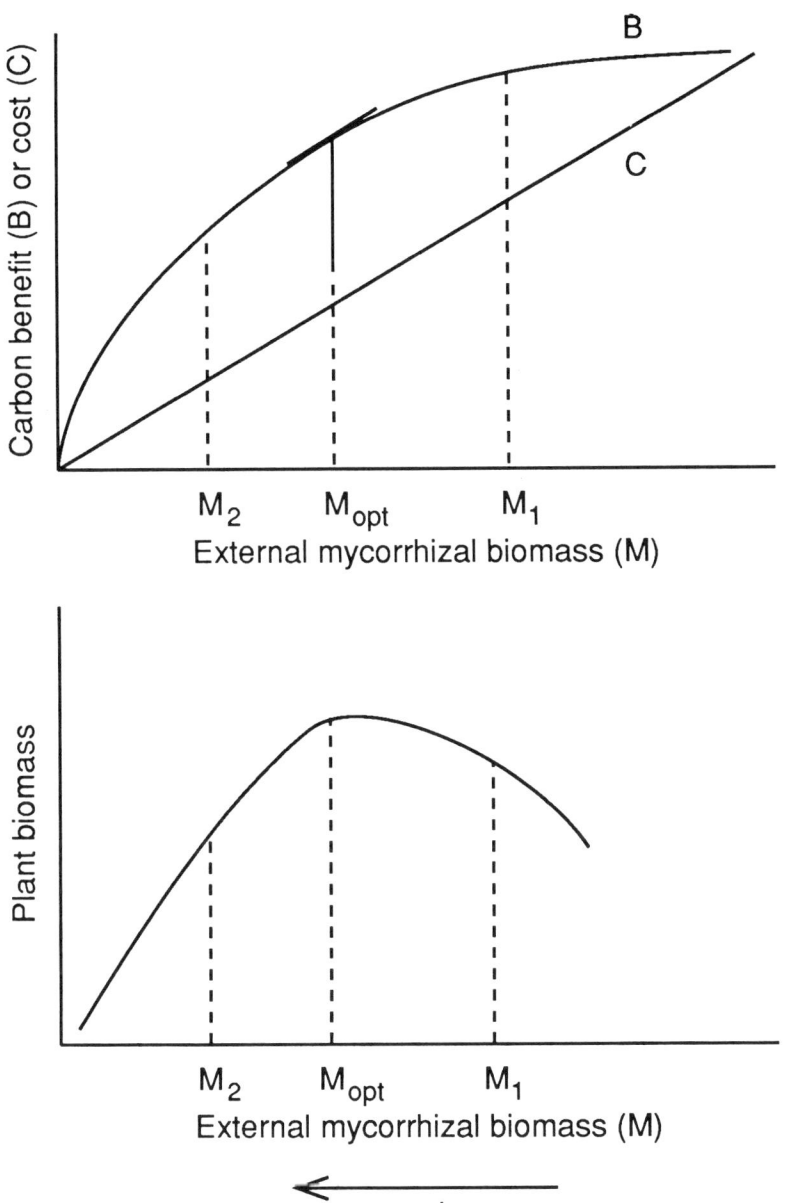

Figure 10.4. Diagrams to illustrate cost–benefit analysis of mycorrhizal grazing. (a) Carbon cost (C) is assumed to be a linear function of external mycorrhizal biomass. Carbon benefit (B) is the additional carbon fixed because of increased P uptake. The optimum mycorrhizal biomass is found where $dB/dM = dC/dM$. (b) Grazing reduces external mycorrhizal biomass from its initial value (M_1) through M_{opt} to a lower value (M_2), with consequent effects on plants growth.

Field Experiments

Pot experiments that include soil animals represent a significant increase in realism over the traditional mycorrhizal pot experiment. One development from these would be the construction of microcosms containing gnotobiotic systems in which to study mycorrhizal responses. This has not yet been done, though workable microcosm systems have been used to investigate decomposition and nutrient cycling. An alternative approach is to move directly into the field and to measure mycorrhiza–fauna interactions directly, either by observation or manipulation.

The most obvious manipulation to make is the application of a biocide to reduce or eliminate the populations of fungivores. This is a very crude manipulation, since no biocides are specific for fungivores, and indeed nearly all those used to date have very broad-spectrum activity. One of the first such studies was that of Finlay (1985), who applied the insecticide "chlorfenvinphos" (diethyl 1-(2', d'-dichlorophenyl)-2-chlorovinyl phosphate) and the fungicide "benomyl" to soil in plots in a field that had been fumigated with methyl bromide 10 months previously. All plots received an inoculum of 2 kg/m^2 of *Glomus occultum* inoculum, and were sown with red clover *Trifolium pratense*. Benomyl, which is an effective anti-VAM agent (Hale & Sanders, 1982; Fitter & Nichols, 1988), reduced shoot mass, P contents, and P uptake rates (expressed as rate of P accumulation per unit shoot mass).

The insecticide treatment reduced collembola numbers by 80%; other soil animals were not examined. Plants in insecticide treated plots had higher shoot mass, shoot P concentration, shoot P content, and shoot P accumulation rates (Fig. 10.5). The increases were not due to changes in P availability, and the results are consistent with the notion that reducing collembola numbers permitted better mycorrhizal function. The increase in shoot mass following insecticide treatment could be due to reductions in root-feeding insects, but the effect on P accumulation rate is not so obviously explained in this way.

This study was performed in a relatively P-deficient soil under agricultural conditions likely to maximize the potential contribution of mycorrhizal fungi to plant growth. In natural and seminatural communities the evidence for VAM benefits to plants is much less convincing (Fitter, 1989), and so the chances of obtaining positive responses to the reduction of grazing pressure are less good. Nevertheless, one of the most obvious possible reasons for the failure of VAM to produce good responses in such communities is grazing (Fitter, 1985), so it is important to extend these studies. McGonigle and Fitter (1988) applied insecticide, again chlorfenvinphos as in Finlay's (1985) study, to a seminatural, species-rich grassland at Wheldrake Ings, Yorkshire (England). They monitored the phosphate nutrition of the grass *Holcus lanatus*, an abundant species in the community. As previously, insecticide reduced collembola density 3-fold and significantly increased shoot mass, especially at the end of the experiment, 69

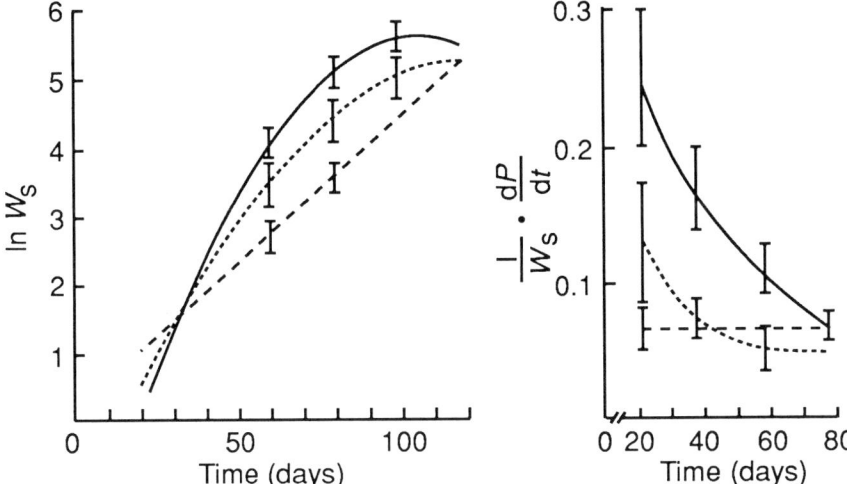

Figure 10.5. Changes in (a) mean shoot mass m^{-2} and (b) rate of phosphorus accumulation into shoots in *Trifolium pratense* plants grown in field plots treated with insecticide (——), fungicide (— —), or untreated (···). Bars represent 95% confidence intervals to fitted curves. From Finlay (1985).

days after first application. At this time there was also a large increase in P flux, from 70 fmol/cm^2/sec in the controls to 230 in the insecticide-treated plots. P fluxes (uptake rate per unit root surface area) were calculated rather than inflows (uptake rate per unit root length) because the very fine roots of *H. lanatus* gave very low inflow values.

This experiment provides evidence similar to that of Finlay (1985), but for a seminatural community. Various explanations are possible, but the increased P uptake rate per unit root length or area is hard to explain by any mechanism other than one involving mycorrhizas. The increased P fluxes in *H. lanatus* coincided with the onset of flowering, which is likely to increase P demand by the plant considerably. It is possible that plants growing in the field under natural conditions normally have low P demands that can be satisfied by diffusive flux through soil, thus rendering the mycorrhizal association superfluous, but that at certain times in the life-cycle, of which sexual reproduction is an obvious example, P demand is much greater, rendering the mycorrhizal flux highly beneficial. This interpretation is supported by the data of Dunne and Fitter (1989) for cultivated strawberry: they measured P inflows of 13 fmol/cm/sec in the vegetative phase rising to 435 fmol/cm/sec at fruiting. Such high values could not possibly be achieved by diffusive flux. If converted to a flux across the root, the highest inflow for strawberry represents 6.9 pmol/cm^2/sec, higher than reported for any other mycorrhizal plant (McGonigle & Fitter, 1988) and higher than the maximum P fluxes reported for solution culture experiments (Barber, 1984; p.216).

These data are consistent with the hypothesis that soil animals do consume significant amounts of VAM fungal hyphae under realistic field conditions and so render the association less effective. Since VAM often give little obvious benefit to plants in the field, possibly because of inherently low P demand, it is not surprising that another attempt to manipulate a natural ecosystem using biocides, that by Ingham et al. (1986), produced few mycorrhizal responses. In this study various biocides were added to undisturbed prairie vegetation; carbofuran, a nematicide, increased VAM infection in roots of *Bouteloua gracilis,* but without altering shoot P concentrations. This agrees with the study by Fitter (1986) in an alpine grassland, in which reducing VAM infection in the roots of various species actually increased shoot P concentration. In neither case were plants receiving an obvious quantitative benefit from infection.

Other experiments currently in progress may give further insights into this complex system. For example, Brown and Gange (1989), studying the role of herbivory in succession, found very large changes in plant species composition following the application of insecticide to soil. Whether these might be related to effects on mycorrhizal grazers remains to be seen. What is urgently needed is a study in which the interaction between the soil fauna and mycorrhizal fungi is studied throughout the life-cycle of a plant species. Until this is achieved, it will remain difficult to offer more than conjecture in this area, although there seems little doubt that mycorrhizal fungi are an important food resource for soil fungivores, and that these animals can radically alter the functioning of the symbiosis. Any attempt to understand the role of mycorrhizas in either natural or agricultural systems must take grazing into account.

Acknowledgments

We are grateful to Dr. Roger Finlay for helpful comments on the manuscript and to the British Ecological Society for permission to reproduce Figures 10.2–10.4.

References

Anderson, J. M. (1978). Competition between two unrelated species of soil Cryptostigmata (Acari) in experimental microcosms. *Journal of Animal Ecology* **44,** 475–495.

Anderson, J. M., & Ineson, P. (1984). Interactions between microorganisms and soil invertebrates in nutrient flux pathways of forest ecosystems. In: *Invertebrate-Microbial Interactions* (Ed. by J. M. Anderson, A. D. M. Rayner, & D. W. H. Walton), pp. 59–88. Cambridge University Press, Cambridge.

Barber, S. A. (1984). *Soil Nutrient Bioavailability.* Wiley, New York.

Bethlenfalvay, G. J., Brown, M. S., & Pacovsky, R. S. (1982). Relationships between host and endophyte development in mycorrhizal soybeans. *New Phytologist* **90,** 537–543.

Brown, V. K., & Gange, A. C. (1989). Differential effects of above- and below-ground insect herbivory during early plant succession. *Oikos* **54**, 67–76.

Brundrett, M. C., & Kendrick, B. (1988). The mycorrhizal status, root anatomy and phenology of plants in a sugar maple forest. *Canadian Journal of Botany* **66**, 1153–1173.

Coleman, D. C. (1985). Through a ped darkly: An ecological assessment of root-soil-microbial-faunal interactions. In: *Ecological Interactions in Soil* (Ed. by A. H. Fitter, D. Atkinson, D. J. Read, & M. B. Usher), pp. 1–21. Special publication no. 4 of the British Ecological Society, Blackwell Scientific Publications, Oxford.

Coleman, D. C., Cole, C. V., Hunt, H. W., & Klein, D. A. (1978). Trophic interactions in soil as they affect energy and nutrient dynamics I: Introduction. *Microbial Ecology* **4**, 345–349.

Cromack, K., Fichter, A. M., Moldenke, A. M., Entry, J. A., & Ingham, E. R. (1988). Interactions between soil animals and ectomycorrhizal fungal mats. *Agricultural Ecosystems and Environment* **24**, 161–168.

Dunne, M. J., & Fitter, A. H. (1989). The phosphorus budget of a field-grown strawberry (*Fragaria* × *ananassa* cv. Hapil) crop: Evidence for a mycorrhizal contribution. *Annals of Applied Biology* **114**, 185–193.

Elliott, E. T., Coleman, D. C., Ingham, R. E., & Trofymow, J. A. (1984). Carbon and energy flow through microflora and microfauna in the soil subsystem of terrestrial ecosystems. In: *Current Perspectives in Microbial Ecology* (Ed. by M. J. King & C. A. Reddy), pp. 424–433. American Society for Microbiology, Washington, D.C.

Finlay, R. D. (1985). Interactions between soil micro-arthropods and endomycorrhizal associations of higher plants. In: *Ecological Interactions in Soil* (Ed. by A. H. Fitter, D. Atkinson, D. J. Read, & M. B. Usher), pp. 319–331. Special publication no. 4 of the British Ecological Society, Blackwell Scientific Publications, Oxford.

Fitter, A. H. (1985). Functioning of vesicular-arbuscular mycorrhizas under field conditions. *New Phytologist* **99**, 257–265.

Fitter, A. H. (1986). Effect of benomyl on leaf phosphorus concentration in alpine grasslands: A test of mycorrhizal benefit. *New Phytologist* **103**, 767–776.

Fitter, A. H. (1989). The ecology of vesicular-arbuscular mycorrhizas in temperate ecosystems. *Agriculture, Ecosystems and Environment* **29**, 137–151.

Fitter, A. H., & Nichols, R. (1988). The use of benomyl to control infection by vesicular-arbuscular mycorrhizal fungi. *New Phytologist* **110**, 201–206.

Fogel, R. (1985). Roots as primary producers in below-ground ecosystems. In: *Ecological Interactions in Soil* (Ed. by A. H. Fitter, D. Atkinson, D. J. Read, & M. B. Usher), pp. 23–36. Special publication no. 4 of the British Ecological Society, Blackwell Scientific Publications, Oxford.

Fogel, R., & Hunt, G. (1979). Fungal and arboreal biomass in a western Oregon Douglas-fir ecosystem. *Canadian Journal of Forest Research* **9**, 245–256.

Fogel, R., & Hunt, G. (1983). Contribution of mycorrhizae and soil fungi to nutrient cycling in a Douglas-fir ecosystem. *Canadian Journal of Forest Research* **13**, 219–232.

Fogel, R., & Trappe, J. M. (1978). Fungus consumption (mycophagy) by small mammals. *Northwest Science* **52**, 1–31.

Hale, M. G., & Sanders, F. E. (1982). Effects of benomyl on vesicular-arbuscular mycorrhizal infections of red clover (*Trifolium pratense* L.) and consequences for phosphorus inflow. *Journal of Plant Nutrition* **5**, 1355–1367.

Hanlon, R. D. G. (1981). Influence of grazing by Collembola on the activity of senescent fungal colonies grown on media of different nutrient concentration. *Oikos* **36**, 362–367.

Harris, K. K., & Boerner, R. E. J. (1990). Effects of belowground grazing by collenbola on growth, mycorrhizal infection and P uptake of *Geranium robertianum*. *Plant and Soil*, **129**, 203–210.

Hunt, H. W., Coleman, D. C., Ingham, E. R., Ingham, R. E., Elliot, E. T., Moore, J. C., Reid, C. C. P., & Morley, C. R. (1987). The detrital food web in a shortgrass prairie. *Biology and Fertility of Soil* **3**, 57–68.

Hussey, R. S., & Roncadori, R. W. (1981). Influence of *Aphelenchus avenae* on vesicular-arbuscular endomycorrhizal growth response in cotton. *Journal of Nematology* **13**, 48–52.

Ingham, E. R. (1988). Interactions between nematodes and vesicular-arbuscular mycorrhizae. *Agricultural Ecosystems and Environment* **24**, 169–182.

Ingham, E. R., Trofymow, J. A., Ames, R. N., Hunt, H. W., Morley, C. R., Moore, J. C., & Coleman, D. C. (1986). Trophic interactions and nitrogen cycling in a semiarid grassland. Part II. System responses to removal of different groups of soil microbes or fauna. *Journal of Applied Ecology* **23**, 615–630.

Koch, K. E., & Johnson, C. R. (1984). Photosynthate partitioning in split-root seedlings with mycorrhizal and non-mycorrhizal root systems. *Plant Physiology* **75**, 26–30.

McGonigle, T. P. (1987). Vesicular-arbuscular mycorrhizas and plant performance in a semi-natural grassland. D.Phil. thesis, University of York, U.K.

McGonigle, T. P., & Fitter, A. H. (1988). Ecological consequences of arthropod grazing on VA mycorrhizal fungi. *Proceedings of the Royal Society of Edinburgh* **94B**, 25–32.

McIlveen, W. D., & Cole, H. Jr. (1976). Spore dispersal of Endogonaceae by worms, ants, wasps and birds. *Canadian Journal of Botany* **54**, 1486–1489.

Moore, J. C. (1988). The influence of microarthropods on symbiotic and non-symbiotic mutualism in detrital-based below-ground food webs. *Agricultural Ecosystems and Environment* **24**, 147–159.

Moore, J. C., St. John, T. V., & Coleman, D. C. (1985). Ingestion of vesicular-arbuscular mycorrhizal hyphae and spores by soil microarthropods. *Ecology* **66**, 1979–1981.

Newell, K. (1984a). Interactions between two decomposer basidiomycetes and a collembolan under sitka spruce: Distribution, abundance and selective grazing. *Soil Biology and Biochemistry* **16**, 227–233.

Newell, K. (1984b). Interactions between two decomposer basidiomycetes and a collembolan under sitka spruce: Grazing and its potential effects on fungal distribution and leaf litter. *Soil Biology and Biochemistry* **16**, 235–239.

Parkinson, D., Visser, S., & Whittaker, J. B. (1979). Effects of collembolan grazing on fungal colonization of leaf litter. *Soil Biology and Biochemistry* **11**, 529–535.

Parton, W. J., Anderson, D. W., Cole, C. V., & Stewart, J. W. B. (1984). Simulation of soil organic matter formation and mineralization in semiarid ecosystems. In: *Nutrient Cycling in Agricultural Ecosystems* (Ed. by R. Lowrance, R. L. Todd, L. E. Asmussen, & R. A. Leonard), pp. 533–550. University of Georgia College of Agriculture Publication 23.

Rabatin, S. C., & Stinner, B. R. (1985). Arthropods as consumers of vesicular-arbuscular fungi. *Mycologia* **77**, 320–322.

Rabatin, S. C., & Stinner, B. R. (1988). Indirect effects of interactions between VAM fungi and soil-inhabiting invertebrates on plant processes. *Agricultural Ecosystems and Environment* **24**, 135–146.

Rich, J. R., & Schenck, N. C. (1981). Seasonal variations in populations of plant-parasitic nematodes and vesicular-arbuscular mycorrhizae in Florida field corn. *Plant Disease* **65**, 804–807.

Riffle, J. W. (1967). Effect of an *Aphelenchoides* species on the growth of a mycorrhizal and a pseudomycorrhizal fungus. *Phytopathology* **57**, 541–544.

Rosendahl, S., Rosendahl, C. L., & Sochting, U. 1989. Distribution of VA mycorrhizal endophytes amongst plants from a Danish grassland community. *Agriculture, Ecosystems and Environment* **29**, 329–335.

Salawu, E. O., & Estey, R. H. (1979). Observations on the relationships between a vesicular-arbuscular fungus, a fungivorous nematode and the growth of soybeans. *Phytoprotection* **60**, 99–102.

Sanders, F. E., Tinker, P. B., Black, R. L., & Palmerley, S. M. (1977). The development of endomycorrhizal root systems I. Spread of infection and growth promoting effects with four species of vesicular-arbuscular mycorrhizas. *New Phytologist* **78**, 257–268.

Seastedt, T. R., James, S. W., & Todd, T. C. (1988). Interactions among soil invertebrates, microbes and plant growth in a tallgrass prairie. *Agricultural Ecosystems and Environment* **24**, 219–228.

Shaw, P. J. A. (1985). Grazing preferences of *Onychiurus armatus* (Insecta: Collembola) for mycorrhizal and saprophytic fungi in pine plantations. In: *Ecological Interactions in Soil* (Ed. by A. H. Fitter, D. Atkinson, D. J. Read, & M. B. Usher), pp. 333–337. Special publication of the British Ecological Society no. 4, Blackwell Scientific Publications, Oxford.

Shaw, P. J. A. (1988). A consistent hierarchy in the fungal feeding preferences of the collembola *Onychiurus armatus*. *Pedobiologia* **31**, 179–187

Sieverding, E., Toro, S., & Mosquera, O. (1989). Biomass production and nutrient concentrations in spores of VA mycorrhizal fungi. *Soil Biology and Biochemistry* **21**, 69–72.

Smith, S. E., & Gianinazzi-Pearson, V. (1988). Physiological interactions between symbionts in vesicular-arbuscular mycorrhizal plants. *Annual Review of Plant Physiology and Plant Molecular Biology* **39**, 221–244.

Snellgrove, R. C., Splitstoesser, W. E., Stribley, D. P., & Tinker, P. B. (1982). The distribution of carbon and the demand of the fungal symbiont in leek plants with vesicular-arbuscular mycorrhizas. *New Phytologist* **92,** 75–87.

Stuart, J. W. B., & McKercher, R. B. (1982). Phosphorus cycle. In: *Experimental Microbial Ecology* (Ed. by R. G. Burns & J. H. Slater), pp. 221–238. Blackwell Scientific Publications, Oxford.

Sutherland, J. R., & Fortin, J. A. (1968). Effect of the nematode *Aphelenchus avenae* on some ectotrophic mycorrhizal fungi and on a red pine mycorrhizal relationship. *Phytopathology* **58,** 519–523.

Usher, M. B. (1976). Aggregation responses of soil arthropods in relation to the soil environment. In: *The Role of Terrestrial and Aquatic Organisms in Decomposition Processes* (Ed. by J. M. Anderson & A. MacFadyen), pp. 61–94. Blackwell Scientific Publications, Oxford.

Visser, S. (1985). Role of the soil invertebrates in determining the composition of soil microbial communities. In: *Ecological Interactions in Soil* (Ed. by A. H. Fitter, D. Atkinson, D. J. Read, & M. B. Usher), pp. 297–317. Special Publications of the British Ecological Society no. 4. Blackwell Scientific Publications, Oxford.

Walter, D. E., Hudgens, R. A., & Freckman, D. W. (1986). Consumption of nematodes by fungivorous mites, *Tyrophagus* spp. (Acarina: Astigmata: Acaridae). *Oecologia* **70,** 357–361.

Warnock, A. J., Fitter, A. H., & Usher, M. B. (1982). The influence of a springtail *Folsomia candida* (Insecta, Collembola) on the mycorrhizal association of leek *Allium porrum* and the vesicular-arbuscular mycorrhizal endophyte *Glomus fasciculatus*. *New Phytologist* **90,** 285–292.

Zak, B. (1965). Aphids feeding on mycorrhizae of Douglas-fir. *Forest Science* **11,** 410–411.

SECTION 3
System Dynamics and Application

11

Specificity Phenomena in Mycorrhizal Symbioses: Community-Ecological Consequences and Practical Implications

Randy Molina, Hugues Massicotte, and James M. Trappe

SUMMARY. Most land plants form mycorrhizae, and the species of mycorrhizal fungi number in the thousands. Many host–fungus combinations are incompatible, however, and breadth of host range or fungus associates differs strongly among both fungus and plant symbionts. Evaluation of the occurrence and function of linkages between plants via shared mycorrhizal fungi requires a full understanding of compatibility and specificity phenomena. We recognize and describe six phenomena that bear on this problem: (1) *dependency vs. independency*—defines whether plants form or do not form mycorrhizae; (2) *facultative vs. obligate symbionts*—defines the ability or inability of symbionts to complete life cycles in the absence of mycorrhiza formation; (3) *fidelity to a class of mycorrhiza*—recognizes that most symbionts typically form one class of mycorrhizae (e.g., vesicular-arbuscular vs. ectomycorrhizae), but that some can form two or more classes; (4) *host range of mycorrhizal fungi*—describes a spectrum that varies from a narrow range, typically genus restricted, to an intermediate range, often restricted to a plant family, to a broad range, typically extending across diverse plant families and orders; (5) *host receptivity*—defines the numbers and diversity of mycorrhizal fungi accepted by a particular host, also ranging from narrow (low number) to broad (high number); and (6) *ecological specificity*—the influence of biotic and abiotic factors on ability of plants to form functional mycorrhizae with particular fungi in natural soils.

Specificity phenomena in mycorrhizal associations directly affect plant community development in a variety of ways. Mycorrhizal dependence, fidelity to a mycorrhizal class, and mycobiont propagule availability are prime determinants of early stages of community development and affect long-term plant dominance. During community succession, shared compatibility for mycorrhizal fungal associates determines linkage potential for those plants and influences plant interactions. Ecological linkages of plants via mycorrhizal interactions may contribute to the formation of functional guilds of plants and fungi. Shared compatibility for mycorrhizal symbionts also influences the ability of plants to migrate into new locations, and so may affect plant migrations during periods of rapid global climate change.

Much research is needed to determine how specificity phenomena determine linkage potential between plants and subsequent effects on community development. In this chapter we document those needs and provide a framework for integrating our understanding of mycorrhizal specificity phenomena into the broader context of ecosystem function and stability.

Introduction

Over the last two decades, mycorrhiza research has taken a mostly reductionist approach to study plant–fungus mycorrhizal interactions. Practical applications in agriculture and forestry have driven mycorrhiza research, focusing on host response. This approach may logically follow the "green revolution" strategy to increase plant yields through modern technology, but management of resources today calls for an ecosystem-level of understanding. Long-term productivity is measured in terms of sustainable ecosystem processes that maintain fertility and system "health" as well as recovery after adverse disturbance. This requires detailed understanding of key biological linkages that determine ecosystem trajectory (Perry et al., 1987, 1989a, 1990). Mycorrhizal associations function as dynamic biological linkages, yet we understand poorly how they operate in the soil and affect the outcome of community development. Understanding the extent, diversity, and function of these linkages is vital to predict future productivity and resiliency.

Linkages involve direct or indirect connections between biological components in space and time. Although mycorrhizae act at small rhizosphere sites, they profoundly affect community development and ecosystem functioning across landscapes. Perry et al. (1989a) hypothesize that such linkages have created the formation of social "guilds" of host plants that share compatibility for belowground mutualists. Such guilds function in a holistic manner; positive interactions and feedbacks (linkages) between individuals promote benefit for the whole. We hypothesize that linkages developed during early plant successional seres differ from those of more stable late seral plant communities. We need to learn which biotic components are connected and which are not, how and by what they are connected, and how these connections regulate composition of a given plant community.

Although mycorrhizal associations are the norm for most land plants, the several mycorrhiza classes, i.e., vesicular-arbuscular mycorrhiza (VAM), ectomycorrhiza (EM), ectendomycorrhiza, ericoid mycorrhiza, etc., and the tremendous array of mycorrhizal fungal species lead to a vast number of possible host–fungus combinations. Thus, understanding the functional nature of mycorrhizal linkages requires a comprehensive knowledge and appreciation for the formation of compatible mycorrhizae resulting from chance encounters of plant roots and fungi.

Much of the discussion about compatibility in mycorrhizal associations has centered around recognition phenomena and comparing mycorrhizae to host specificity of plant pathogens. This helps in understanding the evolution of plant–fungus relationships, but it clouds the ecological significance of mutualism and the widely disparate processes yielding mutualism versus parasitism. Reviews of mycorrhizal specificity commonly emphasize a "lack of specificity" in mycorrhizal associations, i.e., there is no known example of gene-for-gene level of host fungus specificity. This generalization promotes the impression that most mycorrhizal fungi are promiscuous soil fungi and redundant in function. This masks their complex ecological diversity. We contend that as many if not more mycorrhizal fungus–host associations are specific as are nonspecific, and only by describing this entire range of compatible interactions will we discover their potential "linkage" functions in diverse terrestrial ecosystems.

Range and Definitions of Specificity Phenomena

Historical Perspective

Europeans have recognized for centuries the close association of many mushrooms with particular tree species, especially prized edible mushrooms and truffles (Trappe, 1962). This represents the first recognition of mycorrhizal specificity and continues to attract attention today.

Around the turn of the twentieth century, several investigators (see Rayner, 1927) documented the variety of mycorrhizae found among plants. A general understanding of relationships between plant families and their classes of mycorrhizae began to emerge. It became clear that the endotrophic (later VAM) mycorrhiza was the most widespread class, but ectotrophic mycorrhizae (later ectomycorrhizae) prevailed among the Pinaceae, Fagaceae, and Betulaceae, and other classes are unique to the Ericales and Orchidaceae. Some families, such as the Salicaceae or Myrtaceae, could form two or more of these classes. These observations were a critical basis for discussing specificity and compatibility phenomena. Indeed, fidelity to a particular mycorrhizal class is a basic compatibility phenomenon with considerable evolutionary and ecological consequences.

Identifying mycorrhizal symbionts and determining the functions of these root associations paved the way for further understanding of specificity phenomena. Melin's (1922, 1923, 1925) elegant pure culture syntheses of EM conclusively demonstrated differences in mycotrophy between soil fungi (some were mycorrhizal and some were not) and in compatibility between fungi and plants as seen in the ability of fungi to form EM with some plant species but not others. He also noted differences in how "actively" the fungi colonized tree roots and developed concepts of virulence that have now largely fallen into disuse. During this same period "pseudomycorrhizae" were also described and discussed to indicate the widespread nature of mycorrhizal-like root colonization. This type of root colonization still needs reexamination in regard to mycorrhizal specificity phenomena.

By the mid-twentieth century, improved physiological techniques not only demonstrated the nutritional benefits of mycorrhizae to plants but also showed that fungi differed strongly in this regard and that environmental factors, particularly soil conditions, affected formation and function of mycorrhizae. Culture and experimental manipulation of different mycorrhizal fungi also pointed out strong differences in carbon dependency and substrate utilization. Mycorrhizal plants and fungi were seen to interact and develop dependencies in many ways. As appreciation of the ecology of mycorrhizal symbioses has increased during the last two decades, we now realize the fundamental importance of mycorrhizal interactions, including specificities, in influencing plant ecology. Finally, the question of specificity and compatibility is becoming better resolved with modern techniques of biochemistry, electron microscopy, and molecular genetics. These allow examination of mechanisms of plant–fungus recognition and indicators of functional compatibility.

Dependency versus Independency

Our discussion of the levels of mycorrhizal specificities begins with the realization that not all plants and fungi form mycorrhizae. In an ecological sense, a plant can be termed independent of mycorrhizae if it can survive as an individual, compete in a community, and reproduce without them (Trappe, 1989). This may seem a simple yes-or-no phenomenon, but it quickly blends to gray as we expand the concept into one of facultative vs. obligate mycorrhizal symbionts. Trappe (1987) recently reviewed this subject in detail for angiosperms.

General reviews of mycorrhizae often begin by stating something like "90% of the world's land plants are mycorrhizal," or as corrected by Trappe (1987), "belong to families that are commonly mycorrhizal." Despite the low nonmycorrhizal percentage, the actual number is certainly in the tens of thousands of plant species. Gerdemann (1968) lists the possibly nonmycorrhizal or rarely mycorrhizal plant families: Aizoaceae, Amaranthaceae, Caryophyllaceae, Chenopodiaceae, Comelinaceae, Cruciferae, Cyperaceae, Fumariaceae, Juncaceae, Nyctaginaceae, Phytolaccaceae, Polygonaceae, Portulacaceae, and Urticaceae. However, many exceptions occur within these families, as well as within families typically mycorrhizal (Trappe, 1987; Newman & Reddell, 1987). Newman and Reddell (1987) point out that in the typically nonmycorrhizal plant families listed by Gerdemann (1968), the percentage of mycorrhizal species is now known to range from 13 to 63%. Environmental factors can influence whether a plant is mycotrophic or not in some settings as can presence of companion plants. Nevertheless, the point is that some plants are routinely independent of mycorrhizae—Trappe (1987) indicates 18% of angiosperms examined so far—and this phenomenon has many ecological ramifications.

Conversely by our estimate only about 10% of the tens of thousands of soil fungus species are mycorrhizal, perhaps some 5000 to 6000 species in all. Mycorrhizal fungi are in the Ascomycotina, Basdiomycotina, and Zygomycotina (Table 11.1).

Table 11.1. *Estimated Numbers (in Parentheses) of Ectomycorrhizal and Ectendomycorrhizal Fungus Species by Fruiting Habit, Class, Family, and Genus*

Habit, class, family	Genera
Epigeous Habit (4450)	
Ascomycotina (135)	
Ascobolaceae (5)	*Sphaerosoma* (3), others (2)
Geoglossaceae (20)	*Geoglossum* (5), *Leotia* (5), *Trichoglossum* (5), others (5)
Helvellaceae (10)	*Helvella* (10)
Pezizaceae (55)	*Peziza* (50), *Sarcosphaera* (1), others (4)
Pyronemataceae (30)	*Geopora* (3), *Humaria* (5), *Sphaerosporella* (4), *Trichophaea* (5), *Wilcoxina* (3), others (10)
Sarcoscyphaceae (15)	*Plectania* (10), *Pseudoplectania* (5)
Basidiomycotina (4225)	
Amanitaceae (200)	*Amanita* (200)
Astraeaceae (5)	*Astraeus* (5)
Boletaceae (410)	*Boletus* (150), *Gyrodon* (20), *Gyroporus* (20), *Leccinum* (40), *Phylloporus* (20), *Pulveroboletus* (40), *Suillus* (50), *Tylopilus* (50), others (20)
Cantharellaceae (110)	*Cantharellus* (70), *Craterellus* (20), others (20)
Clavariaceae (150)	*Clavaria* (20), *Clavariadelphus* (10), *Ramaria* (100), others (20)
Corticiaceae (70)	*Amphinema* (5), *Byssocorticium* (5), *Byssoporia* (5), *Piloderma* (5), others (50)
Cortinariaceae (1400)	*Cortinarius* (900), *Cuphocybe* (5), *Dermocybe* (85), *Descolea* (15), *Hebeloma* (120), *Inocybe* (210), *Leucocortinarius* (2), *Naucoria* (50), *Rozites* (10), *Stephanopus* (3)
Entolomataceae (200)	*Clitopilus* (20), *Entoloma* (160), *Leptonia* (20)
Gomphaceae (30)	*Gomphus* (20), others (10)
Gomphidiaceae (30)	*Cystogomphus* (1), *Chroogomphus* (20), *Gomphidius* (9)
Hydnaceae (150)	*Bankera* (5), *Dentinum* (5), *Hydnellum* (20), *Hydnum* (100), *Phellodon* (20)
Hygrophoraceae (380)	*Camarophyllus* (30), *Hygrocybe* (100), *Hygrophorus* (250)
Lycoperdaceae (50)	*Lycoperdon* (50)
Paxillaceae (20)	*Paxillus* (15), others (5)
Pisolithaceae (5)	*Pisolithus* (5)
Polyporaceae (20)	*Albatrellus* (20)
Russulaceae (700)	*Lactarius* (200), *Russula* (500)
Sclerodermataceae (25)	*Scleroderma* (25)
Strobilomycetaceae (110)	*Austroboletus* (30), *Boletellus* (60), *Strobilomyces* (20)
Thelephoraceae (40)	*Boletopsis* (2), *Thelephora* (38)
Tricholomataceae (210)	*Catathelasma* (5), *Laccaria* (30), *Leucopaxillus* (25), *Tricholoma* (150)

(continued)

Table 11.1. (continued)

Habit, class, family	Genera
Hypogeous Habit (965)	
Ascomycotina (280)	
Ascobolaceae (1)	*Sphaerosoma* (1)
Balsamiaceae (15)	*Balsamia* (6), *Picoa* (9)
Elaphomycetaceae (50)	*Elaphomyces* (50)
Geneaceae (30)	*Genabea* (5), *Genea* (25)
Helvellaceae (20)	*Barssia* (3), *Fischerula* (2), *Hydnotrya* (15)
Pezizaceae (25)	*Amylascus* (5), *Hydnotryopsis* (10), *Peziza* (3), *Pachyphloeus* (6), *Ruhlandiella* (1)
Pyronemataceae (35)	*Geopora* (10), *Hydnocystis* (2), *Labyrinthomyces* (1), *Dingleya* (5), *Reddellomyces* (9), *Sphaerozone* (1), *Stephensia* (7)
Terfeziaceae (34)	*Choiromyces* (5), *Hydnobolites* (2), *Terfezia* (27)
Tuberaceae (70)	*Paradoxa* (1), *Tuber* (69)
Basidiomycotina (680)	
Astraeaceae (10)	*Pyrenogaster* (1), *Radiigera* (9)
Boletaceae (185)	*Alpova* (15), *Gastroboletus* (15), *Rhizopogon* (150), *Truncocolumella* (5)
Cortinariaceae (110)	*Cortinarius* (5), *Destuntzia* (5), *Hymenogaster* (75), *Setchelliogaster* (4), *Thaxterogaster* (24)
Chondrogastraceae (5)	*Chondrogaster* (5)
Cribbeaceae (1)	*Mycolevis* (1)
Entolomataceae (5)	*Richoniella* (5)
Gomphidiaceae (2)	*Brauniellula* (1), *Gomphogaster* (1)
Hysterangiaceae (40)	*Hysterangium* (40), *Trappea* (2)
Leucogastraceae (25)	*Leucogaster* (21), *Leucophleps* (4)
Melanogastraceae (20)	*Melanogaster* (20)
Mesophelliaceae (45)	*Castoreum* (5), *Malajczukia* (10), *Mesophellia* (20), others (10)
Octavianinaceae (30)	*Octavianina* (15), *Sclerogaster* (15)
Russulaceae (140)	*Archangeliella* (10), *Elasmomyces* (10), *Gymnomyces* (30), *Macowanites* (25), *Martellia* (40), *Zelleromyces* (25)
Sclerodermataceae (6)	*Scleroderma* (6)
Sedeculaceae (1)	*Sedecula* (1)
Strobilomycetaceae (50)	*Austrogautieria* (10), *Chamonixia* (10), *Gautieria* (25), *Wakefieldia* (5)
Tricholomataceae (5)	*Hydnangium* (5)
Zygomycotina (5)	
Endogonaceae (5)	*Endogone* (5)
Total, all taxa (5415)	

Facultative versus Obligate Symbionts

In the strictest sense, facultative plants or fungi are those capable of forming mycorrhizae but also able to complete their life cycles without forming mycorrhizae. Trappe (1987) regards plants as facultative mycotrophs if they function without mycorrhizae in some natural situations and with mycorrhizae in others. Many orchid mycorrhizal fungi are facultative mycobionts; several are well known saprophytes or plant parasites. Ability of some plants to function without mycorrhizae but also form functional mycorrhizae, depending on the ecological setting and associated plants, is exemplified by chenopods in desert habitats (Miller, 1979). Trappe's (1987) data indicate that about 12% of the angiosperms are facultative.

Obligate mycotrophs cannot complete their life cycles without mycorrhizae. Many field and greenhouse experiments indicate that most mycotrophs are obligately mycorrhizal. If about 12% of examined angiosperm species are facultatively mycorrhizal and 18% are typically nonmycorrhizal, then about 70% are obligate mycotrophic plants (Trappe, 1987).

Fidelity to a Class of Mycorrhiza

Mycorrhizal fungi typically form only one class of mycorrhiza. For example, VAM fungi do not form EM and EM fungi do not form VAM. Exceptions would include some ascomycetes and basidiomycetes that can form both ecto- and ectendo (arbutoid) mycorrhizae, but this simply reflects differing host morphological responses to colonization. The main classes of mycorrhizae are illustrated in Figure 11.1.

Most plants also show strong fidelity to a particular mycorrhiza class. In the Angiosperms examined so far, 65% of the species are reported to form only one class of mycorrhiza, whereas only 5% form more than one class (Trappe, 1987). Examples are shown in Table 11.2. Species of *Alnus, Salix, Populus,* and *Eucalyptus* can form VAM either solely or simultaneously with EM, sometimes in the same rootlet (Chilvers et al., 1987). Similarly some typically ericoid plants have also been shown to occasionally form EM (Largent et al., 1980); some ericaceous plants in Hawaii form both ericoid and VAM (Koske et al., 1990). Ecto- and ectendomycorrhizae are commonly found on the same host and in Australia some unusual EM fungi can also form orchidoid mycorrhizae (Warcup, 1988). If we also classify the less well-known dark-septate fungi and other miscellaneous root endophytes as mycorrhizal classes the fidelity question is less clear.

Host Range of Mycorrhizal Fungi

Host range refers to the diversity of plant species with which a fungus species can form mycorrhizae. This character varies between fungal species and mycorrhiza

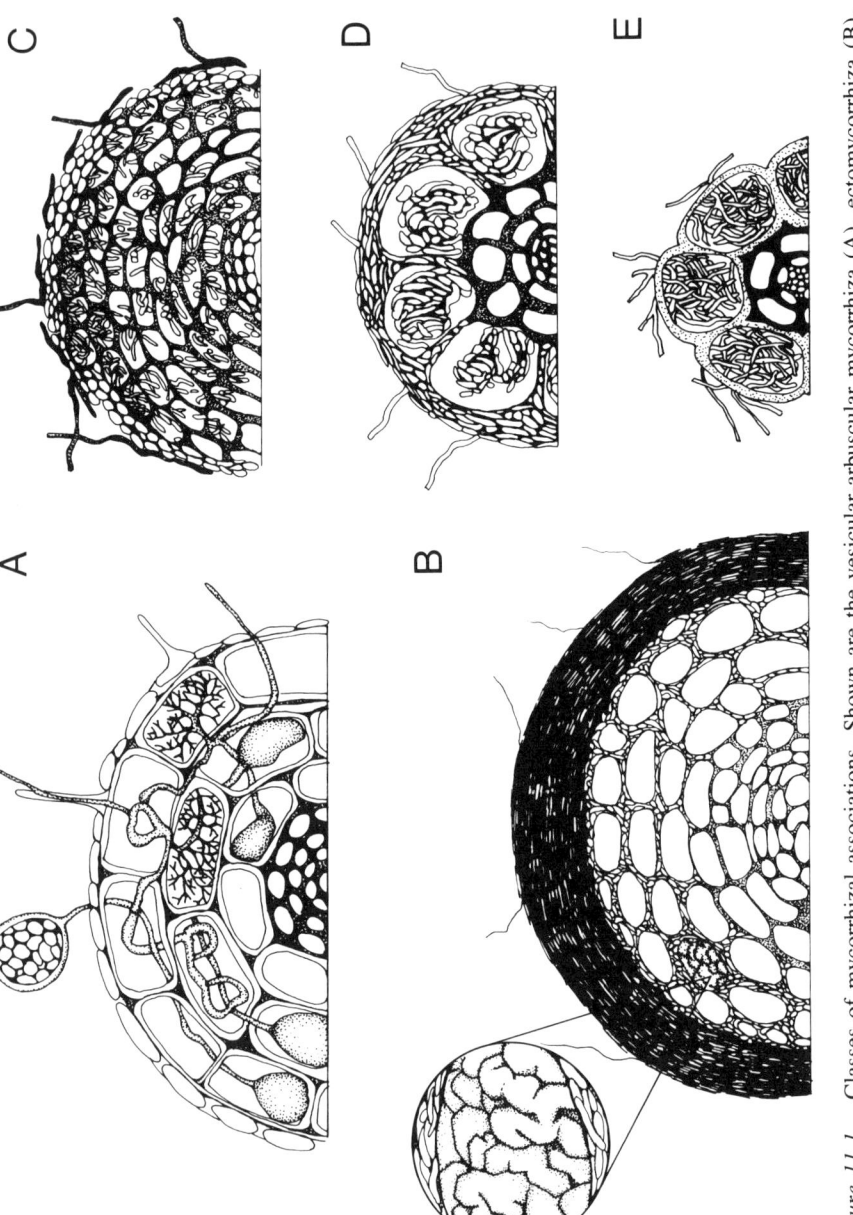

Figure 11.1. Classes of mycorrhizal associations. Shown are the vesicular-arbuscular mycorrhiza (A), ectomycorrhiza (B), ectendomycorrhiza (C), arbutoid mycorrhiza (D), and ericoid mycorrhiza (E). Each of these classes of mycorrhizae has varying levels of specificity.

Table 11.2. Hosts Reported to Form Both EM and Ericoid Mycorrhizae and VAM by Order, Family, and Genus[a]

Order	Family	Genus
Asterales	Asteraceae	Crepis, Helichrysum, Helipterum, Homogyne, Lactuca, Leptorhynchos, Podolepis, Pogonolepis, Rutidosis, Waitzea
Campanulales	Campanulaceae	Campanula, Lobelia
Capparales	Brassicaceae	Cochlearia
Caryophyllales	Caryophyllaceae	Silene
	Nyctaginaceae	**Neea**
Casuarinales	Casuarinaceae	**Casuarina**
Celastrales	Aquifoliaceae	**Ilex**
Coniferae	Cupressaceae	Cupressus, Juniperus
	Pinaceae	Abies, Larix, Pinus, Pseudotsuga, Tsuga
	Taxaceae	Taxus
	Taxodiaceae	**Cryptomeria**, Sciadopitys
Cornales	Cornaceae	**Cornus**
Dipsacales	Caprifoliaceae	Diervilla, **Sambucus**
Ericales	Ericaceae	**Vaccinium**
	Pyrolaceae	Pyrola
Eufilicales	Aspidiaceae	**Dryopteris, Polystichum**
	Blechnaceae	**Blechnum**
	Hypolepidaceae	Histiopteris
	Onocleaceae	Onoclea
	Polypodiaceae	**Adiantum**, Pteridium
	Sinopteridaceae	Pellaea
Euphorbiales	Buxaceae	Buxus
	Caesalpiniaceae	Cercis
	Fabaceae	Brachysema, Eutaxia, Kennedia, Mirbellia, Oxylobium, Platylobium, Pultenaea, **Robinia**, Vicia, Viminaria
	Mimosaceae	**Acacia**, Inga
Fabales	Betulaceae	**Alnus**, Betula
	Fagaceae	Castanea, Fagus, **Quercus**
Gnetales	Gnetaceae	Gnetum
Hamamelidales	Hamamelidaceae	Hamamelis
	Platanaceae	Platanus
Juglandales	Juglandaceae	**Carya, Juglans**
Lamiales	Lamiaceae	**Thymus**
Laurales	Lauraceae	Ocotea, **Sassafras**
Magnoliales	Magnoliaceae	Liriodendron
Malvales	Sterculiaceae	Lasiopetalum
	Tiliaceae	Tilia
Myricales	Myricaceae	**Myrica**
Myrtales	Lythraceae	Lythrum
	Myrtaceae	**Eucalyptus, Leptospermum, Melaleuca**
	Thymelaeaceae	Daphne, Pimetea
Papaverales	Papaveraceae	Sanguinaria
Polygonales	Polygonaceae	**Coccoloba, Polygonum**
Proteales	Elaeagnaceae	Elaeagnus

(continued)

Table 11.2. (continued)

Order	Family	Genus
Rhamnales	Rhamnaceae	Pomaderris, **Rhamnus**
	Vitaceae	Vites
Rosales	Grossulariaceae	**Ribes**
	Rosaceae	**Cercocarpus, Crataegus,** Dryas, Fragaria, **Malus,** Potentilla, Poterium, **Prunus, Rosa, Rubus, Sorbus**
Rubiales	Rubiaceae	**Galium,** Rubia
Salicales	Salicaceae	**Populus, Salix**
Sapindales	Aceraceae	Acer
	Anacardiaceae	**Rhus**
	Zygophyllaceae	Peganum
Scrophulariales	Bignoniaceae	Jacaranda
	Oleaceae	**Fraxinus,** Olea, Osmanthus
	Scrophulariaceae	Melampyrum, Pedicularis
Urticales	Cecropiaceae	Cecropia
	Ulmaceae	**Ulmus**
Violales	Cistaceae	**Helianthemum**

^aGenera in boldface type are well documented and have sure ecological significance. Other genera are poorly documented or the ecological significance is slight or unknown.

classes (Table 11.3). A continuum of host ranges exists, but for simplicity we can divide this character into three categories: narrow, intermediate, and broad.

Narrow Host Range

The narrowest host range would be a fungal association with only one species of plant (analogous to many well-known rust pathogens), or even a fungal race to plant species. Only one known possibility exists for this degree of specificity. Smits (1983) suggests that EM fungi for some dipterocarp species in Malaysia are restricted to single host species. But numerous ectomycorrhizal fungi form mycorrhizae or produce sporocarps only in association with a specific plant genus (see Table 11.4). Such fungi may sometimes be induced in pure culture to form mycorrhizae with other genera, but the sporocarp association in nature shows a more strongly specialized relationship. There are also examples of narrow host range at the fungus strain level. For example, although overall the species *Boletus edulis* has a broad host range, some strains are confined to gymnosperm associations and others to angiosperms (Trappe, 1962).

Intermediate Host Range

Host ranges of many fungi are neither narrow (restricted) nor broad (mostly unrestricted). For example, many intermediate range ectomycorrhizal fungi are

11.3. *Estimated Percentages of Species of Selected Basidiomycota with Narrow, Intermediate, or Broad Host Restriction by Fruiting Habit, Family, and Genus*[a]

Habit, family	Genus	Narrow %	Intermediate %	Broad %
Epigeous Habit				
Amanitaceae	*Amanita*	5	35	60
Boletaceae	*Suillus*	90	10	
	Boletus	20	35	45
	Leccinum	80	20	
	Small genera	30	30	40
Cortinariaceae	*Cortinarius*	20	55	25
	Dermocybe	30	45	25
	Hebeloma	25	45	30
	Inocybe	20	45	35
	Naucoria	65	35	
Gomphidiaceae	*Gomphidius*	65	35	
	Chroogomphus	60	40	
Hygrophoraceae	*Hygrophorus*	35	50	15
Russulaceae	*Lactarius*	40	35	25
	Russula	25	45	30
Sclerodermataceae	*Scleroderma*	25	45	30
Tricholomataceae	*Laccaria*		20	80
	Tricholoma	20	45	35
Hypogeous Habit				
Boletaceae	*Alpova*	40	60	
	Rhizopogon	50	50	
Cortinariaceae	*Hymenogaster*	50	50	
	Thaxterogaster	70	30	
Gomphidiaceae	*Brauniellula*	100		
	Gomphogaster	100		
Russulaceae	*Gymnomyces*	50	50	
	Macowanites	50	50	
	Martellia	60	40	
Sclerodermataceae	*Scleroderma*	25	75	
Tricholomataceae	*Hydnangium*	100		

[a]Narrow, restricted to a host genus; intermediate, restricted to angiosperms, gymnosperms, or a single host family; broad, associates with many diverse ectomycorrhizal hosts.

limited to a single family of hosts, e.g., *Suillus* and *Rhizopogon* spp. with the Pinaceae, or to angiosperms or gymnosperms (Table 11.5).

Broad Host Range

Fungi in this category can form mycorrhizae with diverse host plants, typically crossing between plant families, orders, and even classes. VAM fungi typify this category and are rarely limited in ability to form VAM with a typically VAM

Table 11.4. Examples of Ectomycorrhizal Fungal Species That Associate with Only One Genus of Hosts (Narrow Host Range) by Fruiting Habit, Class, Family, and Genus

Habit, class, family	Genus	Species	Host
Epigeous Habit			
Ascomycotina			
Basidiomycotina			
Amanitaceae	Amanita	diemii	Nothofagus (Fagaceae)
		hyperborea	Salix (Salicaceae)
		princeps	Shorea (Dipterocarpaceae)
		umbrinella	Nothofagus (Fagaceae)
Boletaceae	Boletus	betulicola	Betula (Betulaceae)
		carpinaceus	Carpinus (Betulaceae)
		gabretae	Picea (Pinaceae)
		guadeloupae	Coccoloba (Polygonaceae)
		lentistipitatus	Nothofagus (Fagaceae)
		leptospermi	Leptospermum (Myrtaceae)
		loyo	Nothofagus (Fagaceae)
		lupinus	Fagus (Fagaceae)
		mirabilis	Tsuga (Pinaceae)
		novae-zelandiae	Leptospermum (Myrtaceae)
		pseudorubinus	Pinus (Pinaceae)
		rawlingsii	Leptospermum (Myrtaceae)
		rubinus	Quercus (Fagaceae)
		speciosus	Fagus (Fagaceae)
	Gyrodon	lividus	Alnus (Betulaceae)
		monticola	Alnus (Betulaceae)
	Leccinum	duriusculum	Populus (Salicaceae)
		quercinum	Quercus (Fagaceae)
		salicicola	Salix (Salicaceae)
		scabrum	Betula (Betulaceae)
		testaceoscabrum	Betula (Betulaceae)
	Psiloboletinus	lariceti	Larix (Pinaceae)
	Pulveroboletus	hemichrysus	Pinus (Pinaceae)
		shoreae	Shorea (Dipterocarpaceae)
	Suillus	acerbus	Pinus, 3-needle (Pinaceae)
		americanus	Pinus, 5-needle (Pinaceae)
		caerulescens	Pseudotsuga (Pinaceae)
		cavipes	Larix (Pinaceae)
		granulatus	Pinus (Pinaceae)
		grevillei	Larix (Pinaceae)
		lakei	Pseudotsuga (Pinaceae)
		placidus	Pinus, 5-needle (Pinaceae)
		tomentosus	Pinus (Pinaceae)
		variegatus	Pinus (Pinaceae)
	Tubosaeta	brunneosetosa	Paramacrolobium (Caesalpiniaceae)
	Veloporphyrellus	pantoleucus	Quercus (Fagaceae)
Cantharellaceae	Craterellus	botacensis	Quercus (Fagaceae)

(continued)

Table 11.4. (continued)

Habit, class, family	Genus	Species	Host
Epigeous Habit			
Basidiomycotina			
Clavariaceae	Ramaria	holorubella	Nothofagus (Fagaceae)
		patagonica	Nothofagus (Fagaceae)
		sinapicolor	Eucalyptus (Myrtaceae)
		velocimutans	Tsuga (Pinaceae)
		stuntzii	Tsuga (Pinaceae)
Cortinariaceae	Cortinarius	albonigrellus	Salix (Salicaceae)
		alnetorum	Alnus (Betulaceae)
		alneus	Alnus (Betulaceae)
		alpinus	Salix (Salicaceae)
		archeri	Eucalyptus (Myrtaceae)
		austroacutus	Nothofagus (Fagaceae)
		betulinus	Betula (Betulaceae)
		boyacensis	Quercus (Fagaceae)
		citrinofulvescens	Picea (Pinaceae)
		claricolor	Picea (Pinaceae)
		dilutus	Alnus (Betulaceae)
		geosmus	Nothofagus (Fagaceae)
		herculeus	Cedrus (Pinaceae)
		humicola	Fagus (Fagaceae)
		levipileus	Dryas (Rosaceae)
		lundelli	Picea (Pinaceae)
		orellanoides	Quercus (Fagaceae)
		percavus	Dryas (Rosaceae)
		radicatus	Eucalyptus (Myrtaceae)
		saporatus	Pinus (Pinaceae)
		scotoides	Salix (Salicaceae)
		subnotatus	Fagus (Fagaceae)
		triumphans	Betula (Betulaceae)
	Cuphocybe	alborosea	Nothofagus (Fagaceae)
	Dermocybe	alnophila	Alnus (Betulaceae)
		cinnabarina	Fagus (Fagaceae)
		fervida	Pinus (Pinaceae)
		uliginosa	Salix (Salicaceae)
	Descolea	antarctica	Nothofagus (Fagaceae)
	Hebeloma	candidipes	Picea (Pinaceae)
		album	Cistus (Cistaceae)
		eburneum	Cedrus (Pinaceae)
		kuehneri	Salix (Salicaceae)
		moseri	Nothofagus (Fagaceae)
		ochroalbum	Populus (Salicaceae)
		sinuosum	Fagus (Fagaceae)
	Inocybe	aurea	Pinus (Pinaceae)
		chilensis	Nothofagus (Pinaceae)
		eucoblema	Picea (Pinaceae)
		granulosipes	Eucalpytus (Myrtaceae)
		maculipes	Dryas (Rosaceae)

(continued)

Table 11.4. (continued)

Habit, class, family	Genus	Species	Host
Epigeous Habit			
Basidiomycotina			
Cortinariaceae	*Inocybe*	*orbata*	*Cedrus* (Pinaceae)
		rufula	*Cedrus* (Pinaceae)
		submaculipes	*Pinus* (Pinaceae)
		tristis	*Salix* (Salicaceae)
		vaccina	*Picea* (Pinaceae)
		vulpinella	*Salix* (Salicaceae)
	Naucoria	*alnetorum*	*Alnus* (Betulaceae)
		cephalescens	*Salix* (Salicaceae)
		escharoides	*Alnus* (Betulaceae)
		fellea	*Betula* (Betulaceae)
		salicia	*Salix* (Salicaceae)
		striatula	*Alnus* (Betulaceae)
	Stephanopus	*caeruleus*	*Nothofagus* (Fagaceae)
Gomphidiaceae	*Chroogomphus*	*flavipes*	*Larix* (Pinaceae)
		rutilus	*Pinus*, 3-needle (Pinaceae)
		sibiricus	*Pinus*, 5-needle (Pinaceae)
		vinicolor	*Pinus*, 3-needle (Pinaceae)
	Gomphidius	*gracilis*	*Larix* (Pinaceae)
		maculatus	*Larix* (Pinaceae)
		roseus	*Pinus* (Pinaceae)
Hydnaceae	*Bankera*	*fuligineo-alba*	*Pinus* (Pinaceae)
Hygrophoraceae	*Hygrophorus*	*flavodiscus*	*Pinus* (Pinaceae)
		gliocyclus	*Pinus* (Pinaceae)
		lindtneri	*Corylus* (Betulaceae)
		melizeus	*Betula* (Betulacae)
		piceae	*Picea* (Pinaceae)
		poetarum	*Fagus* (Fagaceae)
		queletii	*Larix* (Pinaceae)
		speciosus	*Larix* (Pinaceae)
Paxillaceae	*Paxillus*	*boletinoides*	*Nothofagus* (Fagaceae)
		defibulatus	*Nothofagus* (Fagaceae)
		filamentosus	*Alnus* (Betulaceae)
Russulaceae	*Lactarius*	*azonites*	*Quercus* (Fagaceae)
		blennius	*Fagus* (Fagaceae)
		bresadolianus	*Picea* (Pinaceae)
		caribaeus	*Coccoloba* (Polgonaceae)
		deterrimus	*Picea* (Pinaceae)
		dryadophilus	*Dryas* (Rosaceae)
		insulsus	*Quercus* (Fagaceae)
		nebulosus	*Coccoloba* (Polygonaceae)
		obscuratus	*Alnus* (Betulaceae)
		porninsus	*Larix* (Pinaceae)
		pseudouvidus	*Salix* (Salicaceae)
		pubescens	*Betula* (Betulaceae)
		pusillus	*Alnus* (Betulaceae)
		sanguifluus	*Pinus* (Pinaceae)
		torminosus	*Betula* (Betulaceae)

(continued)

Table 11.4. (continued)

Habit, class, family	Genus	Species	Host
Epigeous Habit			
Basidiomycotina			
Russulaceae	Russula	alpina	*Salix* (Salicaceae)
		alnetorum	*Alnus* (Betulaceae)
		azurea	*Picea* (Pinaceae)
		carpini	*Carpinus* (Betulaceae)
		coerulea	*Pinus* (Pinaceae)
		cremeolilacina	*Coccoloba* (Polygonaceae)
		fellea	*Fagus* (Fagaceae)
		flava	*Betula* (Betulaceae)
		fuegiana	*Nothofagus* (Fagaceae)
		nigrodisca	*Polygonum* (Polygonaceae)
		maior	*Nothofagus* (Fagaceae)
		nitida	*Betula* (Betulaceae)
		pascua	*Dryas* (Rosaceae)
		puellula	*Fagus* (Fagaceae)
		straminea	*Quercus* (Fagaceae)
		torulosa	*Pinus* (Pinaceae)
		viscida	*Picea* (Pinaceae)
Strobilomycetaceae	Austroboletus	festivus	*Hymenaea* (Caesalpiniaceae)
		heterospermus	*Abies* (Pinaceae)
	Boletellus	alveolatus	*Quercus* (Fagaceae)
		cyanescens	*Nothofagus* (Fagaceae)
Tricholomataceae	Leucopaxillus	barbarus	*Quercus* (Fagaceae)
		lentus	*Picea* (Pinaceae)
	Tricholoma	bakamatsutake	*Quercus* (Fagaceae)
		coarctatum	*Eucalyptus* (Myrtaceae)
		diemii	*Nothofagus* (Fagaceae)
		populinum	*Populus* (Salicaceae)
		robustum	*Pinus* (Pinaceae)
Hypogeous Habit			
Ascomycotina			
Ascobolaceae	Sphaerosoma	fuscescens	*Eucalyptus* (Myrtaceae)
Helvellaceae	Barssia	oregonensis	*Pseudotsuga* (Pinaceae)
Pezizaceae	Amylascus	herbertianus	*Eucalyptus* (Myrtaceae)
	Hydnotryopsis	setchelli	*Quercus* (Fagaceae)
Tuberaceae	Tuber	gibbosum	*Pseudotsuga* (Pinaceae)
Basidiomycotina			
Boletaceae	Alpova	alexsmithii	*Tsuga* (Pinaceae)
		diplophloeus	*Alnus* (Betulaceae)
	Rhizopogon	colossus	*Pseudotsuga* (Pinaceae)
		hawkeri	*Pseudotsuga* (Pinaceae)
		luteolus	*Pinus* (Pinaceae)
		nigrescens	*Pinus* (Pinaceae)
		occidentalis	*Pinus* (Pinaceae)
		villosulus	*Pseudotsuga* (Pinaceae)
		vinicolor	*Pseudotsuga* (Pinaceae)

(continued)

Table 11.4. (continued)

Habit, class, family	Genus	Species	Host
Hypogeous Habit			
Basidiomycotina			
Boletaceae	*Truncocolumella*	*citrina*	*Pseudotsuga* (Pinaceae)
		rubra	*Tsuga* (Pinaceae)
Cortinariaceae	*Cortinarius*	*globuliformis*	*Eucalyptus* (Myrtaceae)
	Hymenogaster	*albus*	*Eucalyptus* (Myrtaceae)
		alnicola	*Alnus* (Betulaceae)
		mcmurphyi	*Quercus* (Fagaceae)
	Setchelliogaster	*tenuipes*	*Eucalyptus* (Myrtaceae)
	Thaxterogaster	*pingue*	*Abies* (Pinaceae)
Chondrogastraceae	*Chondrogaster*	*pachysporus*	*Eucalyptus* (Myrtaceae)
Gomphidiaceae	*Brauniellula*	*albipes*	*Pinus* (Pinaceae)
	Gomphogaster	*leucosarx*	*Pinus* (Pinaceae)
Hysterangiaceae	*Hysterangium*	*separabile*	*Quercus* (Fagaceae)
Mesophelliaceae	*Castoreum*	*radicatum*	*Eucalyptus* (Myrtaceae)
	Mesophellia	*arenaria*	*Eucalyptus* (Myrtaceae)
		castanea	*Eucalyptus* (Myrtaceae)
		ingratissima	*Eucalyptus* (Myrtaceae)
		pachythrix	*Eucalyptus* (Myrtaceae)
		parvispora	*Eucalyptus* (Myrtaceae)
Russulaceae	*Arcangeliella*	*crassa*	*Abies* (Pinaceae)
	Macowanites	*americanus*	*Pinus* (Pinaceae)
	Martellia	*gilkeyae*	*Pseudotsuga* (Pinaceae)
	Zelleromyces	*cinnabarinus*	*Pinus* (Pinaceae)
Sclerodermataceae	*Scleroderma*	*paradoxum*	*Eucalyptus* (Myrtaceae)
Strobilomycetaceae	*Austrogautieria*	*clelandii*	*Eucalyptus* (Myrtaceae)
		manjimupana	*Eucalyptus* (Myrtaceae)
	Gautieria	*novae-zelandiae*	*Nothofagus* (Fagaceae)
Tricholomataceae	*Hydnangium*	*carneum*	*Eucalyptus* (Myrtaceae)

plant in experimental confrontations. Many ectomycorrhizal fungi fall into this category (Table 11.6).

Host Receptivity

Plant species (or genera) can range from being narrowly to broadly receptive toward fungal symbionts. The numbers of mycorrhizal fungi that associate with a particular plant indicate breadth of receptivity. Most mycorrhizal plants can form mycorrhizae with numerous fungus species. Still, the spectrum of host receptivity goes from narrow (low number of fungal associates) to broad (high number). No plant species is known to form mycorrhizae with only a single fungus species. However, some host genera such as *Alnus* have few fungal associates as compared to ectomycorrhizal conifers in the same plant communities (Molina, 1979, 1981). Similarly, ectomycorrhizal fungus associates for some dipterocarp species and other tropical EM hosts (Smits, 1983; Janos, 1985) are

Table 11.5. Examples of Ectomycorrhizal Fungal Species Restricted to Angiosperms, Gymnosperms, or a Single Host Family (Intermediate Host Range) by Fruiting Habit, Class, Family, and Genus

Habit, class, family	Genus	Species	Host
Epigeous Habit			
Ascomycotina			
Basidiomycotina			
Amanitaceae	*Amanita*	*beckeri*	Angiosperms
		mairei	Angiosperms
		porphyria	Pinaceae
		smithiana	Pinaceae
		spreta	Angiosperms
		subalpina	Pinaceae
		submembranacea	Pinaceae
Boletaceae	*Boletus*	*aereus*	Fagaceae
		amarellus	Pinaceae
		fechtneri	Fagaceae
		fragrans	Angiosperms
		impolitus	Angiosperms
		moravicus	Angiosperms
		ornatipes	Angiosperms
		pinicola	Pinaceae
		piperatus	Pinaceae
		queletii	Angiosperms
		rhodoxanthus	Fagaceae
		satanus	Angiosperms
		spadiceus	Fagaceae
		zelleri	Pinaceae
	Gyrodon	*rompelii*	Angiosperms
	Leccinum	*aurantiacum*	Angiosperms
		crocipodium	Angiosperms
		griseum	Angiosperms
		holopus	Betulaceae
		quercinum	Fagaceae
		subleucophaeum	Betulaceae
		vulpinum	Pinaceae
	Pulveroboletus	*auriporus*	Angiosperms
	Suillus	*bovinus*	Pinaceae
		brevipes	Pinaceae
		imitatus	Pinaceae
		luteus	Pinaceae
		pictus	Pinaceae
		punctatipes	Pinaceae
	Tylopilus	*formosus*	Angiosperms
		potamogeton	Angiosperms
Cantharellaceae	*Cantharellus*	*cinnabarinus*	Angiosperms
		guyanensis	Caesalpiniaceae
	Craterellus	*cornucopioides*	Angiosperms

(continued)

Table 11.5. (continued)

Habit, class, family	Genus	Species	Host
Epigeous Habit			
Basidiomycotina			
Clavariaceae	*Ramaria*	*bataillei*	Angiosperms
		brunnea	Pinaceae
		fennica	Pinaceae
		fumigata	Angiosperms
		testaceo-flava	Pinaceae
Corticiaceae	*Amphinema*	*byssoides*	Pinaceae
	Byssoporia	*terrestris*	Pinaceae
Cortinariaceae	*Cortinarius*	*aremoricus*	Fagaceae
		australiensis	Myrtaceae
		caesiogriseus	Angiosperms
		cotoneus	Angiosperms
		isabellinus	Pinaceae
		ignicolor	Pinaceae
		melinus	Pinaceae
		polymorphus	Angiosperms
		pulchripes	Betulaceae
		tenebricus	Angiosperms
		variegatus	Pinaceae
	Cuphocybe	*ferruginea*	Fagaceae
	Dermocybe	*carpineti*	Angiosperms
		holoxantha	Pinaceae
		sanguinea	Pinaceae
		squamulosa	Pinaceae
		splendida	Myrtaceae
	Descolea	*recedens*	Angiosperms
	Hebeloma	*clavulipes*	Angiosperms
		fastibile	Pinaceae
		sacchariolens	Angiosperms
		velutipes	Angiosperms
	Inocybe	*calamistrata*	Pinaceae
		confusa	Angiosperms
		gymnocarpa	Pinaceae
		lateritia	Fagaceae
		lucifuga	Pinaceae
		perbrevis	Angiosperms
		pudica	Pinaceae
		squammata	Angiosperms
		tigrina	Pinaceae
	Leucocortinarius	*bulbiger*	Pinaceae
	Rozites	*similis*	Fagaceae
Entolomataceae	*Entoloma*	*clypeatum*	Rosaceae
		sinuatum	Angiosperms
Gomphaceae	*Gomphus*	*clavatus*	Pinaceae
		floccosus	Pinaceae

(continued)

Table 11.5. (continued)

Habit, class, family	Genus	Species	Host
Epigeous Habit			
Basidiomycotina			
Gomphidiaceae	Chroogomphus	helveticus	Pinaceae
		tomentosus	Pinaceae
	Gomphidius	glutinosus	Pinaceae
		oregonensis	Pinaceae
		roseus	Pinaceae
Hydnaceae	Bankera	carnosa	Pinaceae
	Hydnellum	caerulum	Pinaceae
		suaveolens	Pinaceae
Hygrophoraceae	Camarophyllus	borealis	Pinaceae
		virgineus	Angiosperms
	Hygrophorus	agathosmus	Pinaceae
		barbatulus	Fagaceae
		eburneus	Angiosperms
		erubescens	Pinaceae
		leucophaeus	Angiosperms
		olivaceoalbus	Pinaceae
		penarius	Angiosperms
		ponderatus	Angiosperms
		purpurascens	Pinaceae
		russula	Fagaceae
Paxillaceae	Paxillus	muelleri	Angiosperms
Polyporaceae	Albatrellus	confluens	Pinaceae
		ellissii	Pinaceae
		flettii	Pinaceae
		ovinus	Pinaceae
Russulaceae	Lactarius	acris	Fagaceae
		aspideus	Angiosperms
		chrysorrheus	Fagaceae
		controversus	Salicaceae
		deliciosus	Pinaceae
		lapponicus	Betulaceae
		lignyotus	Pinaceae
		mairei	Angiosperms
		picinus	Pinaceae
		zonarius	Angiosperms
	Russula	adusta	Pinaceae
		brunneoviolacea	Fagaceae
		decolorans	Pinaceae
		integra	Pinaceae
		laeta	Fagaceae
		mustelina	Pinaceae
		nauseosa	Pinaceae
		pectinata	Angiosperms
		sororia	Angiosperms
		velenovskyi	Angiosperms

(continued)

Table 11.5. (continued)

Habit, class, family	Genus	Species	Host
Epigeous Habit			
Basidiomycotina			
Sclerodermataceae	*Scleroderma*	*albidum*	Angiosperms
Strobilomycetaceae	*Austroboletus*	*cookei*	Angiosperms
	Boletellus	*badiovinosus*	Fagaceae
Tricholomataceae	*Catathelasma*	*imperiale*	Pinaceae
	Leucopaxillus	*mirabilis*	Pinaceae
	Tricholoma	*colossum*	Pinaceae
		focale	Pinaceae
		fulvocastaneum	Fagaceae
		impolitum	Angiosperms
		magnivelare	Pinaceae
		matsutake	Pinaceae
		pessundatum	Pinaceae
		vaccinum	Pinaceae
		virgatum	Pinaceae
Hypogeous Habit			
Ascomycotina			
Balsamiaceae	*Balsamia*	*nigrens*	Pinaceae
	Picoa	*carthusiana*	Pinaceae
Elaphomycetaceae	*Elaphomyces*	*aculeatus*	Angiosperms
		leucosporus	Angiosperms
		leveillei	Angiosperms
		morrettii	Pinaceae
Geneaceae	*Genea*	*hispidula*	Angiosperms
		klotzschii	Angiosperms
		verrucosa	Angiosperms
Helvellaceae	*Fischerula*	*subcaulis*	Pinaceae
	Hydnotrya	*cerebriformis*	Pinaceae
		cubispora	Pinaceae
		michaelis	Pinaceae
		variiformis	Pinaceae
Pezizaceae	*Ruhlandiella*	*berolinensis*	Myrtaceae
Pyronemataceae	*Geopora*	*cooperi*	Pinaceae
	Labyrinthomyces	*varius*	Angiosperms
Terfeziaceae	*Hydnobolites*	*cerebriformis*	Angiosperms
	Terfezia	*arenaria*	Angiosperms
		boudieri	Cistaceae
		claveryi	Cistaceae
Tuberaceae	*Tuber*	*magnatum*	Angiosperms
		macrosporum	Angiosperms
Basidiomycotina			
Astraeaceae	*Radiigera*	*atrogleba*	Pinaceae

(continued)

Table 11.5. (continued)

Habit, class, family	Genus	Species	Host
Hypogeous Habit			
Basidiomycotina			
Boletaceae	*Alpova*	*olivaceoniger*	Pinaceae
	Gastroboletus	*subalpinus*	Pinaceae
		turbinatus	Pinaceae
	Rhizopogon	*ellenae*	Pinaceae
		fuscorubens	Pinaceae
		rubescens	Pinaceae
		subcaerulescens	Pinaceae
		truncatus	Pinaceae
Cortinariaceae	*Cortinarius*	*bigelowii*	Pinaceae
	Destuntzia	*fusca*	Pinaceae
	Hymenogaster	*arenarius*	Angiosperms
		boozeri	Angiosperms
		sublilacinus	Pinaceae
Cribbeaceae	*Mycolevis*	*siccigleba*	Pinaceae
Hysterangiaceae	*Hysterangium*	*cistophilum*	Fagaceae
		clathroides	Angiosperms
		coriaceum	Pinaceae
		crassirhachis	Pinaceae
	Trappea	*darkeri*	Pineceae
Leucogastraceae	*Leucogaster*	*badius*	Angiosperms
		rubescens	Pinaceae
	Leucophleps	*magnata*	Pinaceae
		spinispora	Pinaceae
Octavianinaceae	*Octavianina*	*asterosperma*	Angiosperms
		tasmanica	Angiosperms
Russulaceae	*Macowanites*	*chlorinosmus*	Pinaceae
		luteus	Pinaceae
	Martellia	*brunnescens*	Pinaceae
		ellipsospora	Pinaceae
		maculata	Pinaceae
		oregonensis	Pinaceae
		subochracea	Pinaceae
	Zelleromyces	*gilkeyae*	Pinaceae
Strobilomycetaceae	*Chamonixia*	*caespitosa*	Pinaceae
	Gautieria	*monticola*	Pinaceae
		morchelliformis	Angiosperms
Zygomycotina			
Endogonaceae	*Endogone*	*flammicorona*	Pinaceae

Table 11.6. Examples of Ectomycorrhizal Fungal Species with Little Host Restriction (Broad Host Range) by Fruiting Habit, Class, Family, and Genus

Habit, class, family	Genus	Species
Epigeous Habit		
Ascomycotina		
Basidiomycotina		
Amanitaceae	Amanita	aspera, fulva, gemmata, inaurata, muscaria, pantherina, phalloides, rubescens, solitaria, spissa, strobiliformis, vaginata, verna, virosa
Astraeaceae	Astraeus	hygrometricus, pteridus
Boletaceae	Boletus	appendiculatus, armeniacus, badius, calopus, chrysenteron, edulis, erythropus, luridus, minatioolivaceus, pulverulentus, regius, rubellus, spadiceus, subtomentosus, truncatus
	Gyroporus	castaneus, cyanescens
	Phylloporus	rhodoxanthus
	Pulveroboletus	ravenelii
	Tylopilus	chromapes, felleus, gracilis, porphyrosporus
Cantharellaceae	Cantharellus	cibarius, infundibuliformis, tubiformis
Clavariaceae	Clavaria	argillacea
	Ramaria	aurea, botrytis, flava, formosa, mairei, subbotrytis
Corticiaceae	Byssocorticium	atrovirens
	Byssoporia	sublutea
	Piloderma	bicolor, byssinum, sulphureum
Cortinariaceae	Cortinarius	acutus, anomalus, bicolor, bivelus, cinnamomeus, everneus, hemitrichus, leucophanes, mucosus, multiformis, obtusus, phrygianus, saniosus
	Dermocybe	anthracina, cinnamomea, malicoria, palustris, phoenicea
	Hebeloma	crustuliniforme, cylindrosporum, hiemale, longicaudum, mesophaeum, minus, pumilum, sinapizans
	Inocybe	asterospora, bongardii, brunnea, cincinnata, dulcamera, fastigata, jurana, lacera, lanuginella, petiginosa, terrigena, umbrina
	Rozites	caperata
Entolomataceae	Clitopilus	prunulus
	Entoloma	nidorosum, prunuloides, rhodopolium, sericeum, turbidum
Hydnaceae	Dentinum	repandum
	Hydnellum	velutinum
	Hydnum	imbricatum, rufescens, scabrosum
Hygrophoraceae	Hygrophorus	capreolarius, camarophyllus, chrysodon, discoideus, hypothejus, karstenii, marzuolus, pudorinus
Lycoperdaceae	Lycoperdon	perlatum, pyriforme, umbrinum
Paxillaceae	Paxillus	involutus
Pisolithaceae	Pisolithus	tinctorius
Polyporaceae	Albatrellus	cristatus

(continued)

Table 11.6. (continued)

Habit, class, family	Genus	Species
Epigeous Habit		
Basidiomycotina		
Russulaceae	Lactarius	decipiens, fuliginosus, helvus, necator, piperatus, repraesentaneus, rufus, scrobiculatus, spinosulus, uvidus, vellereus, volemus
	Russula	aeruginea, albonigra, amoena, anthracina, cyanoxantha, densifolia, emetica, foetens, heterophylla, lutea, nigricans, ochroleuca, odorata, olivacea, paludosa, palumbina, parazurea, vesca, virescens, xerampelina
Sclerodermataceae	Scleroderma	bovista, cepa, citrinum, laeve, polyrhizum, verruosum
Strobilomycetaceae	Boletellus	betula, chrysenteroides
	Strobilomyces	floccopus
Thelephoraceae	Thelephora	anthocephala, atrocitrina, penicillata, terrestris
Tricholomataceae	Catathelasma	ventricosum
	Laccaria	amethystina, bicolor, laccata, montana, proxima
	Leucopaxillus	cerealis
	Tricholoma	caligatum, columbetta, flavobrunneum, flavovirens, myomyces, saponaceum, sulphureum
Hypogeous Habit		
Ascomycotina		
Balsamiaceae	Balsamia	magnata, platyspora, vulgaris
Elaphomycetaceae	Cenococcum	geophilum
	Elaphomyces	anthracinus, granulatus, muricatus, mutabilis, reticulatus, variegatus
Geneaceae	Genabea	cerebriformis
	Genea	gardneri, harknessii, intermedia
Helvellaceae	Hydnotrya	tulasnei
Pezizaceae	Pachyphloeus	citrinus, ligericus, melanoxanthus
Terfeziaceae	Choiromyces	alveolatus, venosus
Tuberaceae	Tuber	aestivum, borchii, brumale, californicum, excavatum, melanosporum, puberulum, rapaeodorum, rufum
Cortinariaceae	Hymenogaster	bulliardii, calosporus, citrinus, decorus, lilacinus, luteus, olivaceus, populetorum, tener, vulgaris
Hysterangiaceae	Hysterangium	membranaceum
Leucogastraceae	Leucogaster	nudus
Melanogastraceae	Melanogaster	ambiguus, broomeianus, euryspermus, intermedius, tuberiformis, variegatus
Russulaceae	Elasmomyces	mattirolianus
	Zelleromyces	stephensii
Sclerodermataceae	Scleroderma	hypogaeum
Strobilomycetaceae	Gautieria	graveolens, mexicana, otthii
Zygomycotina		
Endogonaceae	Endogone	lactiflua

reported as low in number, although this may be an artifact of inadequate study of those systems. The number of fungal associates of ericoid plants seems low, but again the associations are inadequately studied. Alternatively, Molina and Trappe (1982a) experimentally demonstrated that *Arbutus* and *Arctostaphylos* spp. are receptive in pure culture to a broad range of ectomycorrhizal fungi, more so than many ectomycorrhizal hosts.

Ecological Specificity

The concept of ecological specificity phenomena is critical to evaluation of the ecological consequences of compatibilities in mycorrhizal associations. Harley and Smith (1983) coined the term "ecological specificity" to emphasize that mycorrhizal associations induced in laboratory experiments may not occur under field conditions and that competition in natural habitats may increase expression of specificity. We expand the definition to include all environmental biotic and abiotic factors that affect the ability of plants to form functional mycorrhizae with particular fungi. Examples within this category include interactions of roots and fungal propagules (e.g., germination phenomena); interactions between plants, between plants and fungi, and between fungi that restrict or enhance growth or function of either symbiont in particular habitats; soil microniches that favor or restrict certain associations; and rhizosphere microorganisms that influence differential development of host–fungus associations. Ecological specificities are critical: these interactions yield the final outcome of associations (what is mycorrhizal with what) and thus the potential for mycorrhizal linkages in terrestrial ecosystems.

Recognition Phenomena and Cellular Compatibility

The ultimate resolution of compatibility phenomena lies within the ultrastructural, biochemical, and genetic levels of interactions between fungus and plant. Initial recognition phenomena, alterations in cellular structure, and establishment of functional interfaces are all vital to balanced relationships and exchanges of metabolites. Considerable research is in progress on this subject and is reviewed elsewhere (Anderson, 1988; Duddridge, 1986a, 1987; Gianinnazzi-Pearson, 1983). We will refer to this level of specificity/compatibility only in passing or when we envisage an ecological consequence.

Specificity versus Compatibility

Specificity and compatibility are terms that are often interchanged, but can have different meanings. In plant pathology, host specificity refers to limits in host range for a particular pathogen. Pathogenic relationships with strong host–fungus specialization are emphasized, particularly at the species-for-species level or even race level. Brian (1976) recognizes that levels of specificities exist, from pathogens to nonpathogens. Harley and Smith (1983) begin their discussion of mycorrhiza specificity similarly. Duddridge (1986a) used the term specificity to denote "selectivity" rather than differential effectiveness in enhancing host nutrition. Yet, selectivity is not further defined. Selectivity is appropriate for discussion of specificity, because it indicates processes that determine whether fungus–root associations form functional mycorrhizae in field situations.

"Compatibility" is typically used to express structural characters or operational mechanisms that define a particular state of symbiosis. We use the term "compatibility" in the sense of Gianinazzi-Pearson (1983) to emphasize the merging of symbionts to form a structurally defined mycorrhiza in which physiological activity and exchange of metabolites between the partners improve nutrition of host and fungus. Ecological benefits toward one or both partners indicate functional compatibility, particularly with reference to mycorrhizal linkages in ecosystems. Incompatibility produces deleterious effects on fungus or host or both. For example, the production of phenolics in root tissue or tissue disorganization in response to fungal penetration may be considered signs of incompatibility. In the context of this review, we do not use "compatibility" in the sense of genetic interactions, such as complementary mating types.

Levels of Specificity within Different Mycorrhiza Classes

Vesicular-Arbuscular Mycorrhizae (Fig. 11.1A)

Vesicular-arbuscular mycorrhizal associations generally lack specificity. We know of no examples of VAM fungi that form mycorrhizae only with a restricted group of hosts as is true of other mycorrhiza classes. Individual species of VAM fungi can associate with diverse plant groups, from herbs to long-lived woody perennials. Similarly, VAM plants often associate with many VAM fungi. For example, Molina et al. (1979) found *Festuca* spp. associated with 11 species of Endogonaceae in habitats scattered over the western United States.

Repeated failures to isolate and grow axenic cultures of VAM fungi emphasize the obligate nature of VAM fungi. VAM plants, on the other hand, range widely from facultative to obligate dependency on mycorrhizal association. Many facultative plants are VAM in some ecological settings but may lack mycorrhizae in others (Trappe, 1987). Some commonly reported facultative plants include members of the genera *Brassica, Equisetum, Saxifraga,* and *Trisetum*. Several of these examples belong to families of mostly nonmycorrhizal plants.

Determining the obligate vs. facultative nature of mycorrhizal associations is difficult because we lack information on how the symbiosis affects the reproductive fitness of the plant or fungus. A truly obligate associate would need mycorrhizae sometime in its lifetime to successfully reproduce under natural field conditions. Data on reproductive fitness are lacking in most ecological studies of mycorrhizal associations. However, plants that cannot survive to the reproductive stage, as is the case with virtually all terrestrial woody perennials, clearly are obligate.

Caution is needed in generalizing the lack of specificity in VAM associations. Most plant species and habitats remain unexamined for VAM, and only a few experimental cross-inoculations between VAM fungi and hosts have been tried. Geographically isolated habitats or ecologically unique VAM associations may yet reveal VAM specificity.

Phenomena of ecological specificity and functional compatibility are barely explored for VAM associations in natural settings. Although mechanistic barriers to recognition may be lacking, ecological and physiological interactions between fungi and hosts may play major roles in determining selectivity or preference between symbionts for associations. For example, McGonigle and Fitter (1990) compared several grass species for VAM fungus associates in the field. They report that *Holcus lanatus,* unlike other grass species, was preferentially infected by the "fine edophyte" *Glomus tenue* rather than by other coarse endophytes. Different VAM fungi colonize root systems of the same or different hosts in varying degree; VAM fungi also differ strongly in nutrient translocation to hosts and host carbon utilization. Regardless of the lack of recognition barriers to VAM colonization in experimental inoculations, fungus–fungus, plant–fungus, and plant–plant interactions as mediated by environmental conditions in natural settings will affect the outcome or selectivity of particular VAM associations. For example, companion plants can influence whether neighboring plants form VAM and the degree of VAM colonization, and influence competitive interactions (Miller, 1987). Similarly, interactions of VAM fungi and other rhizosphere microorganisms, e.g., *Rhizobium* strains and VAM fungus species on legumes (Azcón et al., 1991), can also influence the functional compatibility of the symbioses. It is within this setting of ecological specificity and functional compatibility that the impact of VAM on plant community dynamics, particularly the linking of plants by shared fungi, will be best understood and useful to natural resource managers.

Ectomycorrhizae (Fig. 11.1B)

Experimental and ecological data document differences among EM associations in their degree of compatibility and specificity. Most EM fungi produce large, showy fruiting structures whose constancy of host association in the field can be repeatedly recorded. Many EM fungi can be isolated and grown in pure culture and their ability to form EM with known hosts can be experimentally tested. Many EM fungi also form distinctive EM morphologies (types) that can be visually identified by color, mantle structure, and external mycelium (Agerer, 1986; Trappe, 1967; Zak, 1973). This permits comparison of *in vivo* field associations with *in vitro* experiments, confirmation of sporocarp associations with specific mycorrhiza types, and measurement of abundance and dominance of different EM fungi on root systems.

Fidelity of Ectomycorrhizal Plants and Fungi

EM hosts typically demonstrate strong EM fidelity, especially Pinaceae and Fagaceae, but many EM plants can also form VAM, either simultaneously or independently. Commonly cited examples are listed in Table 11.3. The ecological setting or presence and abundance of VAM vs. ectomycorrhizal fungal propagules typically determine the sequence of mycorrhizal formation. For example, in *Eucalyptus* initially high inoculum potential of VAM fungus propagules yield early colonization by VAM over EM; however, EM fungi, once established, dominate VAM in secondary root colonizations as humus builds up (Chilvers et al., 1987).

Many EM plants also form ectendomycorrhizae, again either exclusively or simultaneously with EM. Ectendomycorrhizae are close structural relatives of EM, so their occurrence on EM hosts does not represent strong infidelity compared to the taxonomically and morphologically distant VAM (see the section on "Ectendomycorrhizae"). Many ericaceous plants that typically form arbutoid mycorrhizae also form typical EM lacking intracellular fungal penetration of the epidermis. Arbutoid mycorrhizae are also close kin to EM for reasons discussed under ericoid mycorrhizae and thus also do not represent infidelity on EM hosts. Largent et al. (1980) also report that other ericoid genera form occasional EM; given the flexibility of mycorrhizal relations shown within the Ericales, we need to more accurately assess their mycorrhizal status in natural settings and experimental confrontations.

It is simpler to discuss the fidelity of EM fungi than hosts. We know of no examples where EM fungi form VAM or ericoid mycorrhizae. Many fungi that otherwise form EM produce arbutoid mycorrhizae on *Arbutus* and *Arctostaphylos* species (Zak, 1976a,b; Molina & Trappe, 1982a). For this reason, we perceive that arbutoid mycorrhizae are a type of EM. A unique attribute of the ericoid mycorrhizal class is fungal intracellular penetration in the epidermis. Arbutoid hosts allow development of this character with EM fungi. Some EM fungi also form monotropoid mycorrhizae (see the section on "Monotropoid Mycorrhizae"). We know, however, that the fungi simultaneously form EM with nearby EM hosts and that carbohydrates move from tree to fungus to achlorophyllous plant (Björkman, 1960; Furman & Trappe, 1971). Unusual or unexpected associations of EM fungi are occasional on nonectomycorrhizal hosts. For example, *Cenococcum geophilum* has been reported on species of *Dryopteris, Galium,* and *Lactuca* (Trappe, 1962). Clearly, we need to experimentally demonstrate how environmental conditions and organism interactions (e.g., competition, interference, dominance) determine the functional compatibility of these unusual relationships.

Facultative versus Obligate Ectomycorrhizal Associations

Most EM plants obligately depend on an EM association. Probably all Pinaceae fall into this category as demonstrated by the repeated failures to establish exotic plantations until proper EM fungi were introduced (Mikola, 1970, 1973), or by the stunting of nonectomycorrhizal seedlings in fumigated nursery soil (Trappe & Strand, 1969). We know of no routinely EM plant that exists in the field without mycorrhizae. Seedlings of *Tsuga heterophylla* may initially survive (but grow slowly) free of EM while growing in rotten wood but usually become EM by the second growing season (Christy et al., 1982). Data on mycorrhizal dependency in various ecological settings are needed to clarify the obligate nature of EM hosts.

Defining and clarifying facultative vs. obligate EM fungi are more difficult than for the host; we poorly understand the natural ecology and physiology of the thousands of EM fungi. If facultative EM fungi are defined as being able to live free from EM association, then no truly facultative EM fungus is known (Harley & Smith, 1983). The fungus *Sphaerosporella brunnea* may prove the exception to this generalization. Danielson (1984) found saprophytic strains isolated from burnt wood able to form EM with *Pinus banksiana* and function similarly to root isolates of *S. brunnea*. Ectomycorrhizal fungi can grow freely in the rhizosphere of non-EM hosts under experimental conditions (Theodorou & Bowen, 1971), but the ecological significance of this is unknown. Some EM fungi possibly can grow somewhat in soil and rhizospheres without forming EM but be unable to complete their life cycle, i.e., produce sporocarps in the absence of mycorrhizal association.

An inappropriate facultative designation for several EM fungi has been used in describing ecological, physiological, and host range differences between EM fungi. For example, common putative "early stage" fungi. e.g., *Laccaria* and *Thelephora* spp., have been termed facultative EM fungi, as are other fungi with broad host ranges. This distinction typically entails an assumption of reduced dependency on the host for carbon or other essential host metabolites necessary for fungus growth and reproduction, e.g., ability to utilize soil carbon substrates. Ectomycorrhizal fungi vary widely in substrate utilization and host carbon dependency (Durall et al., 1992), but such physiological differences should not be confused with facultativeness, i.e., the ability to grow and reproduce completely in the absence of EM development in the field.

The rareness with which EM fungi produce sporocarps in the absence of EM hosts strongly indicates an obligate relationship in a reproductive fitness context. The growth of many EM fungi on artificial media and the inability of others to grow on such media indicate differences among these obligate fungi. An index of EM fungus dependency on host metabolites for growth and reproduction similar to the concept of mycorrhizal dependency developed for plants is needed to bring ecological bearing on the ecosystematic roles of the diverse EM fungus flora.

Host Range of EM Fungi

Host range and compatibility evaluations of EM fungi have been based on two criteria, and we are currently developing a third. The most readily available and abundant data come from observations of sporocarp–host associations in the field. Many EM fungi consistently fruit with particular host genera while others fruit with a wide variety of hosts. The EM themselves can sometimes be directly linked with the sporocarps to confirm the association. Lack of sporocarp association does not necessarily imply inability to form mycorrhizae with hosts in the field, however (Trappe, 1962; Molina & Trappe, 1982b).

The potential of fungi and hosts to form EM is typically demonstrated in pure culture syntheses; aseptically (Molina & Palmer, 1982) or semiaseptically (Fortin et al., 1980) grown seedlings are inoculated with single isolates of known fungi. Although these tests are artificial, the results can be useful in documenting compatible and incompatible responses, particularly if exogenous sugars are not abundant in the rooting substrate. In greenhouse studies, vegetative mycelium or spores of known fungi can be used to inoculate seedlings, and we have gathered host range data for a limited number of fungi from such experiments. Recently, we have grown multiple host species in a common soil, either individually or in mixtures, and recorded the simultaneous EM development on the various hosts. Similarly, a mixture of plants grown in a common sterile soil can be inoculated with known fungi. In these experiments we assess compatibility differences between hosts in regard to individual fungi, information extremely useful for documenting host specificity, guild concepts, and potential for interplant connections.

Narrow or Restricted Host Range

Contrary to conventional wisdom, narrowly restricted host ranges are widespread among EM fungi and occur among taxonomically diverse hosts, fungi, and ecological habitats. The most common restriction is expressed at the host genus level. Fungi that fruit exclusively with *Betula, Pinus,* and *Larix* are commonly cited in the European literature because of the long history of collecting mushrooms in those forests. Table 11.4 acknowledges these well-known associations but also emphasizes the phenomenon among other widespread hosts. For example, *Pseudotsuga menziesii* in western North America has one of the larger assemblages of genus specific fungi. We estimate that approximately 250 fungi are specific to *Pseudotsuga,* primarily in the genera *Cortinarius, Gomphidius, Rhizopogon,* and *Suillus. Eucalyptus* is another genus with many species of genus-specific fungi, particularly of hypogeous basidiomycetes. Recent data from mycological expeditions in Australia (Trappe, Castellano, Bougher, & Malajczuk, unpublished) indicate that *Eucalyptus* and related Myrtaceae may have the richest flora of genus-specific EM fungi in the world, a possibility not surprising given the diversity of *Eucalyptus* species and their dominance on an isolated continent. Another host genus likely to have many genus-specific fungi is *Nothofagus* in the southern hemisphere. The extensive collections and descriptions by Singer and Moser (1965), Horak (1983), and Garrido (1988) in South America and the South Pacific region document numerous fungus species known to occur only with *Nothofagus*. In Malaysia, Becker (1983) and Smits (1983) report the occurrence of EM fungi that may associate only with individual dipterocarp species. If such proves true, these dipterocarp EM associations may truly represent our first known examples of EM fungi restricted to a single host species. Study of other isolated groups of EM hosts such as the Ceasalpiniodeae and other Dipterocarpaceae in the tropics may add to the growing list of restricted-host-range EM fungi.

Tables 11.3 and 11.4 indicate the major genera of fungi that express narrow range (at the host genus level). The phenomenon is perhaps most strongly expressed in the Gomphidiaceae and Boletceae (*Chroogomphus, Gomphidius, Suillus, Leccinum,* and their hypogeous relatives *Brauniellula, Gomphogaster, Rhizopogon, Truncocolumella, Austrogautieria, Alpova*). But examples are known from at least 30 genera in 15 families. Within fungal genera such as *Rhizopogon* and *Suillus,* some subgenus groupings also show restricted host ranges. For example, Molina and Trappe (unpublished data) found the section *Villosuli* of *Rhizopogon* restricted to *Pseudotsuga* and the section *Rhizopogon* predominantly associated with pines.

These genus-level specificities are based primarily on sporocarp–host associations in the field. In laboratory pure culture syntheses, some EM fungi with genus-restricted host–sporocarp associations form EM with unexpected hosts. Molina and Trappe (1982b) found some such unexpected EM had unusual morphological characters or were infrequent or less developed on unexpected hosts compared to sporocarp associates. Also, as demonstrated by Duddridge (1986a,b), high glucose levels in pure culture synthesis can induce abnormalities in EM structure and lend greater colonizing or parasitic potential to a fungus than if the host is the primary source of carbon to initiate and develop functional EM as in the field.

These results prompted Harley and Smith (1983) to develop the concept of ecological specificity and hypothesize that the host ranges of EM fungi are likely more limited in the field than in pure culture host–fungus confrontations. Our research causes us to agree. In pure culture syntheses, Molina and Trappe (1982b) demonstrated the ability of *Rhizopogon vinicolor,* a Douglas-fir sporocarp associate, to form traces of EM on *Tsuga heterophylla* and questionable EM on *Picea sitchensis* and *Pinus monticola*. In a subsequent greenhouse experiment, Massicotte and Molina (unpublished data) inoculated several conifer species grown singly or in mixtures with spores of various *Rhizopogon* species. *Rhizopogon vinicolor* formed EM only with Douglas-fir. When Douglas-fir were grown together with *Tsuga heterophylla* or *Pinus,* the abundant EM of *R. vinicolor* on

Douglas-fir produced prolific external mycelium and rhizomorphs that ramified extensively through all root systems of neighboring conifers but formed no EM. Massicotte and Molina (unpublished data) report the same phenomenon for other *Rhizopogon* species of known sporocarp–host specificity, e.g., pine *Rhizopogon* did not cross over to Douglas-fir and vice versa. From those results and other recent soil bioassays for host associations (Molina, unpublished data), we hypothesize that sporocarp associations are indeed a strong indicator of restricted specificity.

Extensive field experimentation and examination of mycorrhizal potential when hosts are grown in mixtures under nonsterile conditions are needed to address this hypothesis. These must be coupled with physiological tests of functional compatibility between host–fungus combinations, particularly for unusual combinations. For example, Finlay (1989) reported that the larch associates *Suillus cavipes* and *Suillus grevillei*, although capable of forming EM with *Pinus sylvestris* in microcosms, were inferior to the pine fungus *Suillus bovinus* in nutrient uptake and translocation to the pine. As stressed by Finlay (1989), we need to study the functional nature of these various degrees of compatibilities among mycorrhizal associates to understand the role played by mycorrhizal specificity in plant community development.

Although genus level restriction in host range is most common among EM fungi, restricted host ranges follow other patterns. Table 11.4 provides examples of narrow host ranges based primarily on constancy of sporocarp–host associations. Many fungus species, genera, or even families can be restricted to plant families or plant orders. For example, the genera *Suillus* and *Rhizopogon* are mostly restricted to Pinaceae. Other fungi appear to associate only with genera of angiosperms such as *Boletus leptospermi* with *Leptospermum* and *Leccinum scabrum* with *Betula*.

Strains of an EM fungus species can also differ in host range. *Boletus edulis* is reported to have varieties separated taxonomically on sporocarp characters that occur only with pines, firs, or oaks, respectively (Trappe, 1962). Strains of EM fungi often differ significantly in pH optima, growth rates on synthetic media, phosphorus uptake, auxin and enzyme production, drought resistance, etc., so it is not surprising that differences in host compatibility might be expressed at subspecific or ecotypic levels. Considerable research on experimental confrontations is needed to illuminate this level of specificity. Tonkin et al. (1989) report that *Pisolithus tinctorius* isolated originally from sporocarps associated with *Eucalyptus* in Australia initiated EM more quickly than a pine isolate of *P. tinctorius* from the United States. Although both isolates were structurally compatible on *Eucalyptus*, Tonkin et al. (1989) state that the speed of initiation may reflect differences in the host–fungus recognition process.

In a related study of EM formation on micropropagated eucalyptus seedlings, Tonkin et al. (1989) found that one of two *Pisolithus tinctorius* isolates formed EM only on clonal lines from mature trees while the other isolate formed EM on seedlings as well as on clonal lines of juvenile and mature trees. The *P. tinctorius* isolate restricted to clonal lines of mature trees showed lack of Hartig net and a build-up of phenolics on the seedlings and juvenile clones, indications of true incompatibility. Tonkin et al. (1989) state that specificity phenomena such as these may influence successional development of fungi in maturing forests. Deacon and Fleming (Chapter 8, this volume) likewise hypothesize that plant age will affect receptivity and so influence fungal succession in maturing plant communities.

Intermediate to Broad Host Range

Although EM host–fungus association records are far from complete, Trappe's (1962) compilation and data compiled since then (Trappe, unpublished) indicate that many, perhaps most, EM fungi are intermediate to broad host ranging. Tables 11.5 and 11.6 list common examples of EM fungi of intermediate and broad host range, based primarily on host–sporocarp associations but also on experimental syntheses. Note that these fungi are taxonomically diverse and represent most families and genera with predominantly EM members. Indeed, some of the best known EM fungi are of broad host range because they fruit abundantly around the world. *Cenococcum geophilum* is perhaps the most broadly host ranging of known EM fungi. Trappe (1964) lists *C. geophilum* in association with about 150 hosts from nearly all the represented EM genera listed. *Cenococcum geophilum* associates not only with most known EM hosts, including some herbs, but also with some typically VAM plants. Other well known EM fungi of broad host range include *Amanita muscaria, Hebeloma crustuliniforme, Laccaria laccata, Pisolithus tinctorius,* and *Thelephora terrestris,* three of which have been used in forest tree inoculation programs. In a pure culture synthesis experiment, Molina and Trappe (1982b) found that of 28 fungus species tested on 7 species of Pinaceae (including 5 genera), 11 formed EM with all hosts, and for the most part EM were well developed and abundant. Half of those were with fungi with known broad host–sporocarp association, and the other half were fungi previously untested for host range. EM fungi that do not express some level of restricted sporocarp–host association in nature are likely broad or intermediate in their host range.

Molina and Trappe (1982b) described a category of intermediate host range fungi typically with some host–sporocarp specificity showing a wider host range in pure culture synthesis. However, the EM formed in those unexpected associations were usually low in abundance and showed occasional incompatibility in structure, probably due to exogenous sugar as pointed out by Duddridge (1986a,b). Although some fungi may be able to form EM with many hosts in pure culture, their restricted sporocarp–host associations in the field may indicate that ecological specificity is operating. As noted for restricted-host-range fungi, we need to design experiments to evaluate physiological and ecological conditions that determine the expression of such ecological specificity.

Floristic studies by Lee and Kim (1987) of EM sporocarp associations in stands of *Abies, Betula, Castanea, Larix, Picea, Pinus, Populus,* and *Quercus* in Korea yielded groupings of narrow, intermediate, and broad-host-ranging fungi. The largest grouping was the broad host range category with common fungus species in *Laccaria, Amanita,* and *Russula*. However, nearly all hosts had components of genus-restricted fungi. Groupings of narrow, intermediate and broad host range fungi may also be apparent in studies where compatibility is explored by growing multiple hosts in a common soil. For example, Massicotte and Molina (unpublished data) found that of 17 EM types recovered from a soil bioassay with four different hosts, 8 formed on all hosts (broad host ranging), 4 formed on only one host (genus specific), and 5 others formed on 2–3 hosts (intermediate). Data collected in this format will be absolutely necessary to determine how the levels of specificities operating in forest soils influence the formation and potential function of EM guilds.

Host Receptivity, Selectivity, and Ecological Specificity

EM plants vary widely in receptivity toward EM fungi and receptivity may be related to plant age (Tonkin et al., 1989). Data from known or suspected fungus associates provide a starting point in describing differences. As seen in Trappe's (1962) compilation and subsequent data, the number of fungal associates differs strongly among EM hosts. The genus *Alnus* is perhaps our best known example for restricted receptivity based on both sporocarp associations and experimental inoculations. Only a few hundred fungus species are likely sporocarp associates. In a large pure culture synthesis with *Alnus rubra*, Molina (1979) found that of 28 confirmed EM fungi, only 4 formed EM with *A. rubra*. Seven fungi that were widely compatible with other hosts (Pinaceae) failed to form EM with *A. rubra*. This pattern of restricted receptivity was confirmed in pure culture synthesis tests with four other species of *Alnus* (Molina, 1981). The entire genus *Alnus* may have evolved strong specialization toward EM fungi by restricted receptivity. Miller et al. (1991) confirmed this strong specialization toward limited fungus species for *A. rubra* by examining the roots of red alder growing in a variety of ecological habitats and in experimental soil bioassays; they found only 11 types of EM on red alder, a remarkable limit given the thousands of EM fungi occurring with Pinaceae and Fagaccae hosts in the surrounding forests.

We suspect other examples of restricted receptivity occur among EM hosts, particularly for plants that may be geographically isolated from other EM hosts or common to particular ecological habitats. For example, Janos (1985) reported the predominance of a single white EM type on *Neea lactivirens* (Nyctaginaceae) in Costa Rica. Similarly, EM fungus diversity can be low in some dipterocarp forests. Smit (1983) states that dipterocarp species in Malaysia select out specific

EM fungi. He found that soil from beneath the EM tree *Dipterocarpus grandiflorus* failed to provide EM fungi for *Shorea obtusa* of the same family. Becker (1983) found that *S. leprosulans* and *S. maxwelliana* seedlings growing in the same vicinity only shared 2 of 20 EM types. As we examine some of the less well known EM associations, such as the desert truffle mycorrhizae of *Helianthemum* (Alsheikh & Trappe, 1983), the unusual EM of Australian herbs (Warcup, 1988), or scattered species in families typically VAM, additional examples of restricted receptivity will appear.

Further examination of Tables 11.1–11.6 reveals the commonly cited diversity and abundance of EM host–fungus associations, particularly among dominant forest trees. For example, Douglas-fir from western North America alone may have as many as 2000 associates (Trappe, 1977a) and the pines as many. These data are crude in reflecting receptivity and are based primarily on host–sporocarp association. To understand how host receptivity operates in a guild concept, i.e., the ecological importance of neighboring plants to enter into a shared EM network, we need to explore the overlap in compatibility.

We have performed this exercise for *Pseudotsuga* and *Pinus* in the Pacific Northwest of the United States. Douglas-fir and pines are dominant tree genera throughout much of this region. Of the approximately 2000 fungus associates for both hosts, we estimate the overlap of compatible fungi to be in the order of 1800 species or 72%. Thus, although both Douglas-fir and pines have developed strong genus level specificity with some genera such as *Rhizopogon* and *Suillus,* they still share an overwhelming number of potentially compatible "guild" fungi, many of which are broadly host ranging with a diversity of other EM hosts in the region and worldwide.

Differences in receptivity between hosts will determine the degree to which they operate within a guild of shared fungi. These differences must be demonstrated either in the field or in experimental associations wherein different host species are grown in close proximity, allowing roots to contact common fungi. We have completed two studies that evaluate such potential guild structure.

Miller, Koo, and Molina (unpublished data) grew Douglas-fir and red alder separately in six forest soils taken from pure red alder sites as well as from various conifer sites and examined roots for EM types in common. They found that of the 12 total EM types recovered over all forest soil types, only 1 type was shared in common between alder and Douglas-fir. Although this study evidences the restricted nature of red alder EM fungi and that red alder and Douglas-fir select different fungus species, it does not examine how fungi of Douglas-fir or alder react when they contact roots of neighboring hosts. To examine this interaction, the hosts must be grown in the same soil so that their roots and associated fungi make contact.

Massicotte and Molina (unpublished data) used such an approach with four EM tree species from different genera, both singly and in mixtures in common soils. They found that of the 17 total EM types recovered by all hosts, 8 were shared by all hosts, 4 occurred on only one host, and 5 occurred on two or three hosts. Furthermore, when grown in monoculture, the individual hosts formed EM with different EM fungi than when grown in host mixtures, i.e., companion plants influenced EM formation of neighbors and thus overall EM diversity.

Newton (1991) conducted similar seedling bioassays in Great Britain to compare overlap in EM fungus associates between *Quercus robur* and *Betula pendula* when grown in a variety of soils. The degree of overlap was similar to that reported above for Douglas-fir and red alder: of 41 total EM types recovered, the two tree species shared only three, *Paxillus involutus, Scleroderma citrinum,* and *Cenococcum geophilum.* In one field planting, birch was dominated by *Paxillus involutus* and oak was dominated by *Scleroderma citrina.* Newton (1991) states, "the fact that birch and oak developed different mycorrhizal floras when grown on the same soil type, indicates that mycorrhizal fungi differ in their ability to colonise and proliferate on the root systems of these species." He further emphasizes the role of "host preference" in determining the distinct EM fungus floras of oaks and birch, a phenomenon we categorize under ecological specificity and host selectivity. Regardless of terminology, these studies provide critical data on overlap of ecological compatibility so that the functional operation of guild concepts can be experimentally evaluated.

Broad host ranges for the EM fungi and broad receptivity by EM hosts both have been successful adaptations. For example, Zak (1976a,b) and later Molina and Trappe (1982a) demonstrated *Arctostaphylos uva-ursi* and *Arbutus menziesii* (Ericaceae) could form both arbutoid mycorrhizae and EM with a diversity of EM fungi in pure culture syntheses, including several fungi with otherwise restricted sporocarp–host associations. *Arbutus* and *Arctostaphylos* showed greater receptivity than other EM hosts in the Pinaceae (Molina & Trappe, 1982b) tested for EM formation with the same fungi. Those pure culture syntheses did include exogenous sugar in the substrate, so caution must be exercised in extending these results to the field where ecological specificity may determine host receptivity. In field studies, Dalberg (1990) reports that *Arctostaphylos urv-ursi* may provide EM fungus inoculum for pine and spruce in Swedish clearcut forest sites and the same phenomenon occurs in clearcuts of Southwest Oregon and British Columbia (M. Amaranthus and G. Hunt, personal communication). We are currently testing shared compatibility between arbutoid hosts and coniferous and fagaceous hosts grown in mixtures in a variety of forest soils.

Ectendomycorrhizae (Fig. 11.1C)

Ectendomycorrhizae have a Hartig net, a usually reduced mantle, and intracellular colonization of epidermal and cortical cells (Mikola, 1965).

Only a few fungal taxa are known to form ectendomycorrhizae. For instance, the genus *Wilcoxina* (the "E-strain") has three taxa confirmed as such: *W. mikolae* var. *mikolae, W. mikolae* var. *tetraspora,* and *W. rehmii* (Yang & Korf, 1985).

These fungi are widely distributed in Finland and North America (Mikola, 1965, 1988; Laiho, 1965; Wilcox et al., 1974, 1983; Danielson, 1982). Other confirmed ectendomycorrhizal fungi include *Phialophora finlandia, Chloridium paucisporum* (Wang & Wilcox, 1985; Wilcox & Wang, 1987a,b), and *Sphaerosporella brunnea* (Egger & Paden, 1986).

E-strain fungi appear to form ectendomycorrhizae only on *Pinus* and *Larix* species, although they form EM with a wide range of other hosts, such as *Picea abies, P. glauca, P. engelmannii, P. pungens, P. sitchensis, Abies lasiocarpa, A. procera, Pseudotsuga menziesii, Tsuga heterophylla* and *Betula verrucosa* (Laiho, 1965), *Picea pungens* and *P. glauca* (Danielson & Pruden, 1989), *Shorea parvifolia* (Louis, 1988), and *Lithocarpus densiflora, Abies grandis,* and *Pseudotsuga menziesii* (Massicotte et al., unpublished data). In addition, an E-strain fungus, probably *Wilcoxina* sp., forms arbutoid mycorrhizae on *Arbutus menziesii* (Massicotte et al., unpublished data).

Overall, the ectendomycorrhizal fungi appear to be broad host ranging in EM associations but intermediate host ranging with *Pinus* and *Larix* in ectendomycorrhizal relationships. These different responses of different hosts to a single fungus are analogous to the arbutoid mycorrhizae of *Arbutus* nd *Arctostaphylos* formed by fungi that are ectomycorrhizal on other hosts.

Arbutoid and Pyroloid Mycorrhizae (Fig. 11.1D)

Arbutoid mycorrhizae are those typical of two genera of Ericaceae, *Arbutus* and *Arctostaphylos*. These mycorrhizae resemble ectendomycorrhizae in that hyphae colonize both intercellularly and intracellularly. However, the Hartig net and intracellular coils are restricted to the root epidermis as opposed to ectendomycorrhizae, where hyphal colonization may extend to the endodermis. We regard the distinction as pedantic for lack of evidence on functional differences between the two.

The fungi that form arbutoid mycorrhizae on *Arbutus* and *Arctostaphylos* form EM on other hosts (Molina & Trappe, 1982a; Zak, 1976a,b). Two groups of mycobionts appear to interact with arbutoid hosts to form arbutoid mycorrhizae: the typical EM fungi, being either broad host ranging or restricted (Zak, 1973, 1974, 1976a,b; Molina & Trappe, 1982a), and an E-strain ascomycete, likely *Wilcoxina* (Massicotte et al., unpublished data). We do not known of fungi restricted only to arbutoid hosts, although this may occur for some *Leccinum* species (Thiers, 1975; Acsai & Largent, 1983a,b). Because most mycobionts of arbutoid hosts are not unique to them, the potential of linkages between plants is enormous.

Ericoid mycorrhizae and EM also may occur on arbutoid hosts (Largent et al., 1980; Mejstrik & Hadac, 1975; Trappe, 1964; Zak, 1973, 1974). However, recent laboratory studies found that *Arbutus unedo* inoculated with ericoid mycorrhizal fungi failed to form ericoid mycorrhizae; inoculated seedlings remained stunted or died (Giovannetti & Lioi, 1990). They also found *A. unedo* incapable of forming VAM after inoculation with six VAM fungi.

Mycorrhizae on *Pyrola* (pyroloid) resemble arbutoid mycorrhizae anatomically, having an epidermal Hartig net, a disorganized fungal sheath, and intracellular penetration restricted to the epidermis (Harley & Smith, 1983; Robertson & Robertson, 1985). Both ascomycetes and basidiomycetes appear to associate with *Pyrola* (Robertson & Robertson, 1985). We can think of no reason to regard "pyroloid" mycorrhizae as a class distinct from arbutoid.

Ericoid Mycorrhizae (Fig. 11.1E)

Ericoid mycorrhizae develop on most Ericales (Ericoideae, Vacciniodeae, and Rhododendroidae of most Ericaceae, as well as Epacridaceae and Empetraceae). A septate fungus forms intracellular coils, restricted to the epidermal cells. Each epidermal cell is colonized individually with little change in root morphology.

Only a few endophytes are known; ascomycetes such as *Hymenoscyphus ericae, Gymnascella dankaliensis, Myxotrichum setosum,* and *Pseudogymnoascus roseus* (Dalpé, 1989), hyphomycetes such as *Oidiodendron griseum, O. cerealis,* and *O. rhodogenum* (Dalpé, 1986; Couture et al., 1983), *O. maius* (Douglas et al., 1989), dark sterile mycelia such as *Mycelium radicis myrtillis* (Friesleben, 1936), and basidiomycetes such as *Clavaria argillacea* and *C. oronoensis* (Petersen & Litten, 1989; Englander & Hull, 1980; Seviour et al., 1973).

These fungi seem to be broad host ranging among ericaceous hosts but restricted to them. Interplant linkages between Ericales and nonericaceous hosts thus appear unlikely. However, Largent et al. (1980) found some typically ericoid plants ectomycorrhizal in the field, Johnson (1981) produced VAM on *Rhododendron* in the greenhouse, and Koske et al. (1990) reported VAM on Ericaceae in Hawaii.

Monotropoid Mycorrhizae

Monotropoid mycorrhizae form on achlorophyllous members of the Ericaceae (the Monotropoideae). They resemble EM, with a Hartig net restricted to epidermis and a thick fungal sheath. In addition, fungal pegs develop in the epidermal cell walls, at least in *Monotropa, Sarcodes,* and *Pterospora* (Lutz & Sjolund, 1973; Duddridge & Read, 1982; Robertson & Robertson, 1985). As in the case of arbutoid and ectendomycorrhizae, the monotropoid type seems to be a hostmediated variant of EM.

No one has succeeded in germinating seed of the Monotropoideae, so experimental evidence on the identity of the mycobionts is lacking and host–fungus associations are based on field observations. Early reports indicate that *Boletus* species may be involved (Francke, 1934; Björkman, 1960; Khan, 1972). Castellano and Trappe (1985) found monotropoid mycorrhizae with the distinctive characteristics of *Elaphomyces muricatus, Cenococcum geophilum, Truncocolu-*

mella citrina, and *Rhizopogon vinicolor* EM in the Pacific Northwest. In addition, *Hymenoscyphus monotropae,* a discomycete, has been described fruiting on *Monotropa uniflora* root clusters (Kernan & Finocchio, 1983). Monotropoid mycorrhizae often form with narrow-host ranging-fungi. Fungal isolates from monotropoid mycorrhizae genetically close to *Rhizopogon* have been identified using polymerase chain reaction techniques (T. Bruns, personal communication). Monotropoid hosts thus appear to share wide receptivity similar to arbutoid hosts, most certainly an ecological advantage given the proven dependence of achlorophyllous plants on carbon of mycorrhizally linked EM plants (Björkman, 1960; Furman & Trappe, 1971).

Orchidoid Mycorrhizae

Specificity patterns in orchid mycorrhizae are complicated in that two types of colonization may occur: primary, involving the germinating seed and seedling, and secondary, involving new roots (Harley & Smith, 1983). A fungus efficient in the adult phase may not be so for the seedling germination phase. Patterns of specificity, if existing, should consequently reflect different selective pressures for different growth phases of orchids.

Another level of complication is the number of orchid species involved, well in excess of 17,500 species according to Mabberley (1989), and the number of fungal species interacting with them, both ascomycetes and basidiomycetes. A comprehensive and updated list of demonstrated and putative orchid endophytes is highly desirable if patterns of specificity are to emerge, but is hampered by the difficulty of growing many orchid species *in vitro* and by the omnipresent task of isolating and linking anamorph and teleomorph stages for numerous endophytes. At least 30 species of orchid endophytic *Rhizoctonia* species have been described. Recent work by Currah (1987), Currah et al. (1987, 1988, 1990), and Moore (1987) detail recent changes occurring in orchid endophyte taxonomy.

Ecological relationships with other plants, e.g., parasitism and saprophytism, of many orchid mycobionts further compound the problem of recognizing specificity patterns. For instance, some orchid mycobionts are parasitic (*Armillaria mellea, Rhizoctonia solani, R. cerealis,* see Alconero, 1969; Smreciu & Currah, 1989; Harley & Smith, 1983), saprophytic (*Coriolus versicolor,* see Harley & Smith, 1983), and even symbiotic with EM hosts, such as *Sebacina vermifera* (teleomorph basidiomycete) on *Melaleuca uncinata* (Warcup, 1988).

In Australian orchids, Warcup (1981, 1971) indicated that orchid specificity occurs at different levels, from species to subtribe; he notes some reservations about the subtribe level because of difficult orchid classification. Clearly, it may be premature to assign ranges of specificity that would apply to this vast group of orchid mycobionts. However, given the spectrum of restricted habitats throughout the world, many close functional relationships (with restricted host range) may have evolved.

Reproductive Biology in Relation to Host Range

Asexual versus Sexual Propagules

Probably all mycorrhizal fungi propagate asexually by hyphae growing from living mycorrhizae to colonize other, receptive rootlets. Trappe and Fogel (1977) reported a single hypha of *Cenococcum geophilum* that extended 2 m from a mycorrhiza of *Pseudotsuga menziesii* and had at least 43 branches that connected with other mycorrhizae, some of nearby *Tsuga heterophylla*. The microcosms used by Finlay and Read (1986a,b) vividly demonstrate vegetative propagation by hyphal growth from one host to another. Some mycorrhizal fungi produce far-ranging strands or rhizomorphs that can effectively spread the colony. *Pisolithus tinctorius* exemplifies this capability particularly well: it will sometimes fruit on road shoulders on the opposite side of the road from the host tree, the rhizomorphs growing under the road surface from the mycorrhizae on one side to the sporocarp on the other.

Specialized reproductive structures generally relate more to degree of host restriction than does mycelial growth. Asexual structures such as sclerotia, chlamydospores, or conidia seem to be more common with broad-host-ranging mycorrhizal fungi than with the narrow-host-ranging ones. And, although sexual fruiting structures are produced by nearly all EM fungi, some forms are typically associated with broad-host-ranging fungi, whereas certain others appear related more to intermediate to narrow restriction of host range.

Cenococcum geophilum, perhaps the most broad host ranging of the EM fungi (Trappe, 1964), has sclerotia as its primary propagules. Sclerotia are common to other broad-host-ranging fungi such as *Pisolithus tinctorius* and *Hebeloma* spp. Knowledge of sclerotium formation by EM fungi is limited to only a few species, however, so we cannot generalize about how commonly they are formed by narrowly-host-ranging species.

Presumptive chlamydospores, asexual spores, are formed by VA mycorrhizal fungi in the genera *Glomus* and *Sclerocystis,* which contain very broad-host-ranging species. They also are produced by some of the ectendomycorrhizal fungi, which seem to be similarly unrestricted as to hosts. Conidia, another type of asexual spores, are particularly formed by ascomycetes, including some of the broad-host-ranging ecto-and ectendomycorrhizal fungi. So far as is known, these asexual spore types are rare or lacking with narrowly restricted EM fungi.

Most ecto-, ectendo-, ericoid, arbutoid, and orchiduid mycorrhizal fungi, whether broadly or narrowly host ranging, produce sexual spores: basidiospores, ascospores, or zygospores. Narrow host restriction of mycorrhizal fungi seems to occur mostly, perhaps exclusively, among species that form sexual spores.

Host versus Environmental Regulation of Propagule Formation

Vegetative propagation by growth of hyphae, strands, and rhizomorphs is regulated by the particular interactions of each fungus with its host and the environment. For fungi that can be grown in culture, the host serves mostly as a source of energy and perhaps growth regulators such as thiamin. Temperature, moisture, and substrate may each influence the rate and form of vegetative growth. The VAM fungi and many EM fungi do not grow in axenic culture by techniques attempted to date. Among the EM fungi, some broad-host-ranging species and some extremely narrowly restricted species grow well in axenic culture, others of both categories do not grow at all. So far as is known, the nature and success of vegetative propagation by mycorrhizal fungi are not related to the breadth or narrowness of their host ranges.

The sclerotia of the broad-host-ranging *Cenococcum geophilum* can form whenever a usable carbon source is available and environmental conditions are right (Massicotte et al., 1991). In nature, the most usual carbon source would be the host, but *Cenococcum* will form sclerotia in axenic culture as well. The host seems not to regulate sclerotium function other than to provide energy. Data on sclerotium formation by other EM fungi are unavailable so far.

The asexual spores of VA fungi form only in connection with host roots. The host provides the needed energy (Ho & Trappe, 1973) and evidently other compounds needed for the fungus to grow and sporulate. These fungi generally appear able to sporulate with any host, in keeping with their broad host range. They can also sporulate with annual as well as perennial plants and under a variety of environmental conditions. Environmental regulation of spore formation may be mostly the usual factors of temperature and moisture coupled with availability of food reserves in host roots. Sometimes the VA mycelium associated with roots of annuals such as *Bromus tectorum* or *Zea mays* sporulates abundantly just before the host dies, a phenomenon akin to the "stress crop" of seeds produced by many vascular plants just before death.

Sexual fruiting of the best known of the broad-host-ranging ecto- and ectendo-mycorrhizal fungi, such as *Cantharellus cibarius*, *Hebeloma* spp., *Inocybe* spp., *Laccaria bicolor*, *L. laccata*, *Pisolithus tinctorius*, *Sphaerosporella brunnea*, and *Thelephora terrestris*, appears to be more regulated by environment than host. These fungi may fruit in greenhouse, nursery, plantation, or forest with hosts of all ages, often whenever moisture and temperature are suitable. Such opportunistic fruiting appears to be regulated as much or more by environment than by host. In contrast, narrow-host-ranging mycobionts such as many species of *Cortinarius*, *Rhizopogon*, *Suillus*, and *Truncocolumella* fruit under apparently much stricter control of the host. This is demonstrated in the coastal fog-belt forests of the Pacific Northwestern North America. Most years these *Picea–Tsuga* forests can be cool and moist throughout the growing season. Some broad-host-ranging species may fruit any time in summer or autumn. Other broad-host-ranging species and most host-restricted species, on the other hand, retain strong seasonality regardless of weather; most fruit in late October through December, no matter how favorable moisture and temperature may have been during summer.

Substrate may also regulate occurrence and fruiting of certain EM fungi. Some broad-host-ranging species such as *Pisolithus tinctorius* typically fruit in soils of low organic matter. More narrow-host-ranging species often fruit only in high organic substrates. The epitome of this latter mode is represented by *Boletus mirabilis,* which forms mycorrhizae and fruits only with rootlets of *Tsuga* spp. growing in brown-cubical-rooted wood.

Spore Dispersal Strategies

The least specialized form of propagule dispersal is the totally passive movement of soil-borne propagules with movement of soil. This strategy characterizes VA mycorrhizal fungi and *Cenococcum geophilum,* the broadest host ranging of the mycorrhizal fungi. The propagules are produced among roots and may be moved within soil by tunneling animals (including insects, worms, reptiles, and mammals). Or they may be exposed to the surface by erosion or animal activity (Allen & MacMahon, 1988). Wind or water may move the spore-containing soil (Allen, 1988), or it may be ingested by or cling to the feet or fur of traveling animals such as deer or elk to be later deposited elsewhere (Allen, 1987; Cazares & Trappe, unpublished data).

With few exceptions, the broad-host-ranging ecto-, ectendo-, arbutoid, ericoid, and orchidoid mycorrhizal fungi fruit as puffballs, mushrooms, or cup fungi that discharge spores to the air for wind dispersal. The discharge may be passive but with sporocarps structured for spore ejection, as with puffballs, or active, as with mushrooms and cup fungi that forceably discharge spores from basidia or asci. Air currents may not be random, but spore dispersal by air may be regarded as essentially random in respect to delivering spores to potential host roots. The combination of broad host range and random spore distribution seems well adapted for establishing mycorrhizal colonies over broad areas without discrimination about climate or soil.

Intermediate- to narrow-host-ranging fungi with only rare exception either form mushrooms or cup fungi with active spore discharge mechanisms or hypogeous (truffle and truffle-like) species that fruit belowground and depend on animal mycophagy for spore dispersal. The latter strategy is the most complex: it involves three-way interdependency of host, fungus, and animal (Trappe & Maser, 1977). The host provides energy and probably other vital compounds to the fungus and habitat and niches for the mycophagist animals. The fungus provides nutrients, water, growth regulators, and protection against fine-root pathogens to the tree and food to the animals. The animals disperse the spores of the fungus and thereby provide the tree with opportunity for mycorrhiza formation with host specific fungi. Hypogeous fungi are far more abundant as EM symbionts than is generally realized. They have been found in all EM forests where they have been intensively sought. Data are limited, but in western North American coniferous forests and Australian eucalypt forests the frequency of hypogeous fruiting bodies often exceeds that of the mushrooms and puffballs.

Evolution of Mycorrhizal Specificity

Origins and Development over Time

The fossil record yields nothing about origins of terrestrial plants. Pirozynski and Malloch (1975) hypothesize that terrestrial plants derived from a symbiosis of aquatic or semiaquatic algae and fungi. In this scenario, the algae could capture solar energy through photosynthesis but could not extract nutrients from soil or organic matter. The fungi, in turn, could not photosynthesize but could get nutrients from soil. Each would require the complementary capabilities of the other to survive on soil. By this hypothesis the first truly terrestrial plants would have been a kind of primitive lichen.

Over the eons, land plants evolved into more complex and organized forms. The earliest rooting structures in the fossil record were lycopods, found in cherts dating from about 350 million years before present (bp) (Kidston & Lang, 1921). These show structures closely resembling vesicles and chlamydospores of present-day VA mycorrhizae. Putative fossil VA mycorrhizae have been detected in younger deposits (Pirozynski, 1981; Stubblefield & Taylor, 1988), but the earliest totally convincing VA mycorrhiza is of a cycad in the Triassic, some 220 million years bp (Stubblefield et al., 1987).

The VA strategy—lack of host specificity and dispersal of asexual spores mostly by movement of soil—is simple and inelegant but highly successful, judging from the predominance of VA mycorrhizal communities over much of the earth today. This fact may seem to support Vanderplank's (1978) assertion: "In mutualistic symbiosis the host loses by mutation to resistance because this ends the symbiosis. Mutations to resistance in mycorrhizal plants are eliminated by selection, because they are disadvantageous; and the elimination also eliminates a major source of specificity." Harley and Smith (1983) espouse this view as an explanation for their contention that "close specificity is not a common characteristic of either ectomycorrhizal hosts or fungi." Despite the assumed "disadvantage" of resistance to a mutualistic symbiont, such resistance has indeed evolved. One may argue the terms "close" or "common," but it appears that about 20% of the angiosperms resist all mycorrhizal fungi and 75% resist at least some types of symbionts (Trappe, 1987). Only 5% appear receptive to a broad range of endophytes. In the gymnosperms, none is known to resist all symbionts but virtually all either resist VA endophytes or asco- and basidiomycete mycobionts (Trappe, unpublished data).

Clearly, selection *has* operated to produce resistance to certain mutualistic symbionts, in other words to increase specificity. The fossil record so far has not revealed when this resistance appeared or whether it was gradual or abrupt. We can infer only from leaf and wood fossils of present day VA-resistant host families, such as the Pinaceae, that resistance to VA colonization began no sooner than the early Cretaceous, perhaps a 130 million years bp (Miller, 1977). Disjunct distributions of hypogeous EM fungi suggest that EM were firmly entrenched before 45 million years bp (Trappe, 1977b). The presence now of hosts receptive to both VA and EM fungi, i.e., *Salix, Alnus,* and *Eucalyptus* spp., suggests that the EM habit replaced VA gradually. This is also supported by seemingly atavistic VA structures in strongly EM hosts such as *Quercus* (Grand, 1969) or *Pseudotsuga* (Cazares, personal communication) or in isolated populations of ericoid mycorrhiza formers (Koske et al., 1990).

The fossil record is even more bereft of fleshy fungal material than of roots, so no direct evidence on the timing of their evolution is available. Trappe (1977b) suggested that disjunct distributions of many hypogeous fungi, dependent on animals for spore dispersal, can best be explained by their presence in the Laurasian supercontinent before tectonic events separated land migration routes between Europe and North America some 50 million years bp. Assuming that mushroom-forming species are ancestral to hypogeous forms (Trappe & Cazares, 1991), the EM mushroom formers would have originated earlier than that.

The southern hemispheric continents separated even earlier, 70 million years or more bp. The southern EM forests are characterized by endemic hypogeous species (Trappe, Castellano, & Malajczuk, unpublished data), suggesting their evolution from mushroom-forming species since continental separation. One may speculate, then, that hypogeous fungi appeared <70 million years bp. This would place the origin of ancestral mushroom formers at >70 million years bp, into the period when the EM hosts first appear in the fossil record.

Lack of specificity at the host species level but its common appearance at the host genus level may also provide clues to the timing of evolution of specificity. For example, *Suillus cavipes* and numerous other EM fungi are restricted to the genus *Larix* and probably associate with all its species. However, the distributions of the world's *Larix* species rarely, if ever, overlap. The same can be said for *Cedrus, Pseudotsuga,* and *Picea* species. Specificity evidently evolved with ancestral hosts and persisted as the tree genera speciated and their species became separated over time. This possibility opens large possibilities for inferring the timing of evolution of specificity, but we are not prepared to pursue them here.

Plants of the present that lack mycorrhizae have evolved strong resistance to most or all mycorrhizal fungi and also tend to be highly evolved. The Cyperaceae and Polygonaceae, for example, are regarded as advanced families and are mostly nonmycorrhizal. Interesting exceptions include the genus *Kobresia* in the Cyperaceae and several *Polygonum* species in the Polygonaceae, which are strongly ectomycorrhizal. The evolutionary advance thus seems to come from primitive lichens to VA mycorrhizae, then either to EM or absence of mycorrhizae: from broad receptiveness to partial resistance to total resistance. This progression conflicts directly with Vanderplank's (1978) concept that resistance in mycorrhizal plants is disadvantageous and thus not a source of selectivity by the fungi.

Selection Pressures toward Specificity

Practically nothing is known about the physiology of mycorrhizal host specificity. Horan and Chilvers (1990) discovered a chemotropic effect exerted by eucalypt roots on their mycobionts but not on mycobionts of birch and pine. VAM host roots produced no such effect. They hypothesized that "such chemotropism could provide the signal which initiates the ectomycorrhizal infection process in the root cap region, and guides the inward growth of the hyphae through that tissue during the early stage of sheath formation." More than chemotropism is clearly involved in specific host–mycobiont reactions, however, as evidenced by tannin deposition or a hypersensitive reaction when roots of one host are confronted with a mycobiont specific to another (Malajczuk et al., 1984).

That host specificity is an advanced character as opposed to a broad-host-ranging habit is consonant with specificity's prevalence with both the advanced, EM hosts, and the hypogeous fungi. No cases of specificity by an EM mycobiont to a single host species have been documented, although conceivably some could exist with monotypic EM genera such as *Keteleeria* (Pinaceae) or for isolated species of dipterocarps in Malaysian forests (Smits, 1983). The apparent general lack of evolution of specificity from a host genus level to a species level suggests that specificity at the genus level met the demands of selection pressure. In an ecological context, however, specificity at the genus level may often operate in a species-specific manner. A *Larix* × *Suillus cavipes* association, for example, would have the ecological function of a host-specific association, because only one species of *Larix* will be present in most, if not all, natural stands. On the other hand, clusters of closely related fungal species with many intermediates suggest rampant evolutionary activity in some present host-specific mycorrhizal fungi, the genus *Rhizopogon* being a good example (Smith & Zeller, 1966). In genera with isolated species such as *Larix* or *Pseudotsuga*, this activity could be headed toward species specificity.

The selection pressures that fomented evolution of host specificity (or, in Vanderplank's sense, advantageous mutations) in mycorrhizae can only be hypothesized at present for lack of understanding of the physiology involved. Malloch et al. (1980), Pirozynski (1981), and Trappe (1989) have outlined differences between forests with nonspecific VA mycorrhizae and those of the more specific EM associations. The VAs tend to form communities that are host-species rich but low in mycobiont diversity and biomass; they predominate on moderate to highly productive sites with rapid organic matter decomposition; they typically have animal pollinated flowers, fleshy fruits for animal dispersal of seed, and mycobiont spores that are dispersed by soil movement, often by animals, but relatively few leaf toxins that minimize herbivory. The EM hosts dominate communities that are host-species poor but high in mycobiont diversity and biomass; they predominate on sites seasonally limited by temperature extremes or available moisture, relatively low in nutrients and with slow organic matter decomposition; they typically have wind-pollinated flowers and dry seeds that are wind-dispersed or eaten by animals, mycobiont spores that are dispersed by wind or animal mycophagy, and relatively abundant leaf toxins to minimize herbivory.

The most advanced and most mycobiont-resistant plants show three major patterns of adaptation in the absence of mycobionts: they may be parasites on other plants, they may be perennials that grow in wet to ponded soils with roots immersed in a soil solution or in marginal and often dry soils with a large part of their biomass dedicated to an ultrafine, profusely branched root system with abundant root hairs, or they may be short-lived annuals of wastelands that respond quickly to seasonally available moisture and produce seed before drought limits their reproductive capability (Trappe, 1987).

Exceptions to these generalities can easily be cited, but the trend to increasing host resistance to mycobionts, i.e., increasing specificity or elimination of the mycobiont, parallels increasing ability to disperse pollen and spores for long distances, to colonize marginal sites, and to produce stands of a single or only a few species. Selection pressures for these traits could develop with continental drift and global climate change, in which marginal habitats are continuously created by new extremes of temperature and moisture, or by environmental catastrophy such as volcanic activity or fire denuding vast tracts.

Marginal habitats typically have strong seasonality in moisture and temperature. In forest zones they are mostly occupied by EM forests. These habitats would produce additional selection pressure for mycobiont diversity: different mycobionts would be selected for physiological activity at different times of year or different niches in soil as organic matter accumulates. The increase in specificity characteristic by EM over VAM would be accompanied by an increase in diversity. In pure stands, that diversity could include rather narrow host specificity if a particular host–mycobiont combination increased the fitness of both symbionts.

In the case of the hypogeous fungi, which commonly express relatively narrow host specificity, the association with extensive pure stands of a host would likely be prerequisite to success. Most commonly, the distance of propagule dispersal by animals is limited, especially because small rodents of limited range are the most common agents of dispersal. Moreover, the inoculum is delivered to a point location in a fecal pellet rather than broadcast broadly. Accordingly, success of that strategy requires that spores be deposited on roots of compatible hosts. The huge pure stands of Douglas-fir, pines, eucalypts, and southern beech abound with diverse populations of host-specific, hypogeous fungi as testimony to that success.

Quite likely many physiological selection pressures have led to the narrowest host specificities. For example, *Alnus rubra* appears to host the most strongly specific fungi and resist colonization by many broad-host-ranging fungi (Molina, 1979). It also strongly resists root pathogens, probably in large part because of the abundance of phenolics in its roots (Trappe et al., 1973). Such compounds might limit penetration by most EM fungi, providing a selection pressure in favor of particular fungi that mutated to a tolerance of those compounds, the alder-specific fungi. Similar biochemical phenomena might be the case with eucalypts. Tolerance to terpenes and other coniferous compounds might separate the EM fungi adapted to Pinaceae from those restricted to angiosperms. Other physiological adaptations between particular hosts and their specific fungi can be hypothesized virtually without limit. The mutation rate of fungi is relatively high, and nature has probably conducted countless experiments over the eons. The few that worked are with us today.

Community Level Considerations of Specificity Phenomena

Successional Relationships

Specificity phenomena in mycorrhizal associations directly affect plant community development, particularly early stages of plant establishment. During primary succession, the ecology of fungal propagule dispersal determines fungal migration and inoculum availability. Long distance wind or animal dispersal of EM fungi can give EM plants a colonizing advantage over obligate VAM plants as evident in retreating glacier forefronts (Trappe, 1988). Thus, fidelity to a mycorrhiza class together with presence or absence of kinds of mycorrhizal fungus propagules immediately affect plant establishment and direction of community succession (Trappe, 1989).

The impact of specificity phenomena on secondary succession is more difficult to assess than for primary succession. Composition of past host species, soil development and properties (pH, fertility, organic matter), and type and severity of disturbance all influence mycorrhizal fungus inoculum potential (Perry et al., 1987); separating biotic from abiotic influences on plant and mycorrhizal succession requires careful field observation and experimentation. In severely disturbed ecosystems, such as degraded mine spoils or those experiencing long-term disturbance (e.g., extended periods of cattle grazing), the phenomena of dependence and facultativeness influence successional patterns. Nonmycorrhizal and facultatively mycorrhizal plants are well known early colonizers of disturbed systems. Trappe (1987) lists many of the world's worst weeds as nonmycorrhizal or facultatively mycorrhizal plants. The severity and duration of disturbance also affect the potential long-term dominance of nonmycorrhizal or facultatively mycorrhizal plants (Miller, 1987). In disturbed grasslands or forested range lands, facultative VAM annual grasses can replace perennial, obligately VAM shrubs (Reeves et al., 1979). After disturbance in tropical forest ecosystems, many early invaders are nonmycorrhizal or facultatively VAM plants (Janos, 1980). In semiarid shrub-steppe communities, replacement of VAM grasses by nonmycorrhizal or facultatively mycorrhizal Chenopodiaceae significantly reduces VAM fungus propagules (Allen & Allen, 1990). Several additional examples are cited by Trappe (1987). We emphasize here that the mycorrhizal specificity phenomena of fidelity/infidelity, dependence/independence, and facultativeness/obligateness along with mycobiont propagule availability are prime determinants of early stages of community development and affect long-term plant dominance.

Popular concepts of "early stage" and "late stage" fungi have been developed from recurring patterns of EM fungal succession with birch in Great Britain (see Deacon & Fleming, Chapter 8, this volume). Although such patterns and concepts nicely characterize successional changes associated with birch, particularly in plantations, most forest ecosystems remain unexplored for this phenomenon. EM fungal succession during periods of secondary plant succession will be strongly influenced by the severity of disturbance, the EM plants remaining on and adjacent to the site, and the suite of environmental factors noted previously. For example, in Pacific northwestern forests of North America, frequent fires create a complex mosaic of forest structures and species composition (Franklin & Dyrness, 1973). Long-lived Pinaceae in the region (Douglas-fir and ponderosa pine can live hundreds of years) can often survive repeated fires. Their EM mycobionts persist on tree roots as active mycelium, a highly effective inoculum for establishing conifer seedlings. In the arid forest ecosystems of the Siskiyou Mountains of southern Oregon/northern California, many EM or arbutoid shrubs or understory trees in genera such as *Arctostaphylos, Arbutus,* or the Fagaceae quickly sprout from root crowns or stems, again providing functional EM fungi in the soil for later establishment by Pinaceae seedlings. In these examples, ability of hosts to associate with common EM fungi, i.e., compatibility and overlap in EM fungi, will strongly determine patterns of fungal succession; "early stage" and "late stage" concepts are irrelevant to these ecological settings.

Last et al. (1987) propose a model of EM fungal succession in which "very early stages of succession are characterized by a small proportion of broad-host-ranging fungi which probably increases in diversity until tree canopy closure. The EM fungal diversity decreases thereafter with a larger proportion of fungi with narrow host ranges." Our experience with Pacific Northwest forests contradicts this model. In several soil bioassays (Amaranthus & Perry, 1989; Borchers & Perry, 1990; Miller & Molina, unpublished data; Schoenberger & Perry, 1982; Massicotte, Molina, & Smith, unpublished data) and outplanting trials of inoculated seedlings (Castellano & Molina, 1989), the predominate EM fungi on Douglas-fir seedlings are genus-specific fungi, primarily *Rhizopogon vinicolor* and related *Rhizopogon* species in the subgenus *Villosuli*. In fact, Douglas-fir-specific *Rhizopogon* species prevail throughout the life history of Douglas-fir; sporocarps of these *Rhizopogon* species commonly abound in mature to old-growth Douglas-fir stands (600–1000 years of age). Danielson and Visser (1989) cite pine-specific *Rhizopogon* and *Suillus* species as aggressive EM colonizers on outplanted pine seedlings. *Rhizopogon rubescens* was a more aggressive colonizer of jack pine than *Laccaria laccata* or *Pisolithus tinctorius* on pine seedlings planted into natural jack pine stands (McAfee & Fortin, 1986). Even in fumigated nurseries, in addition to common broad-host-ranging *Thelephora, Laccaria,* and ectendomycorrhizal fungi, we have observed 2- to 3-year-old seedlings strongly mycorrhizal with *Rhizopogon* and producing abundant fruitings of hypogeous sporocarps.

The successional sequence of broad-host-ranging vs. restricted-range fungi also differs according to host plant. For example, western hemlock (*Tsuga heterophylla*) typically replaces Douglas-fir and becomes climax in the west temperate forests of western Oregon and Washington. Kropp and Trappe (1982) found that EM fungi of western hemlock are typically broad host ranging. They propose that selection pressures would favor this because western hemlock typically establishes in the Douglas-fir understory and remains an understory tree for much of its life history; accordingly it would be advantageous to associate with already present and functioning EM fungi. Douglas-fir, on the other hand, associates with a wide assemblage of genus-specific fungi and is a well known pioneer on disturbed forest sites. Kropp and Trappe (1982) hypothesize that host specific fungi are more common with pioneering plants than late successional plants. Seedling assays by Schoenberger and Perry (1982) of variously disturbed forest soils with Douglas-fir and western hemlock seedlings support this hypothesis. They found that EM fungus associates of Douglas-fir (predominantly *Rhizopogon*) were less sensitive to burning disturbance than were fungus associates of western hemlock.

In the Douglas-fir–western hemlock association, host specificity appears advantageous to the overstory Douglas-fir, because it protects part of the root system from epiparasitism by understory hemlocks via shared mycorrhizal mycelium. Lack of specificity, in contrast, is advantageous to the understory hemlock, for it can plug into the established mycorrhizal system and perhaps even be a sink for the Douglas-fir photosynthate source. The system is among the more productive of temperate forests, so these countervailing habits have reached a healthy balance.

In natural forests, a complex of biotic and abiotic factors, including competitive interactions between fungi, interactions of hosts and fungi, influence of rhizosphere microbes on mycorrhizal development, and soil properties influence root colonization and later replacement, i.e., succession of EM fungi. Danielson and Visser (1989) stress that ecological adaptation of EM fungi to specific soil types ("soil specificity") varies widely among EM fungi and may play a more important role than competitive replacement in determining sequence of EM colonization. Furthermore, Garbaye and Bowen (1987, 1989) showed that soil type and soil–plant interactions selectively stimulate beneficial rhizosphere microbes, which in turn influence development of particular EM associations. Thus, in natural forest ecosystems, particularly those experiencing periodic disturbance, diverse habitat types and host assemblages, and long-lived dominant tree species, a simplified model of early stage–late stage fungus succession proves inadequate.

Specificity Phenomena in Exotic Plantations of EM Host Trees

Several decades of planting EM tree species outside their natural ranges yield unexpectedly rich data on the ecological consequences of specificity phenomena in EM associations. These plantations serve as uncontrolled experiments that examine the migration of EM fungi as influenced by host relationships, compatibility of exotic EM hosts with previously geographically isolated EM fungi, and ecological roles of broad-host-ranging vs. narrow-host-ranging fungi in plantation establishment and performance.

The absence of EM fungi typically results in plantation failure of obligate EM hosts. However, even if EM fungi are present with native plants, incompatibility with introduced EM hosts can still result in plantation failure or delay in establishment. Such was the case in the introduction of *Pinus radiata* into Australia and elsewhere (Mikola, 1970). Even though abundant EM fungal associates of *Eucalyptus* were present in Australia, pine seedlings did not flourish until "pine fungi" were introduced into nurseries (Mikola, 1970).

Expression of incompatibility between *Eucalyptus* and *Pinus* continues in the extensive pine plantations throughout Australia. Malajczuk (1987) reports that the diversity of EM fungi in radiata pine stands is low, typically less than 20 species; in contrast, the eucalypt EM flora is rich, numbering in the thousands. Yet, even if pine and eucalypt stands are adjacent, eucalypt fungi do not move into pine stands. Similarly, Hilton et al. (1988) report that pine-specific *Suillus* spp. are never found in eucalypt forests. Malajczuk (1987) also states that although developing eucalypt forests show a recognizable succession of EM fungi as stands age, the comparatively impoverished pine EM fungus flora changes little; often EM fungi from the nursery persist throughout the pine rotation.

Theodorou and Bowen (1987) found that incompatibility between pine and eucalyptus was expressed during spore germination; germination of pine-specific *Rhizopogon luteolus, Suillus luteus,* and *S. granulatus* responded only to pine roots and not roots of eucalyptus or VAM plants. Horan and Chilvers (1990) reported that mycobionts of eucalypts experience a chemotropic attraction to eucalypt roots, but pine and birch mycobionts showed no such response. However, the patterns of incompatibility between eucalyptus and pine go beyond differences in genus-restricted fungi. Malajczuk (1987) points out that strain specificity within typically broad-host-ranging fungi operates between these two EM host genera. He states that in spite of the broad host range of typical early stage fungi in eucalyptus stands such as *Laccaria laccata* and *Pisolithus tinctorius*, these fungi remain absent in adjacent pine plantations. In a recent synthesis study, Tonkin et al. (1990) demonstrated difference in speed of colonization and incompatibility responses between strains of *P. tinctorius* native to pine vs. eucalyptus.

Rhizosphere microorganisms can also influence patterns of incompatibility between pine and eucalypts. Garbaye and Bowen (1987) report that soil microbial suspensions from a variety of soils differentially affected EM development on *Pinus radiata;* the pine fungus *Rhizopogon luteolus* only formed EM in pine soil or with the microflora from a pine nursery. Such fungus strain incompatibilities (Tonkin et al., 1990) and "selective stimulation" by soil type or soil–plant interaction (Garbaye & Bowen, 1987) express ecological specificity wherein the final outcome of EM development and associations is determined by interacting biotic and abiotic factors in the field.

Infidelity by hosts to a mycorrhiza class and compatibility with broad-host-ranging native fungi are often credited for the successful establishment of exotic trees (Meyer, 1973). The ability to form both VAM and EM is certainly an advantage for hosts at the establishment stage if propagules of only one or the other are available. Hilton et al. (1989) point out that VAM colonization is important in the early establishment of eucalypts in Australia; EM typically prevail soon after establishment. The plantation success of eucalypts worldwide may relate to its functional compatibility with VAM and EM fungi. When eucalypts are planted in regions dominated by other EM hosts, Malajczuk et al. (1982) hypothesize that compatible broad-host-ranging fungi are important for early plant establishment; they demonstrated the ability in pure culture for several broad-host-ranging mycobionts from Oregon to form EM with a range of eucalypt species.

Although broad-host-ranging fungi are important for successful exotic plantations, that ecological role should not overshadow the importance of restricted-range EM fungi in the mycorrhizal functioning of these plantations. Numerous genus-restricted EM fungi have migrated with their specific hosts into exotic locations worldwide, often becoming a dominant EM component. In Chile, introduced species of *Pinus, Larix,* and *Eucalyptus* have conspicuous components of genus-specific fungi such as *Suillus luteus* and *S. granulatus* on pine and *S. grevillei* on *Larix* (Garrido, 1986). *Pinus* and *Eucalyptus* species are the most widely planted of exotics; nearly all pine plantations feature pine-specific *Rhizopogon* and *Suillus* species and nearly all eucalypt plantations feature eucalypt-specific *Hydnangium* and *Hymenogaster* species (Trappe 1962; Mikola 1973; Molina, Castellano, & Trappe, unpublished data). Trappe, Molina, Castellano, and Alvarez (unpublished data) recently collected the Douglas-fir-specific *Rhizopogon parksii* in Douglas-fir plantations in northeastern Spain; this fungis is native to Douglas-fir forests of the Pacific Northwest. Many *Rhizopogon* species have been used successfully in seedling inoculation programs (Castellano & Molina, 1989).

Ectomycorrhizal studies in plantations of exotic trees in New Zealand document not only migration of host specific fungi but emphasize the abundance of these fungi there. Chu-Chou (1979) report that of 7000 attempted root isolations from *Pinus radiata* growing in nurseries and plantations 32% yielded the pine-specific *Rhizopogon luteolus* and *R. rubescens;* the next highest percentage was for *Amanita muscaria* at 3%. In a similar study of exotic Douglas-fir EM fungi, Chu-

Chou and Grace (1981a) report 14% of all Douglas-fir root isolates yielded *Rhizopogon vinicolor* (host specific to Douglas-fir), followed by 6% for *Amanita muscaria*. In a field survey of Douglas-fir EM types in plantations, Chu-Chou and Grace (1983a) report that of the 13 types found, Douglas-fir-specific *R. vinicolor* and *R. parksii* were present in all root samples and in the highest proportions.

A strong analogy is apparent for *Eucalyptus* plantations in New Zealand. Chu-Chou and Grace (1981b,c, 1982) report the widespread occurrence of *Eucalyptus*-specific hypogeous fungi in the genera *Hymenogaster*, *Hydnangium*, and *Mesophellia*; *Hydnangium* was most frequently isolated from eucalypt roots. Sporocarps of *Hydnangium carneum* are also regularly collected in *Eucalyptus* plantations in northern California as are eucalypt-specific species in the genera *Chondrogaster*, *Hymenogaster*, *Hysterangium*, and *Setchelliogaster*, all hypogeous genera (Castellano & Trappe, unpublished data).

Determining the ecological roles and relative importance of broad-host-range vs. restricted-range fungi in exotic plantations requires consideration of a variety of factors, including exotic host compatibility with EM fungi of native trees, length of time exotics have been cultivated in the region, proximity to established plantations, and interactions between fungi and environmental factors, i.e., functional outcomes of ecological specificity. Malajczuk et al. (1982) hypothesize that the success of eucalypts in exotic plantations can be attributed in part to their compatibility with broad-host-ranging fungi of native trees. Eucalypt-specific fungi may be introduced with the planted stock or migrate into the new plantations. Castellano (unpublished data) finds that *Eucalyptus* stands in northern California become rapidly colonized by eucalypt-specific fungi, usually within 2-4 years of planting or immediately if planted in close proximity to established eucalypts. Small mammal dispersal of the rich hypogeous EM flora of the region is strongly developed there, so it is not surprising that the hypogeous, eucalypt-specific fungi are widespread.

Abundance of EM fungal species can indicate their ecological importance, but separating differences between broad-host-ranging and restricted-host-ranging fungi in how they benefit host growth and survival requires experimentation. Chu-Chou and Grace (1985) found that *Rhizopogon rubescens* and *R. luteolus* were more effective in promoting growth and nutrient uptake of *Pinus radiata* than the broad-host-ranging fungi *Laccaria laccata* or *Hebeloma crustuliniforme*. Similarly, Douglas-fir-specific *R. vinicolor* provides greater drought resistance to Douglas-fir seedlings than the broad-host-ranging *L. laccata* or *Pisolithus tinctorius* (Parke et al., 1983), higher photosynthesis rates than *L. laccata* or *H. crustuliniforme* (Dosskey et al., 1990), and is also superior to these broad-host-ranging fungi in enhancing survival and growth following seedling outplanting (Castellano & Molina, 1989). Broad-host-ranging fungi such as *Pisolithus* or *Laccaria* can also enhance tree performance in exotic plantations (Marx 1980; LeTacon et al., 1988). Thus, broad-host-ranging and narrow-host-ranging fungi can both provide nutritional and seedling performance advantages, and the prevalence of one group of fungi over the other will depend on host and ecological setting.

Specificity Phenomena, Mycorrhizal Linkages, and Guilds

Shared compatibility between plants for common mycorrhizal fungus associates determines linkage potential of those plants. Newman (1988) lists five ecological/ecosystematic consequences of mycorrhizal linkages between plants: (1) seedlings quickly link into an established network of mycorrhizal fungi, (2) carbon from overstory plant is "donated" to seedlings, enhancing survival, (3) linkages influence balance of competition between plants, (4) minerals pass between plants, and (5) nutrients from dying plants pass on to linked plants. Carbon and nutrients flow between plants linked by shared fungi in microcosms (Read, 1984; Read & Finlay, 1985), yet Newman (1988) concludes that evidence on the ecological significance in natural ecosystems is inconclusive. He emphasizes the lack of evidence that proximity to plants with the same mycorrhizal type enhances survival or growth of that new plant.

Since that review of mycorrhizal linkages, several studies in southwestern Oregon have begun to build better information on the ecosystematic importance of mycorrhizal linkages in forests. Zak (1976a,b) and Molina and Trappe (1982a) synthesized arbutoid mycorrhizae onto *Arctostaphylos uva-ursi* and *Arbutus menziesii* with a wide range of EM fungi associated with Pinaceae. *Arbutus menziesii* and about 10 species of *Arctostaphylos* are widespread through southwestern Oregon–northern California, as are many Fagaceae (*Castanopsis, Lithocarpus,* and *Quercus* spp.). Molina and Trappe (1982a) hypothesized a successional role for arbutoid plants in maintaining EM fungi in these forest soils following disturbance, thereby facilitating the establishment of valuable Pinaceae timber species. Amaranthus and Perry (1989) planted Douglas-fir seedlings into adjacent sites of recently cleared *Arctostaphylos viscida,* grass, and grass plus *Quercus garryana* fields. An additional treatment of soil taken from beneath *Arbutus menziesii* was added to planting holes for some seedlings on all sites. Growth and survival were significantly greatest on the cleared *Arctostaphylos* site; poor survival and growth resulted in grass and *Quercus* sites. The addition of *Arbutus menziesii* soil significantly increased the number of EM root types in the *Arctostaphylos* site only, suggesting a synergistic interaction between the arbutoid hosts. Borchers and Perry (1990) followed this study by growing Douglas-fir seedlings in soils taken from immediately under to 4 m away from *Lithocarpus densiflora, Quercus chrysolepis,* and *Arbutus menziesii* shrubs growing in a 5-year-old clearcut forest habitat; seedlings grown in soil beneath the hardwood crowns (i.e., in immediate rhizosphere soil) had 60% greater height, and twice the weight and number of EM tips than seedlings grown in soil 4 m away. Ectomycorrhizal types also differed with distance from the hardwood shrubs; *Cenococcum geophi-*

lum and *Rhizopogon* types dominated in soil near the shrubs and a brown type was most common distant from the shrubs. In a survey for conifer survival associated with proximity to hardwood shrubs, Amaranthus et al. (1990) report five times greater conifer regeneration beneath *Arbutus menziesii* compared to beneath VAM shrub species or in open areas; conifer seedlings growing in the immediate rooting zone of *Arbutus menziesii* had the greatest number of EM tips. In a multiple host soil bioassay in this same forest type, Massicotte and Molina (unpublished data) found that of a total 17 EM types recovered, Douglas-fir, *Pinus ponderosa,* and *Abies grandis* shared 8 types in common with *Lithocarpus densiflora* when grown together. Perry et al. (1989a) describe these ecologically linked EM hosts and fungi of southwestern Oregon as functional guilds: "Associations for mutual aid and the promotion of common interests," in this case as defined by common belowground mycobionts.

Defining functional guilds in other mycorrhizal systems will require extensive observation of natural successional patterns and experimental confirmation of shared compatibilities for belowground mycobionts. For example, in African rain forests, EM trees in the Caesalpinaceae (*Afzelia* and *Gilbertiodendron* spp.) typically occur in clumps among VAM tree species. Newbery et al. (1988) hypothesize that this clumping is due to a combination of inefficient seed dispersal, sufficient EM fungus inoculum potential beneath established EM trees, and possible shared carbon between overstory trees and seedlings. Interestingly, an EM tree species in the Euphorbiaceae, *Uapaca* spp., typically accompanies the EM legumes in these clumped distributions (Newbery et al., 1988). Yet, of 43 putative EM fungi associated with these trees, *Uapaca* and *Afzelia* shared only 6 species in common (Thoen & Ba, 1989). Furthermore, in an EM synthesis study, Ba and Thoen (1990) found all *Uapaca* EM isolates failed to form EM on *Afzelia*. Clearly, further experimentation is needed in these potential African EM guilds to unravel facilitative interactions between plants and environmental factors affecting their development.

Although floristic studies of EM fungal communities in relation to forest community development are rare, they are absolutely essential to document overlap in EM fungal species between different tree species and develop understanding of guild relationships. Bills et al. (1986) compared EM basidiomycete communities between adjacent red spruce and heterogeneous EM angiosperm communities in the northeastern United States. Of 54 fungus species, 19 were found exclusively in spruce, 27 exclusively in hardwoods, and only 8 in both tree communities. Spruce was dominated by a few ubiquitous EM fungi; *Laccaria, Lactarius,* and *Russula* spp. were common in nearly all plots. Lee and Kim (1987) collected 196 EM fungal species from pure stands of EM trees in the genera *Abies, Betula, Castanea, Larix, Picea, Pinus, Populus,* and *Quercus* in Korea. The fungi ranged from narrow to broad host ranging, with the broadhost-ranging group being largest; *Laccaria, Amanita,* and *Russula* species were common to most tree species. Such floristic studies provide only a glimpse in potential overlap in ability to associate with particular EM fungi. To develop guild concepts of facilitated interactions between plants as mediated by shared mycobionts, the floristic studies must be conducted in forests where mixtures of EM hosts occur, particularly in successional time sequences.

Guild structure and function in natural ecosystems are complex and multilayered in plant–fungus interactions through time; a simple vision of interconnected plants is unrealistic. Unraveling functional aspects of guilds such as competitive interactions and outcomes in relation to specificity phenomena pose a great challenge for mycorrhizal ecologists. Harley and Smith (1983) discuss the ecological advantage of hosts to associate with the same broad-host-ranging fungi and the selective pressure favoring this adaptation. They state that plants capable of forming mycorrhizae with the same fungi will be able to "compete" for these fungi to obtain nutrients; lack of specificity in mycorrhizal associations enhances this chance of competition and access to nutrients, i.e., enhances fitness. Such seems the case in VAM ecosystems that express little or no incompatibility between hosts and fungi. Janos (1980, 1985) theorizes that lack of specificity among VAM associates contributes to synergistic interactions within communities of VAM plants in tropical ecosystems, reducing competition and favoring coexistence.

In EM forest ecosystems, however, many host–fungus associations are restricted and that degree of specialization must have been ecologically advantageous just as lack of specificity has been. Janos (1985) states that specialization in host–fungus association, i.e., closer specificity, can enhance fitness of a mycobiont if more resources are obtained through such specialization: "Mycobiont specialization must increase occupancy of the roots of the specific hosts sufficiently to compensate for decreased or no association with other plant species." A similar argument can be made for the host with regard to ecological advantages of associating with specific fungi. We hypothesize that host-specific fungi provide a biological mechanism to partition soil resources, thereby providing an exclusive avenue for different plant species to obtain nutrients and enhance their fitness. The EM guild is benefited by reduced competition for shared mutualists. Perry et al. (1989b) found that competition between Douglas-fir and ponderosa pine seedlings was reduced when seedlings were grown in soils inoculated with a mixture of genus-specific *Rhizopogon* species and broad-host-range fungi (*Laccaria laccata, Hebeloma crustuliniforme,* and *Thelephora terrestris*) compared to growth only in the presence of *Thelephora terrestris*. Experiments of this nature are critical to discover the ecological function of specificity phenomena in mycorrhizal guilds and plant community dynamics.

Management Implications and Future Considerations

Specificity phenomena, particularly ecological specificity, must be integrated into future efforts to develop mycorrhizal inoculation strategies. In developing VAM fungus inoculum, wide host range is tauted as an advantage for broad applicability. Yet, numerous studies have shown strong differences between and within VAM fungus species in inoculum effectiveness in different ecological settings.

Searching for fungus isolates that show superior effectiveness through specialization for host or habitat may prove more fruitful than developing a strain for broad application.

The same argument applies for application of EM fungus inoculum (Trappe, 1977a). The wide host ranges of fungi such as *Pisolithus tinctorius* and *Laccaria laccata* give them broad applicability (Hung & Molina, 1986; Marx, 1980; Molina, 1982; Molina & Chamard, 1983). Both fungus species have proven successful in some application trials (Marx, 1980; Marx et al., 1984; LeTacon et al., 1988), but they have yielded poor results in others (Castellano & Molina, 1989). Mikola (1972) hypothesized that host-specific EM fungi, being evolutionarily specialized with particular hosts, would provide greater benefit than cosmopolitan host generalists. He points out the success of the pine symbiont *Suillus plorans* in benefiting *Pinus cembra* in high elevations in the Alps, and the numerous examples of pine-specific fungi in exotic pine plantations. Likewise, in Oregon and Washington, our greatest success in enhancing outplanting performance of Douglas-fir and pine has come from inoculations with narrow-host-ranging fungi in the genus *Rhizopogon;* broad-host-ranging fungi such as *Laccaria laccata* and *Hebeloma crustuliniforme* have performed comparatively poorly in field trials in this region (Bledsoe et al., 1982; Castellano & Molina, 1989). Interestingly enough, however, an Oregon isolate of *Laccaria laccata* that failed to enhance outplanting performance of inoculated Douglas-fir in Oregon has been extremely beneficial for outplanted Douglas-fir in France (LeTacon, personal communication). Thus, although host range data are critical in developing mycorrhizal applications, trial-and-error experimentation can still reveal good fungus isolates for specific ecological settings.

Perry et al. (1987) revisited mycorrhizal applications in forestry and emphasized that although inoculation may be necessary in certain reforestation or afforestation situations (e.g., severe, long-term degradation), maintaining the biological diversity of mycorrhizal fungi and other soil microbes through wise resource management may be our best way to apply mycorrhizal technology. Perry et al. (1989a) hypothesize that diversity in both plant and microbial communities stabilizes the plant–soil system during environmental fluctuations, particularly following adverse disturbance: "Physiologically diverse mutualists are likely to buffer the plant-soil system by extending the range of environments in which the plant is able to maintain net photosynthesis." However, mycorrhizal fungi are rich not only in species number and physiological activities; the many host–fungus combinations multiply the diversity far beyond the simple numbers of fungi and hosts alone. Specificity phenomena in mycorrhizal associations act as primary sources of "functional biodiversity" in terrestrial ecosystems. Functional biodiversity is likely to be a more relevant indicator of ecosystem health than sheer numbers of organisms.

Perry et al. (1989a) refer to the positive feedbacks in soil–plant linkages as "bootstrapping in ecosystems." They state: "The redundancy provided by shared microflora means that one guild member is able to maintain soil organisms required for all. In most ecosystems, variability in the nature of timing of disturbances leads to uncertainty in the composition of the pioneering community and such redundancy benefits both individual plants and the community as a whole." Maintaining overall biodiversity of soil microorganisms is critical to ecosystem health, but we caution overemphasizing redundancy in function of soil organisms. We have described considerable differences between plants and fungi in terms of expressed specificity phenomena and their ecological consequences. Even if many of the mycorrhizal fungi within the spectrum of narrow to broad host ranges perform similar functions, they can have strikingly different ecologies that are important at different times of the year or over long successional periods.

Specificity phenomena in mycorrhizal associations will also play an important role in determining plant species migration during periods of rapid global climate change. Perry et al. (1990) state "Compatibility between belowground mutualists of resident species and needs of immigrant species will strongly influence the successful transition from one perennial plant community to another during climate change." Climate change is a widescale disturbance, and specificity aspects of fidelity, dependence/independence, and facultativeness/obligateness will immediately be factors in determining new community composition. But as plant species migrate, the overlap in shared mutualists and plant ranges will determine the ease of plant community transition (Perry et al., 1990). For example, in regions with sharp ecotones between EM and VAM dominated ecosystems (e.g., forests-grasslands), EM hosts may have difficulty migrating into VAM ecosystems and vice versa. On the other hand, in broadly contiguous mountainous regions dominated by diverse EM tree species, considerable overlap in compatible root mutualists may facilitate migration across mountain ranges or altitudes for EM hosts. But how much overlap in compatible mutualists and geographic host range is enough to facilitate smooth transitions of species migrations? How will dispersal ecologies of mycorrhizal fungi and other important rhizosphere organisms likewise affect future community transitions? We must also consider the narrow specificities that have evolved for many EM associations of pioneering plants such as Douglas-fir. How will the migration of these narrow-host-ranging mycobionts affect the successful migration of their specific hosts?

To answer such complex question requires considerable more research on the functional aspects of specificity phenomena and linkages in mycorrhizal associations, and exploring guild concepts at scales ranging from community to landscape levels. From such studies we can build a better understanding of the role mycorrhizae play in community development and ecosystem stability, and provide knowledgeable tools for resource managers to protect and utilize the belowground ecosystem.

Acknowledgments

The concepts presented in this chapter were developed during research funded by the Indo-U.S. Science and Technology Initiative, National Science Foundation

Grants BSR 8505972 and 8717427, and the National Research Council of Canada. We thank Drs. Steven Miller and David Janos for the many hours of stimulating discussions on these concepts. We wish to dedicate this chapter to the memory of Dr. Jack Harley. Dr. Harley provided early encouragement to the senior author to conduct research on mycorrhizal specificity and provided stimulating discussion of the subject in his writings and lectures.

References

Acsai, J., & Largent, D. L. (1983a). Fungi associated with *Arbutus menziesii*, *Arctostaphylos uva-ursi* in central and northern California. *Mycologia* **75**, 544–547.

Acsai, J., & Largent, D. L. (1983b). Mycorrhizae of *Arbutus menziesii* Pursh and *Arctostaphylos manzanita* Parry in northern California. *Mycotaxon* **16**, 519–536.

Agerer, R. (1986). Studies on ectomycorrhizae II. Introducing remarks on characterization and identification. *Mycotaxon* **26**, 473–492.

Alconero, R. (1969). Mycorrhizal synthesis and pathology of *Rhizoctonia solani* in vanilla orchid roots. *Phytopathology* **59**, 426–430.

Allen, M. F. (1987). Reestablishment of mycorrhizas on Mount St. Helens: Migration vectors. *Transactions of the British Mycological Society* **88**, 413–417.

Allen, M. F. (1988). Re-establishment of VA mycorrhizas following severe disturbance: Comparative patch dynamics of a shrub desert and a subalpine volcano. *Proceedings of the Royal Society of Edinburgh* **94B**, 63–71.

Allen, M. F., & Allen, E. B. (1990). Carbon source of VA mycorrhizal fungi associated with Chenopodiaceae from semiarid shrub-steppe. *Ecology* **71**, 2019–2021.

Allen, M. F., & MacMahon, J. A. (1988). Direct VA mycorrhizal inoculation of colonizing plants by pocket gophers. (*Thomomys talpoides*) on Mount Saint Helens. *Mycologia* **80**, 754–756.

Alsheikh, A. M., & Trappe, J. M. (1983). Desert truffles: The genus *Tirmania*. *Transactions of the British Mycological Society* **81**, 83–90.

Amaranthus, M. P., & Perry, D. A. (1989). Interaction effects of vegetation type and Pacific madrone soil inocula on survival, growth, and mycorrhiza formation of Douglas-fir. *Canadian Journal of Forest Research* **19**, 550–556.

Amaranthus, M. P., Molina, R., & Perry, D. A. (1990). Soil organisms, root growth and forest regeneration. *Proceedings of the Society of American Foresters*, pp. 89–93. National Convention, Spokane, Washington, September, 1989.

Anderson, A. J. (1988) Mycorrhizae—host specificity and recognition. *Phytopathology* **78**, 375–378.

Azcón, R., Rubio, R., & Barea, J. M. (1991). Selective interactions between different species of mycorrhizal fungi and *Rhizobium meliloti* strains, and their effects on growth, N_2-fixation (^{15}N) and nutrition of *Medicago sativa* L. *New Phytologist* **117**, 399–404.

Ba, A. M., & Thoen, D. (1990). First syntheses of ectomycorrhizas between *Afzelia africana* Sm. (Caesalpinioideae) and native fungi for West Africa. *New Phytologist* **114**, 99–103.

Becker, P. (1983). Ectomycorrhizae on *Shorea* (Dipterocarpaceae) seedlings in a lowland Malaysian rainforest. *Malaysian Forester* **46**, 146–170.

Bills, G. F., Holtzmann, G. I., & Miller, O. K., Jr. (1986). Comparison of ectomycorrhizal Basidiomycete communities in red spruce versus northern hardwood forests of West Virginia. *Canadian Journal of Botany* **64**, 760–768.

Björkman, E. (1960). *Monotropa hypopitys* L.—an epiparasite on tree roots. *Physiologia Plantarum* **13**, 308–327.

Bledsoe, C. S., Tennyson, K., & Lopushinsky, W. (1982). Survival and growth of outplanted Douglas-fir seedlings inoculated with mycorrhizal fungi. *Canadian Journal of Forest Research* **12**, 720–723.

Borchers, S. L., & Perry, D. A. (1990). Growth and ectomycorrhiza formation of Douglas-fir seedlings grown in soils collected at different distances from pioneering hardwoods in southwest Oregon clear-cuts. *Canadian Journal of Forest Research* **20**, 712–721.

Brian, P. W. (1976). The phenomenon of specificity in plant diseases. In: *Specificity in Plant Diseases* (Ed. by R. K. S. Wood & A. Graniti), pp. 15–22. Plenum, New York.

Castellano, M. A., & Molina, R. (1989). Mycorrhizae. In: *The Container Tree Nursery Manual, Volume 5, Agricultural Handbook 674* (Ed. by T. D. Landis, R. W. Tinus, S. E. McDonald, & J. P. Barnett), pp. 101–167. United States Department of Agriculture, Forest Service, Washington, D.C.

Castellano, M. A., & Trappe, J. M. (1985). Mycorrhizal associations of five species of Monotropoideae in Oregon. *Mycologia* **77**, 499–502.

Chilvers, G. A., Lapeyrie, F. F., & Horan, D. P. (1987). Ectomycorrhizal vs endomycorrhizal fungi within the same root systems. *New Phytologist* **107**, 441–448.

Christy, E. J., Sollins, P., & Trappe, J. M. (1982). First-year survival of *Tsuga heterophylla* without mycorrhizae and subsequent ectomycorrhizal development on decaying logs and mineral soil. *Canadian Journal of Botany* **60**, 1601–1605.

Chu-Chou, M. (1979). Mycorrhizal fungi of *Pinus radiata* in New Zealand. *Soil Biology and Biochemistry* **11**, 557–562.

Chu-Chou, M., & Grace, L. J. (1981a). Mycorrhizal fungi of *Pseudotsuga menziesii* in the North Island of New Zealand. *Soil Biology and Biochemistry* **13**, 247–249.

Chu-Chou, M., & Grace, L. J. (1981b). *Hydnangium carneum*, a mycorrhizal fungus of *Eucalyptus* in New Zealand. *Transactions of the British Mycological Society* **77**, 650–651.

Chu-Chou, M., & Grace, L. J. (1981c). *Hymenogaster albus*—a mycorrhizal fungus of *Eucalyptus* in New Zealand. *New Zealand Journal of Forestry Research* **11**, 186–190.

Chu-Chou, M., & Grace, L. J. (1982). Mycorrhizal fungi of *Eucalyptus* in the North Island of New Zealand. *Soil Biology and Biochemistry* **14**, 133–137.

Chu-Chou, M., & Grace, L. J. (1983a). Characterization and identification of mycorrhizas of Douglas fir in New Zealand. *Sonderdruck aus European Journal of Forest Pathology* **13**, 251–260.

Chu-Chou, M., & Grace, L. J. (1983b). Hypogeous fungi associated with some forest trees in New Zealand. *New Zealand Journal of Botany* **21**, 183–190.

Chu-Chou, M., & Grace, L. J. (1985). Comparative efficiency of the mycorrhizal fungi *Laccaria laccata*, *Hebeloma crustuliniforme* and *Rhizopogon* species on growth of radiata pine seedlings. *New Zealand Journal of Botany* **23**, 417–424.

Couture, M., Fortin, J. A., & Dalpé, Y. (1983). *Oidiodendron griseum* Robak: An endophyte of ericoid mycorrhiza in *Vaccinium* spp. *New Phytologist* **95**, 375–380.

Currah, R. S. (1987). *Thanatephorus pennatus* sp. nov. isolated from mycorrhizal roots of *Calypso bulbosa* (Orchidaceae) from Alberta. *Canadian Journal of Botany* **65**, 1957–1960.

Currah, R. S., Hambleton, S., & Smreciu, A. (1988). Mycorrhizae and mycorrhizal fungi of *Calypso bulbosa*. *American Journal of Botany* **75**, 739–752.

Currah, R. S., Sigler, L., & Hambleton, S. (1987). New records and new taxa of fungi from the mycorrhizae of terrestrial orchids of Alberta. *Canadian Journal of Botany* **65**, 2473–2482.

Currah, R. S., Smreciu, A., & Hambleton, S. (1990). Mycorrhizae and mycorrhizal fungi of boreal species of *Platanthera* and *Coeloglossum*. *Canadian Journal of Botany* **68**, 1171–1181.

Dahlberg, A. (1990). Effect of soil humus cover on the establishment and development of mycorrhiza on containerised *Pinus sylvestris* L. and *Pinus contorta* spp. *latifolia* Engelm, after outplanting. *Scandinavian Journal of Forest Research* **5**, 103–112.

Dalpé, Y. (1986). Axenic synthesis of ericoid mycorrhiza in *Vaccinium angustifolium* Ait. by *Oidiodendron* species. *New Phytologist* **103**, 391–396.

Dalpé, Y. (1989). Ericoid mycorrhizal fungi in the Myxotrichaceae and Gymnoasaceae. *New Phytologist* **113**, 523–527.

Danielson, R. M. (1982). Taxonomic affinities and criteria for identification of the common ectendomycorrhizal symbiont of pines. *Canadian Journal of Botany* **60**, 7–18.

Danielson, R. M. (1984). Ectomycorrhiza formation by the operculate discomycete *Sphaerosporella brunnea* (Pezizales). *Mycologia* **76**, 454–461.

Danielson, R. M., & Pruden, M. (1989). The ectomycorrhizal status of urban spruce. *Mycologia* **81**, 335–341.

Danielson, R. M., & Visser, S. (1989). Host response to inoculation and behaviour of introduced and indigenous ectomycorrhizal fungi of jack pine grown on oil-sands tailings. *Canadian Journal of Forest Research* **19**, 1412–1421.

Dosskey, M. G., Linderman, R. G., & Boersma, L. (1990). Carbon-sink stimulation of photosynthesis in Douglas fir seedlings by some ectomycorrhizas. *New Phytologist* **115**, 269–274.

Douglas, G. C., Heslin, M. C., & Reid, C. (1989). Isolation of *Oidiodendron maius* from *Rhododendron* and ultrastructural characterization of synthesized mycorrhizas. *Canadian Journal of Botany* **67**, 2206–2212.

Duddridge, J. A. (1986a). Specificity and recognition in mycorrhizal associations. In: *Physiological and Genetical Aspects of Mycorrhizae* (Ed. by V. Gianinazzi-Pearson & S. Gianinazzi), pp. 45–58. INRA, Paris.

Duddridge, J. A. (1986b). The development and ultrastructure of ectomycorrhizas. III.

Compatible and incompatible interactions between *Suillus grevillei* (Klotzsch) Sing. and 11 species of ectomycorrhizal hosts *in vitro* in the absence of exogenous carbohydrate. *New Phytologist* **103**, 457–464.

Duddridge, J. A. (1986c). The development and ultrastructure of ectomycorrhizas. IV. Compatible and incompatible interactions between *Suillus grevillei* (Klotzsch) Sing. and a number of ectomycorrhizal hosts *in vitro* in the presence of exogenous carbohydrate. *New Phytologist* **103**, 465–471.

Duddridge, J. A. (1987). Specificity and recognition in ectomycorrhizal associations. In: *Fungal Infection of Plants* (Ed. by G. F. Pegg & P. G. Ayres), pp. 25–44. Cambridge University Press, Cambridge.

Duddridge, J. A., & Read, D. J. (1982). An ultrastructural analysis of the development of mycorrhizas in *Monotropa hypopitys* L. *New Phytologist* **92**, 203–214.

Durall, D. M., Todd, A. W., & Trappe, J. M. (1992). Decomposition of ^{14}C-labeled substrates by ectomycorrhizal fungi in association with Douglas fir. *New Phytologist, in press*.

Egger, K. N., & Paden, J. W. (1986). Biotrophic associations between lodgepole pine seedlings and postfire ascomycetes (Pezizales) in monoxenic culture. *Canadian Journal of Botany* **64**, 2719–2725.

Englander, L., & Hull, R. J. (1980). Reciprocal transfer between ericaceous plants and a *Clavaria* sp. *New Phytologist* **84**, 661–667.

Finlay, R. D. (1989). Functional aspects of phosphorus uptake and carbon translocation in incompatible ectomycorrhizal associations between *Pinus sylvestris* and *Suillus grevillei* and *Boletinus cavipes*. *New Phytologist* **112**, 185–192.

Finlay, R. D., & Read, D. J. (1986a). The structure and function of the vegetative mycelium of ectomycorrhizal plants. I. Translocation of ^{14}C-labelled carbon between plants interconnected by a common mycelium. *New Phytologist* **103**, 143–156.

Finlay, R. D., & Read, D. J. (1986b). The structure and function of the vegetative mycelium of ectomycorrhizal plants. II. The uptake and distribution of phosphorus by mycelial strands interconnecting host plants. *New Phytologist* **103**, 157–165.

Fortin, J. A., Piché, Y., & Lalonde, M. (1980). Technique for the observation of early morphological changes during ectomycorrhiza formation. *Canadian Journal of Botany* **58**, 361–365.

Francke, H. L. (1934). Beiträge zur Kenntnis der Mykorrhiza von *Monotropa hypopitys* L. Analyse und Synthese der Symbiose. *Flora* **129**, 1–52.

Franklin, J. F., & Dyrness, C. T. (1973). *Natural Vegetation of Oregon and Washington*. United States Department of Agriculture, Forest Service, General Technical Report, PNW-8.

Friesleben, R. (1936). Weitere Untersuchungen uber die Mycotrophie der Ericaceen. *Jahrbuch für Wissenschaftliche Botanik* **82**, 413–459.

Furman, T. E., & Trappe, J. M. (1971). Phylogeny and ecology of mycotrophic achlorophyllous angiosperms. *Quarterly Review of Biology* **46**, 219–225.

Garbaye, J., & Bowen, G. D. (1987). Effect of different microflora on the success of ectomycorrhizal inoculation of *Pinus radiata*. *Canadian Journal of Forest Research* **17,** 941–943.

Garbaye, J., & Bowen, G. D. (1989). Stimulation of ectomycorrhizal infection of *Pinus radiata* by some microorganisms associated with the mantle of ectomycorrhizas. *New Phytologist* **112,** 383–388.

Garrido, N. (1986). Survey of ectomycorrhizal fungi associated with exotic forest trees in Chile. *Nova Hedwigia* **43,** 423–442.

Garrido, N. (1988). Agaricales s.l. und irhe Mykorrhizen in den *Nothofagus*-Wäldern Mittelchiles. *Bibliotheca Mycologica* **120,** 1–528.

Gerdemann, J. W. (1968). Vesicular-arbuscular mycorrhiza and plant growth. *Annual Review of Phytopathology* **6,** 397–418.

Gianinazzi-Pearson, V. (1983). Host-fungus specificity, recognition and compatibility in mycorrhizae. In: *Genes Involved in Microbe Plant Interactions* (Ed. by E. S. Dennis, B. Hohn, T. Hohn, P. King, J. Schell, & D. P. S. Verma), pp. 225–253. Springer-Verlag, Vienna.

Giovannetti, M., & Lioi, L. (1990). The mycorrhizal status of *Arbutus unedo* in relation to compatible and incompatible fungi. *Canadian Journal of Botany* **68,** 1239–1244.

Grand, L. F. (1969). A beaded endotrophic mycorrhiza of northern and southern red oak. *Mycologia* **61,** 408–409.

Harley, J. L., & Smith, S. E. (1983). *Mycorrhizal Symbiosis*. Academic Press, London.

Hilton, R. N., Malajczuk, N., & Pearce, M. H. (1988). Larger fungi in the jarrah forest: An ecological and taxonomic survey. In: *The Jarrah Forest: A Complex Mediterranean Ecosystem* (Ed. by B. Dell, J. J. Havel, & N. Malajczuk), pp. 89–110. Kluwer Academic Publishers, The Netherlands.

Ho, I., & Trappe, J. M. (1973). Translocation of ^{14}C from *Festuca* plants to their endomycorrhizal fungi. *Nature (London)* **224,** 30–31.

Horak, E. (1983). Mycogeography in the South Pacific region: Agaricales, Boletales. *Australian Journal of Botany Supplemental Series* **1D,** 1–41.

Horan, D. P., & Chilvers, G. A. (1990). Chemotropism—the key to ectomycorrhizal formation? *New Phytologist* **116,** 297–301.

Hung, L. L., & Molina, R. (1986). Use of the ectomycorrhizal fungus *Laccaria laccata* in forestry. III. Effects of commercially produced inoculum on container-grown Douglas-fir and ponderosa pine seedlings. *Canadian Journal of Forest Research* **16,** 802–806.

Janos, D. P. (1980). Mycorrhizae influence tropical succession. *Biotropica* **12**(2, Suppl), 56–64.

Janos, D. P. (1985). Mycorrhizal fungi: Agents or symptoms of tropical community composition? In: *6th North American Conference on Mycorrhizae* (Ed. by R. Molina), pp. 98–103. Oregon State University, Forest Research Laboratory, Corvallis, OR.

Johnson, C. R. (1981). Benefits of mycorrhizae in woody container production. *Florida Nurseryman* **90–91,** 98–102.

Kernan, M. J., & Finocchio, A. F. (1983). A new discomycete associated with the roots of *Monotropa uniflora* (Ericaceae). *Mycologia* **75,** 916–920.

Khan, A. G. (1972). Mycorrhizae in the Pakistan ericales. *Pakistan Journal of Botany* **4,** 183–194.

Kidston, R., & Lang, W. H. (1921). On old red sandstone plants showing structure from the Rhynie Chert Bed, Aberdeenshire. Pt. IV. *Transactions of the Royal Society of Edinburg* **52,** 855–902.

Koske, R. E., Gemma, J. N., & Englander, L. (1990). Vesicular-arbuscular mycorrhizae in Hawaiian Ericales. *American Journal of Botany* **77,** 64–68.

Kropp, B. R., & Trappe, J. M. (1982). Ectomycorrhizal fungi of *Tsuga heterophylla*. *Mycologia* **74,** 479–488.

Laiho, O. (1965). Further studies on the ectendotrophic mycorrhiza. *Acta Forestalia Fennica* **79,** 1–35.

Largent, D. L., Sugihara, N., & Wishner, C. (1980). Occurrence of mycorrhizae on ericaceous and pyrolaceous plants in northern California. *Canadian Journal of Botany* **58,** 2274–2279.

Last, F. T., Dighton, J., & Mason, P. A. (1987). Successions of sheathing mycorrhizal fungi. *Trends in Ecology and Evolution* **2,** 157–161.

Lee, K. J., & Kim, Y. S. (1987). Host specificity and distribution of putative ectomycorrhizal fungi in pure stands of twelve tree species in Korea. *Korean Journal of Mycology* **15,** 48–69.

Le Tacon, F., Garbaye, J., Bouchard, D., Chevalier, G., Olivier, J. M., Guimberteau, J., Poitou, N., & Frochot, H. (1988). Field results from ectomycorrhizal inoculation in France. In: *Canadian Workshop on Mycorrhizae in Forestry* (Ed. by M. Lalonde & Y. Piché), pp. 51–74. Université Laval.

Louis, I. (1988). Ecto- and ectendomycorrhizae in the tropical dipterocarp, *Shorea parvifolia*. *Mycologia* **80,** 845–849.

Lutz, R. W., & Sjolund, R. D. (1973). *Monotropa uniflora:* Ultrastructural details of its mycorrhizal habitat. *American Journal of Botany* **60,** 339–345.

Mabberley, D. J. (1989). *The Plant-Book*. Cambridge University Press, Cambridge.

Malajczuk, N. (1987). Ecology and management of ectomycorrhizal fungi in regenerating forest ecosystems in Australia. In: *Mycorrhizae in the Next Decade—Practical Applications and Research Priorities, 7th NACOM* (Ed. by D. M. Sylvia, L. L. Hung, & J. H. Graham), pp. 118–120. University of Florida, Gainesville, FL.

Malajczuk, N., Molina, R., & Trappe, J. M. (1982). Ectomycorrhiza formation in *Eucalyptus*. I. Pure culture synthesis, host specificity and mycorrhizal compatibility with *Pinus radiata*. *New Phytologist* **91,** 467–482.

Malajczuk, N., Molina, R., & Trappe, J. M. (1984). Ectomycorrhiza formation in *Eucalyptus*. II. The ultrastructure of compatible and incompatible mycorrhizal fungi and associated roots. *New Phytologist* **96,** 43–53.

Malajczuk, N., Lapeyrie, F., & Garbaye, J. (1990). Infectivity of pine and eucalypt isolates of *Pisolithus tinctorius* on roots of *Eucalyptus urophylla in vitro*. I. Mycorrhiza formation in model systems. *New Phytologist* **114**, 627–631.

Malloch, D. W., Pirozynski, K. A., & Raven, P. H. (1980). Ecological and evolutionary significance of mycorrhizal symbioses in vascular plants, a review. *Proceedings of the National Academy of Science U.S.A.* **77**, 2113–2118.

Marx, D. H. (1980). Ectomycorrhizal fungus inoculations: a tool for improving forestation practices. In: *Tropical Mycorrhiza Research* (Ed. by P. Mikola), pp. 11–71. Clarendon Press, Oxford.

Marx, D. H., Cordell, C. E., Kenney, D. S., Mexal, J. G., Artman, J. D., Riffle, J. W., & Molina, R. J. (1984). Commercial vegetative techniques of *Pisolithus tinctorius* and inoculation techniques for development of ectomycorrhizae on bare-root tree seedlings. *Forest Science Monograph* **25**, 1–101.

Massicotte H. B., Trappe, J. M., Peterson, R. L. & Melville, L. H. (1991). Studies on *Cenococcum geophilum*. II. Sclerotium morphology, germination, and formation in pure culture and growth pouches. *Canadian Journal of Botany* **70**, 125–132.

McAfee, B. J., & Fortin, J. A. (1986). Competitive interactions of ectomycorrhizal mycobionts under field conditions. *Canadian Journal of Botany* **64**, 848–852.

McGonigle, T. P., & Fitter, A. H. (1990). Ecological specificity of vesicular-arbuscular mycorrhizal associations. *Mycological Research* **94**, 120–122.

Mejstrik, V., & Hadac, E. (1975). Mycorrhizas of *Arctostaphylos uva-ursi*. *Pedobiologia* **15**, 336–342.

Melin, E. (1922). Untersuchungen über die *Larix* mykorrhiza I. Synthese der mykorrhiza in reinkulture. *Svensk Botanisk Tidskrift* **16**, 161–196.

Melin, E. (1923). Experimentelle Untersuchungen über die Birkin un Espenmycorrhizen und ihre Pilzsymbionten. *Svensk Botanisk Tidskrift* **17**, 479–520.

Melin, E. (1925). *Untersuchungen über die Bedeutung der Baummykorrhiza. Eine ökologisch-physiologische Studie*. G. Fisher, Jena.

Meyer, F. H. (1973). Distribution of ectomycorrhizae in native and man-made forests. In: *Ectomycorrhizae—Their Ecology and Physiology* (Ed. by G. C. Marks & T. T. Kozlowski), pp. 79–105. Academic Press, New York.

Mikola, P. (1965). Studies on the ectotrophic mycorrhiza of pine. *Acta Forestalia Fennica* **79**, 1–56.

Mikola, P. (1970). Mycorrhizal inoculation in afforestation. *International Review of Forestry Research* **3**, 123–196.

Mikola, P. (1973). Application of mycorrhizal symbiosis in forestry practice. In: *Ectomycorrhizae—Their Ecology and Physiology* (Ed. by G. C. Marks & T. T. Kozlowski), pp. 383–406. Academic Press, New York.

Mikola, P. (1988). Ectendomycorrhiza of conifers. *Silva Fennica* **22**, 19–27.

Miller, C. N., Jr. (1977). Mesozoic conifers. *Botanical Review* **43**, 217–280.

Miller, R. M. (1979). Some occurrences of vesicular-arbuscular mycorrhiza in natural and disturbed ecosystems of the Red Desert. *Canadian Journal of Botany* **57**, 619–623.

Miller, R. M. (1987). The ecology of vesicular-arbuscular mycorrhizae in grass- and shrublands. In: *Ecophysiology of VA Mycorrhizal Plants* (Ed. by G. R. Safir), pp. 135–170. CRC Press, Boca Ratan, FL.

Miller, S. L., Koo, C. D., & Molina, R. (1991). Characterization of red alder ectomycorrhizae: A preface to monitoring belowground ecological responses. *Canadian Journal of Botany* **69**, 516–531.

Molina, R. (1979). Pure culture synthesis and host specificity of red alder mycorrhizae. *Canadian Journal of Botany* **57**, 1223–1228.

Molina, R. (1981). Ectomycorrhizal specificity in the genus *Alnus*. *Canadian Journal of Botany* **59**, 325–334.

Molina, R. (1982). Use of the ectomycorrhizal fungus *Laccaria laccata* in forestry. I. Consistency between isolates in effective colonization of containerized conifer seedlings. *Canadian Journal of Forest Research* **12**, 469–473.

Molina, R., & Chamard, J. (1983). Use of ectomycorrhizal fungus *Laccaria laccata* in forestry. II. Effects of fertilizer forms and levels on ectomycorrhizal development and growth of container-grown Douglas-fir and ponderosa pine seedlings. *Canadian Journal of Forest Research* **13**, 89–95.

Molina, R., & Palmer, J. G. (1982). Isolation maintenance, and pure culture manipulation of ectomycorrhizal fungi. In: *Methods and Principles of Mycorrhizal Research* (Ed. by N. C. Schenck), pp. 115–129. American Phytopathological Society, St. Paul, MN.

Molina, R., & Trappe, J. M. (1982a). Lack of mycorrhizal specificity in the ericaceous hosts *Arbutus menziesii* and *Arctostaphylos uva-ursi*. *New Phytologist* **90**, 495–509.

Molina, R., & Trappe, J. M. (1982b). Patterns of ectomycorrhizal host specificity and potential among Pacific Northwest conifers and fungi. *Forest Science* **28**, 423–458.

Molina, R. J., Trappe, J. M., & Strickler, G. S. (1978). Mycorrhizal fungi associated with *Festuca* in the western United States. *Canadian Journal of Botany* **56**, 1691–1695.

Moore, R. T. (1987). The genera of *Rhizoctonia*-like fungi: *Ascorhizoctonia, Ceratorhiza* gen. nov., *Epulorhiza* gen. nov., *Moniliopsis*, and *Rhizoctonia*. *Mycotaxon* **29**, 91–99.

Newbery, D. M., Alexander, I. J., Thomas, D. W., & Gartlan, J. S. (1988). Ectomycorrhizal rain-forest legumes and soil phosphorus in Korup National Park, Cameroon. *New Phytologist* **109**, 433–450.

Newman, E. I. (1988). Mycorrhizal links between plants: Functioning and ecological significance. *Advances in Ecological Research* **18**, 243–270.

Newman, E. I., & Reddell, P. (1987). The distribution of mycorrhizas among families of vascular plants. *New Phytologist* **106**, 745–752.

Newton, A. C. (1991). Mineral nutrition and mycorrhizal infection of seedling oak and birch. III. Epidemiological aspects of ectomycorrhizal infection, and the relationship to seedling growth. *New Phytologist* **117**, 53–60.

Parke, J. L., Linderman, R. G., & Black, C. H. (1983). The role of ectomycorrhizas in drought tolerance of Douglas-fir seedlings. *New Phytologist* **95**, 83–95.

Perry, D. A., Molina, R., & Amaranthus, M. P. (1987). Mycorrhizae, mycorrhizospheres, and reforestation: Current knowledge and research needs. *Canadian Journal of Forest Research* **17**, 929–940.

Perry, D. A., Amaranthus, M. P., Borchers, J. G., Borchers, S. L., & Brainerd, R. E. (1989a). Bootstrapping in ecosystems. *BioScience* **39**, 230–237.

Perry, D. A., Margolis, H., Choquette, C., Molina, R., & Trappe, J. M. (1989b). Ectomycorrhizal mediation of competition between coniferous tree species. *New Phytologist* **112**, 501–511.

Perry, D. A., Borchers, J. G., Borchers, S. L., & Amaranthus, M. P. (1990). Species migrations and ecosystem stability during climate change; the belowground connection. *Conservation Biology* **4**, 266–274.

Petersen, R. H., & Litten, W. (1989). A new species of *Clavaria* fruiting with *Vaccinium*. *Mycologia* **81**, 325–327.

Pirozynski, K. A. (1981). Interactions between fungi and plants through the ages. *Canadian Journal of Botany* **59**, 1824–1827.

Pirozynski, K. A., & Malloch, D. W. (1975). The origin of land plants: A matter of mycotrophism. *BioSystems* **6**, 153–164.

Rayner, M. C. (1927). Mycorrhiza. *New Phytologist* Reprint #15, Wheldon & Wesley, London.

Read, D. J. (1984). The structure and function of vegetative mycelium of mycorrhizal roots. In: *Ecology and Physiology of the Fungal Mycelium* (Ed. by D. H. Jennings & A. D. M. Rayner), pp. 215–240. Cambridge University Press, Cambridge.

Read, D. J., Francis, R., & Finlay, R. D. (1985). Mycorrhizal mycelia and nutrient cycling in plant communities. In: *Ecological Interactions in Soil* (Ed. by A. H. Fitter, D. Atkinson, D. A. Read, & M. B. Usher), pp. 193–217. Blackwell Scientific Publications, Palo Alto, CA.

Reeves, F. B., Wagner, D., Moorman, T., & Kiel, J. (1979). The role of endomycorrhizae in revegetation practices in the semi-arid west. I. A comparison of incidence of mycorrhizae in severely disturbed vs. natural environments. *American Journal of Botany* **66**, 6–13.

Robertson, D. C., & Robertson, J. A. (1985). Ultrastructural aspects of *Pyrola* mycorrhizae. *Canadian Journal of Botany* **63**, 1089–1098.

Schoenberger, M. M., & Perry, D. A. (1982). The effect of soil disturbance on growth and ectomycorrhizae of Douglas-fir and western hemlock seedlings: A greenhouse bioassay. *Canadian Journal of Forest Research* **12**, 343–353.

Seviour, R. J., Willing, R. R., & Chilvers, G. A. (1973). Basidiocarps associated with ericoid mycorrhizas. *New Phytologist* **72**, 381–385.

Singer, R., & Moser, M. (1965). Forest mycology and forest communities in South America. *Mycopathologia Mycologia Applicata* **27**, 129–190.

Smith, A. H., & Zeller, S. M. (1966). A preliminary account of the North American species of *Rhizopogon*. *Memoirs of the New York Botanical Gardens* **14**, 1–178.

Smits, W. Th. M. (1983). VIII. Dipterocarps and mycorrhiza, an ecological adaptation and a factor in forest regeneration. *Flora Malesiana Bulletin* **36**, 3926–3937.

Smreciu, E. A., & Currah, R. S. (1989). Symbiotic germination of seeds of terrestrial orchids of North America and Europe. *Lindleyana* **4**, 6–15.

Stubblefield, S. P., & Taylor, T. N. (1988). Recent advances in palaeomycology. *New Phytologist* **108**, 3–25.

Stubblefield, S. P., Taylor, T. N., & Trappe, J. M. (1987). Fossil mycorrhizae: A case for symbiosis. *Science* **237**, 59–60.

Theodorou, C., & Bowen, G. D. (1971). Effects of non-host plants on growth of mycorrhizal fungi of radiata pine. *Australian Forestry* **35**, 17–22.

Theodorou, C., & Bowen, G. D. (1987). Germination of basidiospores of mycorrhizal fungi in the rhizosphere of *Pinus radiata* D. Don. *New Phytologist* **106**, 217–223.

Thiers, H. D. (1975). *California Mushrooms, a Field Guide to the Boletes*. Hafner Press, New York.

Thoen, D., & Ba, A. M. (1989). Ectomycorrhizas and putative ectomycorrhizal fungi of *Afzelia africana* Sm. and *Uapaca guineensis* Müll. Arg. in southern Senegal. *New Phytologist* **113**, 549–559.

Tonkin, C. M., Malajczuk, N., & McComb, J. A. (1989). Ectomycorrhizal formation by micropropagated clones of *Eucalyptus marginata* inoculated with isolates of *Pisolithus tinctorius*. *New Phytologist* **111**, 209–214.

Trappe, J. M. (1962). Fungus associates of ectotrophic mycorrhizae. *Botanical Review* **28**, 538–606.

Trappe, J. M. (1964). Mycorrhizal hosts and distribution of *Cenococcum graniforme*. *Lloydia* **27**, 100–106.

Trappe, J. M. (1967). Pure culture synthesis of Douglas-fir mycorrhizae with species of *Hebeloma, Suillus, Rhizopogon,* and *Astraeus*. *Forest Science* **13**, 121–130.

Trappe, J. M. (1977a). Selection of fungi for ectomycorrhizal inoculation in nurseries. *Annual Review of Phytopathology* **15**, 203–222.

Trappe, J. M. (1977b). Biogeography of hypogeous fungi: Trees, mammals and continental drift. *2nd International Mycological Congress* (Abstract), p. 675.

Trappe, J. M. (1987). Phylogenetic and ecologic aspects of mycotrophy in the angiosperms from an evolutionary standpoint. In: *Ecophysiology of VA Mycorrhizal Plants* (Ed. by G. R. Safir), pp. 2–25. CRC Press, Boca Ratan, FL.

Trappe, J. M. (1988). Lessons from alpine fungi. *Mycologia* **80**, 1–10.

Trappe, J. M. (1989). The meaning of mycorrhizae to plant ecology. In: *Mycorrhizae for Green Asia* (Ed. by M. Mahadevan, N. Ramen, & K. Natarayan), pp. 347–349. Centre for Advanced Studies, University of Madras, India.

Trappe, J. M., & Cazares, E. (1991). Evolucion y ecologia de los hongos hipogeos. *Revista Mexicana de Micologia* **6**, 33–40.

Trappe, J. M., & Fogel, R. D. (1977). Ecosystematic functions of mycorrhizae. *Colorado State University Range Science Department Sciences Series* **26**, 205–214.

Trappe, J. M., & Maser, C. (1977). Ectomycorrhizal fungi: Interactions of mushrooms and truffles with beasts and trees. In: *Mushrooms and Man, an Interdisciplinary Approach to Mycology* (Ed. by T. Walters), pp. 165–179. Linn-Benton Community College, Albany, OR.

Trappe, J. M., & Strand, R. F. (1969). Mycorrhizal deficiency in a Douglas-fir region nursery. *Forest Science* **15**, 381–389.

Trappe, J. M., Li, C. Y., Lu, K. C., & Bollen, W. B. (1973). Differential response of *Poria weirii* to phenolic acids from Douglas-fir and red alder roots. *Forest Science* **19**, 191–196.

Vanderplank, J. E. (1978). *Genetic and Molecular Basis of Plant Pathogenesis*. Springer-Verlag, New York.

Wang, C. J. K., & Wilcox, H. E. (1985). New species of ectendomycorrhizal and pseudomycorrhizal fungi: *Phialophora finlandia, Chloridium paucisporum,* and *Phialocephala fortinii*. *Mycologia* **77**, 951–958.

Warcup, J. H. (1971). Specificity of mycorrhizal association in some Australian terrestrial orchids. *New Phytologist* **70**, 41–46.

Warcup, J. H. (1981). The mycorrhizal relationship of Australian orchids. *New Phytologist* **87**, 371–382.

Warcup, J. H. (1988). Mycorrhizal associations of isolates of *Sebacina vermifera*. *New Phytologist* **110**, 227–231.

Wilcox, H. E., & Wang, C. J. K. (1987a). Ectomycorrhizal and ectendomycorrhizal associations of *Phialophora finlandia* with *Pinus resinosa, Picea rubens,* and *Betula alleghaniensis*. *Canadian Journal of Forest Research* **17**, 976–990.

Wilcox, H. E., & Wang, C. J. K. (1987b). Mycorrhizal and pathological associations of dematiaceous fungi in roots of 7-month-old tree seedlings. *Canadian Journal of Forest Research* **17**, 884–889.

Wilcox, H. E., Neumann, R., Ganmore, R., & Wang, C. J. K. (1974). Characteristics of two fungi producing ectendomycorrhizae in *Pinus resinosa*. *Canadian Journal of Botany* **52**, 2279–2282.

Wilcox, H. E., Yang, C. S., & LoBuglio, K. F. (1983). Responses of pine roots to E-strain ectendomycorrhizal fungi. *Plant and Soil* **71**, 293–297.

Yang, C. S., & Korf, R. P. (1985). A monograph of the genus *Tricharina* and a new, segregate genus, *Wilcoxina* (Pezizales). *Mycotaxon* **24**, 467–531.

Zak, B. (1973). Classification of ectomycorrhizae. In: *Ectomycorrhizae—Their Ecology and Physiology* (Ed. by G. C. Marks & T. T. Kozlowski), pp. 43–78. Academic Press, New York.

Zak, B. (1974). Ectendomycorrhiza of Pacific madrone (*Arbutus menziesii*). *Transactions of the British Mycological Society* **62**, 201–204.

Zak, B. (1976a). Pure culture synthesis of bearberry mycorrhizae. *Canadian Journal of Botany* **54**, 1297–1305.

Zak, B. (1976b). Pure culture synthesis of Pacific madrone ectendomycorrhizae. *Mycologia* **68**, 362–369.

12

Physiological Ecology of Ectomycorrhizae: Implications for Field Application

Caroline S. Bledsoe

SUMMARY. **In forest ecosystems, virtually all plant roots are mycorrhizal and enhanced nutrition and growth of forest trees are the result. However, less is known about the different effects of specific fungal species forming these mycorrhizas. Studies on small trees or seedlings have shown that certain fungal species may be more beneficial to the host than other species. How these results apply to larger trees in a natural forest environment is uncertain. In forests, many different fungi may infect roots of a single tree, and these associations change over time with season and with age of the stand. Using new, molecular methods, mycorrhizal scientists are learning to identify these root-associated fungi. As we increase our understanding of the importance of the changing mycorrhizal population, forest management practices are also changing—to enhance mycorrhizal inoculum and diversity. Future research should focus on the physiological roles of different fungi, in order for future foresters to choose wisely the particular fungal partner for their trees.**

Introduction

It is well known that mycorrhizal fungi and plants can form a mutually beneficial, symbiotic relationship. Simply stated, fungi, in intimate association with plant roots, receive a supply of energy, as carbohydrates, from the host. Plants receive mineral nutrients, in amounts greater than that which the plant roots would acquire alone; enhanced plant growth results. There are other benefits to both fungus and plant that are not discussed here, in order to focus on the major benefit to plants—enhanced nutrient acquisition.

This nutritional benefit to plants is of interest to plant growers, including foresters interested in the enhancement of tree growth. Foresters, forest biologists, and ecologists would like to understand whether the choice of a particular fungal species as a partner for a given tree species on a given reforestation site can significantly affect the survival, growth, and nutrition of the tree.

This chapter deals with this question of fungal-specific effects by considering what is known about the ecology and physiology of mycorrhizal fungi associated with forest tree species, particularly conifers. Further, the chapter discusses what aspects of mycorrhizal ecology may apply to field forestry—the planting of seedlings, growth of trees, and reforestation of a harvested site. In the following three sections, we discuss (1) current knowledge of mycorrhizal ecology and physiology of nutrient acquisition, (2) current practices in field applications, and (3) new research and future opportunities for mycorrhizal applications.

Current Knowledge: Mycorrhizal Ecophysiology and Nutrition

Fungal Effects on the Host

A number of laboratory, greenhouse, and nursery studies have attempted to discover how different mycorrhizal species affect the physiology of a host tree. For example, if fungal species differ in their phosphate uptake rates, will ectomycorrhizal roots also have uptake rates similar to that of the fungal partner? This topic is difficult to address in mature trees in forests since one cannot control the choice of the fungal partner. Thus, although there are numerous limitations in seedling research, virtually all of our information comes from this type of study.

These lab/greenhouse/nursery studies, a few of which are described below, generally demonstrate that fungi do differ in some of their physiological characteristics, although the differences are generally not large. Further, the choice of the fungal partner can affect the physiological characteristics of the mycorrhizal root or seedling, although the differences are, again, not large.

When considering only the fungus alone, fungal species do differ in anatomical and physiological characteristics, both of which may affect the mycorrhizal association and the host itself. These differences may be at the family, genus, species, or strain level. For example, at the strain level, different strains of *Laccaria bicolor* differed in degree of mycorrhizal formation as measured by mantle thickness, density, and degree of Hartig net penetration (Wong et al., 1989). At higher levels of taxonomic organization, many authors have reported differences among fungi in their ability to (1) withstand water stress (Mexal & Reid, 1973; Coleman et al., 1989), (2) take up nitrogen (Littke et al., 1984), (3) take up phosphorus (Harley & Smith, 1983; Findlay & Read, 1986, and many others), (4) produce phosphatases (Ho & Zak, 1979), and many other characteristics.

When considering the mycorrhizal association, differences are also found, in water stress (Parke et al., 1983), nitrogen uptake (Rygiewicz et al., 1984a,b), phosphorus uptake (Harley & Smith, 1983; Findlay & Read, 1986), and phosphatase activity (Bartlett & Lewis, 1973; Antibus et al., 1981; Kropp, 1989).

Thus, a number of studies of mycorrhizal fungi alone or of mycorrhizal associations have demonstrated differences in the anatomy or physiology, which are due to differences in the fungal species. However, since all these studies are laboratory, greenhouse, or seedling studies, it is not clear whether these differences may be of significance in a forest environment. The next section discusses this topic.

Mycorrhizal Functions in the Forest

Little is known about how mycorrhizae function in a mature forest environment. There are two major questions: which fungal species are present on roots, and do fungal species change (undergo succession) over time?

Presence of Fungal Species

It is difficult to determine the identity of fungal species on tree roots, since it is very difficult to identify fungi on roots. Fungal identification is based on characteristics of the fruiting body or sporocarp. The sporocarps are generally found aboveground, although some fruit belowground (e.g., truffles and false truffles). Although sporocarps must be linked via hyphae and rhizomorphs to tree roots that supply carbohydrates for sporocarp growth, these hyphal connections are difficult to trace. The appearance of fungal fruit bodies is often assumed to be presumptive evidence of functioning mycorrhizae on nearby roots.

Mycorrhizal researchers have long wanted to be able to identify fungal species on roots. Until recently, mycorrhizal roots were characterized on the basis of visible, morphological characters—such as color of hyphae, degree of branching, and thickness of the mantle. With the development of some molecular techniques (see the Section on "Promising New Research"), we now have a tool for much more positive identification of these root-associated mycorrhizal fungi (see papers by Lee & Taylor, 1989, Rogers et al., 1989; Bruns et al., 1990, & others). These methods allow fungi to be identified to genus, sometimes species.

Change of Fungal Species

There is increasing evidence that, just as a forest undergoes succession, there may be a succession of mycorrhizal fungi (Last et al., 1984, 1985; Dighton & Mason, 1985). As forests age, many conditions change. Forest soils are spatially very heterogeneous; physical, chemical, and biological conditions change rapidly over very small distances. Microsites are simply small three-dimensional volumes in soils within which microorganisms affect and are affected by the physical and chemical environment—soil moisture, fertility, aeration, etc. Coleman (1985) discusses soil microsites in detail.

These "microsites" in soils affect the growth and function of soil microorganisms, including mycorrhizal fungi. As plant species change over forest succession and tree stands age, the soil environment is also altered. For example, in coniferous forests, soils may become more acidic, litter layers may increase, and coarse woody debris accumulates. Carbon/nitrogen ratios change, altering availability of nutrients for microbes and plants. It is likely that fungal species also change in response to changing soil environment and plant species.

Mycorrhizal species are strongly affected by environmental gradients—such as soil type, ecosystem type, and altitude (Read, 1984). Thus it is likely that changing environmental gradients in soils also cause a succession of mycorrhizal fungi. This presumed fungal succession should be documented in a number of different ecosystems, including understanding of how mycorrhizae interact with other soil biota (Fogel, 1988).

There have been few studies that systematically observed either the fruit bodies in a forest stand or the types of ectomycorrhizal roots with time. The work in Scotland (Dighton & Mason, 1985; Last et al., 1985) is an outstanding exception. They documented changing patterns of both fruit bodies and root-associated fungi in stands of varying ages and in the same stand over a period of several years. These scientists suggest that there may be "early" stage, "mid" stage, and "late" stage fungi, corresponding to early, mid, and late-successional stages of the forest. More data from other groups are needed to determine whether this observation is a general condition.

The diversity of fungal symbionts is high in coniferous forests, whereas the diversity of the host species is low. However, we can only speculate why. Perry et al. (1987) suggested that this fungal diversity may be a buffer against disturbance. We can only guess at the different ecological and physiological functions of these various species.

The previous examples demonstrate the need for more information on the fungi that are functioning mycorrhizae in a forest, as well as on the changing fungal species over time. Once we have this understanding, then we can begin to study how mycorrhizal fungi function in the forest.

Current Field Applications: Mycorrhizal Inoculation/Retention

Seedlings

Most of the demonstrated benefits of mycorrhizae for forest trees are improved nutrition for phosphorus (primarily) and nitrogen (secondarily). In considering whether to inoculate forest tree seedlings, one must ask whether there is a practical, economic benefit to inoculation? As Stribley (1989) noted, the economics of fertilization should be an inextricable part of any experimental design that focuses on mycorrhizal inoculation. The interaction of mycorrhizae and fertility is complex, but remains a subject that needs examination.

If one determines that inoculation is desirable, then the single most important decision regarding inoculation of seedlings is which fungal species should be used. Although there are other decisions to be made, such as what method of inoculation and what season, if the chosen fungus does not form mycorrhizae, prosper in the field environment, and benefit the seedling, then all other choices matter very little.

This question was discussed by Trappe (1977), who emphasized the need for more information about the physiological and ecological characteristics of mycorrhizal fungal species. Unfortunately, in the past 13 years, although many outplanting studies have been conducted, the answers to this question are largely unknown. Since the choice of a fungus depends on many factors, the answer is rarely simple. Unfortunately we rarely know enough about the ecology of individual mycorrhizal fungi to make this choice.

In the southeastern United States, the fungus *Pisolithus tinctorius* (Pt) in combination primarily with *Pinus* spp. has been used in over 100 bare root nursery tests in 38 states (Marx et al., 1984) and many field outplantings. Generally Pt enhanced seedling survival and growth, although results were variable and the difficulties and costs of inoculation may not equal the benefits (Cordell et al., 1987). With oak seedlings, 11 different isolated were tested, with varying success (Dixon et al., 1984; Mitchell et al., 1984).

In the northwestern United States, several different fungal species have been tested, with modest and variable success. In harsh dry sites, *Heboloma crustuliniforme* and *Laccaria laccata* did not compete successfully with native fungal species or increase Douglas-fir seedling growth and survival (Bledsoe et al., 1982). In field and nursery tests, these same two fungi as well as Pt failed to readily colonize conifer seedlings or enhance plant growth (Castellano, 1987; Hung, 1985; Molina, 1982). The fungus *Cenococcum geophilum* was more promising and stimulated seedling top growth (Kropp et al., 1985). *Rhizopogon vinicolor*-inoculated seedlings also benefited from mycorrhizal association (Castellano & Trappe, 1985).

In Australia, the benefits from mycorrhizal inoculation are more evident, since a major commercial timber species, *Pinus radiata,* is an exotic, does not regenerate naturally, and has no native fungal partners (Tommerup et al., 1987). Mycorrhizal fungi have been a forest management tool for more than 60 years. As in the United States, in Australia, mycorrhizal scientists are searching for more suitable fungal species with enhanced benefits for the forest tree species.

In France, different fungal species when mycorrhizal with pine seedlings produced different height growth responses (Mousain et al., 1987). Current programs are in progress to screen and select wild mycorrhizal strains according to their enzymatic activities (nitrate reductase, phosphatase, etc.).

In tropical forests, where exotic seedling species are used, inoculation is essential (Mikola, 1980).

Thus, there are indications that certain species may be more "beneficial" than others in enhancing tree growth, but there are also many negative results and inconsistent results. Clearly, we need to know much more about the physiology of fungal species and their abilities to form efficient associations with a particular tree species under field conditions.

Many methods have been developed for inoculation, including use of soil, roots of inoculated trees, spores and spore pellets, and mycelia (see the following references for extensive discussion of these methods: Marx & Kenny, 1982; Riffle & Maronek 1982; Marx et al., 1984). In the opinion of this author, the question of which fungus or fungi to use must be answered first, before additional work is needed on methods of inoculation.

Commercial Forest Management

Once a mycorrhizal seedling is outplanted, it seems that little is done to manage the mycorrhizal association throughout the life of the tree. It may be necessary to alter current forest management practices to ensure the development and maintenance of beneficial mycorrhizal fungi in seedlings, young trees, mature trees, and in harvested areas under preparation for replanting.

How long does the mycorrhizal fungus persist on the roots? In many cases, the existing fungus is rapidly replaced by native fungi. For example, on an arid site in the western United States, two introduced species, *Hebeloma* and *Laccaria*, were replaced within 6 months after outplanting by an unidentified native fungus (Bledsoe et al., 1982). In another case, the original fungi remained 3 years after outplanting of mycorrhizal *Quercus rubra* seedlings, although there were differences among the different fungal species used for inoculation (Beckjord & McIntosh, 1983). And in a third example, although the original fungus remained on the roots, new roots that grew outward into the soil were infected by the indigenous fungus, not by the original fungus (Ruehle, 1983). Thus, persistence is a very site-specific phenomenon.

Use of herbicides and fertilizers may alter ectomycorrhizae on roots of young trees. High fertility regimes are known to decrease mycorrhizal numbers. In older stands, there is increasing evidence that mycorrhizal associations change as forest stands age (Dighton & Mason 1985). However, there is little effort by forest managers to promote mycorrhizal diversity in stands in order to provide appropriate mycorrhizal fungi throughout a forest rotation. This lack of appropriate fungal species may affect forest growth. Perhaps the nutrient uptake capabilities are different among early- and late-stage mycorrhizal fungi; in older forest stands, if the fungi are not adapted to nutrient uptake in soils with deep litter layers, forest growth may be decreased. Clearly, we need to know a great deal more about the physiological abilities of mycorrhizal fungi and whether there is a true distinction between early- and late-stage fungi.

In later stages of forest growth, during the harvest and site-preparation stages, the practice of removal of litter and woody debris, as well as slash burning, may reduce inoculum potential (Perry et al., 1987) and make reforestation difficult. In one case in the central Oregon Cascade Mountains, burning of sites reduced the mycorrhizal inoculum from *Cenococcum,* an effect that may last for more than 20 years (Schoenberger & Perry, 1982).

Reforestation—Disturbed, Severe Sites

Numerous studies have shown that mycorrhizal inoculation improves survival and growth of seedlings outplanted in such harsh sites as mine spoils, borrow pits, and high-elevation sites (Marx, 1975; Marx & Artman, 1979; Ruehle, 1980; Grossnickle & Reid, 1983). Mycorrhizae may also provide some protection against heavy metals in soils (Dixon & Buschena, 1988). In most cases, the difficulties of establishment in these harsh sites are so great that the choice of the fungal partner does not seem to be as important as in sites where there is a greater source of natural inoculum.

The degree and type of mycorrhizae on new seedlings can be modified by clearcutting, as Pilz and Perry (1984) demonstrated in the western United States, where clearcutting increased mycorrhizal formation. The presence of roots of older trees may also affect mycorrhizal development. Fleming (1984) trenched birch trees and noted that some "late-succession fungi" colonized seedlings in the trenched areas. Perhaps in clearcut areas, these older roots may support fungi that would not ordinarily occur on seedling roots.

Exotic, Introduced Species

In New Zealand, South America, Australia, Africa, and other parts of the world, new forests have been established using exotic, nonnative tree species. The successful establishment of these tree seedlings often depends on the simultaneous introduction of the mycorrhizal symbiont. In natural tropical forests, about 95% of the tree species are endomycorrhizal (LeTacon et al., 1987). In tropical forestry, the principal ectomycorrhizal species used are pines, eucalypts, and casuarinas. When these species are introduced, inoculum must be included, as stunted, chlorotic seedlings have resulted from the lack of inoculum (LeTacon et al., 1987). As tropical forests are cut and replanted to exotic species, there is an increasing demand for information about the choice of fungal mycorrhizal partner. Unfortunately, little is known about the natural inoculum, or the physiological ecology of introduced fungi (see LeTacon et al., 1987), for an extensive review of this subject).

Future Opportunities: Mycorrhizal Applications in Forestry

Forestry of the Future

Forestry of the 1990s is undergoing a revolution. Over this decade, forest management practices will change more drastically than in any other decade in the past. This revolution can be seen in the current controversy in the northwestern United States over the cutting of old growth forests and the indicator species, the spotted owl. Other countries, particularly West Germany, France, Britain, and the

Scandanavian countries, have dealt with this conflict sooner, primarily because these countries have a smaller land base and a greater awareness by their population of some of the negative aspects of forest harvest. Nevertheless, many countries and people are coming to the recognition that we have a single, global environment, and that forest management and harvest practices must be ecologically sound.

How will "new forestry" practices affect the application of mycorrhizal associations? These practices will require managed, productive forests under a changing global climate, and mycorrhizae will be increasingly important in maintaining productive forests. Although mycorrhizae are a small part of the biomass of a forest, they are central to nutrient cycling and primary productivity. For example, in a 180-year-old *Abies amabilis* stand in western Washington, mycorrhizae were less than 1% of the biomass, but mycorrhizae and fine roots contributed about 75% of the net primary production (Vogt et al., 1982). Clearly, mycorrhizae are a crucial part of forest management.

Promising, New Research

The following areas of mycorrhizal research are suggested as critical in the next 10–20 years:

1. *Maintenance of ecological diversity* of mycorrhizal fungi in natural stands, as a source of fungal species for research.
2. Development of techniques (such as molecular methods) for *identification of fungi on roots*.
3. Field studies that alter soil conditions and evaluate the *changes in mycorrhizal species*.
4. *Genetic engineering* of mycorrhizal fungi, with the goal of selecting certain characteristics (e.g., increased nutrient uptake ability, or saprobic ability) to benefit the host.
5. Large-scale *manipulative experiments* to determine the effects of *increased soil temperature and increased atmospheric carbon dioxide* on the functions of the mycorrhizal association.
6. Increased study of the *spatial arrangement* of mycorrhizae and associated soil biota in the rhizosphere.
7. Detailed studies of the *physiology and metabolism of ectomycorrhizal associations,* in order to determine which genes and which metabolic pathways are operative in the association.
8. Development of an on-line, searchable *computerized database of information on mycorrhizae*—ecology, physiology, and anatomy, for use in selecting mycorrhizal characteristics for future research.

Of these areas of research, I have selected three for particular attention, because I think these three areas are unusually timely, are a current bottleneck in our understanding of mycorrhizae, and are amenable to rapid progress due to the development of new methods.

1. Characterization of ectomycorrhizal roots at various sites, as a first step in selecting fungi for inoculation.

Just as forest ecologists have categorized the species distribution, abundance, biomass, etc. in various forest stands, similar attention is required by the belowground root and rhizosphere system. For a variety of forest stands, we lack basic information on the ectomycorrhizal associations. Studies of the type conducted by Brundrett et al. (1990) and Doudrick et al. (1990) are excellent examples of this type of research.

Brundrett et al. (1990) surveyed the mycorrhizal root structure of 20 important Ontario tree genera, beginning the collection of data that can form the base for future physiological studies. Doudrick et al. (1990) surveyed the distribution and environmental factors associated with presumed ectomycorrhizal fungi of black spruce in northern Minnesota. They noted fruit body production and collected mycorrhizal roots. As they noted, this study "serves as the first step in selecting ecologically adapted fungi for inoculating black spruce seedlings destined for planting in the field" (p. 830). Such background studies are needed for many other forest stands.

2. Identification of root-associated fungi using molecular techniques.

To make any progress on understanding how mycorrhizae function in a forest, we must know what fungi are present on roots and how they change over time. Until the recent developments in molecular biology, it was almost impossible to identify these fungi, based on vegetative hyphae on roots. The available data described the ectomycorrhizae on the basis of color of hyphae, texture, degree of branching, etc. Recently, several groups of mycorrhizal researchers have applied molecular methods to identify fungi on roots (Fortin and co-workers in Quebec; Taylor and Bruns at the University of California, Berkeley; Rogers at the State University of New York, Syracuse; Martin and co-workers at the Forestry Research Institute in Nancy, France; Sen in Finland; Rygiewicz at the Environmental Protection Agency in Corvallis, Oregon).

By using either isozyme analysis (Sen, 1990), restriction fragment length polymorphisms (RFLP) analysis, or DNA amplification methods, it is possible to develop a "library" of information about a number of different fungal species. Using this "library," one can then identify unknown fungi associated with roots. Rogers et al. (1989) identified root-associated fungi from a 20-year-old Douglas-fir stand in the western United States.

Gardes et al. (1991) used polymerase chain reaction (PCR) to amplify a region of the nuclear ribosomal DNA, specifically the internal transcribed spacer. Variation allowed distinguish among *Laccaria* isolates, as well as among other genera.

These new molecular techniques offer a powerful tool for understanding which fungal species are active on tree roots, and how these fungal associations change over time.

3. Understanding of the genes involved in mycorrhizal expression.

In recent work Hilbert and Martin (1988) and others attempt to determine which genes are "turned on" during the mycorrhizal phase. "Mycorrhiza-specific" polypeptides have been identified, providing a starting point for understanding which genes are important during the mycorrhizal process. This kind of work can ultimately give us clues about the physiological functions of different fungal species and how host and fungus interact during their symbiotic association.

References

Antibus, R. K., Croxdale, J. G., Miller, O. K., & Linkins, A. E. (1981). Ectomycorrhizal fungi of *Salix rotundifolia* III. Resynthesized mycorrhizal complexes and their surface phosphatase activities. *Canadian Journal of Botany* **59**, 2458–2465.

Bartlett, E. M., & Lewis, D. H. (1973). Surface phosphatase activity of mycorrhizal roots of beech. *Soil Biology and Biochemistry* **5**, 249–257.

Beckjord, P. R., & McIntosh, M. S. (1983). Growth and fungal retention by field-planted *Quercus rubra* seedlings inoculated with several ectomycorrhizal fungi. *Bulletin of the Torrey Botanical Club* **110**, 353–359.

Bledsoe, C. S., Tennyson, K., & Lopushinsky, W. (1982). Survival and growth of mycorrhizal Douglas-fir seedlings outplanted on burned-over sites in eastern WA. *Canadian Journal of Forest Research* **12**, 720–723.

Bledsoe, C. S., Brown, D., Coleman, M., Littke, W., Rygiewicz, P., Sangwanit, U., Rogers, S., & Ammirati, J. (1989). Physiology and metabolism of ectomycorrhizae. In: *Forest Tree Physiology* (Ed. by E. Dreyer, G. Aussenac, M. Bonnet-Masimbert, P. Dizengremel, J. Favre, J. Garrec, F. LeTacon, & F. Martin), *Annales des Sciences Forestiere*, **46**, 697–705.

Brundrett, M., Murase, G., & Kendrick, B. (1990). Comparative anatomy of roots and mycorrhizae of common Ontario trees. *Canadian Journal of Botany* **68**, 551–578.

Bruns, T. D., Fogel, R., & Taylor, J. W. (1990). Amplification and sequencing of DNA from fungal herbarium specimens. *Mycologia* **82**, 175–184.

Castellano, M. A. (1987). Ectomycorrhizal inoculum production and utilization in the Pacific Northwestern U.S.—A glimpse at the past, a look to the future. In: *Mycorrhizae in the Next Decade, Practical Applications and Research Priorities* (Ed. by D. M. Sylvia, L. L. Hung, & J. H. Graham), pp. 290–292. University of Florida, Gainesville, FL.

Castellano, M. A., & Trappe, J. (1985). Ectomycorrhizal formation and plantation performance of Douglas-fir nursery stock inoculated with *Rhizopogon* spores. *Canadian Journal of Forest Research* **15**, 613–617.

Coleman, D. C. 1985. Through a ped darkley: an ecological assessment of root-soil-microbial-faunal interactions. In: *Ecological Interactions in Soil: Plants, Microbes and Animals* (Ed. by D. Atkinson, D. J. Read, & M. B. Usher), pp. 1–22. Blackwell Scientific Publications, London.

Coleman, M., Bledsoe, C. S., & Lopushinsky, W. (1989). Pure culture response of ectomycorrhizal fungi to imposed water stress. *Canadian Journal of Botany* **67**, 29–39.

Cordell, C. E., Marx, D. H., Maul, S. B., & Owen, J. H. (1987). Production and utilization of ectomycorrhizal fungal inoculum in the eastern United States. In: *Mycorrhizae in the Next Decade, Practical Applications and Research Priorities* (Ed. by D. M. Sylvia, L. L. Hung, & J. H. Graham), pp. 287–289. University of Florida, Gainesville, FL.

Dighton, J., & Mason, P. (1985). Mycorrhizal dynamics during forest tree development. In: *Developmental Biology of Higher Fungi* (Ed. by D. Moore, L. A. Casselton, D. A. Wood, & J. C. Frankland), pp. 117–139. Cambridge University Press, Cambridge.

Dixon, R. K., & Buschena, C. A. (1988). Response of ectomycorrhizal *Pinus banksiana* and *Picea glauca* to heavy metals in soils. *Plant and Soil* **105**, 265–271.

Dixon, R. K., Garrett, H. E., Cox, G. S., Marx, D. H., & Sander, I. L. (1984). Inoculation of three *Quercus* species with eleven isolates of ectomycorrhizal fungi. I. Inoculation success and seedling growth relationships. *Forest Science* **30**, 364–372.

Doudrick, R. L., Stewart, E. L., & Alm, A. A. (1990). Survey and ecological aspects of presumed ectomycorrhizal fungi associated with black spruce in northern Minnesota. *Canadian Journal of Botany* **68**, 825–831.

Findlay, R. D., & Read, D. J. (1986). The uptake and distribution of phosphorus in ectomycorrhizal mycelium. In: *Physiological and Genetical Aspects of Mycorrhizae* (Ed. by V. Gianinazzi-Pearson & S. Gianinazzi), pp. 351–355. INRA, Paris, France.

Fleming, L. V. (1984). Effects of soil trenching and coring on the formation of ectomycorrhizas on birch seedlings grown around mature trees. *New Phytologist* **98**, 143–153.

Fogel, R. (1983). Root turnover and productivity of coniferous forests. In: *Tree Root Systems and Their Mycorrhizas* (Ed. by D. Atkinson, K. Bhat, M. Coutts, P. Mason, & D. Read), pp. 75–86. Junk, Boston.

Fogel, R. (1988). Interactions among soil biota in coniferous ecosystems. *Agriculture, Ecosystems and Environment* **24**, 69–85.

Gardes, M., White, T. J., Fortin, J. A., Bruns, T. D., & Taylor, J. W. (1991). Identification of indigenous and introduced symbiotic fungi in ectomycorrhizae by amplification of nuclear and mitochondrial ribosomal DNA. *Canadian Journal of Botany* **69**, 180–190.

Grossnickle, S. C., & Reid, C. P. P. (1983). Ectomycorrhiza formation and root development patterns of conifer seedlings on a high-elevation mine site. *Canadian Journal of Forest Research* **13**, 1145–1158.

Harley, J. L., & Smith, S. E. (1983). *Mycorrhizal Symbiosis*. Academic Press, London.

Hilbert, J. L., & Martin, F. (1988). Regulation of gene expression in ectomycorrhizas. I. Protein changes and the presence of ectomycorrhiza-specific popypeptides in the *Pisolithus-Eucalyptus* symbiosis. *New Phytologist* **110**, 339–346.

Ho, I., & Zak, B. (1979). Acid phosphatase activity of six ectomycorrhizal fungi. *Canadian Journal of Botany* **57**, 1203–1205.

Hung, L.-L. L. (1985). Ectomycorrhiza inoculation of Douglas-fir plug and +1 seedlings with commercially produced inoculum. In: *Proceedings of the 6th North American Conference on Mycorrhizae* (Ed. by R. Molina), p. 210. College of Forestry, Oregon State University, Corvallis, OR.

Kropp, B. R. (1989). Variation in acid phosphatase activity among progeny from controlled crosses in the ectomycorrhizal fungus *Laccaria bicolor*. *Canadian Journal of Botany* **68**, 864–866.

Kropp, B. R., Castellano, M. A., & Trappe, J. M. (1985). Performance of outplanted western hemlock (*Tsuga heterophylla* (Raf.) Sarg.) seedlings inoculated with *Cenococcum geophilum*. *Tree Planter's Notes* **36**, 13–16.

Last, F. T., Mason, P. A., Ingleby, K., & Fleming, L. V. (1984). Succession of fruit bodies of sheathing mycorrhizal fungi associated with *Betula pendula*. *Forest Ecology and Management* **9**, 229–234.

Last, F. T., Mason, P. A., Wilson, J., Ingleby, K., Munro, R. C., Fleming, L. V., & Deacon, J. W. (1985). Epidemiology of sheathing (ecto-) mycorrhizas in unsterile soils: A case study of *Betula pendula*. *Proceedings of the Royal Society of Edinburgh* **85B**, 299–315.

Lee, S. B., & Taylor, J. W. (1989). DNA Isolation from fungi for use with the polymerase chain reaction. In: *PCR—Protocols and Applications—A Laboratory Manual* (Ed. by N. Innis, D. Gelfand, J. Sninsky, & T. White). Academic Press, New York.

LeTacon, F., Garbaye, J., & Carr, G. (1987). The use of mycorrhizas in temperate and tropical forests. *Symbiosis* **3**, 179–206.

Littke, W., Bledsoe, C. S., & Edmonds, R. L. (1984). Nitrogen uptake and growth *in vitro* by *Hebeloma crustuliniforme* and other Pacific northwest mycorrhizal fungi. *Canadian Journal of Botany* **62**, 647–652.

Marx, D. H. (1975). Mycorrhizae and establishment of trees on strip-mined land. *Ohio Journal of Science* **75**, 288–297.

Marx, D. H., & Artman, J. D. (1979). *Pisolithus tinctorius* ectomycorrhizae improve survival and growth of pine seedlings on acid coal spoils in Kentucky and Virginia. *Reclamation Review* **2**, 23–31.

Marx, D. H., & Kenney, D. S. (1982). Production of ectomycorrhizal fungus inoculum. In: *Methods and Principles of Mycorrhizal Research* (Ed. by N. C. Schenck), pp. 131–146. The American Phytopathological Society, St. Paul, NM.

Marx, D. H., Cordell, C. E., Kenney, D. S., Mexal, J. G., Artman, J. D., Riffle, J. W., & Molina, R. J. (1984). Commercial vegetative inoculum of *Pisolithus tinctorius* and inoculation techniques for development of ectomycorrhizae on bare-root seedlings. *Forest Science Monographs* **25**, 1–101.

Mexal, J., & Reid, C. P. P. (1973). The growth of selected mycorrhizal fungi in response to induced water stress. *Canadian Journal of Botany* **51,** 1579–1588.

Mikola, P. (1980). *Tropical Mycorrhiza Research.* Clarendon Press, Oxford, England.

Mitchell, R. J., Cox, G. S., Dixon, R. K., Garrett, H. E., & Sander, I. L. (1984). Inoculation of three *Quercus* species with eleven isolates of ectomycorrhizal fungi. II. Foliar nutrient content and isolate effectiveness. *Forest Science* **30,** 563–572.

Molina, R. (1982). Use of the ectomycorrhizal fungus *Laccaria laccata* in forestry. I. Consistency between isolates in effective colonization of containerized conifer seedlings. *Canadian Journal of Forest Research* **12,** 469–473.

Mousain, D., Galconnet, G., Gruez, J., Chevalier, G., Tillard, P., Bousquet, N., Plassard, C., & Cleyet-Marel, J. C. (1987). Controlled ectomycorrhizal development of mediterranean forest seedlings in the nursery. First results and prospects. In: *Mycorrhizae in the Next Decade, Practical Applications and Research Priorities* (Ed. by D. M. Sylvia, L. L. Hung, & J. H. Graham), p. 129. University of Florida, Gainesville, FL.

Parke, J. L., Linderman, R. G., & Black, C. H. (1983). The role of ectomycorrhizas in drought tolerance of Douglas-fir seedlings. *New Phytologist* **95,** 83–95.

Perry, D. A., Molina, R., & Amaranthus, M. P. (1987). Mycorrhizae, mycorrhizospheres, and reforestation: Current knowledge and research needs. *Canadian Journal of Forest Resources* **17,** 929–940.

Pilz, D. P., & Perry, D. A. (1984). Impact of clearcutting and slash burning on ectomycorrhizal associations of Douglas-fir seedlings. *Canadian Journal of Forest Research* **14,** 94–100.

Read, D. J. (1984). The structure and function of the vegetative mycelium of mycorrhizal roots. In: *The Ecology and Physiology of the Fungal Mycelium* (Ed. by D. H. Jennings & A. D. M. Rayner), pp. 215–240. Cambridge University Press, London.

Riffle, J. W., & Maronek, D. M. (1982). Ectomycorrhizal inoculation procedures for greenhouse and nursery studies. In: *Methods and Principles of Mycorrhizal Research* (Ed. by N. C. Schenck), pp. 147–156. The American Phytopathological Society, St. Paul, MN.

Rogers, S. O., Rehner, S., Bledsoe, C., Mueller, G., & Ammirati, J. (1989). Extraction of DNA from basidiomycetes for ribosomal DNA hybridizations. *Canadian Journal of Botany* **67,** 1235–1243.

Ruehle, J. L. (1980). Growth of containerized loblolly pine with specific ectomycorrhizae after 2 years on an amended borrow pit. *Reclamation Review* **3,** 95–101.

Ruehle, J. L. (1983). The relationship between lateral-root development and spread of *Pisolithus tinctorius* ectomycorrhizae after planting of container-grown loblolly pine seedlings. *Forest Science* **29,** 519–526.

Rygiewicz, P., Bledsoe, C. S., & Zasoski, R. J. (1984a). Effects of ectomycorrhizae and solution pH on 15-N ammonium uptake by coniferous seedlings. *Canadian Journal of Forest Research* **14,** 885–892.

Rygiewicz, P., Bledsoe, C. S., & Zasoski, R. J. (1984b). Effects of ectomycorrhizae and solution pH on 15-N nitrate uptake by coniferous seedlings. *Canadian Journal of Forest Research* **14,** 893–899.

Schoenberger, M. M., & Perry, D. A. (1982). The effect of soil disturbance on growth and ectomycorrhizae of Douglas-fir and western hemlock seedlings: A greenhouse bioassay. *Canadian Journal of Forest Research* **12,** 343–353.

Sen, R. (1990). Isozymic identification of individual ectomycorrhizas synthesized between Scots pine (*Pinus sylvestris* L.) and isolates of two species of *Suillus*. *New Phytologist* **114,** 617–626.

Stribley, D. P. (1989). Present and future value of mycorrhizal inoculants. In: *Microbial Inoculation of Crop Plants* (Ed. by R. Campbell, & R. M. MacDonald), Vol. 25, pp. 49–65. Special Publication of the Society for General Microbiology, IRL Press, Oxford University, London.

Tommerup, I. C., Kuek, C., & Malajczuk, N. (1987). Ectomycorrhizal inoculum production and utilization in Australia. In: *Mycorrhizae in the Next Decade, Practical Applications and Research Priorities* (Ed. by D. M. Sylvia, L. L. Hung, & J. H. Graham), pp. 293–295. University of Florida, Gainesville, FL.

Trappe, J. M. (1977). Selection of fungi for ectomycorrhizal inoculation in nurseries. *Annual Review of Phytopathology* **15,** 203–222.

Vogt, K. A., Grier, C. C., Meier, C. E., & Edmonds, R. L. (1982). Mycorrhizal role in net primary production and nutrient cycling in *Abies amabilis* ecosystems in western Washington. *Ecology* **63,** 370–380.

Wong, K. K. Y., Piche, Y., Montpetit, D., & Kropp, B. R. (1989). Differences in the colonization of *Pinus banksiana* roots by sib-monokaryotic and dikaryotic strains of ectomycorrhizal *Laccaria bicolor*. *Canadian Journal of Botany* **67,** 1717–1726.

13

The Application of VA Mycorrhizae to Ecosystem Restoration and Reclamation

R. M. Miller and J. D. Jastrow

SUMMARY. The contributions of mycorrhizal research to ecosystem restoration have been considerable. However, the value of mycorrhizae to restoration is not as a panacea, but as a focus that has allowed us to better comprehend a system. By concentrating on this association, we have developed a better understanding of the restoration process, and we have learned much about the ecology of mycorrhizal plants and fungi. The contributions of mycorrhizae to restoration are discussed for three different restoration applications: surface mined lands, coastal dune habitats, and the recreation of soil structure. The role of vesicular-arbuscular mycorrhizae (VAM) in restoration of surface mined lands is especially important in semiarid to arid ecosystems where management of the mycorrhizal fungus is necessary because of the depauperate propagule reserve. Disturbance of this reserve affects the trajectory of revegetation. When the success of mycorrhizal inoculation programs for coastal dune habitats is evaluated on the basis of host production, the results are equivocal; however, inoculation may speed the establishment of outplanted grass tillers used in shore dune stabilization programs. Lately, an important but much overlooked role of the mycorrhizal association is its contribution to soil structure. Roots and mycorrhizal hyphae are involved in the creation of water-stable soil aggregates via several potential mechanisms. The formation of aggregates is an important prelude to soil stabilization and the recreation of a nutrient reserve. Thus, mycorrhizae are important in linking the restoration of vegetation to the reestablishment of soil processes crucial to the formation of soil structure and to the redevelopment of nutrient cycles.

Introduction

It would not be an overstatement to suggest that past promises of mycorrhizal researchers concerning the applied value of mycorrhizal fungi in agriculture, forestry, and horticulture have been more rhetorical than deliverable. In most cases, mycorrhizae have been promoted as a natural fertilizer substitute, primarily for phosphorus and trace minerals. Attempts to pursue this application have often focused on the selection of superior species or strains and the development of an inoculation program. However, unless the economics of using phosphate fertilizers change considerably, we should take another approach to marketing this important symbiosis and avoid using the rhizobia paradigm.

The applications of vesicular-arbuscular mycorrhizae in horticulture, agriculture, and forestry have been the subject of numerous reviews (e.g., Jeffries, 1987; LeTacon et al., 1987; Gianinazzi et al., 1989) and served as the theme for the Seventh North American Conference on Mycorrhizae in 1987 (Sylvia et al., 1988). Similarly, thorough reviews on the physiology of VAM plants (Nelson, 1987; Stribley, 1987; Smith & Gianinazzi-Pearson, 1988) and inoculation programs (Stribley, 1989) have been presented. The intent of this chapter is to demonstrate that mycorrhizae have immediate implications for solving applied problems at the ecosystem scale. Specifically, our aim is to show how the mycorrhizal association has been used to identify and solve problems associated with land reclamation and restoration. We believe that a mycorrhizae paradigm can be used to improve our management practices, especially as related to the contributions of mycorrhizae to the creation of soil structure and the development of nutrient cycles. Thus, we will attempt to show the value of mycorrhizae not as a panacea, but as an approach to gaining a better understanding of the soil system.

Conceptual Background

The importance of the belowground component of terrestrial ecosystems has been overlooked during the development of most restoration strategies (Miller, 1985, 1987a). Commonly used practices place a premium on aboveground production or diversity at the expense of belowground function. Nevertheless, in recent years land managers have come to recognize the importance of soil biota. This invisible portion of the system is crucial to the life of a community because soil organisms control the formation of soil structure and nutrient cycling (Elliott & Coleman, 1988; Jastrow & Miller, 1991). Land managers must understand the roles of soil biota to successfully reconstruct stable systems across a variety of climatic and edaphic conditions.

In recent years, several technological fixes have been proposed to aid in the reconstruction of degraded lands. These range from the use of new equipment for seeding or furrowing to genetically engineered plants and microbes that are expected to provide such factors as improved drought and salt tolerance (M. F. Allen, 1988a). However, for most degraded lands, we see a technology ahead of the science. Our current understanding of whole plant physiology and belowground habitats is not at a point where these advanced technologies can be implemented in any predictable fashion. For example, even the survival potential of introduced plants and microbes in the environment is not known. For biotechnology to successfully contribute to the restoration of degraded landscapes, a greater emphasis will have to be given to research on basic plant and soil biology.

Of the myriad of organisms found in soil, it is the mycorrhizal fungus that provides a direct physical link between primary producers and decomposers. The mycorrhizal association influences plant community production and structure directly via its roles in plant nutrition (St. John & Coleman, 1983; Miller, 1987b) and water relations (Allen & Allen, 1986, 1988), and through its influence on soil structure development (Tisdall & Oades, 1982; Bethlenfalvay et al., 1988; Miller & Jastrow, 1990). A fundamental role for mycorrhizal fungal hyphae may be to bridge the annular space, producing a physical connection between the root surface and surrounding soil particles (Miller, 1987b). In creating such bridges, the hyphae increase the effective surface area of the root and decrease resistance to water flow to the root surface by allowing closer contact with the soil. This physical relationship among the root surface, hyphae, and soil matrix could be especially important to plants growing in soils of high conductive resistance or where drought is common. This connection would also allow continued uptake of nutrients from the soil solution during a drought cycle.

Any discussion regarding the applications of mycorrhizae to ecosystem restoration must recognize the complexity of responses that are expressed by both plants and VAM fungi. For plants this complexity of response is exhibited in their growth forms, life history traits, and physiology, all of which may influence a plant's survival and its competitive ability within a community. Furthermore, these responses may be influenced by the mycorrhizal association. For the fungus, responses will be influenced by both soil and plant. These factors can influence such fungus traits as modes of infection, the kinds of propagules produced, and the allocation of intra- and extraradical fungal hyphae. Thus, mycorrhizae can contribute to restoration via responses of the plant, the fungus, or both.

To discern the different roles of the mycorrhizal association in restoration, studies have been conducted at a variety of organizational scales (Table 13.1). For example, at one level investigators have examined the production and location of extraradical hyphae (e.g., Nicolson, 1959; Allen & MacMahon, 1985), the kinds of roots that possess mycorrhizae (Reinhardt & Miller, 1990), and their role in creating soil aggregates and stabilizing dunes through the physical entanglement of soil or sand particles (e.g., Tisdall & Oades, 1980; Koske & Polson, 1984; Rose, 1988; Miller & Jastrow, 1990). Other studies have investigated how mycorrhizae may influence the growth and physiology of whole plants (e.g., Janos, 1980b; Call & McKell, 1985; Allen & Allen, 1986; Sylvia, 1989). At another level, studies have investigated the importance of mycorrhizae in the composition of successional seres (Allen & Allen, 1990; Gange et al., 1990). For example, many early successional plants in arid and tropical environments

Table 13.1. *Examples of Direct and Indirect Expression of the Mycorrhizal Association on Processes Occurring at Various Hierarchical Organizations of System Structure*

Structural organization	Processes
Mycorrhizal roots and extraradical hyphae	Ion uptake, soil aggregation, and rhizosphere
Whole plants	Nutrition, water relations, growth, reproduction, and disease resistance
Community	Richness and stability
Guilds	Ecto-, endo-, and ericoids can determine community type

are nonmycorrhizal, but species characteristic of later seres typically possess mycorrhizal fungi (Miller, 1979; Reeves et al., 1979; Allen & Allen, 1980; Janos, 1980a). On a much larger scale, plant communities or even ecosystems can possess different kinds or guilds of mycorrhizae. Many temperate and almost all boreal forest trees possess ectomycorrhizae, whereas grassland and shrubland plants are endomycorrhizal (Read, 1984; Miller, 1987b). The type of mycorrhizae has important implications for restoration because ecto- and endomycorrhizal fungi have different modes and rates of propagule recruitment to severely disturbed systems.

Finally, from a practical standpoint, any contributions of mycorrhizae to ecosystem restoration depend on the use of strategies that promote or preserve the mycorrhizal association. To achieve this goal, selection and inoculation may be useful or necessary under some circumstances (e.g., in "sterile" substrates). However, the use of restoration strategies that create or preserve the habitat in which the mycorrhizal association occurs may often be more efficient, economical, and effective. In other cases, both approaches may be necessary; the appropriate habitat must be created and supplied with inoculum.

The Value of Mycorrhizae in Disturbed Ecosystems

Much of our basic knowledge of VAM ecology comes from studies conducted on disturbed ecosystems (Danielson, 1985; Miller, 1987a). The degree of disturbance investigated ranges from minor [disked and fallow fields (e.g., Black & Tinker, 1979; Allen & Allen, 1980; Schenck & Kinlock, 1980; Medve, 1984; Thompson, 1987)] to severe [volcanic activity (Allen et al., 1984), tailings associated with metal ores (e.g., Jasper et al., 1988; Johnson & McGraw, 1988a,b), retorted oil shale and extracted oil sands (Call & McKell, 1982; Danielson et al., 1983; Stahl & Williams, 1986), and the overburden associated with surface mining and collery waste (e.g., Daft & Nicolson, 1974; Lambert & Cole, 1980; Kiernan et al., 1983; Jasper et al., 1987)]. These kinds of studies provide basic data including lists of plants that form mycorrhizae (e.g., Miller, 1979; Reeves et al., 1979; Trappe, 1981; Rothwell & Vogel, 1982), the characteristics of a VAM fungal propagule (e.g., Rives et al., 1980; Evans & Miller,

1988; Jasper et al., 1989a,b), and the vectors by which mycorrhizal fungal propagules are disseminated (e.g., Warner et al., 1987; M. Allen, 1988a,b; Maser et al., 1988).

A major contribution to our understanding of the restoration process that came from studies of the mycorrhizal association in disturbed environments in semiarid deserts and tropical forests is the concept of a functional continuum based on the host's need for mycorrhizal fungi (Reeves et al., 1979; Janos, 1980a). This hypothesis suggests that the responsiveness of a host to mycorrhizae can be either facultative or obligate depending on the importance of symbiont-supplied minerals. These types of studies have shown that mycorrhizae are absent or infrequent during the early stages of succession. Many plants that occur during early seral stages are either nonmycorrhizal or facultatively mycorrhizal. In many tropical systems, the later stages of recovery from disturbance are characterized by plants whose need for mycorrhizal fungi is considered obligate (Janos, 1980b).

As compelling as this functional continuum approach is for providing insight into the restoration process, when used alone it has limited application because of its emphasis on host rather than fungal dynamics. Perhaps a more fruitful approach would be one based on applying the concepts of patch dynamics alone or in combination with the functional continuum approach (Allen, 1989). Accumulated data suggest that landscape disturbances, such as those associated with surface mining, can cause relatively homogeneous systems to become patchy or heterogeneous rather quickly. Larger scale disturbances like those associated with volcanic activities can remain relatively homogeneous for long timespans (M. F. Allen, 1988b; Allen, 1989). For both of these systems, the creation of environmental heterogeneity may be the key to understanding how to accomplish their restoration. The contributions of the mycorrhizal association in creating both vegetation and soil heterogeneity are important components of this process.

The suggestion that the rate of succession or restoration may be hastened by inoculation or manipulation of the mycorrhizal fungus population (Reeves et al., 1979; Janos, 1980a) may be true under conditions where VAM fungal propagule densities are very low. However, a closer examination of the data suggests that mycorrhizal fungi probably do not facilitate the skipping of successional stages. Rather, the importance of mycorrhizal fungi more likely lies in preventing the stagnation of community development. For example, protracted occupation of a site by nonmycotrophic colonizers could influence competitive interactions with host plants (Allen & Allen, 1988), aid in seedling establishment (Read & Birch, 1988), or simply reduce propagule reserves (Miller, 1984, 1987a; Biondini et al., 1985). In any case, these will likely delay the establishment of more obligate mycotrophic plants. A protracted delay in the colonization of a site by mycorrhizal plant species might also result in a reduction in mycorrhizal fungal propagule reserves and, hence, the continued occupation of a site by nonmycotrophic species (Allen & Allen, 1980, 1988; Miller, 1987a). Such sites are usually characterized by having low species richness and often poor productivity. Hence, the primary goal of any restoration program should be to establish and maintain viable populations of mycorrhizal fungi.

Applications of Mycorrhizae to Ecosystem Restoration

The goal of ecosystem restoration is to use human resources to return degraded biological communities to their original state (Jordan et al., 1988). In the following three examples of applications of VA mycorrhizae to ecosystem restoration, we have taken a rather broad view of the term restoration and have included research that could instead be called reclamation. Restoration usually means the recreation of entire communities on the basis of natural models, whereas reclamation is a deliberate attempt to return a damaged ecosystem to a condition short of restoration per se.

Although much of what we know about the restoration process is anecdotal, its success should be judged by criteria such as sustainability, invasibility, productivity, nutrient retention, and biotic interaction rather than by species composition and physiognomy alone (Ewel, 1987). Even though mycorrhizae do not always contribute directly, they are somehow involved with each criterion because they are an association that integrates the establishment of vegetation and the development of soil systems. Conditions where mycorrhizae may be a limiting factor during the restoration process are identified in Table 13.2.

Our first example of restoration applications shows how mycorrhizae have contributed to understanding the consequences of different reclamation practices used in restoring drastically disturbed sites. The kinds of studies cited have taught us much about both the basic and applied ecology of the mycorrhizal association, as well as the contribution of this symbiosis to restoration. The second example illustrates how mycorrhizae can play an important role in stabilizing shoreline dune systems and presents a rationale for the use of mycorrhizal inoculants. Our last example demonstrates the importance of mycorrhizae, especially the extraradical hyphae, in the creation of soil structure and shows how mycorrhizae may be important in restoring nutrient cycles.

Table 13.2. Conditions in Which Management of Mycorrhizae Can Become Important in Ecosystem Restoration

Density of mycorrhizal inoculum low
Form of inoculum mainly colonized roots or extraradical hyphae
Potential for extended stages dominated by nonmycorrhizal plants
Site is either extremely arid or hydric during some portion of the growing season
Cold temperatures when moisture is adequate for plant growth
Low mineral nutrient content of soils
Potential for toxicity of soil to plant growth
Optimization of biodiversity a primary goal

Restoration of Surface Mined Lands

Mycorrhizae become an important management issue in surface mining because three factors inherent to the mining process can decrease the viable populations of mycorrhizal fungi (Jasper et al., 1987). These factors are the removal of growing plants, soil disturbance, and topsoil storage. When surface soils (i.e., "topsoils") are removed and reapplied after mining, reduced mycorrhizal fungal populations in the reapplied soils could affect not only the establishment of grass, forb, and shrub seedlings (Visser et al., 1984b; Miller, 1984, 1987a; Jasper et al., 1987; Stark & Redente, 1987; Jasper et al., 1989b) but also the functional aspects of the developing community (Biondini et al., 1985; Allen & Allen, 1986, 1988). This is particularly likely because large-scale surface mining often occurs in nutrient or moisture-limited environments where the existence of certain species depends on mycorrhizae.

If VAM fungi could be easily reintroduced after mining, restoration might be simplified. However, field inoculation with VAM fungi on a large scale is not currently practical (Stribley, 1989), and success with containerized systems to date has been equivocal (e.g., Call & McKell, 1985; Roskoski et al., 1986; Call & Davies, 1988; Stahl et al., 1988). (Containerized seedlings inoculated with indigenous VAM fungi are available from *Tree of Life Nursery,* San Juan Capistrano, CA.) Hence, because the problem is so large and viable mycorrhizal fungus populations must be maintained, revegetation efforts have had to rely on indigenous populations of VAM fungi and their natural dispersal mechanisms. Thus, considerable attention has been given to the fate of VAM fungi during the mining and reclamation process.

In surface mining and in many landscaping operations, surface soils are commonly stripped and stockpiled in large mounds until they are needed for reapplication during restoration. Many studies have demonstrated that this practice is detrimental to a soil's microbial and physicochemical constituents (e.g., Abdul-Kareem & McRae, 1984; Visser et al., 1984a; Miller et al., 1985; Harris et al., 1989). In addition to the obvious direct effect of compaction by heavy equipment, stockpiling of topsoils has been shown to reduce the effectiveness of biologically mediated processes such as soil aggregation (Abdul-Kareem & McRae, 1984) and litter decomposition (Miller, 1984; Visser et al., 1984a). If topsoil storage practices reduce the viability of VAM populations, there could be long-term ramifications for restoration.

Numerous studies have shown that stockpiling reduces the mycorrhizal fungal density in soils (e.g., Rives et al., 1980; Gould & Liberta, 1981; Liberta, 1981; Abdul-Kareem & McRae, 1984; Miller et al., 1985; Harris et al., 1987; Jasper et al., 1989b). For example, Rives et al. (1980) found a significant reduction in the mycorrhizal infection potential (MIP) of soil stored for 3 years. When this same stockpile was sampled 15 months later, an even greater reduction in inoculum was evident (Gould & Liberta, 1981). Much of the loss of inoculum during

Table 13.3. Mycorrhizal Infection Potential (MIP) for Undisturbed and Stockpiled Topsoils Obtained from Different Ecosystems

Site	Mycorrhizal infection potential (%)[a] of soil dilutions				Length of storage (years)
	Undisturbed		Stored		
	1/1	1/4	1/1	1/4	
North Dakota[b]	91.7	57.7	81.7	7.0	3
North Dakota[c]	92.8	79.6	50.4	30.4	4
Wyoming[d]	35.0	27.2	52.5	32.9	2
Wyoming[d]			28.9	17.2	5
England[e]	79.3	50.0	66.6	5.4	7
Illinois[f]	85.3	80.0	57.3	45.3	1
Illinois[f]			12.0	4.0	3

[a]Moorman and Reeves (1979).
[b]Rives et al. (1980).
[c]Gould and Liberta (1981).
[d]Miller et al. (1985).
[e]Abdul-Kareem and McRae (1984).
[f]Liberta (1981).

storage appeared to be due to the death of colonized roots capable of acting as propagules. This reduction in MIP levels associated with storage appears to be common to most mycorrhizal populations investigated (Table 13.3).

In a more mechanistic approach, Miller et al. (1985) found that the responses of VAM fungi to soil storage were dependent on soil moisture content during stockpiling. When soils were stored in a relatively moist condition (> -2 MPa), a positive linear relationship occurred between soil moisture tension and MIP, i.e., the drier the soil, the higher the mycorrhizal infection percent (MIP). However, when soils were stored in a relatively dry condition (< -2 MPa), the length of storage time was the determining factor in survival of VAM fungal propagules. The results of this study suggest that to preserve the fungus, soil stockpiles should be constructed to sizes and depths that are appropriate for regional climatic conditions. In semiarid and arid environments, relatively large stockpiles can be constructed with little impact to mycorrhizal fungus populations, if soils are stockpiled when they are relatively dry. However, in more mesic climates where the potential for recharging soil moisture is much greater, stockpile depths should not exceed the rooting depth of the cover vegetation. Alternatively, stockpiles should be designed to minimize moisture recharge. Studies from mesic climates have substantiated the above conclusion as soils from lower depths within stockpiles have a reduced MIP (Harris et al., 1987; Jasper et al., 1987).

Although maintaining viable VAM fungal populations is not the primary reason for salvaging topsoils, it is a major concern in establishment and survival of

Table 13.4. *The Effects of Various Degrees of Soil Disturbance on Mycorrhizal Infection Potential Over Time in a Sagebrush Community.*[a, b]

	Postdisturbance mycorrhizal infection potential (%)		
Disturbance treatment	1 year	4 years	6 years
Vegetation mechanically removed with minimal disturbance to soil	55.0	66.5	67.0
Vegetation removed; topsoil scarified to depth of 30 cm	44.5	52.5	67.5
Topsoil and subsoil removed to depth of 1 m, mixed together, and replaced	13.0	11.0	36.0
Topsoil and subsoil removed separately and replaced in reverse order	11.5	20.0	69.0

[a]MIP levels were 65% for the sagebrush community soil prior to disturbance (Schwab & Reeves, 1981).

[b]Adapted from Doerr et al. (1984) and Biondini et al. (1985).

many plants and is affected by topsoil handling procedures. The influence of mycorrhizal fungi on the revegetation of reapplied topsoil is affected by whether a soil was stockpiled or directly reapplied (Miller, 1984, 1987a; Jasper et al., 1987, 1989b). In semiarid communities, the decrease in inoculum potential associated with topsoil storage appears to be greater than the decrease associated with the disturbance of topsoil alone. Nevertheless, the disturbance associated with the removal and direct reapplication of topsoil may still significantly influence the inoculum of a site (Miller, 1984; Doerr et al., 1984; Biondini et al., 1985; Jasper et al., 1987, 1989b).

To better understand the effects of soil disturbance on the viability of mycorrhizal fungal propagules, the effects of different types of disturbances in a sagebrush community in Northwestern Colorado were tested (Doerr et al., 1984; Biondini et al., 1985). The degree of disturbance had profound influences on MIP and soil biological activity. Simply removing the aboveground vegetation reduced the MIP but the decline was slight, and recovery to predisturbance inoculum levels was almost immediate (Table 13.4). If disturbance was increased by scarifying the topsoil to a depth of 30 cm, greater reduction in MIP was observed and the recovery time lagged. Removal of the surface 1 m of soil, mixing it, and replacing it had the greatest effect on MIP; 6 years after disturbance, the MIP was only about 55% of the predisturbance level. When topsoil and subsoil were removed to a depth of 2 m without mixing and were replaced in reverse order (subsoil first), MIP levels were severely reduced and recovered slowly but were virtually identical to undisturbed levels after 6 years. The large reductions in MIP after soil removal and replacement occurred because the propagule bank is quite small for semiarid communities and the inoculum, like the vegetation, has a patchy distribution (Allen & MacMahon, 1985). Furthermore, the handling of surface soils by heavy equipment not only breaks up the roots and the associated mycorrhizal hyphal network of the existing vegetative community but also mixes the soil, which dilutes the propagule bank in surface layers. Since MIP levels decrease with increasing soil depth (Schwab & Reeves, 1981), the potential for eventual recovery appears to be greater if dilution is minimized by keeping the topsoil and subsoil separate. Unless deleterious soil characteristics are present, surface soils now are often salvaged in two lifts. The first lift typically removes A horizons, and the second lift commonly includes B and C horizons.

In the Darling Plateau of Western Australia, disturbance of soils significantly affects mycorrhizal propagules (Jasper et al., 1989b). In this system, the infectivity of VAM fungal propagules in topsoil was destroyed, even when the soil was stripped and immediately reapplied without stockpiling (Jasper et al., 1989b). Most infectivity was lost within 3 weeks of clearing the vegetation and before the soil was disturbed. The loss of infectivity in these soils occurs because spores are not the main propagule source. Rather, the reduction of infectivity was related to the loss of colonized root fragments, which serve as the primary source of propagules. The apparent loss of propagules was not found for another site at Weipa, Queensland, where only a two-thirds reduction in infectivity was observed for soils after stockpiling (Jasper et al., 1987). The findings from the latter site are more in line with those observed for most North American sites.

In the Red Desert of Wyoming, the effects of different topsoil handling procedures on mycorrhizal inoculum viability can be considerable (Miller, 1984, 1987a, 1988). The mycorrhizal infectivity of the soils determined immediately after soil disturbance and 2 years later were similar to those for soil of the undisturbed community. Nevertheless, the establishment of VAM plant species on these disturbed soils was delayed even though no apparent loss of inoculum density occurred. At this site, soil moisture is limiting to plant growth for much of the growing season. When moisture is available, soil temperatures can be below 10°C. This combination of factors affects seed germination and plant growth and also tends to slow or inhibit spore germination and root colonization by VAM fungi. Apparently, the low moisture and temperature conditions found at this site combined with the dilution of fungal propagules caused by mixing of horizons during topsoil handling to delay the establishment of VAM plants. Thus, nonmycorrhizal plant forms predominated in the early years after reapplication of topsoil, as they do during the early stages of succession in this region. Nonmycorrhizal plants may not be detrimental, however, because they may contribute to establishment of the mycorrhizal community by providing safe sites for establishment of mycorrhizal species (Miller, 1987a; E. B. Allen, 1988; Schmidt & Reeves, 1984).

The distribution and number of VAM fungal propagules within the soil profile vary with plant community type and are often influenced by the distribution of fine roots (White et al., 1989). Thus, edaphic factors that control root distributions can influence VAM fungus populations as well. The data presented in Table 13.4

as well as other data obtained from semiarid and arid sites (Jasper et al., 1989b) demonstrate that even the simple disturbance caused by removing the aboveground portion of the vegetation can reduce VAM fungal populations in the short term.

Studies of the relationship between mycorrhizal infectivity and edaphic factors for replaced topsoils in the Red Desert indicated that factors associated with soil moisture and aeration were the best predictors of mycorrhizal inoculum (Miller, 1984, 1987b). Plant community composition was also related to inoculum levels. Although fungal propagule levels were not associated with mycorrhizal plant cover, nonmycorrhizal cover was negatively correlated with mycorrhizal inoculum in reapplied soil.

Potential trajectories of succession for semiarid and arid environments in relation to mycorrhizal inoculum levels, microclimatic conditions as measured by the interrelationship of precipitation and available soil moisture, and nonmycorrhizal weed density are summarized in Figure 13.1. For sites that occur in these environments, the establishment of mycorrhizal grasses and shrubs is facilitated by a nonmycorrhizal plant colonization stage. In North America, this nonmycorrhizal stage is composed of introduced weeds, including *Halogeton, Salsola,* and *Lepidium,* but also endemic shrubs of the Chenopodeaceae, such as *Atriplex* spp. (Schmidt & Reeves, 1984; Miller, 1987a; Allen & Allen, 1988). At the more mesic end of a continuum of semiarid sites, the successional trend is for a nonmycorrhizal weedy stage to give rise to a mycorrhizal grass stage (E. B. Allen, 1988; Allen & Allen, 1988). At more xeric sites, like the Red Desert, nonmycorrhizal shrubs play an important intermediary role as safe sites for grass seedling establishment (Miller, 1987a). The mechanism appears to be associated with an increase in soil moisture (R. M. Miller, unpublished data) and a potential spore buildup associated with the shrub architecture (M. F. Allen, 1988a). If several repeated drought years occur, the nonmycorrhizal shrub stage increases in its importance as a "facilitator." Also, some typically nonmycorrhizal shrubs can take on mycorrhizal colonization in the presence of mycorrhizal hosts (Miller et al., 1983), hence increasing mycorrhizal inoculum. Even after 10 years of recovery, VAM shrubs still have not developed at the Red Desert site. This "facilitator" plant effect may be also occurring with *Myrica cerifere,* a plant commonly associated with both early and later successional sites on unreclaimed spoil from phosphate mining in Florida (Poole & Sylvia, 1990).

The importance of topsoil replacement in the reestablishment of mycorrhizae cannot be overstated; topsoils are the primary source of mycorrhizal fungus inoculum. However, topsoil salvaging and replacement do not guarantee mycorrhiza development, especially after stockpiling. Nevertheless, without topsoil the development of mycorrhizae on spoil or tailings can be quite slow (Zak & Parkinson, 1983; Waaland & Allen, 1987). If topsoil is not available for replacement over spoil or tailings, the addition of organic amendments may increase the rate of mycorrhiza development (Johnson & McGraw, 1988a). Conversely, the

Application of VA Mycorrhizae to Ecosystem Restoration / 449

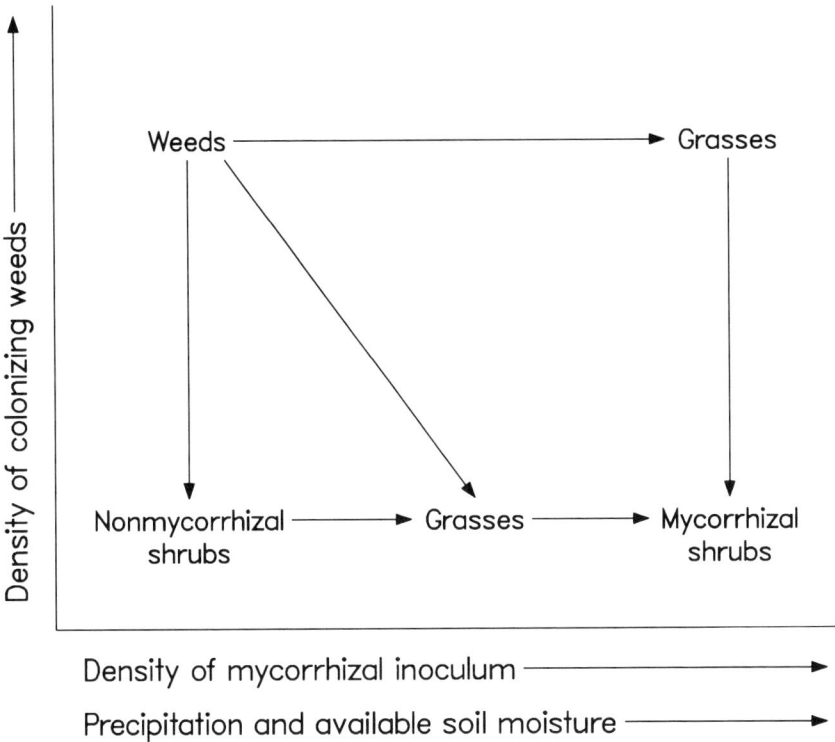

Figure 13.1. Possible trajectories of succession in a shrubland–grassland as affected by introducing weeds, nonmycorrhizal shrubs, density of mycorrhizal inoculum, and site available moisture characteristics.

high levels of inorganic fertilizers often suggested for reclamation generally inhibit mycorrhiza establishment (Zak et al., 1982; Zak & Parkinson, 1982, 1983; Johnson & McGraw, 1988b).

Apart from the passage of time and increasing root densities, factors that influence VAM fungus colonization of disturbed soils remain undetermined. Although the harshness of the site and the severity of disturbance can determine the plant's need for mycorrhizae (Miller, 1979), other factors such as inoculum type and density, edaphic conditions, plant cover and physiognomy, host genotype, and time are critical parameters affecting mycorrhizal formation. These examples are primarily from semiarid ecosystems where recovery of mycorrhizae appears to be rather slow. In more mesic environments where propagules are more ubiquitous (higher root densities), the recovery of mycorrhizal fungi can be quite rapid provided the substrate is not toxic. The addition of topsoil can significantly hasten both the establishment of mycorrhizae and vegetation (Lambert & Cole, 1980), although viable mycorrhizae can occur without topsoil replacement provided organic amendments are applied (Johnson & McGraw, 1988a,b).

Restoration of Coastal Dune Habitats

A major concern for many coastal areas of the world is shoreline erosion. Nature's defense against shoreline erosion is the formation of coastal foredunes that protect the coast by absorbing energy from wind, tide, and wave actions (McHarg, 1972). Shoreline and dune systems are dynamic; there is continual movement of sand, and beach erosion and accretion are both natural processes. However, the introduction of human structures and activities can accelerate the loss of dune systems. The restoration of foredunes may require replenishment of sand that is often obtained offshore at great expense, costing several million dollars per mile (Sylvia & Will, 1988).

An alternative approach is to preserve foredunes by planting them with beach grasses to establish a developing root and rhizome system that will stabilize the dunes. The plants of choice for dune revegetation and stabilization in North America are *Ammophilia breviligulata* Fern. for the mid- and north Atlantic coast and *Uniola paniculata* L. for the south Atlantic and Gulf coasts (Koske & Polson, 1984; Sylvia, 1989; Gemma & Koske, 1989). Because of the nutrient-poor conditions of the substrate (Read, 1989), the role played by mycorrhizal fungi in both plant survival and dune stabilization can be considerable.

The roots of major dune-colonizing species typically form mycorrhizae (Nicolson, 1960; Nicolson & Johnston, 1979; Koske & Halvorson, 1981; Giovannetti & Nicolson, 1983; Ernst et al., 1984). The extent of VAM colonization in roots is apparently associated with both seasonal and successional processes, the latter process being related to the position of the plant within the dune system (Read, 1989). In the coastal dune systems of Britain, the lowest levels of colonization were associated with the most mobile part of the system. Roots were more highly colonized in fixed dunes and late in the growing season (Nicolson & Johnston, 1979). A similar trend apparently occurs along the Atlantic coast of the United States; plants found on the more stable dunes have the highest levels of mycorrhizal fungus colonization (Koske & Polson, 1984). Fixed maritime sand dunes along the Mediterranean coast, however, did not follow this trend. In contrast, they exhibited the highest levels of colonization and spore density in early summer (Giovannetti, 1985). Surprisingly, there is very little similarity between the species of VAM fungi occurring in these Mediterranean systems and those found in the British Isles and North American dune systems.

Few studies have evaluated the influence of mycorrhizal fungi on the growth of beach grasses. In the few glasshouse experiments conducted, plants responded positively to inoculation (Nicolson & Johnston, 1979; Koske & Polson, 1984; Sylvia & Burks, 1988), although the growth response of the host depended on the location within the dune system from which the endophyte was isolated (Nicolson & Johnston, 1979).

The results of field inoculation programs to enhance seedling establishment appear to be inconclusive if success is judged on the basis of host response. Inoculation increased tillering in *A. breviligulata* and biomass in *U. paniculata* (Gemma & Koske, 1989; Sylvia, 1989). The advantage conferred by inoculation was short lived; by the second growing season, inoculated and noninoculated plants had similar growth characteristics and mycorrhizal colonization.

Nevertheless, the prompt establishment of aboveground vegetation decreases the erosional force of wind and water, and allows for the physical stabilization of sand particles by roots, mycorrhizal fungi, and other soil microbes. As important as a productive host is to stabilizing foredunes, it is a host's contribution to the development of a functioning belowground community that is critical to the stabilization process. The goal of any dune stabilization program is the creation of aggregates that can withstand wind erosion.

An important factor in the creation of these aggregates is the production of extraradical hyphae of VAM fungi (Koske et al., 1975; Sutton & Sheppard, 1976; Nicolson & Johnston, 1979; Forster & Nicolson, 1981a; Jehne & Thompson, 1981; Rose, 1988). Extraradical hyphal lengths of up to 12 m/g of sand or 592 m/cm of the colonized length of *U. paniculata* roots have been reported in foredunes (Sylvia, 1986). These hyphae act as a mechanical network holding sand in place. The formation of these initial aggregates of sand may be the dominant agent in the aggregating processes leading to dune stabilization (Read, 1989). In later stages of aggregation and soil formation, the aggregates are apparently stabilized by filaments of saprophytic fungi and by the gums and glues of bacteria and algae (Forster & Nicolson, 1981a) that predominate in the later successional dunes (Forster & Nicolson, 1981b; Rose, 1988). The formation of aggregates appears to be greater in the presence of roots than in sand without roots (Forster & Nicolson, 1981a).

A limiting factor to field inoculation with mycorrhizal fungi is the availability of inoculum. Unlike many systems where root fragments are the primary means of infection (Read et al., 1976; Rives et al., 1980), spores appear to be the major source of inoculum in dune systems. However, both spores and vegetative fragments are apparently long-distance codispersal agents for the tropical Pacific dune systems (Koske & Gemma, 1990). In more temperate dune systems, inoculum potential and spore densities can vary considerably over the growing season (Gemma & Koske, 1988; Gemma et al., 1989). Furthermore, a cold dormancy period may be necessary for spore germination to occur (Gemma & Koske, 1988). Such seasonal variability for spores does not appear to occur in Florida dune systems (Sylvia, 1986).

A major problem for incorporating mycorrhizae into field outplanting programs is the development of a mycorrhizal inoculum protocol (Jeffries, 1987; Sylvia & Burks, 1988). The procedures currently used for outplanting grass seedlings are often not conducive to the development of vigorous mycorrhizae. For example, present outplanting practices for *Ammophilia* usually use tillers with very little if any root stock attached (Gemma & Koske, 1989). An additional problem is that the most effective isolates of VAM fungi may not sporulate easily (Sylvia & Burks, 1988). Provided these problems can be solved, the use of mycorrhizal inoculants could prove to be a useful strategy in dune stabilization and restoration programs.

Restoration of Soil Structure

If the number of papers published in each area of mycorrhizal research indicated the importance of the association to ecosystem restoration, the nutrition and growth response of a host would top the list. However, this criterion would overlook a very important role of mycorrhizae that is crucial to restoration, the role of the extraradical hyphae of VAM fungi in forming soil aggregates. The formation of soil aggregates that are relatively resistant to breakdown is known to prevent wind and water erosion. Both the quality and size distribution of soil aggregates can also affect a soil's pore size distribution (Emerson et al., 1986; Elliott & Coleman, 1988). Together, these properties constitute what is termed soil structure and can influence soil physical, chemical, and biological processes via their effects on the accessibility of food, shelter, water, oxygen, and nutrients to soil biota (Coleman, 1986; Elliott & Coleman, 1988).

Depending on the degree of degradation of a site, an important component of a successful restoration strategy is the reestablishment of nutrient cycling. In other words, the most critical aspect of creating a self-sustaining system is the establishment of a nutrient reserve (Bradshaw et al., 1982). This reserve is usually initiated by the combination of atmospheric deposition, weathering, and detrital inputs from vegetation or by additions of organic amendments and fertilizers. Unless these organic inputs are stabilized, accumulations of organic matter and microbial biomass and the concomitant buildup of nutrient reserves in soil systems are usually minimal. Organic residues are generally protected or stabilized within soils through the formation of soil aggregates (Oades, 1984; Emerson et al., 1986; Elliott & Coleman, 1988). Hence, a major goal of restoration should be to establish conditions that favor development of soil aggregates thereby facilitating an important step in the creation of a nutrient reserve.

In soils where organic matter is the major binding mechanism, several types of organic binding agents contribute to aggregate formation and stabilization. Inorganic and relatively persistent organic binding agents are important for the stabilization of microaggregates (< 0.25 mm diameter), but physical entanglement by roots and hyphae of mycorrhizal fungi subsequently appears to be a major mechanism whereby microaggregates are bound together into macroaggregates (> 0.25 mm diameter) (Tisdall & Oades, 1982; Oades, 1984; Muneer & Oades, 1989). Individual mycorrhizal and saprophytic fungal hyphae along with fine roots bind soil particles into larger, aggregated units (Allison, 1968; Tisdall & Oades, 1979, 1982; Oades, 1984; Gupta & Germida, 1988; Miller & Jastrow, 1990). Mucigels and polysaccharides produced by bacteria, fungi, and roots also bind and stabilize these aggregates (Tisdall & Oades, 1982; Oades, 1984; Foster, 1985; Emerson et al., 1986).

There is, however, some controversy regarding the relative importance of intermicroaggregate binding agents associated with the stabilization of macroaggregates. Some researchers (e.g., Cheshire et al., 1983, 1984; Elliott, 1986) have contended that polysaccharides and other organic compounds serve as the most important intermicroaggregate binding agents. Although Tisdall and Oades (1982) and Oades (1984) propose that polysaccharides and other organic materials function as persistent binding agents for microaggregates, they suggest that the role of these agents in the stabilization of macroaggregates is merely transient. A partial resolution to this controversy may come from the work of Sparling and Cheshire (1985). They have suggested that the role of polysaccharides may be more important in nonrhizosphere soils, whereas mycorrhizae may be more important in rhizosphere soils.

Elliott and Coleman (1988) hypothesized that the relative importance of organic binding agents versus physical entanglement depends on whether researchers have studied the degradation of native soil or the restoration of disturbed soil, respectively. They suggest that the process of macroaggregate development is a dynamic one in which roots and fungal hyphae function primarily in the formation phase and organic binding agents are largely responsible for stabilization.

However, there is visual evidence to suggest that saprophytic and VAM fungal hyphae are involved in both physical and chemical binding mechanisms. In addition to the simple hyphal entanglement mechanism that may dominate in an initial aggregative phase, a later stabilization phase apparently involves cementation of fungal hyphae to soil particles by organic or amorphous materials (Tisdall & Oades, 1979; Gupta & Germida, 1988).

From these data, we may hypothesize that the contribution of mycorrhizae to the aggregation process during restoration occurs in three closely related processes. The first process involves the growth of extraradical hyphae into the soil matrix to create the skeletal structure that holds the primary soil particles together via physical entanglement (Tisdall & Oades, 1979; Gupta & Germida, 1988). A second process is the creation, by roots and mycorrhizal hyphae, of the conditions that are conducive to the formation of microaggregate structures. In this phase, the mechanical entanglement process physically brings mineral particles (i.e., silts and clays) and organic debris together in such a way that microaggregates may form (Allison, 1968; Tisdall & Oades, 1982; Oades, 1984; Emerson et al., 1986). These particles are cemented together by physicochemical processes involving various binding agents such as extracellular polysaccharides, persistent aromatic compounds, and polyvalent cation bridges that form complexes with clays and organic matter. These persistent gums and glues are probably derived from the residues of mycorrhizal hyphae, root cells, and saprophytic microflora. Finally, microaggregates are physically enmeshed by extraradical hyphae and roots to create the macroaggregate structure that may be further stabilized by cementation with polysaccharides or other organic compounds (Elliott & Coleman, 1988; Gupta & Germida, 1988).

These three processes occur simultaneously because soil aggregation is dynamic. Some soil systems, however, may never or only slowly develop beyond the first process (e.g., systems where clays and organic matter are absent or very low). In systems that do progress past this step, microaggregates may be formed within the centers of relatively stable macroaggregates (Oades, 1984). In addition, just as macroaggregates are being created, they are degraded, due to various factors including the physical pressures of root growth, decomposition of labile organic binding agents, or changes in management practices that reduce root and hyphal growth (Allison, 1968; Tisdall & Oades, 1982). In contrast, microaggregates, once stabilized, are only slowly degraded because of the greater physical protection and chemical resistance of the organic components (Tisdall & Oades, 1982).

A major factor affecting the type of macroaggregates present in a soil is the state of microaggregate development. If a soil's microaggregate structure is poorly developed, the size and stability of macroaggregates may be limited to physical entanglement of primary soil particles and transient cementation by relatively labile organic binding agents. Furthermore, without significant numbers of microaggregates, the development of intramacroaggregate pores will be minimal, which will likely restrict the activities of soil organisms within these macroaggregates (Elliott & Coleman, 1988).

This not to suggest that physical entanglement of primary soil particles by hyphae and roots in itself is not important. Quite the contrary, for many systems (e.g., sand dunes) physical entanglement is the major event (Forster & Nicolson, 1981a), although it does not produce very stable aggregates because the agents responsible for the stabilization and the development of microaggregate structure are absent. If, however, the additional agents conducive to the creation of microaggregates (e.g., clays, polyvalent cations, organic residues) are also present, mycorrhizae can contribute to the development of microaggregates and better quality macroaggregates. When a relatively stable microaggregate structure is already in place, as in many mollisols degraded by row cropping, the basic building blocks for creating very stable macroaggregates are already present, and the contribution of mycorrhizae to the restoration of macroaggregate structure will be extremely important (Tisdall & Oades, 1980; Jastrow, 1987; Miller & Jastrow, 1990).

In many types of restorations, e.g., minesoils, very little structural development is present until the soil has been revegetated. Without vegetation, few or no macroaggregates develop (Wilson, 1957). Mycorrhizal fungi have been found to be important in stabilizing revegetated mine soil (Rothwell, 1984). The establishment of the mycorrhizal association is the major link between the revegetation and aggregation phases of the restoration process. Both revegetation and aggregation contribute to the accretion of organic matter. Even so, if a mine soil is composed mainly of sand-sized particles or if polyvalent cations are scarce, macroaggregate structure will be poorly developed until the substrate is further weathered. The weathering of mineral particles can occur by freezing and thawing or by the actions of organic acids derived from vegetation or microbial activity. In such systems, restoration may proceed slowly unless organic amendments are used to supplement the inputs of organic debris from vegetation.

The work of Tisdall and Oades (1979, 1980, 1982) provides a good example of ways that mycorrhizal research can help determine the consequences of different crop rotations for soil stability. Their initial study suggested that aggregate stability is related to root and mycorrhizal fungal hyphal lengths (Tisdall & Oades, 1979). Follow-up studies indicated that 50 years of crop rotation decreased the amounts of stable macroaggregates (> 0.25 mm diameter) and simultaneously decreased the lengths of roots and VAM fungal hyphae in the soil (Tisdall & Oades, 1980). They found a positive correlation between the amounts of macroaggregates and the length of extraradical VAM fungal hyphae (Fig. 13.2) that was related to the type of crop rotation and the frequency of fallow conditions. A similar correlation was found between macroaggregates and root length. Other investigators have argued that living roots rather than mycorrhizal hyphae are the important factor affecting aggregation (Reid & Goss, 1981; Stone & Buttery, 1989), but these studies have not quantitatively measured extraradical hyphae. The practical implication of this body of research is that good soil structure can be maintained by crop management practices such as minimizing tillage and rotating with plants possessing extensive mycorrhizal root systems.

Soil aggregates apparently can develop quite rapidly in row crop soil that has been restored to tallgrass prairie vegetation (Jastrow, 1987). The mollisol studied had been under continuous cultivation for over 100 years; however, within 11 years after planting to prairie species, the macroaggregate development was not significantly different from that of a nearby virgin prairie remnant. Organic matter accumulation during restoration was minimal, and thus the rapid recovery was most likely due to a well-developed microaggregate structure that remained relatively intact during cropping. Without disturbance, both total root length and the length of roots colonized by VAM fungi also increased (Cook et al., 1988). Furthermore, both root and extraradical hyphal lengths were associated with increases in the percentage of macroaggregates (Figs. 13.3 and 13.4) and the geometric mean diameter of water-stable aggregates (Miller & Jastrow, 1990). The strength of the mycorrhizal association and its relationship to soil aggregation appeared to vary with root morphology. Plant lifeforms possessing a greater proportion of fine roots (0.2–1.0 mm diameter) than very fine roots (< 0.2 mm diameter) had the strongest association with aggregation development (Miller & Jastrow, 1990). In addition, a major portion of the effect of roots on soil aggrega-

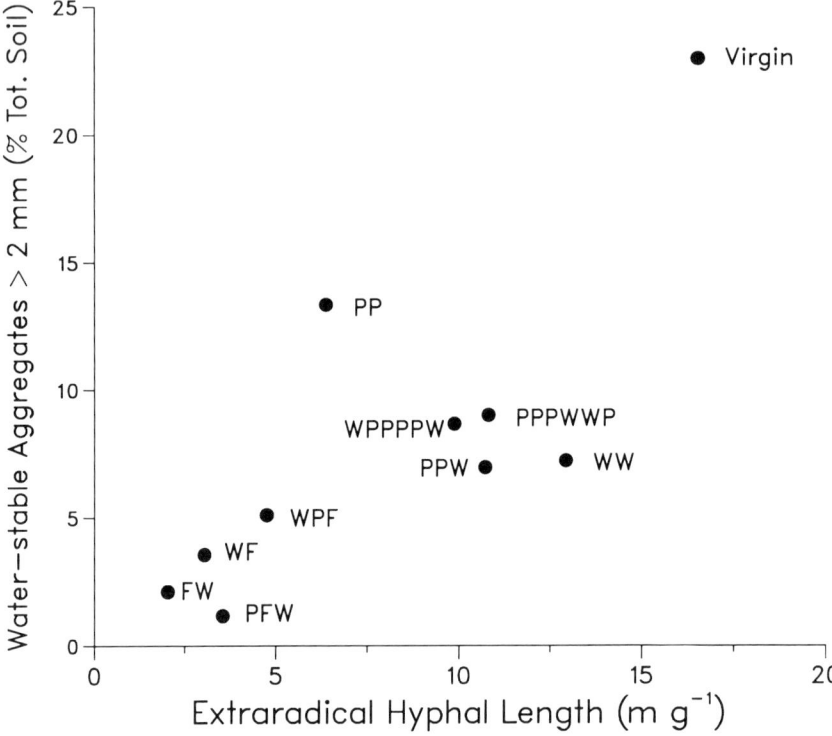

Figure 13.2. The relationship between extraradical hyphal length and percentage water-stable aggregates for a soil under different crop rotations. PP, old pasture; PPW, pasture–pasture–wheat; WW, wheat every year; WPF, wheat–pasture–fallow; PFW, pasture–fallow–wheat; WF, wheat–fallow; FW, fallow–wheat; PPPWWP and WPPPPW, 2 years wheat and 4 years pasture. Data from Tisdall and Oades (1980).

tion occurred indirectly and was due to the roots' association with mycorrhizal fungi.

An area of research much needed in soil restoration is the effects of soil aggregation and resultant pore size distributions on nutrient cycling. Another important and related topic is the role of extraradical hyphae in creating a nutrient reserve. As important as the extraradical hyphae of VAM fungi are to soil aggregation, they are just as important in the selective exploitation of nutrient-rich sites within soil. The studies of St. John et al. (1983a,b) indicate that VAM hyphae proliferate in localized zones of high organic content. Their research suggests that hyphae grow randomly through soil until they reach a nutrient-rich microsite where the hyphae proliferate. An interesting question is whether these sites are associated with or actually are the soil aggregates that the fungal hyphae have helped to create. The encapsulation of organic debris within a soil aggregate

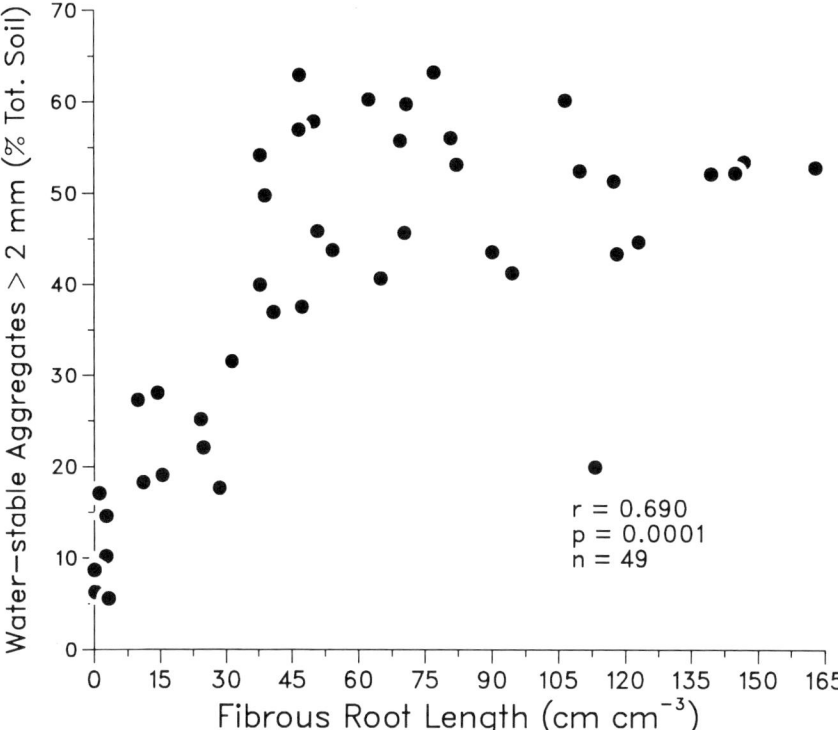

Figure 13.3. The relationship between total fibrous root length and percentage water-stable aggregates in soils from a prairie restoration chronosequence (Miller & Jastrow, unpublished data).

may be a means for creating a more conservative nutrient cycle. Thus, as restoration proceeds, nutrient cycles that are quite open may evolve to where nutrient release is considerably slowed and controlled by the habitats created through the aggregation process.

Conclusions

Contributions of the mycorrhizal association to ecosystem restoration are represented at many levels of organization (Table 13.1). The most obvious role for the mycorrhizae is the linking of primary producers with soil decomposers, allowing for a more closely coupled nutrient cycle. At one scale, mycorrhizal fungi are intimately associated with the creation of a nutrient reserve by their contribution to the development of soil structure. At another scale, mycorrhizae are associated with the day-to-day functioning of a plant via their involvement in whole plant processes such as nutrient uptake, water relations, growth, and

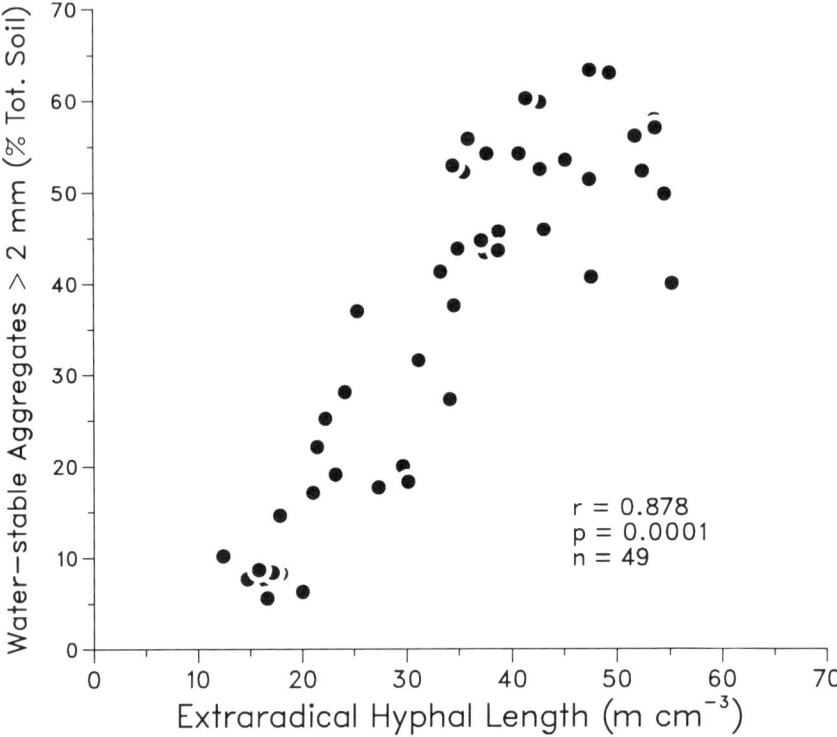

Figure 13.4. The relationship between extraradical hyphae and percentage water-stable aggregates in soils from a prairie restoration chronosequence (Miller & Jastrow, unpublished data).

reproduction, i.e., the integration of whole plant growth. Also, the diversity and stability of the plant community can be affected by any factor that can impact the mycorrhizal propagule reserve. At even larger scales, the kind of mycorrhizae, whether VAM, ectomycorrhizae, or ericoid forms, can influence the type of community that can occupy a site. In many ways the mycorrhizal association can be viewed as a "keystone" feature of a functioning community.

Studies of mycorrhizae have made important contributions to the relatively new field of restoration ecology, which incorporates applied and basic research with land management (Jordan et al., 1987). The examples presented above demonstrate that mycorrhizae can play an important role in developing an understanding of how to restore whole communities. In addition, the examples show that mycorrhizae contribute actively to the restoration process.

In the most simplistic terms, the studies presented emphasize the mycorrhizal mycelial network, i.e., the hyphal links that occur among roots and between roots and soil particles or aggregates, during the restoration process. Furthermore, the degraded state of many communities can be tied to the loss of this network. The significance of the mycelial network has recently been reviewed elsewhere (Read, 1984; Read & Birch, 1988; Newman, 1988), and, as we have demonstrated, its loss can affect mycorrhizal inoculum, seedling establishment, soil structure and belowground habitats, nutrient cycling, and species diversity. Hence, our need for the identification of factors that control the establishment not only of vegetation but of the mycorrhizal mycelial network will be critical to the restoration of whole communities.

Acknowledgments

The preparation of this manuscript was supported by the U.S. Department of Energy, Assistant Secretary for Energy Research, Office of Health and Environmental Research, Ecological Research Division, under contract W-31-109-Eng-38.

References

Abdul-Kareem, A. W., & McRae, S. G. (1984). The effects on topsoil of long-term storage in stockpiles. *Plant and Soil* **76,** 357–363.

Allen, E. B. (1988). Some trajectories of succession in Wyoming sagebrush grassland: Implications for restoration. In: *The Reconstruction of Disturbed Arid Lands: An Ecological Approach,* AAAS Selected Symposium 109 (Ed. by E. B. Allen), pp. 89–112. Westview Press, Boulder, CO.

Allen, E. B. (1989). The restoration of disturbed landscapes with special reference to mycorrhizal fungi. *Journal of Arid Environments* **17,** 279–286.

Allen, E. B., & Allen M. F. (1980). Natural re-establishment of vesicular-arbuscular mycorrhizae following stripmine reclamation in Wyoming. *Journal of Applied Ecology* **17,** 139–147.

Allen, E. B., & Allen, M. F. (1986). Water relations of xeric grasses in the field: Interactions of mycorrhizas and competition. *New Phytologist* **104,** 559–571.

Allen, E. B., & Allen, M. F. (1988). Facilitation of succession by the nonmycotrophic colonizer *Salsola kali* (Chenopodiaceae) on a harsh site: Effects of mycorrhizal fungi. *American Journal of Botany* **75,** 257–266.

Allen, E. B., & Allen, M. F. (1990). The mediation of competition by mycorrhizae in successional and patchy environments. In: *Perspectives on Plant Competition* (Ed. by J. B. Grace & D. Tilman), pp. 367–389. Academic Press, New York.

Allen, M. F. (1988a). Belowground structure: A key to reconstructing a productive arid ecosystem. In: *The Reconstruction of Disturbed Arid Lands: An Ecological Approach,* AAAS Selected Symposium 109 (Ed. by E. B. Allen), pp. 113–135. Westview Press, Boulder, CO.

Allen, M. F. (1988b). Re-establishment of VA mycorrhizas following severe disturbance:

Comparative patch dynamics of a shrub desert and subalpine volcano. *Proceedings of the Royal Society of Edinburgh* **94B**, 63–71.

Allen, M. F., & MacMahon, J. A. (1985). Impact of disturbance on cold desert fungi: Comparative microscale dispersion patterns. *Pedobiologia* **28**, 215–224.

Allen, M. F., MacMahon, J. A., & Andersen, D. C. (1984). Reestablishment of Endogonaceae on Mount St. Helens: Survival of residuals. *Mycologia* **76**, 1031–1038.

Allison, F. E. (1968). Soil aggregation—Some facts and fallacies as seen by a microbiologist. *Soil Science* **106**, 136–143.

Bethlenfalvay, G. J., Thomas, R. S., Dakessian, S., Brown, M. S., & Ames, R. N. (1988). Mycorrhizae in stressed environments: Effects on plant growth, endophyte development, soil stability and soil water. In: *Arid Lands Today and Tomorrow* (Ed. by E. E. Whitehead, C. F. Hutchinson, B. N. Timmermann, & R. G. Varaday), pp. 1015–1029. Westview Press, Boulder, CO.

Biondini, M. E., Bonham, C. D., & Redente, E. F. (1985). Secondary successional patterns in a sagebrush (*Artemisia tridentata*) community as they relate to soil disturbance and soil biological activity. *Vegetatio* **60**, 25–36.

Black, R., & Tinker, P. B. (1979). The development of endomycorrhizal root systems. II. Effect of agronomic factors and soil conditions on the development of vesicular-arbuscular mycorrhizal infection in barley and on the endophyte spore density. *New Phytologist* **83**, 401–413.

Bradshaw, A. D., Marrs, R. H., Roberts, R. D., & Skeffington, R. A. (1982). The creation of nitrogen cycles in derelict land. *Philosophical Transactions of the Royal Society of London Series B* **296**, 557–561.

Call, C. A., & Davies, F. T. (1988). Effects of vesicular-arbuscular mycorrhizae on survival and growth of perennial grasses in lignite overburden in Texas. *Agriculture, Ecosystems and Environment* **24**, 395–405.

Call, C. A., & McKell, C. M. (1982). Vesicular-arbuscular mycorrhizae—a natural revegetation strategy for disposed processed oil shale. *Reclamation and Revegetation Research* **1**, 337–347.

Call, C. A., & McKell, C. M. (1985). Endomycorrhizae enhance growth of shrub species in processed oil shale and disturbed native soil. *Journal of Range Management* **38**, 258–261.

Cheshire, M. V., Sparling, G. P., & Mundie, C. M. (1983). Effect of periodate treatment of soil on carbohydrate constituents and soil aggregation. *Journal of Soil Science* **34**, 105–112.

Cheshire, M. V., Sparling, G. P., & Mundie, C. M. (1984). Influence of soil type, crop and air drying on residual carbohydrate content and aggregate stability after treatment with periodate and tetraborate. *Plant and Soil* **76**, 339–347.

Coleman, D. C. (1986). The role of microfloral and faunal interactions in affecting soil processes, In: *Microfloral and Faunal Interactions in Natural and Agro-Ecosystems* (Ed. by M. J. Mitchell & J. P. Nakas), pp. 317–348. Martinus Nijhoff/Dr. W. Junk, Boston.

Cook, B. D., Jastrow, J. D., & Miller, R. M. (1988). Root and mycorrhizal endophyte development in a chronosequence of restored tallgrass prairie. *New Phytologist* **110**, 355–362.

Daft, M. J., & Nicolson, T. H. (1974). Arbuscular mycorrhizas in plants colonizing coal wastes in Scotland. *New Phytologist* **73**, 1129–1138.

Danielson, R. M. (1985). Mycorrhizae and reclamation of stressed terrestrial environments. In: *Soil Reclamation Processes* (Ed. by R. Tate & D. L. Klein), pp. 173–201. Marcel Dekker, New York.

Danielson, R. M., Visser, S., & Parkinson, D. (1983). Microbial activity and mycorrhizal potential of four overburden types used in the reclamation of extracted oil sands. *Canadian Journal of Soil Science* **63**, 363–375.

Doerr, T. B., Redente, E. F., & Reeves, F. B. (1984). Effects of soil disturbance on plant succession and levels of mycorrhizal fungi in a sagebrush-grassland community. *Journal of Range Management* **37**, 135–139.

Elliott, E. T. (1986). Aggregate structure and carbon, nitrogen, and phosphorus in native and cultivated soils. *Soil Science Society of American Journal* **50**, 627–633.

Elliott, E. T., & Coleman, D. C. (1988). Let the soil work for us. *Ecological Bulletins* **39**, 23–32.

Emerson, W. W., Foster, R. C., & Oades, J. M. (1986). Organomineral complexes in relation to soil aggregation and structure. In: *Interactions of Soil Minerals with Natural Organics and Microbes* (Ed. by P. M. Huang & M. Schnitzer), pp. 521–548. Soil Science Society of America, Madison, WI.

Ernst, W. H. O., Van Duin, W. E., & Oolbekking, G. T. (1984). Vesicular-arbuscular mycorrhiza in dune vegetation. *Acta Biologica Neerlandica* **33**, 151–160.

Evans, D. G., & Miller, M. H. (1988). Vesicular-arbuscular mycorrhizas and the soil-disturbance-induced reduction of nutrient absorption in maize. I. Causal relations. *New Phytologist* **110**, 67–74.

Ewel, J. J. (1987). Restoration is the ultimate test of ecological theory. In: *Restoration Ecology: A Synthetic Approach to Ecological Research* (Ed. by W. R. Jordan, M. E. Gilpin, & J. D. Abers), pp. 31–33. Cambridge University Press, Cambridge.

Forster, S. M., & Nicolson, T. H. (1981a). Aggregation of sand from a maritime embryo sand dune by microorganisms and higher plants. *Soil Biology and Biochemistry* **13**, 199–203.

Forster, S. M., & Nicholson, T. H. (1981b). Microbial aggregation of sand in a maritime dune succession. *Soil Biology and Biochemistry* **13**, 205–208.

Foster, R. C. (1985). *In situ* localization of organic matter in soils. *Quaestiones Entomologicae* **21**, 609–633.

Gange, A. C., Brown, V. K., & Foster, L. M. (1990). A test of mycorrhizal benefit in an early successional plant community. *New Phytologist* **115**, 85–91.

Gemma, J. N., & Koske, R. E. (1988). Seasonal variation in spore abundance and dormancy of *Gigaspora gigantea* and in mycorrhizal inoculum potential of a dune soil. *Mycologia* **80**, 211–216.

Gemma, J. N., & Koske, R. E. (1989). Field inoculation of American beach grass

(*Ammophila breviligulata*) with V-A mycorrhizal fungi. *Journal of Environmental Management* **29**, 173–182.

Gemma, J., Koske, R. E., & Carreiro, M. (1989). Seasonal dynamics of selected species of V-A mycorrhizal fungi in a sand dune. *Mycological Research* **92**, 317–321.

Gianinazzi, S., Trouvelot, A., & Gianinazzi-Pearson, V. (1989). Conceptual approaches for the rational use of VA endomycorrhizae in agriculture: Possibilities and limitations. *Agriculture, Ecosystems and Environment* **29**, 153–161.

Giovannetti, M. (1985). Seasonal variations of vesicular-arbuscular mycorrhizas and endogonaceous spores in a maritime sand dune. *Transactions of the British Mycological Society* **84**, 679–684.

Giovannetti, M., & Nicolson, T. H. (1983). Vesicular-arbuscular mycorrhizas in Italian sand dunes. *Transactions of the British Mycological Society* **80**, 552–557.

Gould, A. B., & Liberta, A. E. (1981). Effects of topsoil storage during surface mining on the viability of vesicular-arbuscular mycorrhiza. *Mycologia* **73**, 914–922.

Gupta, V. V. S. R., & Germida, J. J. (1988). Distribution of microbial biomass and its activity in different soil aggregate size classes as affected by cultivation. *Soil Biology and Biochemistry* **21**, 777–786.

Harris, J. A., Hunter, D., Birch, P., & Short, K. C. (1987). Vesicular-arbuscular mycorrhizal populations in stored topsoils. *Transactions of the British Mycological Society* **89**, 600–603.

Harris, J. A., Birch, P., & Short, K. C. (1989). Changes in the microbial community and physico-chemical characteristics of topsoils stockpiled during opencast mining. *Soil Use and Management* **5**, 161–167.

Janos, D. P. (1980a). Mycorrhizae influence tropical succession. *Biotropica* **12**, 56–64.

Janos, D. P. (1980b). Vesicular-arbuscular mycorrhizae affect lowland tropical rain forest plant growth. *Ecology* **61**, 151–162.

Jasper, D. A., Robson, A. D., & Abbott, L. K. (1987). The effect of surface mining on the infectivity of vesicular-arbuscular mycorrhizal fungi. *Australian Journal of Botany* **35**, 641–652.

Jasper, D. A., Robson, A. D., & Abbott, L. K. (1988). Revegetation in an iron-ore mine—nutrient requirements for plant growth and the potential role of vesicular-arbuscular (VA) mycorrhizal fungi. *Australian Journal of Soil Research* **26**, 497–507.

Jasper, D. A., Abbott, L. K., & Robson, A. D. (1989a). Soil disturbance reduces the infectivity of external hyphae of vesicular-arbuscular mycorrhizal fungi. *New Phytologist* **112**, 93–99.

Jasper, D. A., Abbott, L. K., & Robson, A. D. (1989b). The loss of VA mycorrhizal infectivity during bauxite mining may limit the growth of *Acacia pulchella* R. Br. *Australian Journal of Botany* **37**, 33–42.

Jastrow, J. D. (1987). Changes in soil aggregation associated with tallgrass prairie restoration. *American Journal of Botany* **74**, 1656–1664.

Jastrow, J. D., & Miller, R. M. (1991). Methods for assessing the effects of biota on soil structure. *Agriculture, Ecosystems and Environment* **34**, 279–303.

Jeffries, P. (1987). Use of mycorrhizae in agriculture. *CRC Critical Reviews in Biotechnology* **5**, 319–357.

Jehne, W., & Thompson, C. H. (1981). Endomycorrhizae in plant colonization on coastal sand-dunes at Cooloola, Queensland. *Australian Journal of Ecology* **6**, 221–230.

Johnson, N. C., & McGraw, A. C. (1988a). Vesicular-arbuscular mycorrhizae in taconite tailings. I. Incidence and spread of endogonaceous fungi following reclamation. *Agriculture, Ecosystems and Environment* **21**, 135–142.

Johnson, N. C., & McGraw, A. C. (1988b). Vesicular-arbuscular mycorrhizae in taconite tailings. II. Effects of reclamation practices. *Agriculture, Ecosystems and Environment* **21**, 143–152.

Jordan, W. R., Gilpin, M. E., & Abers, J. D. (1987). *Restoration Ecology: A Synthetic Approach to Ecological Research*. Cambridge University Press, Cambridge.

Jordan, W. R., Peters, R. L., & Allen, E. B. (1988). Ecological restoration as a strategy for conserving biological diversity. *Environmental Management* **12**, 55–72.

Kiernan, J. M., Hendrix, J. W., & Maronek, D. M. (1983). Endomycorrhizal fungi occurring on orphan strip mines in Kentucky. *Canadian Journal of Botany* **61**, 1798–1803.

Koske, R. E., & Gemma, J. N. (1990). VA mycorrhizae in strand vegetation of Hawaii: Evidence for long-distance codispersal of plants and fungi. *American Journal of Botany* **77**, 466–474.

Koske, R. E., & Halvorson, W. L. (1981). Ecological studies of vesicular-arbuscular mycorrhizae in a barrier sand dune. *Canadian Journal of Botany* **59**, 1413–1422.

Koske, R. E. & Polson, W. R. (1984). Are VA mycorrhizae required for sand dune stabilization? *BioScience* **34**, 420–424.

Koske, R. E., Sutton, J. C., & Sheppard, B. R. (1975). Ecology of *Endogone* in Lake Huron sand dunes. *Canadian Journal of Botany* **53**, 87–93.

Lambert, D. H., & Cole, H. (1980). Effects of mycorrhizae on establishment and performance of forage species in mine spoil. *Agronomy Journal* **72**, 257–260.

LeTacon, F., Garbaye, J., & Carr, G. 1987. The use of mycorrhizas in temperate and tropical forests. *Symbiosis* **3**, 179–206.

Liberta, A. E. (1981). Effects of topsoil-storage duration on inoculum potential of vesicular-arbuscular mycorrhizae. In: *Symposium on Surface Mining Hydrology, Sedimentology and Reclamation*, pp. 45–48. University of Kentucky, Lexington, KY.

Maser, C., Maser, Z., & R. Molina. (1988). Small-mammal mycophagy in rangelands of central and southeastern Oregon. *Journal of Range Management* **41**, 309–312.

McHarg, I. (1972). Best shore protection. Nature's own dunes. *Civil Engineering* **42**, 66–71.

Medve, R. J. (1984). The mycorrhizae of pioneer species in disturbed ecosystems in western Pennsylvania. *American Journal of Botany* **71**, 787–794.

Miller, R. M. (1979). Some occurrences of vesicular-arbuscular mycorrhiza in natural and disturbed ecosystems of the Red Desert. *Canadian Journal of Botany* **57**, 619–623.

Miller, R. M. (1984). Microbial ecology and nutrient cycling in disturbed arid ecosystems.

In: *Ecological Studies of Disturbed Landscapes: A Compendium of Results of Five Years of Research Aimed at the Restoration of Disturbed Ecosystems*, Chapter 3, pp. 1–29. DOE/NBM-5009372. Springfield, VA.

Miller, R. M. (1985). Mycorrhizae. *Restoration and Management Notes* **3**, 14–20.

Miller, R. M. (1987a). Mycorrhizae and succession. In: *Restoration Ecology: A Synthetic Approach to Ecological Research* (Ed. by W. R. Jordan, M. E., Gilpin, & J. D. Aber), pp. 205–219. Cambridge University Press, Cambridge.

Miller, R. M. (1987b). The ecology of vesicular-arbuscular mycorrhizae in grass- and shrublands. In: *Ecophysiology of VA Mycorrhizal Plants* (Ed. by G. R. Safir), pp. 135–170. CRC Press, Boca Raton, FL.

Miller, R. M. (1988). The management of VA mycorrhizae in semi-arid environments. In: *Mycorrhizae in the Next Decade: Practical Applications and Research Priorities* (Ed. by D. M. Sylvia, L. L. Hung, & J. H. Graham), pp. 139–141. Institute of Food and Agricultural Research, University of Florida, Gainesville, FL.

Miller, R. M., & Jastrow, J. D. (1990). Hierarchy of root and mycorrhizal fungal interactions with soil aggregation. *Soil Biology and Biochemistry* **22**, 579–584.

Miller, R. M., Moorman, T. B., & Schmidt, S. K. (1983). Interspecific plant association effects on vesicular-arbuscular mycorrhiza occurrence in *Atriplex confertifolia*. *New Phytologist* **95**, 241–246.

Miller, R. M., Carnes, B. A., & Moorman, T. B. (1985). Factors influencing the survival of vesicular-arbuscular mycorrhiza propagules during topsoil storage. *Journal of Applied Ecology* **22**, 259–266.

Moorman, T., & Reeves, F. B. (1979). The role of endomycorrhizae in revegetation practices in the semi-arid west. II. A bioassay to determine the effects of land disturbance on endomycorrhizal populations. *American Journal of Botany* **66**, 14–18.

Muneer, M., & Oades, J. M. (1989). The role of Ca-organic interactions in soil aggregate stability. III. Mechanisms and models. *Australian Journal of Soil Research* **27**, 411–423.

Nelsen, C. E. (1987). The water relations of vesicular arbuscular mycorrhizal systems. In: *Ecophysiology of VA Mycorrhizal Plants* (Ed. by G. R. Safir), pp. 71–91. CRC Press, Boca Raton, FL.

Newman, E. I. (1988). Mycorrhizal links between plants: their functioning and ecological significance. *Advances in Ecological Research* **18**, 243–270.

Nicolson, T. H. (1959). Mycorrhiza in the gramineae. I. Vesicular-arbuscular endophytes, with special reference to the external phase. *Transactions of the British Mycological Society* **42**, 421–438.

Nicolson, T. H. (1960). Mycorrhiza in the gramineae. II. Development in different habitats, particularly sand dunes. *Transactions of the British Mycological Society* **43**, 132–145.

Nicolson, T. H., & Johnston, C. (1979). Mycorrhiza in the Gramineae. III. *Glomus fasciculatus* as the endophyte of pioneer grasses in a maritime sand dune. *Transactions of the British Mycological Society* **72**, 261–268.

Oades, J. M. (1984). Soil organic matter and structural stability: Mechanisms and implications for management. *Plant and Soil* **76**, 319–337.

Poole, B. C., & D. M. Sylvia. (1990). Companion plants affect colonization of *Myrica cerifera* by vesicular-arbuscular mycorrhizal fungi. *Canadian Journal of Botany* **68**, 2703–2707.

Read, D. J. (1984). The structure and function of the vegetative mycelium of mycorrhizal roots. In: *The Ecology and Physiology of the Fungal Mycelium* (Ed. by D. H. Jennings & A. D. Rayner), Symposium of the British Mycological Society 8, pp. 215–240. Cambridge University Press, Cambridge.

Read, D. J. (1989). Mycorrhizas and nutrient cycling in sand dune ecosystems. *Proceedings of the Royal Society of Edinburgh* **96B**, 89–110.

Read, D. J., & Birch, P. (1988). The effects and implications of disturbance of mycorrhizal mycelial systems. *Proceedings of the Royal Society of Edinburgh* **94B**, 13–24.

Read, D. J., Koucheki, H. K., & Hodgson, J. (1976). Vesicular arbuscular mycorrhiza in natural vegetation systems. I. the occurrence of infection. *New Phytologist* **77**, 641–653.

Reeves, F. B., Wagner, D., Moorman, T., & Keil, J. (1979). The role of endomycorrhizae in revegetation practices in the semi-arid west. I. A comparison of incidence of mycorrhizae in severely disturbed vs. natural environments. *American Journal of Botany* **66**, 6–13.

Reid, J. B., & Goss, M. J. (1981). Effect of living roots of different plant species on the aggregate stability of two arable soils. *Journal of Soil Science* **32**, 521–541.

Reinhardt, D. R., & Miller, R. M. (1990). Size classes of root diameter and mycorrhizal colonization in two temperate grassland communities. *New Phytologist* **116**, 129–136.

Rives, C. S., Bajwa, M. I., Liberta, A. E., & Miller, R. M. (1980). Effects of topsoil storage during surface mining on the viability of VA mycorrhiza. *Soil Science* **129**, 253–257.

Rose, S. L. (1988). Above and belowground community development in a marine sand dune ecosystem. *Plant and Soil* **109**, 215–226.

Roskoski, J. P., Pepper, I., & Pardo, E. (1986). Inoculation of leguminous trees with rhizobia and VA mycorrhizal fungi. *Forest Ecology and Management* **16**, 57–68.

Rothwell, F. M. (1984). Aggregation of surface mine soil by interaction between VAM fungi and lignin degradation products of *Lespedeza*. *Plant and Soil* **80**, 99–104.

Rothwell, F. M., & W. G. Vogel. (1982). Mycorrhizae of planted and volunteer vegetation on surface-mined sites. Broomall, Pa. Northeastern Forest Experiment Station; USDA Forest Service General Technical Report NE-66. 12 p.

Schenck, N. C., & Kinlock, R. A. (1980). Incidence of mycorrhizal fungi on six field crops in monoculture on newly cleared woodland site. *Mycologia* **72**, 445–479.

Schmidt, S. K., & F. B. Reeves. (1984). Effect of the non-mycorrhizal pioneer plant *Salsola kali* L. (Chenopodiaceae) on vesicular-arbuscular mycorrhizal (VAM) fungi. *American Journal of Botany* **71**, 1035–1039.

Schwab, S., & Reeves, F. B. (1981). The role of endomycorrhizae in revegetation practices in the semi-arid west. III. Vertical distribution of vesicular-arbuscular (VA) mycorrhiza inoculum potential. *American Journal of Botany* **68,** 1293–1297.

Smith, S., & Gianinazzi-Pearson, V. (1988). Physiological interactions between symbionts in vesicular-arbuscular mycorrhizal plants. *Annual Review of Plant Physiology and Plant Molecular Biology* **39,** 221–244.

Sparling, G. P., & Cheshire, M. V. (1985). Effect of periodate oxidation on the polysaccharide content and microaggregate stability of rhizosphere and non-rhizosphere soils. *Plant and Soil* **88,** 113–122.

Stahl, P. D., & Williams, S. E. (1986). Oil shale process water affects activity of vesicular-arbuscular fungi and *Rhizobium* 4 years after application to soil. *Soil Biology and Biochemistry* **18,** 451–455.

Stahl, P. D., Williams, S. E., & Christensen, M. (1988). Efficacy of native vesicular-arbuscular mycorrhizal fungi after severe soil disturbance. *New Phytologist* **110,** 347–354.

Stark, J. M., & Redente, E. F. (1987). Production potential of stockpiled topsoil. *Soil Science* **144,** 72–76.

St. John, T. V., & Coleman, D. C. (1983). The role of mycorrhizae in plant ecology. *Canadian Journal of Botany* **61,** 1005–1014.

St. John, T. V., Coleman, D. C., & Reid, C. P. P. (1983a). Growth and spatial distribution of nutrient-absorbing organs: Selective exploitation of soil heterogeneity. *Plant and Soil* **71,** 487–493.

St. John, T. V., Coleman, D. C., & Reid, C. P. P. (1983b). Association of vesicular-arbuscular mycorrhizal hyphae with soil organic particles. *Ecology* **64,** 957–959.

Stribley, D. P. (1987). Mineral Nutrition. In: *Ecophysiology of VA Mycorrhizal Plants* (Ed. by G. R. Safir), pp. 59–70. CRC Press, Boca Raton, FL.

Stribley, D. P. (1989). Present and future value of mycorrhizal inoculants. In: *Microbial Inoculation of Crop Plants* (Ed. by R. Cambell & R. M. MacDonald), pp. 49–65. Special Publication of the Society of General Microbiology.

Stone, J. A., & Buttery, B. R. (1989). Nine forages and the aggregation of a clay loam soil. *Canadian Journal of Soil Science* **69,** 165–169.

Sutton, J. C., & Sheppard, B. R. (1976). Aggregation of sand-dune soil by endomycorrhizal fungi. *Canadian Journal of Botany* **54,** 326–333.

Sylvia, D. M. (1986). Spatial and temporal distribution of vesicular-arbuscular mycorrhizal fungi associated with *Uniola paniculata* in Florida foredunes. *Mycologia* **78,** 728–734.

Sylvia, D. M. (1989). Nursery inoculation of sea oats with vesicular-arbuscular mycorrhizal fungi and outplanting performance on Florida beaches. *Journal of Coastal Research* **5,** 747–754.

Sylvia, D. M., & Burks, J. N. (1988). Selection of a vesicular-arbuscular mycorrhizal fungus for practical inoculation of *Uniola paniculata*. *Mycologia* **80,** 565–568.

Sylvia, D. M., & Will, M. E. (1988). Establishment of vesicular-arbuscular mycorrhizal fungi and other microorganisms on a beach replenishment site in Florida. *Applied and Environmental Microbiology* **54,** 348–352.

Sylvia, D. M., Huang, L. L., & Graham, J. H. (1988). *Mycorrhizae in the Next Decade: Practical Applications and Research Priorities* (Proceedings of the 7th North American Conference on Mycorrhizae). University of Florida, Gainesville, FL.

Thompson, J. P. (1987). Decline of vesicular-arbuscular mycorrhizae in long fallow disorder of field crops and its expression in phosphorus deficiency of sunflower. *Australian Journal of Agriculture Research* **38,** 847–867.

Tisdall, J. M., & Oades, J. M. (1979). Stabilization of soil aggregates by the root systems of ryegrass. *Australian Journal of Soil Research* **17,** 429–441.

Tisdall, J. M., & Oades, J. M. (1980). The effect of crop rotation on aggregation in a red-brown earth. *Australian Journal of Soil Research* **18,** 423–433.

Tisdall, J. M., & Oades, J. M. (1982). Organic matter and water-stable aggregates in soils. *Journal of Soil Science* **33,** 141–163.

Trappe, J. M. (1981). Mycorrhizae and productivity of arid and semiarid rangelands. In: *Advances in Food Producing Systems in Arid and Semi-Arid Lands* pp. 581–599. Academic Press, New York.

Visser, S., Fujikawa, J., Griffiths, C. L., & Parkinson, D. (1984a). Effect of topsoil storage on microbial activity, primary production and decomposition potential. *Plant and Soil* **82,** 41–50.

Visser, S., Griffiths, C. L., & Parkinson, D. (1984b). Topsoil storage effects on primary production and rates of vesicular-arbuscular mycorrhizal development in *Agropyron trachycaulum*. *Plant and Soil* **82,** 51–60.

Waaland, M. E., & Allen, E. B. 1987. Relationships between VA mycorrhizal fungi and plant cover following surface mining in Wyoming. *Journal of Range Management* **40,** 271–276.

Warner, N. J., Allen, M. F., & MacMahon, J. A. (1987). Dispersal agents of vesicular arbuscular mycorrhizal fungi in a disturbed arid ecosystem. *Mycologia* **79,** 721–730.

White, J. A., Munn, L. C., & S. E. Williams. (1989). Edaphic and reclamation aspects of vesicular-arbuscular mycorrhizae in Wyoming Red Desert soils. *Soil Science Society of America Journal* **53,** 86–90.

Wilson, H. A. (1957). Effect of vegetation upon aggregation in strip mine spoils. *Soil Science Society Proceedings* **21,** 637–640.

Zak, J. C., & Parkinson, D. (1982). Initial vesicular-arbuscular mycorrhizal development of slender wheatgrass on two amended mine spoils. *Canadian Journal of Botany* **60,** 2241–2248.

Zak, J. C., & Parkinson, D. (1983). Effects of surface amendation of two mine spoils in Alberta, Canada on vesicular-arbuscular mycorrhizal development of slender wheatgrass: A 4-year study. *Canadian Journal of Botany* **61,** 798–803.

Zak, J. C., Danielson, R. M., & Parkinson, D. (1982). Mycorrhizal fungal spore numbers and species occurrence in two amended mine spoils in Alberta, Canada. *Mycologia* **74,** 785–792.

14

Biotechnology and the Future of VAM Commercialization
Timothy Wood and Brian Cummings

SUMMARY. **Despite their potential for use in agriculture and horticulture, vesicular-arbuscular mycorrhizal (VAM) inoculants have not been broadly commercialized. Factors including high inoculum costs, technical problems in inoculum development, constraints on product efficacy, and limited market needs are involved. This chapter describes the biological bases for these problems and discusses three biotechnological approaches toward their solution. The three approaches include development of improved culture systems for VAM fungi, expansion of product efficacy through development of multiagent inoculants, and expansion of product efficacy through genetic transformation of VAM fungi. Each holds promise for enhancing commercial utilization of mycorrhizal fungi and for advancing our understanding of VAM biology.**

Introduction

During the past two decades, major advances in the biological sciences have generated a new set of tools that allows us to more fully unlock, manipulate, and utilize our biological resources. Collectively, these tools are known as biotechnology, and they represent an integration of advances and techniques from biochemistry, cellular biology, molecular biology, genetics, and microbiology. In total these advances are changing the role of biology in commerce, medicine, agriculture, and natural resource management. Importantly, biotechnology has not replaced the applied biological sciences that existed prior to 1970, e.g., plant breeding, pharmaceutical sciences, and pest management. Rather it has integrated them into a technological framework that draws on an improved understanding of fundamental life processes.

One biological tool that is now being integrated into biotechnology is the development of commercial VAM inoculants for use in agriculture, horticulture, forestry, and reclamation. The impetus for commercialization comes largely from a scientific literature replete with reports showing that VAM associations benefit plant growth, development, and vigor. Basic studies have demonstrated that mycorrhizal associations play an important role in plant nutrition, and that plants colonized by mycorrhizal fungi take up greater amounts of phosphorus and trace metals, particularly when those nutrients are sparingly soluble in soils (Abbott & Robson, 1984). Improved resistance to drought, environmental stresses, and some root pathogens have also been shown to result from VAM colonization and development (Nelson & Safir, 1982; Allen & Boosallis, 1983; Bagyaraj, 1984; Graham, 1986).

More applied studies have added to the list of benefits. Under experimental conditions, commercial crops inoculated with VAM fungi have shown increases in growth and yield (Menge, 1981; Powell, 1984), improved uniformity (Biermann & Linderman, 1983), reductions in phosphorus and trace metal fertilizer requirements (Menge et al., 1978c), reduced losses due to environmental stresses and root disease (Nelson & Safir, 1982; Ojala et al., 1983; Hussey & Roncadori, 1982, reduced stunting on fumigated soils (Menge et al., 1977, Timmer & Leyden, 1978; Kleinschmidt & Gerdemann, 1972), improved transplant establishment (Johnson & Crews, 1979; Menge et al., 1978a; Schenck & Tucker, 1974), and reduced cropping times (Sasa et al., 1987). All translate into increased profits for farmers and nurserymen, and all suggest a significant commercial potential for VAM inoculants.

Despite this potential, VAM inoculants have not been widely commercialized. The reasons for this are many, and they involve a complex of factors associated with inoculum cost, the technologies of inoculum development, constraints on product efficacy, and market needs. These factors will be discussed in detail below. Suffice it to say here that for VAM inoculants to become widely used they must address significant grower needs with consistent and broadscale efficacy in a cost-effective manner.

VAM inoculants developed to date have fallen short of this mark. Production techniques for VAM fungi remain expensive, and the costs of inoculum are relatively high. VAM inoculants show significant and consistent efficacy under a fairly narrow spectrum of conditions and relatively inconsistent performance under the majority of horticultural and agricultural conditions (Wood, 1991). And given the current availability of phosphorus and trace metal fertilizers, VAM inoculants do not appear to address significant grower needs across the broad spectrum of agriculture and horticulture.

Given the fundamental importance of mycorrhizal associations for plant structure and function, however, the potential for using VAM inoculants in plant production remains. The purpose of this chapter is to discuss ways in which current advances in biotechnology can be applied toward expanding the commercial utilization of VAM fungi in agriculture and horticulture.

Strategies for Commercial Expansion through Biotechnology

Several groups, including private companies, government agencies, and university scientists, have worked to develop commercial VA mycorrhiza inoculants, or components and prototypes thereof. Superior strains of VAM fungi for given applications have been isolated and identified (Govinda et al., 1983; Powell et al., 1982). Culture systems for these fungi have been devised, and some have been patented (Menge, 1984; Dehne & Backhaus, 1986; Mugneir & Mosse, 1987). Several inoculant formulations have been developed (Powell, 1984; Hall, 1979; Warner, 1985) and techniques for applying inoculants to nursery and agricultural soils have been designed and tested (Hayman et al., 1981; Ferguson & Menge, 1986). In short, many of the traditional technologies necessary for development of VAM inoculants are in place, and reasonably good, first-generation products have been developed and introduced.

The failure of these inoculants to attain widespread commercial use, at least in the United States, is again tied to issues of high inoculum cost, limited efficacy, and limited grower need. In this section, we examine three areas of biotechnology that could impact these shortcomings. The technologies are (1) development of improved culture systems for VAM fungi to reduce inoculum production costs, (2) development of multiagent inoculants to expand breadth of efficacy, and (3) development of genetic transformation systems to introduce new traits of significant commercial interest to VAM fungal strains.

Improved Culture Systems for VAM Fungi

Vesicular-arbuscular mycorrhizal fungi are obligate symbionts. In nature they require association with root cortical tissues for growth, development, and sporulation. In the laboratory, under monoaxenic conditions, spores of VAM fungi can be germinated, and extensive germtube growth can be obtained on specialized media (Hepper, 1984). However, no one to our knowledge has succeeded in subculturing VAM fungal hyphae or in obtaining significant fungal differentiation and sporulation under monoaxenic conditions.

Instead, VAM fungi are typically cultured in association with the roots of living plants. A variety of systems based on this approach have been developed. Pot culture is most widely used both for experimental and commercial inoculum production (Ferguson & Woodhead, 1982; Menge, 1984). Here, VAM fungi are produced on the roots of living plants grown in pots in the greenhouse. The pots contain a solid substrate, and the fungi develop and sporulate both within that substrate and within host roots. Pot culture is an established technique, and it is amenable to commercial inoculum production. However, it carries two significant limitations. First, because the inoculum is produced in open containers, it is subject to contamination by unwanted bacteria and fungi including plant pathogens. Obviously, such contamination is unacceptable, and the problem has been reduced through development of techniques for disinfesting and testing the quality of pot culture inoculum on a routine basis (Dehne & Backhaus, 1986).

The second limitation involves cost. VAM fungi are slow growing, and high levels of sporulation (up to several million spores per liter of culture substrate) are generally achieved only after 4 months or more in pot culture. These long incubation periods make pot culture an expensive process, and they keep the costs of VAM inoculum high relative to other agricultural inputs.

Variations on pot culture involving hydroponics and aeroponics have been developed (Howeler et al., 1982; Elmes & Mosse, 1980; Hung & Sylvia, 1987), but both systems remain subject to external contamination, and both require extended culture periods.

A second approach to production uses axenic root organ cultures as living substrates for VAM fungi (Mosse & Hepper, 1975; Miller-Wideman & Watrud, 1984; Mugnier & Mosse, 1987). Here, tissue cultured roots grown on sterile substrates in closed containers are inoculated with surface sterilized, pregerminated VAM fungal spores. Normal mycorrhizas develop in such systems, and mature, viable fungal spores can form. VAM root organ culture systems are largely experimental in nature, and, to our knowledge, commercial quantities of inoculum have not been produced in this way. The root organ culture approach clearly reduces the contamination problem. However, there is no evidence that VAM fungi grow and develop more rapidly under axenic conditions. As such, inoculants produced in this way are still likely to be expensive.

The magnitude of the cost problem is significant. We calculate that, given current production technologies, the cost of inoculum will average 0.2 cents per plant for many applications. Although this outlay may be acceptable for uses in high-value horticulture, it is too high for use with most major agronomic row crops such as corn, soybean, and cotton. At 50,000 plants per hectare (Chapman & Carter, 1976), inoculum costs for corn would total $100/ha, far higher than farmers are used to paying for phosphate fertilizer, insecticides, *Rhizobium* inoculants, etc. (typically $2–20 per hectare per application).

Furthermore, for a new agricultural amendment to be adopted widely in agriculture, growers must receive a 3 to 4-fold payback on product costs. That is, the economic benefit gained from the product in terms of improved yield or crop quality must be three or four times greater than the cost of that product to the grower. Assuming an average revenue of $750/ha for corn, VAM inoculation would have to improve corn yields by at least 40% on a consistent basis before it became a widely adopted practice. To date these levels of efficacy have not been demonstrated.

Can inoculum costs be dramatically reduced by developing technologies for producing VAM fungi in monoaxenic, large-scale fermentation? Scientists have been exploring such culture systems for decades. The principal approach used to date involves systematic screening of defined and complex media for support of hyphal growth and differentiation. Dramatic levels of germtube growth have been achieved from surface sterilized spores, but again, to our knowledge, extensive fungal differentiation, sporulation, and hyphal subculture have not been achieved on abiotic media, even when those media contain exudates and extracts from the roots of VAM host plants (Hepper, 1984).

The failure of this approach has led to speculation concerning nonnutritional foundations for the obligate symbiosis. Hypotheses include the following: (1) Permeability barriers in VAM germtubes and hyphae prevent nutrient uptake except across arbuscular membranes (Millner, 1988). (2) VAM fungi lack genomic instructions for critical enzymes or biochemical pathways, and thus depend on host cells for specific metabolites or metabolic processes. And (3) VAM fungi are genetically competent, but genes critical for growth and differentiation must be activated through association with the host plant. Activation may involve chemical triggers (e.g., compounds present in, or derived from host cell walls) or physical triggers (e.g., thigmotropic stimuli similar to those reported for rust fungi).

Regardless of the mechanism involved, the obligate biotrophy of VAM fungi poses a central challenge for mycorrhizal research. Unfortunately, that challenge contains an inherent dilemma: it is difficult to investigate and characterize the biology of an organism when you cannot grow it in isolation, and it is difficult to discover how to grow it in isolation when you do not understand its fundamental biology.

Current biotechnologies offer no immediate solutions to the culture problem, but they do provide precise tools that facilitate the study of VAM biology in the absence of a pure culture system. Monoclonal antibody techniques, for example, could be used to study the transport proteins in VAM fungal membranes and to examine the "permeability" factor. Proteins could be extracted from the membranes of saprophytic fungi that are closely related to VAM fungi. Monoclonal antibodies could be produced, tagged, and used to probe the membranes in germtubes, hyphae, and arbuscules of VAM fungi. If significant numbers of those membrane proteins were found in arbuscules but not in hyphae and germtubes, then one might suspect that arbuscules were required for specific aspects of nutrition. Furthermore, proteins of interest could be sequenced and compared with sequences in databanks to search for clues to protein and arbuscular function.

Similarly, RNA and DNA technologies could be used to study the genetic competence of VAM fungi. Messenger RNAs could be extracted from germinated VAM fungal spores and from VAM fungal hyphae collected from intact mycorrhizas (e.g., from VAM root organ cultures). Complimentary DNA strands (cDNAs) could be produced from these RNAs and then cloned for amplification. Hybridization techniques could then be used to identify RNAs that were present, and hence genes that were active, in the symbiotic as opposed to the asymbiotic state. The cDNAs specific to those RNAs could then be sequenced and compared with sequences in gene databanks in an attempt to identify gene products and their functions. That information, in turn, might provide clues as to which genes were being activated as a result of mycorrhiza formation, and which cell functions were lacking in the asymbiotic state.

In addition, cDNA sequences that were differentially expressed in intact mycorrhizas could be used as probes to determine the locations of those genes in the fungal genome, and to explore the presence or absence of associated promoter and repressor sequences.

Techniques from molecular biology could be applied in many more ways to the study of VAM biology. Our purpose here, however, is not to promote "armchair biology" but to point out that the tools of biotechnology offer promise for exploring the obligate nature of VAM symbioses and for eventually developing pure culture systems for VAM fungi.

Expanded Efficacy through Multiagent Inoculants

Despite their broad commercial potential, VAM inoculants appear to perform consistently well in only one small niche market. They improve phosphorus uptake, and hence the growth and yield of VAM-dependent horticultural crops (e.g., citrus and grape) grown on fumigated soils, particularly when those soils contain moderate-to-low levels of phosphorus, and/or when those soils fix significant amounts of phosphorus. Here, efficacy is substantial and consistent, and a true market for inoculum exists (Menge et al., 1977, Timmer & Leyden, 1978, Kleinschmidt & Gerdemann, 1972; Koch et al., 1982; Koepfner et al., 1983; Menge et al., 1983; Strobel et al., 1982; Haas & Krikun, 1985; Schultz et al., 1981; Kormanik et al., 1982; Wood, 1991).

Outside of this market, the efficacy of VAM inoculants declines. Crop species and cultivars vary in the degree to which they depend on VAM associations for nutrient uptake, and not all plants require mycorrhizae to a significant degree (Heckman & Angle, 1987; Krishna et al., 1985; Azcon & Ocampo, 1981; Toth & Castleberry, 1984; Menge et al., 1978b). VA mycorrhizae enhance uptake of phosphorus and trace elements when those nutrients are present but sparingly soluble in soils. However, they have little or no efficacy when those nutrients are abundant and soluble (Abbott & Robson, 1984; Menge et al., 1982). High levels of available phosphorus and host phosphorus sufficiency inhibit VAM development (Ratnayake et al., 1978; Graham et al., 1981; Schwab et al., 1983). And competition with indigenous populations of mycorrhizal fungi in nonfumigated soils can obscure or suppress the benefits of inoculation (Powell, 1984). One or more of these factors come into play in the majority of agricultural and horticultural situations, and, as a result, VAM inoculants tend to provide limited and/or inconsistent benefits most of the time. Challenges exist, then, for expanding the breadth of VAM efficacy.

One approach to this challenge focuses on development of multiagent inoculants, that is, inoculants that contain several different microorganisms, each with a unique role to play in promoting plant growth and plant health. In theory, mixed inoculants would carry a broader spectrum of potential benefits, and thus might show efficacy across a broader range of uses.

Phosphate-solubilizing bacteria and fungi, including species in the genera *Pseudomonas, Agrobacterium, Bacillus, Aspergillus,* and *Penicillium,* are one class of organisms that might be included in multiagent inoculants. Used for years as "bacterial fertilizers" in the U.S.S.R., these microbes enhance the solubilization of chemically bound phosphorus in the rhizosphere (Alexander, 1977). A dual inoculant containing both phosphorus-solubilizing microbes and VAM fungi could have the compound effect of both enhancing the chemical availability of P in the soil and improving plant uptake of that phosphorus.

Interactions between phosphate-solubilizing microbes and VAM fungi, and their combined effects on plant growth, have been studied. Dual inoculants have been shown to stimulate plant growth, and synergies have been demonstrated. For example, Barea et al. (1975) and Raj et al. (1981) both showed that phosphate-solubilizing bacteria maintained higher populations for longer periods in the rhizospheres of mycorrhizal as opposed to nonmycorrhizal plants. In addition, some phosphate-solubilizing bacteria and fungi have been shown to produce hormones, vitamins, and other growth factors that could stimulate mycorrhizal development and plant growth independently of a phosphorus effect (Bagyaraj, 1984). Clearly, multiple interactions are involved.

Symbiotic and free-living nitrogen-fixing bacteria could also be combined with VAM fungi in mixed inoculants. The legume–*Rhizobium*–VAM association has been studied repeatedly, and many reports show that coinoculation of legumes with both VAM fungi and *Rhizobium* bacteria gives improved plant growth over inoculation with either symbiont alone (Barea & Azcon-Aguilar, 1983; Bagyaraj, 1984). Both nodulation and nitrogen fixation are known to be phosphorus-intensive processes (McLachlan & Norman, 1961; Gibson, 1976; Bergersen, 1971), and effective VAM colonization is thought to be important in meeting that high phosphorus requirement. Similar benefits of coinoculation have been reported for free-living nitrogen fixers. Bagyaraj and Menge (1978) inoculated tomato with *Glomus fasciculatum* and *Azotobacter chroococcum* and found beneficial interactions. Mycorrhizal colonization increased and stabilized the *A. chroococcum* population in the rhizosphere, and the bacteria, in turn, enhanced VAM colonization and spore production. Manjunath et al. (1981) inoculated onion with a free-living nitrogen-fixing bacteria, a VAM fungus, and a phosphate-solubilizing fungus, and noted similar additive and synergistic effects on plant growth.

Plant growth-promoting rhizobacteria might also be included in general purpose inoculants. Select strains of *Pseudomonas fluorescens, P. putida,* and *P. syringae,* isolated from root surfaces and rhizosphere soils, have been used successfully as seed and soil inoculants to promote growth and yield of canola, potato, radishes, rice, sugar beet, and wheat (Schroth & Weinhold, 1986; Kloepper et al., 1988, 1989). In field experiments, significant growth and yield responses to inoculation have been seen about 50% of the time. Modes of action are thought to involve suppression of pathogens via the action of siderophores (Leong, 1986), as well as production of plant growth regulators (Kloepper et al., 1988). Synergistic increases in plant growth have been reported to result from coinoculation of soils with VAM fungi and *Pseudomonas putida* (Meyer & Linderman, 1986b).

Biocontrol agents are a fourth category of organisms that could be included in multiagent inoculants. Select bacteria, including strains of *Bacillus subtilis* and *B. cereus*, have been used as seed inoculants to suppress seedling diseases in field crops, e.g., *Sclerotium* on onion, *Phytophthora* on peanut, and *Fusarium* on corn (Weller, 1988; Kloepper et al., 1989). Disease-suppressive fungi, including strains of *Trichoderma* and *Gliocladium*, have shown similar potential for control of diseases of horticultural crops caused by *Rhizoctonia*, *Sclerotium*, *Fusarium*, and *Phytophthora* species (Papavizas, 1985; Sivan et al., 1987; Elad et al., 1984). The principal mode of action for control by *Bacillus* species is thought to involve production of antibiotics (Fravel, 1988). *Trichoderma* species are thought to parasitize their target fungi (Elad et al., 1983). Given these antifungal activities, the feasibility of combining VAM fungi with biocontrol agents in mixed inocula remains in question.

Although it is easy to speculate on the potential uses of mixed inoculants in agriculture and horticulture, difficulties remain in addressing the challenges of formulation, compatibility between agents, efficacy testing, and cost. In theory, formulation of mixed inoculants can be achieved using current technologies. Dry, shelf-stable VAM inoculants have already been developed and marketed. Similar formulations are available for nitrogen-fixing bacteria, plant growth-promoting rhizobacteria, and a variety of biocontrol agents, although shelf lives of 18 months or more (desirable for commercial products) have not been achieved in all cases. Nevertheless, no serious obstacles appear to exist for combining several organisms in a single product.

Functional compatibilities between VAM fungi and biocontrol agents represent an area of growing interest. Given the diversity of such agents and the complexity of the rhizosphere environment, any number of interactions between these organisms is possible. Nevertheless, results to date show no strong and consistent antagonisms between VAM fungi and the biocontrol agents tested. Meyer and Linderman (1986b) found lower populations of fluorescent pseudomonads around the roots of mycorrhizal as opposed to nonmycorrhizal clover plants. In a similar study, Paulitz and Linderman (1990) found temporary suppression of *Pseudomonas putida* by VAM fungi in the rhizosphere of cucumber, but the effect declined after 6–9 weeks. Furthermore, suppression was seen with only one of the two VAM fungi tested. Other workers have shown neutral to positive effects of VAM colonization on the populations and activities of biocontrol agents in soils (Meyer & Linderman, 1986a; Sccilia & Bagyaraj, 1987). Fluorescent pseudomonads, in turn, have been reported to delay germination of VAM fungal spores, but not to reduce germination over the long term (Paulitz & Linderman, 1990). And in two studies, inoculation with fluorescent pseudomonads has had either no effect or a stimulatory effect on rates of VAM colonization (Paulitz & Linderman, 1990; Meyer & Linderman, 1986a). A single study examining interactions between VAM fungi and strains of *Trichoderma hamatum* and *T. harzianum* showed no effects of these two biocontrol fungi against VAM colonization (Kohl & Schlosser, 1989). Studies to date, then, provide grounds for cautious optimism concerning the compatibilities of VAM fungi and biocontrol agents. Nevertheless, results are incomplete, and considerably more work covering a range of microbes, soils, and field conditions is needed before sound conclusions can be drawn.

The potentially high costs of combination inoculants pose serious barriers to commercialization. As discussed above, VAM products are already too expensive to be used in most agricultural situations, and adding further agents to those products should only compound the problem. Clearly, the costs of VAM fungal production will have to be reduced before mixed inoculants offer a realistic solution to problems of limited VAM efficacy.

Finally, questions of true efficacy need to be addressed in detailed field studies. The same inconsistencies in performance that have been found for VAM fungi have been reported for most other individual soil inoculants under field conditions. Furthermore, the benefits and synergies reported for mixed inoculants have been demonstrated largely in greenhouse pot experiments. Field performance has not been characterized in detail. Clearly, there is no way to know at this time whether multiagent inoculants will show improved and consistent performance under routine commercial use. The concept, however, deserves further study.

Expanded Efficacy through Genetic Transformation

Phosphorus is an essential plant nutrient, and phosphate fertilizer is a valued input for most segments of American agriculture and horticulture. Although VAM fungi improve plant uptake of phosphorus and promote more efficient use of applied fertilizers, the need for such improvements is not perceived by the majority of U.S. growers. Phosphorus fertilizer works, it is available, and it is relatively cheap. In short, VAM inoculants do not address significant grower needs, except in the small niche markets discussed earlier.

Nevertheless, VAM fungi hold an intriguing potential for use in agriculture and horticulture. These fungi have exceedingly broad host ranges. Any given strain can colonize the roots of approximately 90% of all plant taxa. If a VAM fungus could be manipulated to perform functions that are broadly valuable to commercial growers, then sizable markets for VAM inoculants, covering a wide variety of crops, could develop. Characteristics that might be amplified or moved into VAM fungi include (1) improved abilities to suppress root disease development, (2) tolerance to specific fungicides, (3) production of the *Bacillus thuringensis* (Bt) endotoxin to suppress root damage caused by insect larvae, (4) production of antinematode agents; and others.

Three approaches to genetically manipulating VAM fungi are discussed below. These are controlled anastomosis, protoplast fusion, and genetic transformation.

Hyphal Anastomosis

The genetics of VAM fungi is poorly understood. Individual spores are known to contain hundreds to thousands of nuclei, but it is not known whether these nuclei represent multiple copies of the same genome or whether they are genetically diverse. Furthermore, these nuclei appear to share a common cytoplasm within a coenocytic thallus. Asexual, or vegetative, reproduction has been well documented in VAM fungi. Sexual life stages and sexual reproduction have not been reported. As such, questions remain as to how VAM fungi are able to recombine traits and maintain genetic diversity. One possible mechanism involves hyphal anastomosis.

When spores of a given strain of a VAM fungus are germinated on filter paper, germtubes arising from different spores can be seen to fuse or anastomose. It is not known at this time whether exchange of cytoplasm and nuclei occur between the two fused organisms, but if such exchanges do occur, anastomosis might provide a natural mechanism for genetic recombination. It might also be used as a tool to "hybridize" VAM fungi and to produce novel strains carrying desirable traits combined from a variety of isolates.

DNA probe technologies, which have been applied recently to VA mycorrhizal fungi (Cummings & Wood, 1990), might be used to explore these possibilities. Consider the following experiment. Two strains of a given VAM fungal species could be assessed for their ability to cross anastomose by cogerminating their spores on filter paper and checking for hyphal fusion. If fusion was found, the two strains could then be subjected to restriction fragment linked polymorphism (RFLP) or polymerase chain reaction (PCR) analysis, and specific molecular markers that differentiated the two strains could be identified. PCR techniques have an advantage over RFLP analyses for VAM research in that only small amounts of DNA are required as starting materials. Next, the two strains could be cocultured in pot cultures where anastomosis would presumably occur, and where, over time, a new generation of spores would be produced. Many single spore isolations could then be made from these mixed cultures, and the isolates could be multiplied and analyzed, again with differential RFLP or PCR probes, for genetic recombination. The presence of "hybrids" would suggest that genetic material had been exchanged between strains. Follow-up studies could be performed to determine whether anastomosis was limited to intraspecific crosses, or whether interspecific and intergeneric recombination were also possible.

If recombination was demonstrated, the preceding protocols could be used to combine traits of commercial interest and to produce novel, superior strains of VAM fungi. For example, isolates of a given VAM fungus, such as *Glomus intraradices,* that had been selected for superior plant growth promotion on acid soils could be hybridized with other strains of the same species that had been selected for superior performance on neutral to basic soils. Few if any natural strains of VAM fungi have been found to provide superior growth stimulation on both acid and neutral to basic soils (Wood, 1991), and production and use of such a hybrid would significantly improve the breadth of inoculant efficacy. Other commercially important characteristics that might be recombined into single superior strains include tolerance to high levels of phosphorus, aggressive colonization, and tolerance to rapid drying (important in formulation).

Protoplast Fusion

Protoplast fusion is another method that can be used to create unique fungal hybrids. In this procedure, mixtures of cell wall degrading enzymes are used to remove hyphal or spore cell walls and to create protoplasts (complete cell wall removal) or spheroplasts (partial cell wall removal) (Selitrennikoff & Bloomfield, 1984; Quigley et al., 1985). Protoplasts of two desired strains are then fused via polyethylene glycol or electrofusion methods. Hybrid organisms expressing the combined genomes are regenerated, typically on selective media that favor their growth and suppress growth of parent cell lines (Anne & Peberdy, 1975; Zimmerman, 1982). Intraspecific, interspecific, and intergeneric protoplast fusions have been reported for fungi (Homolka et al., 1988).

Several technical hurdles must be overcome before protoplast fusion techniques can be routinely used with VAM fungi. The lack of a pure culture system for these fungi poses the most serious obstacle. Hyphae are a preferred starting material for fungal protoplast production, but even under optimized conditions involving robust saprophytic fungi, techniques produce only 1–10% viable protoplasts (Quigley et al., 1987). Successful fusion frequencies producing stable hybrids are even lower, typically less than 1% (Cullen & Leong, 1986). Because VAM fungal hyphae cannot be produced in large amounts in isolation, it will be difficult to obtain sufficient starting material and to produce protoplasts in the numbers necessary to carry out successful cell fusions on a routine basis. VAM fungal spores provide alternative starting materials, but here, wall degradation is problematic given the thickness of the spore walls and the high concentrations of chitin and pectin present within wall layers. Successful spheroplasts have been produced from VAM fungal spores, but quantities have been low, osmotic stability has been poor (Upshall, 1986), and, to our knowledge, such preparations have not been used in successful fusions. In summary, then, while protoplast fusion offers potential for producing hybrid VAM fungi, several technical hurdles must be overcome before those technologies can be realistically applied.

Transformation

Recombinant DNA technologies provide a more controlled approach to the genetic manipulation and genetic enrichment of VAM fungi. This approach allows manipulation at the single gene level and avoids problems associated with the gross fusion of two genomes. Furthermore, recombinant DNA techniques are not limited to intraspecific, interspecific, or intergeneric transfers. Genes from a wide range of prokaryotes and eukaryotes could, in theory, be moved into VAM fungi.

Several general steps are involved in developing transformation systems for a target organism. These include development of an expression vector, development of a DNA delivery method, and development of a system for selecting successful transformants. Although such systems have been successfully tested with a variety of fungi, including ectomycorrhizal fungi (Barrett et al., 1989), unique challenges exist in VAM applications.

In theory, these challenges are not serious in the steps required to construct a VAM fungal expression vector. Generally effective protocols are established. They involve obtaining a suitable plasmid, inserting a dominant selectable marker, and identifying and inserting a strong promoter to drive the marker gene. Plasmids that contain selectable markers for benomyl, hygromycin, and bleomycin resistance (potentially useful in VAM fungal transformation systems) have been developed and are available. Furthermore, standard restriction digestions can be used to produce random DNA fragments from a VAM fungus. These fragments can be cloned into the vector and eventually screened for the presence of strong promoter sequences. Obtaining a functional promoter is often an intricate process. The sequence of the dominant selectable marker must be known, and the random fragments must be inserted in precise locations. Nevertheless, standard protocols exist, and construction of an expression vector for VAM fungal transformations is immediately feasible.

Once a suitable expression vector has been constructed and tested, a gene of commercial interest can be inserted adjacent to the promoter. Again, single gene traits that might be incorporated into VAM fungi include those for fungicide resistance and Bt toxin production. The vector is now ready for delivery.

Three approaches to delivery of the transformation vector into target cells exist. These are protoplast uptake, microinjection, and biolistics. Protoplast uptake has been successfully used in a variety of fungal transformation systems to deliver DNA constructs into target cells (Peberdy, 1987). In conventional use, protoplasts of the target fungus are produced using methods previously described. They are then suspended in osmotically stabile solutions that contain both the expression vector and polyethylene glycol to promote vector uptake. With time, the protoplasts spontaneously absorb the vectors. They are then washed and placed on selective media where successful transformants regenerate. The obstacles described earlier for production and regeneration of VAM fungal protoplasts would pose barriers for applying protoplast delivery techniques in VAM transformation systems.

Microinjection offers a more controlled method for delivering DNA into fungal cells. Here, a 0.1-μm glass pipette is used to penetrate the walls of spores or conidia and to inject transformation vectors into the cytoplasm or nuclei of target cells. In a small proportion of cases, vectors are spontaneously incorporated into cell genomes to effect the transformation. The technique has been used successfully to transform *Neurospora* (Yamamoto & Furusawa, 1979), and it may be applicable to VAM fungi. Biolistic delivery systems represent a "shotgun" approach to microinjection. Here, minute tungsten particles coated with DNA are fired into target cells where, with variable but generally low frequency, that DNA is again spontaneously incorporated into the genome. The technology was developed to circumvent some of the limitations associated with plant transformation systems, and it has been successfully used with fungal spores (Sanford et al., 1991; Klein et al., 1987).

Because microinjection and biolistics can be performed on intact spores, these technologies may be better suited than protoplast uptake for use in VAM fungal transformation systems. Their use avoids the problems associated with protoplast production and regeneration. Low transformation frequencies pose problems for microinjection techniques, as large numbers of spores must be individually treated. These problems are reduced with biolistics as literally thousands of spores can be treated with a single "shot."

The major hurdle associated with all the delivery systems discussed lies in getting expression of desired genes in a VAM fungus. With hundreds to thousands of nuclei per spore, important questions remain as to how many of those nuclei must be successfully transformed before traits are expressed. Answers may lie in part in the selection systems employed.

Systems for selection of successful transformants typically employ selective media to identify cells expressing the dominant selectable marker. Two alternative approaches have been used in most microbial transformations. Auxotrophic markers and selection systems involve the ability of the transformed organism to use novel substrates. These approaches, however, cannot be applied to VAM fungi until their nutritional requirements have been defined. Fungicide and antibiotic resistance systems offer more feasible alternatives. VAM fungal spores, for example, that had been successfully transformed via microinjection or biolistics to express the gene for benomyl resistance could potentially be identified on germination media containing levels of benomyl that inhibited germtube growth from nontransformed individuals but permitted germtube growth from transformed individuals. Alternatively, selection could be attempted in pot cultures where benomyl was applied at rates that allowed only successful transformants to develop and form mycorrhizas.

Again, important questions remain as to how effective such selection systems might be if only a few of the many nuclei in a VAMF spore were transformed. One can only guess at this point that success may require maintaining selective pressures for extensive periods of time to promote selective multiplication of transformed nuclei and to ensure that transformations were stable from generation to generation.

Conclusion

This chapter has discussed three approaches to advancing the commercial use of VAM inoculants in agriculture and horticulture. Development of pure culture systems, development of multiagent mixes, and development of genetic recombination systems have been proposed as avenues for reducing the cost and increasing the value of VAM inoculants. In each case, biotechnology plays a dual role. On the one hand, a new culture system for fermentation of VAM fungi, a mixed biological inoculant, and a transformed VAM fungus represent, in their own rights, biotechnologies with eventual value in improving plant performance in agriculture and horticulture. On the other hand, each development is a product of biotechnology, a result of applying recent advances in our understanding of fundamental life processes.

In this sense, biotechnology provides an interface between basic and applied mycorrhiza science. VAM fungi are fundamentally difficult organisms to work with. Their status as obligate symbionts and their unusual (poorly understood) genetic systems pose significant obstacles to both basic and applied research. Current biotechnological tools, including monoclonal antibody and recombinant DNA techniques, provide a platform for basic studies characterizing the nutritional requirements, developmental processes, and genetic mechanisms of VAM fungi. Advances in these important areas, in turn, will provide foundations for research leading toward enhanced utilization of these potentially valuable biological resources.

References

Abbott, L. K., & Robson, A. D. (1984). The effect of mycorrhizae on plant growth. In: *VA Mycorrhiza* (Ed. by C. L. Powell & D. J. Bagyaraj), pp. 113–130. CRC Press, Boca Raton, FL.

Alexander, M. (1977). *Soil Microbiology*. Wiley, New York.

Allen, M. F., & Boosallis, M. G. (1983). Effects of two species of VA mycorrhizal fungi on drought tolerance of winter wheat. *New Phytologist* **93,** 67–76.

Anne, J., & Peberdy, J. F. (1975). Conditions for induced fusion of fungal protoplasts in polyethylene glycol solutions. *Archives of Microbiology* **105,** 201–205.

Azcon, R., & Ocampo, J. A. (1981). Factors affecting the vesicular-arbuscular infection and mycorrhizal dependency of thirteen wheat cultivars. *New Phytologist* **87,** 677–685.

Bagyaraj, D. J., & Menge, J. A. (1978). Interactions between a VA mycorrhizal fungus and *Azotobacter* and their effects on rhizosphere microflora and plant growth. *New Phytologist* **80,** 567–573.

Bagyaraj, D. J. (1984). Biological interactions with VA mycorrhizal fungi. In: *VA Mycorrhiza* (Ed. by C. L. Powell & D. J. Bagyaraj), pp. 131–154. CRC Press, Boca Raton, FL.

Barea, J. M., & Azcon-Aguilar, C. (1983). Mycorrhizas and their significance in nodulating nitrogen-fixing plants. *Advances in Agronomy* **36**, 1–54.

Barea, J. M., Azcon, R., & Hayman, D. S. (1975). Possible synergistic interactions between *Endogone* and phosphate-solubilizing bacteria in low-phosphate soils. In: *Endomycorrhizas* (Ed. by F. E. Sanders, P. B. Tinker, & B. Mosse), pp. 408–417. Academic Press, London.

Barrett, V., Lemke, P. A., & Dixon, R. K. (1989). Protoplast formation from selected species of ectomycorrhizal fungi. *Applied Microbiology and Biotechnology* **30**, 381–387.

Bergersen, F. J. (1971). Biochemistry of symbiotic nitrogen fixation in legumes. *Annual Review of Plant Physiology* **22**, 121–140.

Biermann, B., & Linderman, R. G. (1983). Increased geranium growth using preplant inoculation with a mycorrhizal fungus. *Journal of the American Society of Horticultural Science* **108**, 972–976.

Chapman, S. R., & L. P. Carter. (1976). *Crop Production, Principles and Practices*. W. H. Freeman, San Francisco.

Cullen, D., & Leong S. (1986). Recent advances in the molecular genetics of industrial filamentous fungi. *Trends in Biotechnology* November, 285–288.

Cummings, B. A., & Wood, T. (1990). Use of RFLP's as a means of examining genetic relatedness in VAM fungi. *Abstracts of the Eighth NACOM*, Jackson, WY.

Dehne, H. W., & Backhaus, G. F. (1986). The use of vesicular-arbuscular mycorrhizal fungi in plant production. 1. Inoculum production. *Z. Pflanzenkrankheiten Pflanzenschutz* **93**, 415–424.

Elad, Y., Chet, I., Boyle, P., & Henis Y. (1983). Parasitism of *Trichoderma* spp. of *Rhizoctonia solani* and *Sclerotium rolfsii*—scanning electron microscipy and fluorescence microscopy. *Phytopathology* **73**, 85–88.

Elad, Y., Barak, R., & Chet, I. (1984). Parasitism of sclerotia of *Sclerotium rolfsii* by *Trichoderma harzianum*. *Soil Biology and Biochemistry* **16**, 381–386.

Elmes, R., & Mosse, B. (1980). Vesicular-arbuscular mycorrhiza: Nutrient film technique. *Rothamsted Experimental Station Annual Report* **1**, 188.

Ferguson, J. J., & Menge, J. A. (1986). Response of citrus seedlings to various field inoculation methods with *Glomus deserticola* in fumigated nursery soils. *Journal of the American Society of Horticultural Science* **111**, 288–292.

Ferguson, J. J., & Woodhead, S. H. (1982). Increase and maintenance of vesicular-arbuscular mycorrhizal fungi. In: *Methods and Principles of Mycorrhizal Research* (Ed. by N. C. Schenck), pp. 47–54. American Phytopathological Society, St. Paul, MN.

Fravel, D. R. (1988). Role of antibiosis in the biocontrol of plant diseases. *Annual Review of Phytopathology* **26**, 75–91.

Gibson, A. H. (1976). Limitation to nitrogen fixation in legumes. In: *Proceedings of the First International Symposium on Nitrogen Fixation, Vol. 2* (Ed. by W. E. Newton & C. J. Nyman), pp. 400–427. Washington State University Press, Pullman.

Govinda Rao, Y. S., Bagyaraj, D. J., & Rai, P. V. (1983). Selection of an efficient

VA mycorrhizal fungus for finger millet. I. Glasshouse screening. *Zenbralblatt für Mikrobiologie* **138**, 409.

Graham, J. H. (1986). Citrus mycorrhizae: Potential benefits and interactions with pathogens. *HortScience* **21**, 1302–1306.

Graham, J. H., Leonard, R. T., & Menge, J. A. (1981). Membrane-mediated decrease in root exudation responsible for phosphorus inhibition of vesicular-arbuscular mycorrhiza formation. *Plant Physiology* **68**, 548–552.

Haas, J. H., & Krikun, J. (1985). Efficacy of endomycorrhizal-fungus isolates tested at standardized inoculum potential levels and the inoculum quantities required for growth response of bell pepper. *New Phytologist* **100**, 613–621.

Hall, I. R. (1979). Soil pellets to introduce vesicular-arbuscular mycorrhizal fungi into soils. *Soil Biology and Biochemistry* **11**, 85–86.

Hayman, D. S., Morris, E. J., & Page, R. J. (1981) Methods for inoculating field crops with mycorrhizal fungi. *Annals of Applied Biology* **99**, 247–253.

Heckman, J. R., & Angle, J. S. (1987). Variation between soybean cultivars in vesicular-arbuscular mycorrhiza fungi colonization. *Agronomy Journal* **79**, 428–430.

Hepper, C. M. (1984). Isolation and culture of VA mycorrhizal (VAM) fungi. In: *VA Mycorrhiza* (Ed. by C. L. Powell and D. J. Bagyaraj), pp. 95–112. CRC Press, Boca Raton, Florida.

Hoepfner, E. F., Koch, B. L. & Covey, R. P. (1983). Enhancement of growth and phosphorus concentrations in apple seedlings by vesicular-arbuscular mycorrhizae. *Journal of the American Society of Horticultural Science* **108**, 207–209.

Homolka, L., Vyskocil, P., and Pilat, P. (1988). Use of protoplasts in the improvement of filamentous fungi. *Applied Microbiology and Biotechnology* **28**, 166–169.

Howeler, R. H., Asher, C. J., & Edwards, D. G. (1982). Establishment of an effective endomycorrhizal association on cassava in flowing solution culture and its effects on phosphorus nutrition. *New Phytologist* **90**, 229–238.

Hung, L. L. & Silvia, D. M. (1987). VAM inoculum production in aeroponic culture. In *Mycorrhizae in the Next Decade*. (Ed. by D. M. Silvia, L. L. Hung, and J. H. Graham) p. 272. University of Florida, Gainesville.

Hussey, R. S. & Roncadori, R. W. (1982). Vesicular-arbuscular mycorrhizae may limit nematode activity and improve plant growth. *Plant Disease* **66**, 9–14.

Johnson, C. R. & Crews, C. E. (1979). Survival of mycorrhizal plants in the landscape. *American Nurseryman* **150**, 15.

Klein, T. M., Wolf, E. D., Wu, R. & Sabford, J. C. (1987). High velocity microprojectiles for delivering nucleic acids into living cells. *Nature* **327**, 70–73.

Kleinschmidt, G. D. & Gerdemann, J. W. (1972). Stunting of citrus seedlings in fumigated nursery soils related to the absence of endomycorrhizae. *Phytophathology* **62**, 1447–1453.

Kloepper, J. W., Lifshitz, R. & Schroth, M. N. (1988). *Pseudomonas* inoculants to benefit plant production. *ISI Atlas of Science: Animal and Plant Sciences*.

Kloepper, J. W., Lifshitz, R. & Zablotowicz, R. M. (1989). Free-living bacterial inocula for enhancing crop productivity. *Tibtech* **7**, 39–44.

Koch, B. L., Covey, R. P. & Larsen, H. J. (1982). Response of apple seedlings in fumigated soil to phosphorus and vesicular-arbuscular mycorrhiza. *HortScience* **17**, 232–233.

Kohl, J. & Schlosser, E. (1989). Effect of two *Trichoderma* spp. on the infection of maize roots by vesicular-arbuscular mycorrhiza. *Zeitschrift fur Pflanzenkrankheiten und Pflanzenschutz* **96**, 439–443.

Kormanik, P. P., Schultz, R. C., & Bryan, W. C. (1982). The influence of vesicular-arbuscular mycorrhizae on the growth and development of eight hardwood tree species. *Forest Science* **28**, 531–539.

Krishna, K. R., Shetty, K. G., Dart, P. J., & Andrews, D. J. (1985). Genotype dependent variation in mycorrhizal colonization and response to inoculation of pearl millet. *Plant and Soil* **86**, 113–125.

Leong, J. (1986). Siderophores: Their biochemistry and possible role in the biocontrol of plant pathogens. *Annual Review of Phytopathology* **24**, 187–209.

Manjunath, A., Mohan, R., & Bagyaraj, D. J. (1981). Interactions between *Beijerinckia mobilis, Aspergillus niger* and *Glomus fasciculatus* and their effects on growth of onion. *New Phytologist* **87**, 723–727.

McLachlan, K. D., & Norman, B. W. (1961). Phosphorus and symbiotic nitrogen fixation in subterranean clover. *Journal of the Australian Institute of Agricultural Science* **27**, 244.

Menge, J. A. (1981). Mycorrhiza agriculture technologies. In: *Background Papers for Innovative Biological Technologies for Lesser Developed Countries* (Ed. by U.S. Congress, Office of Technology Assessment), pp. 383–424. U.S. Government Printing Office, Washington, D.C.

Menge, J. A. (1984). Inoculum production. In: *VA Mycorrhiza* (Ed. by C. L. Powell & D. J. Bagyaraj), pp. 187–203. CRC Press, Boca Raton, FL.

Menge, J. A., Lembright, H., & Johnson, E. L. V. (1977). Utilization of mycorrhizal fungi in citrus nurseries. *Proceedings of the International Society of Citriculture* **1**, 129–132.

Menge, J. A., Davis, R. M., Johnson, E. L. V., & Zentmyer, G. A. (1978a). Mycorrhizal fungi increase growth and reduce transplant injury in avocados. *California Agriculture* **32**, 6–7.

Menge, J. A., Johnson, E. L. V., & Platt, R. G. (1978b). Mycorrhizal dependency of several citrus cultivars under three nutrient regimes. *New Phytologist* **81**, 553–559.

Menge, J. A., Labanauskas, C. K., Johnson, E. L. V., & Platt, R. G. (1978c). Partial substitution of mycorrhizal fungi for phosphorus fertilization in the greenhouse culture of citrus. *Soil Science Society of America Journal* **42**, 926–930.

Menge, J. A., Jarrell, W. M., Labanauskas, C. K., Ojala, J. C., Huszar, C., Johnson, E. L. V., & Silbert, D. (1982). Predicting mycorrhizal dependency of Troyer citrange on *Glomus fasciculatus* in California citrus soils and nursery mixes. *Soil Science Society of America Journal* **46**, 762–768.

Menge, J. A., Raski, D. J., Lider, L. A., Johnson, E. L. V., Jones, N. O., Kissler, J. J., & Hemstreet, C. L. (1983). Interactions between mycorrhizal fungi, soil fumigation and growth of grapes in California. *American Journal of Enology and Viticulture* **34**, 117–121.

Meyer, J. R., & Linderman, R. G. (1986a). Selective influence on populations of rhizosphere or rhizoplane bacteria and actinomycetes by mycorrhizas formed by *Glomus fasciculatum*. *Soil Biology and Biochemistry* **18**, 191–196.

Meyer, J. R., & Linderman, R. G. (1986b). Response of subterranean clover to dual inoculation with vesicular-arbuscular mycorrhizal fungi and a plant growth-promoting bacterium, *Pseudomonas putida*. *Soil Biology and Biochemistry* **18**, 185–190.

Miller-Wideman, M. A., & Watrud, L. S. (1984). Sporulation of *Gigaspora margarita* on root cultures of tomato. *Canadian Journal of Microbiology* **30**, 642–646.

Millner, P. D. (1988). A minireview of biotechnology as applied to vesicular-arbuscular mycorrhizae. *Developments in Industrial Microbiology* **29**, 149–158.

Mosse, B., & Hepper, C. (1975). Vesicular-arbuscular mycorrhizal infections in root organ cultures. *Physiological Plant Pathology* **5**, 215–223.

Mugnier, J., & Mosse, B. (1987). Vesicular-arbuscular mycorrhizal infection in transformed root-inducing T-DNA roots grown axenically. *Phytopathology* **77**, 1045–1050.

Nelson, D. E., & Safir, G. R. (1982). Increased drought tolerance of mycorrhizal onion plants caused by improved phosphorus nutrition. *Planta* **154**, 407–413.

Ojala, J. C., Jarrell, W. M., Menge, J. A., & Johnson, E. L. V. (1983). Influence of mycorrhizal fungi on the mineral nutrition and yield of onion in saline soil. *Agronomy Journal* **75**, 255–259.

Papavizas, G. C. (1985). Trichoderma and Gliocladium: biology, ecology, and potential for biocontrol. *Annual Review of Phytopathology* **23**, 23–54.

Paulitz, T. C., & Linderman, R. G. (1990). Interactions between fluorescent pseudomonads and VA mycorrhizal fungi. *New Phytologist* **113**, 37–45.

Peberdy, J. F. (1987). Developments in protoplast fusion in fungi. *Microbiological Sciences* **4**, 108–111.

Powell, C. L. (1984). Field inoculation with VA mycorrhizal fungi. In: *VA Mycorrhiza* (Ed. by C. L. Powell & D. J. Bagyaraj), pp. 205–234. CRC Press, Boca Raton, FL.

Powell, C. L., Clark, G. E., & Vergerne, N.J. (1982). Growth response of four onion cultivars to several isolates of VA mycorrhizal fungi. *New Zealand Journal of Agricultural Research* **25**, 465–470.

Quigley, D. R., Jabri, E., & Selitrennikoff, C. P. (1985). Protoplasts of *Neurospora crassa*. *Experimental Mycologist* **9**, 254–258.

Quigley, D. R., Taft, C. S., Stark, T., & Selitrennikoff, C. P. (1987). Optimal conditions for the release of protoplasts of *Neurospora* using Novozym 234. *Experimental Mycology* **11**, 236–240.

Raj, J., Bagyaraj, D. J., & Manjunath, A. (1981). Influence of soil inoculation with vesicular-arbuscular mycorrhizas and a phosphate-dissolving bacterium on plant growth and 32 P uptake. *Soil Biology and Biochemistry* **13**, 105–108.

Ratnayake, M., Leonard, R. T., & Menge, J. A. (1978). Root exudation in relation to supply of phosphorus and its possible relevance to mycorrhizal formation. *New Phytologist* **81**, 543–552.

Sanford, J. C., Klein, T. M., Wolf, E. D., & Allen (1991). *Nuclear Particle Science Technology* (in press).

Sasa, M., Zahka, G., & Jakobsen, I. (1987). The effect of pretransplant inoculation with VA mycorrhizal fungi on the subsequent growth of leeks in the field. *Plant and Soil* **97**, 279–283.

Schenck, N. C., & Tucker, D. P. H. (1974). Endomycorrhizal fungi and the development of citrus seedlings in Florida fumigated soils. *Journal of the American Society of Horticultural Science* **99**, 284.

Schroth, M. N., & A. R. Weinhold. (1986). Root-colonizing bacteria and plant health. *HortScience* **21**, 1295–1298.

Schultz, R. C., Kormanik, P. P., & Bryan, W. C. (1981). Effects of fertilization and vesicular-arbuscular myucorrhizal inoculation on growth of hardwood seedlings. *Soil Science Society of America Journal* **45**, 961–965.

Schwab, S. M., Menge, J. A., & Leonard, R. T. (1983). Comparison of stages of vesicular-arbuscular mycorrhiza formation in sudangrass grown at two levels of phosphorus nutrition. *American Journal of Botany* **70**, 1225–1232.

Secilia, J., & Bagyaraj, D. J. (1987). Bacteria and actinomycetes associated with pot cultures of vesicular-arbuscular mycorrhizas. *Canadian Journal of Microbiology* **33**, 1069–1073.

Selitrennikoff, C. P., & Bloomfield, E. C. (1984). Formation and regeneration of protoplasts of wild type *Neurospora crassa*. *Current Microbiology* **11**, 113–118.

Sivan, A., Ucko, O., & Chet, I. (1987). Biological control of fusarium crown rot of tomato by *Trichoderma harzianum* under field conditions. *Plant Disease* **71**, 587–592.

Strobel, N. E., Hussey, R. S., & Roncadori, R. W. (1982). Interactions of vesicular-arbuscular mycorrhizal fungi, *Meloidogyne incognita,* and soil fertility on peach. *Phytopathology* **72**, 690–694.

Timmer, L. W., & Leyden, R. F. (1978). Stunting of citrus seedlings in fumigated soils in Texas and its correction by phosphorus fertilization and inoculation with mycorrhizal fungi. *Journal of the American Society of Horticulture* **103**, 533–537.

Toth, R., Page, T., & Castleberry, R. (1984). Differences in mycorrhizal colonization of maize selections for high and low ear leaf phosphorus. *Crop Science* **24**, 994–996.

Upshall, A. (1986). Filamentous fungi in biotechnology. *Biotechniques* **4**, 158–166.

Warner, A. (1985). Mycorrhizal seed pellets. U.S. Patent No. 4,551,165. U.S. Patent Office, Washington, D.C.

Weller, D. M. (1988). Biological control of soilborne plant pathogens in the rhizosphere with bacteria. *Annual Review of Phytopathology* **26**, 379–407.

Wood, T. E. (1991). VA mycorrhizal fungi: challenges for commercialization. In: *Hand-*

book of Applied Mycology, Vol. 4, Biotechnology (Ed. by D. K. Arora, K. G. Mukerji, & R. P. Elander). Marcel Dekker, New York, in press.

Yamamoto, F., & Furusawa, M. (1979). A simple microinjection technique not employing a micromanipulator. *Experimental Cell Research* **117,** 441–445.

Zimmerman, U. (1982). Electric field mediated fusion and related electrical phenomena. *Biochimica et Biophysica Acta* **694,** 227–277.

15

Mycorrhizae and the Integration of Scales: From Molecules to Ecosystems

Michael F. Allen, Steven D. Clouse, Barbara S. Weinbaum, Sherri L. Jeakins, Carl F. Friese, and Edith B. Allen

SUMMARY. Observations on successional patterns in semiarid habitats indicate that VA mycorrhizae play a major role in the composition of the plant communities, and they subsequently affect the various processes critical to forming a stable ecosystem. Our initial hypothesis was that the patterns could be explained simply by invoking the enhanced growth of mycotrophic species caused by mycorrhizal formation, and that mycorrhizae would increase competitiveness. Our research over the past 15 years has shown that as we studied each hierarchical level, a new set of factors appeared that played a major role in understanding processes at each level.

This chapter presents an overview of the research we have conducted during this period ranging from ecosystem restoration to molecular interactions between mycorrhizal fungi and host and nonhost plants. Included are assessments of nutrient cycling patterns, productivity, competition between plants, growth responses of plants to mycorrhizal fungi, production of specific compounds in specific organs that regulate phosphate availability, responses of individual tissues of plants to the invasion by mycorrhizal fungi, and, finally, mechanisms that trigger the recognition and regulation of the plant–fungus interaction.

We believe that explicit recognition of the differing scales of interaction between mycorrhizal fungi and plants determines the research questions posed to understanding how mycorrhizae affect processes important to both anthropogenic and natural landscapes.

Introduction

During the past 15 years, one of our major research interests has focused on the interactions between nonmycotrophic and mycotrophic plant species and how mycorrhizae may affect, via those species, the successional patterns in the arid steppes of western North America. We started with the simple premise that because mycorrhizae are mutualistic associations, in the presence of mycorrhizal fungi, the mycotrophic plants should be more competitive. In the absence of these fungi, the plants that fail to form these associations should be at a competitive disadvantage. Early studies (e.g., Schramm, 1966) suggested that disturbance reduces or eliminates mycorrhizal fungi from a site, and coastal successions exhibit a trend of increasing mycotrophy from the primary foredune through the stabilized dunes (e.g., Nicolson, 1960). Therefore, to understand the shifts in species and the dynamics of fungal–plant interactions, we should study the ecological importance of mycorrhizae. In particular, we focus on the vesicular-arbuscular (VA) mycorrhizal type that predominates in the systems we have studied.

Needless to say, as with most interesting research, the problem was not as simple as hypothesized. In the course of attempting to understand these relationships, we started with simple competition studies in reclaimed surface mines. These were advantageous in that there were no surviving mycorrhizal fungi, one of the few truly nonmycorrhizal sites, to which we could add inoculum to study the influence of the association. As we proceeded with our studies, the questions led us into areas all the way from assessing the molecular responses of individual plant cells to the invasion by VA mycorrhizal fungal hyphae, to effects of the fungi on ecosystem processes.

We therefore believe that our research represented a unique opportunity to link studies ranging from molecular biology to ecosystem processes in asking questions about a single problem. We present this chapter as an example of linking hierarchical levels to address the functioning of mycorrhizae in a single ecosystem. We believe that this approach has value in addressing many theoretical and applied problems in ecology as well as mycorrhizal biology.

Statement of Problem

Studies over the past 40 years have suggested that increased mycorrhizal activity correlates closely with succession. For example, Nicolson (1960) reported that *S. kali* colonizing sand dunes was nonmycotrophic, that is, did not form mycorrhizae, and was succeeded by VA mycorrhizal grasses in stabilized dunes. We (Allen & Allen, 1980) observed a similar trend in reclaimed surface mines in the high plains and cold desert regions of Wyoming. This combined with similar observations in other disturbances and other regions (e.g., Marx, 1975; Reeves et al., 1979; Janos, 1980; Miller, 1979) led us to hypothesize that in semiarid regions, VA mycorrhizae could be used as a management tool to hasten reclamation of disturbed arid lands (e.g., Fig. 15.1).

Succession is the process whereby lands change in the scale of years to decades following disturbance. The organisms that survive the disturbance and subsequently colonize affect all of the critical ecosystem processes from primary production to nutrient cycling and influence all land uses of a site. Management

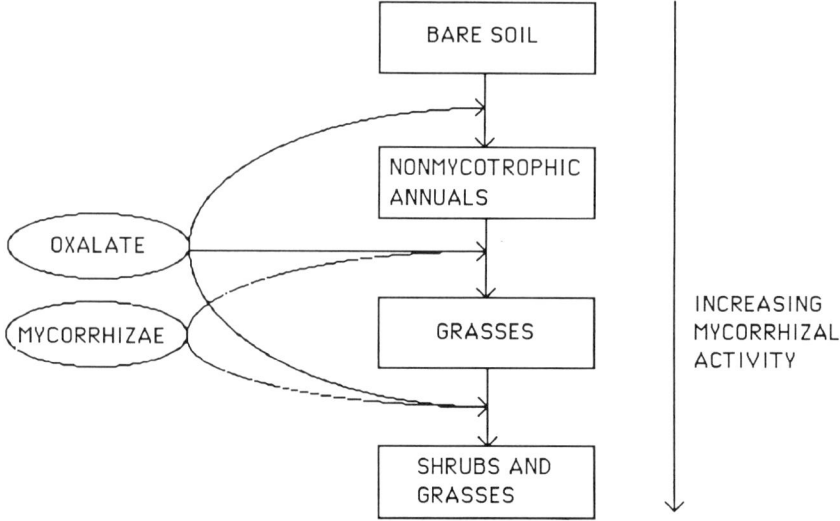

Figure 15.1. Model showing the possible influence of oxalate and VAM fungi on the rate of succession.

of lands is, in essence, a task that requires management of communities of plants to create a conservative set of ecosystem processes. Specifically, Odum (1969) outlined several ecosystem characteristics contrasting late versus early successional sites (Table 15.1). The mature successional characteristics represent those considered desirable in a "reclaimed" or restored system. In particular, a mature ecosystem is characterized by a nutrient conservative system in which nutrients are rapidly cycled between biotic parts and not available to leach out of the system; mycorrhizae are a critical part in that these fungi permeate the soil picking up nutrients and channelling them rapidly to the host plant. Vesicular-arbuscular (VA) mycorrhizal fungi act by forming intimate structures (arbuscules) within

Table 15.1. Some Characteristics of Developing and Mature Ecosystems[a]

Ecosystem attribute	Developing ecosystem	Mature ecosystem
Production/respiration	$>$ or <1	~ 1
Net community production	High	Low
Inorganic nutrients	Extrabiotic	Intrabiotic
Spatial heterogeneity	Poorly organized	Well organized
Mineral cycles	Open	Closed
Nutrient exchange rates	Rapid	Slow
Internal symbiosis	**Undeveloped**	**Developed**
Nutrient conservation	Poor	Good
Stability	Poor	Good

[a]Derived from Odum (1969).

individual plant root cortical cells. Resources are exchanged between symbionts leading to the enhanced growth of the plant via increased mineral or water uptake, and of the fungus via acquisition of energy in the form of carbohydrates; these interactions depend on the recognition between plant and fungus, a molecular process. Thus, it becomes apparent that a clearer understanding of the functioning of mycorrhizae in an ecosystem such as the semiarid shrub–grassland could best be understood by developing a hierarchical model that predicts the ecosystem reactions to the molecular interaction between plant and fungus (e.g., Allen, 1991). In this chapter, we attempt to show how we experimented at a range of scales to develop an understanding of why and when mycorrhizal fungi regulate the successional processes and, secondarily, how the symbiosis may be used as a management tool in restoring desirable communities that have conservative ecosystem properties.

Successional Patterns in a Semiarid Shrub–Steppe

Across a landscape, neither ecosystem processes nor community structure are uniform. Both natural and anthropogenic disturbance agents act to create patches. Interestingly, these patches also indicate that mycorrhizal associations act to regulate both ecosystem and community properties.

In southwestern Wyoming, the most obvious natural disturbances are small animal disturbances, the most numerous of which are ant disks created by harvester ants and badger diggings. Harvester ants create disturbances on the scale of 10 m^2 that are abandoned every 10 to 15 years (e.g., Friese, 1991). Interestingly, the successional patterns on these mounds appear to be restricted to mycotrophic species. Only grasses and shrubs are apparent on these mounds following abandonment. The ants concentrate mycorrhizal inoculum in the form of root clippings with very high concentrations of both hyphal and spore inoculum (Friese, 1991). Badgers were found preferentially to dig into shrub root systems containing high concentrations of spores and leave those diggings in the interspace areas. They also contained high mountain concentrations of mycorrhizal fungi. Plants colonizing both types of sites tend to be facultatively mycotrophic grasses and rarely can nonmycotrophic annuals be observed (M. F. Allen & C. F. Friese, unpublished observations).

Intermixed with these are anthropogenic disturbances ranging from hand-tilled plots to roadsides to large surface mines. In all of these sites, mycorrhizal inoculum is reduced and the dominant early colonizers are nonmycotrophic weeds, primarily members of the Chenopodiaceae and Brassicaceae (Reeves et al., 1979; Miller, 1979; Allen & Allen, 1980, 1990).

There can be two explanations for the different patterns: differential recruitment of plants or differential survival. Seed rain studies show that there are weed, grass, and shrub seeds dispersed onto both small disturbances in native areas and onto a large disturbed mined site (e.g., West & Durham, 1991). Also, during secondary succession, such as when topsoil is retained, weedy nonmycotrophic species are often present or more rapidly colonize than nonmycotrophic perennials. These nonmycotrophic species often persist at least 3 to 4 years during which the perennials must invade, compete, and establish, a formidable task. For this reason, in most managed areas, large disturbances are planted to perennial grasses and shrubs. Despite this planting, the weedy annuals still constitute a large fraction of the cover for several years following revegetation efforts (e.g., Waaland & Allen, 1987).

Thus, we can envision three major pathways for succession to proceed, to nonmycotrophic plants followed by other nonmycotrophic plants, nonmycotrophic plants followed by mycotrophic plants, or directly to mycotrophic plants. We postulated that mycorrhizae were a major player in the differences in the successional patterns and that by understanding the interactions between mycorrhizal fungi and the various plants, we could describe the mechanisms whereby these interactions regulated the vegetation patterns seen in these habitats.

Field Observations: Ecosystem Processes

Several observations at the ecosystem scale suggest that mycorrhizae represent a key difference between a weedy, unstable site where nutrients are predominantly abiotic in form, and a grass or shrub-dominated site that is relatively stable and organic nutrients prevail. Sites in cold deserts including spoil material in the Hanna Basin and Powder River Basin of Wyoming up to 10 years old were dominated by weeds such as *Salsola kali* and *Halogeton glomeratus*. These plants do not form VA mycorrhizae and mycorrhizal fungi were very low in numbers (e.g., Allen & Allen, 1980). Adjacent undisturbed sites dominated by *Artemisia tridentata, Atriplex gardneri, Agropyron smithii,* and *Bouteloua gracilis* were highly mycorrhizal (40–90% of the root length infected). In southwestern Wyoming, undisturbed sites with grasses and shrubs and reclaimed sites that were predominantly covered by grasses and shrubs contained mycorrhizae, whereas sites recently regraded that were dominated by weeds or abandoned sites with very low cover tended to have low mycorrhizal active (Waaland & Allen, 1987). Finally, plant cover was greater on east-facing slopes, where mycorrhizal activity was highest (Waaland & Allen, 1987) and was associated with greater immigration of inoculum (Warner, 1985; M. F. Allen, 1988a).

Nutrient cycling patterns also differ between the nonmycorrhizal early successional systems and the mycorrhizal later successional systems. In sites dominated by the nonmycotrophic annuals such as *S. kali* and *A. rosea,* the organic matter and N content of the soil is low and most of the phosphate is found in the bound inorganic fractions (Duce, 1987). The addition of mycorrhizae to the same soils increased soil respiration, resulting in the mineralization of P (Knight et al.,

1989) and the increased incorporation of P into plant tissue (Duce 1987). In the annual systems, approximately the same amount of N was lost from the system as wind-driven tumbleweed litter, as came onto the site in the precipitation annually (M. Allen, unpublished data). When individual legumes were planted, the N content of the soil did not increase despite the presence of mycorrhizae and *Rhizobium*. Interestingly, the presence of the mycorrhizae increased plant N and increased the flower and seed production of the legume *Hedysarum boreale*, but the increased litter and seed pods were apparently removed by wind (Carpenter & Allen, 1988). However, we predict that if these same plants were located in the wind shadow of a group of shrubs, that N would be retained by the system. Under the mycorrhizal shrubs, organic matter and N increased as the leaf litter remained under the existing canopy in shrubs that were large enough to breach the wind (M. Allen, unpublished data). In fact, the mycorrhizal fungi themselves contained as much as 5–10% of the N in the soil. These fungi as well as the saprophytic hyphal mass could turn over rapidly with a precipitation event, resulting in a rapid pool for plant growth (M. Allen, in preparation). Thus, mycorrhizal associations were present in sites where nutrients tended to be organically bound in living or dead tissue and to be retained through time. In sites dominated by nonmycotrophic species, nutrients were diffuse, inorganic, and subject to erosion.

The third major ecosystem process measured was productivity. Plant production was highly variable with or without mycorrhizae. *Salsola kali* can be highly productive and production was often reduced when mycorrhizal fungi were present (Allen & Allen, 1986). Nevertheless, in shrubs such as *Artemisia tridentata* that often show positive growth responses to mycorrhizae, production declined during years when the mycorrhizal fungi declined (Allen et al., 1987), and the mycorrhizal chenopodiaceous shrub, *A. gardneri* (M. F. Allen, 1983) virtually disappeared at the same time (Allen et al., 1987). These data suggest that mycorrhizae generally tend to increase production in late seral plants but not early ones and is supported experimentally (E. B. Allen, 1984; Allen & Allen, 1990).

Interactions among Plants: Community Processes

Shifts in ecosystem processes such as those described above are generally the result of changes in the species composition of a site. Remembering that a community is composed of plants, animals, and microorganisms, then shifting processes affecting any group can ramify through the entire group of organisms. Thus, the community influences the ecosystem processes and the ecosystem process changes further affect the community. Mycorrhizae represent a component that can simultaneously link plants as well as modify their competitive interactions (see Miller & Allen, Chapter 9, this volume). Therefore, the community level dynamics may be the crucial element in identifying the changes that mycorrhizae can impose on a site.

Mycorrhizal associations are formed by grasses and most forbs. Nonmycotrophic plants tend to be found in a limited number of families such as the Chenopodiaceae and Brassicaceae. This has been well documented for almost a century (Stahl, 1900). These observations were confirmed for several sites in the cold desert to shrub steppes, except that some perennial *Atriplex* species also formed mycorrhizae (e.g., Williams et al., 1976; Miller, 1979; Allen, 1983). In these regions, competition for water and possibly for nutrients between introduced chenopods and the native grasses could be intense, and could affect the rate of succession (Allen, 1982; Allen & Knight, 1984). As mycorrhizae improve water and nutrient acquisition by those plants capable of forming the symbiosis (e.g., Allen, 1984), we postulated that mycorrhizae act at the community level by enhancing the competitive ability of mycotrophic species and by shifting the resources from nonmycotrophic early successional annuals to late-seral mycorrhizal plants.

To test for the influence of mycorrhizae on competitive interaction between plants, a series of glasshouse experiments were conducted. A competition experiment between the nonmycotrophic weed *S. kali* and the mycotrophic grasses *Bouteloua gracilis* and *Agropyron smithii* (Allen & Allen, 1984) showed that the absence or presence of mycorrhizal fungi altered the competitive outcome. Specifically, when mycorrhizal fungi were added, the grasses had relatively greater mass whereas when absent, the *S. kali* grew larger. However, contrary to expectations, when mycorrhizal fungi were added to the *S. kali* alone, both mass and stomatal conductivity were reduced. This implied a direct interaction between the mycorrhizal fungi and this nonmycotrophic plant.

These experiments were followed with a series of field studies designed to study the community-level effects of mycorrhizal fungi on the transition between early seral, nonmycotrophic, and later seral mycotrophic plants. A series of 1/2 m^2 plots were inoculated at several locations across a recontoured surface mined site with little or no mycorrhizal inoculum. The addition of mycorrhizae improved the drought tolerance of the grasses by increasing stomatal conductivity during drought stress and delaying the plant phenology such that these perennial plants could continue to grow an additional 2–3 weeks into the year-end drought. No nutrient responses were found in these fertile sites. Moreover, the mycorrhizal response was enhanced in the face of competition from the nonmycotrophic annuals (Allen & Allen, 1986). However, on observing several sites, mycorrhizae only facilitated succession by reducing competition from weeds at one of the five test sites (E. B. Allen, 1989). At the harshest site, in fact, inoculation with mycorrhizal fungi reduced the density of the *S. kali* by 40% and its cover by 30%. At this site, *S. kali* was found to facilitate the establishment of the grasses and inhibitory effects of the mycorrhizal fungi subsequently reduced the grass density and cover (Allen & Allen, 1988).

The glasshouse and field data on community-level experiments corroborated one another. The addition of mycorrhizae affected the composition of the resulting plant species by enhancing the resource acquisition of the mycotrophic species. However, the fungi also reduced the growth and survival of the chenopodiaceous nonmycotrophs. Based on these results, we began a series of organismal-level studies designed to elucidate direct interactions between mycorrhizal fungi and some host plants of the shrub–steppe biome.

Responses of Plants to VA Mycorrhizal Fungi: Organismal Processes

Over the past 20 years, we and colleagues have studied the responses of several plant species from the Rocky Mountains and the Great Basin to VA mycorrhizae. These include both the growth and resource acquisition studies. We will summarize these here roughly by grouping them according to successional status by early, mid- and late seral groups. We recognize that there is considerable overlap in time and that succession can be arrested or the trajectory can be altered at any point depending on the local conditions (E. B. Allen, 1988).

Early Seral Nonmycotrophic Species

The primary colonizers following severe disturbances (enough to destroy mycorrhizal inoculum) in these regions are members of the Chenopodiaceae and Brassicaceae (e.g., Allen & Knight, 1984), families regarded as nonmycotrophic (Stahl, 1900; Newman & Reddell, 1987). The response of four of these species, *S. kali, Atriplex rosea, Halogeton glomeratus,* and *Brassica nigra,* have been studied. All four species showed either no response, or more commonly in the case of the nonmycotrophic species, a negative response to inoculation with the fungus (Allen, 1984; Benjamin & Allen, 1987; Allen & Allen, 1990, and unpublished data). This includes lowered stomatal conductivity, reduced nutrient uptake, and lower growth rates. Either cover (in field experiments) or mass (in glasshouse studies) could be reduced by as much as 40% with the addition of mycorrhizal fungi.

Other data showed that the addition of mycorrhizal fungi resulted in seedling mortality. Inoculated field plots had 30% lower plant densities than uninoculated plots. In addition, we observed that, on abandoned ant mounds containing high mycorrhizal inoculum, seedlings of these early seral species were generally absent, or, when present, were stunted (C. F. Friese & M. F. Allen, unpublished observations). When this inoculum was placed in pots in the glasshouse and seeded with *S. kali,* those plants in the nonsterile soil containing mycorrhizal inoculum had significantly greater wilting symptoms and greater mortality than in sterile soil (Table 15.2).

Together these data indicate that mycorrhizal fungi directly inhibit the growth of at least some of these plants: an antagonistic relationship may form between plants and fungi in severely disturbed lands, such as abandoned agricultural sites and surface mines. Where mycorrhizal inoculum failed to reestablish, these plants

Table 15.2. Effects of Mycorrhizal Inoculum on Salsola kali Seedlings (% of Germinated Seeds) after 1 Month

	Nonsterile soil	Sterile soil
Experiment 1		
Dead	3.6	1.7
Wilted	2.1	0.8
Experiment 2		
Dead	3.2	0.7
Wilted	4.5	6.0
Experiment 3		
Dead	23.0	12.3
Wilted	0.0	0.0

predominated for long periods (e.g., Allen & Allen, 1980; Waaland & Allen, 1987).

Early Seral Mycotrophic Plants

Following the initial colonization by chenopods and crucifers, a group of annual grasses is often observed to colonize. In the Great Basin and the High Plains east of the Rocky Mountains, cheatgrass, *Bromus tectorum,* shows this habit. In the Central Valley of California and in Southern California, *Avena barbata* and other annual grasses fill this role. Interestingly, both grasses are introduced from habitats where anthropogenic disturbance has been a component of the region for thousands of years and both species form mycorrhizae. However, to date, we have been unable to demonstrate any response of *B. tectorum,* positive or negative, to mycorrhizal fungi although the fungus clearly survives in association (Benjamin & Allen, 1987). *Avena barbata* shows a slight enhancement in leaf growth and seed set and some species of the fungi survive and reproduce readily (Nelson, 1991). These species probably could be categorized as forming a commensalistic (plant 0, fungus +) to a slight mutualistic (+, +) symbiosis. Importantly, when *A. barbata* and *B. tectorum* predominate they appear to select for a group of "weedy" mycorrhizal fungi (*Glomus geosporum, G. aggregatum,* and *G. occultum,* Nelson, 1991; Allen, Allen, & Friese, unpublished data) and, if they become predominant, can arrest succession resulting in a permanent "annualization" of the region (Billings, 1989).

Mid-Seral Plants

Perennial grasses tend to colonize a site following the annuals (Allen & Knight, 1984) or are planted to hasten the reclamation process. These species show a wide range of positive physiological responses to mycorrhizae that would enhance their long-term survival. However, they rarely have major growth responses. These species include various species of *Agropyron, Andropogon, Bouteloua gracilis, Stipa pulcra, S. comata,* and *Koelaria cristata.* Improved drought tolerance is a common mycorrhizal response coupled with delayed phenology, increased photosynthesis, increased tillering, and reduced leaf mortality in both glasshouse and field studies of many species (e.g. Allen, Allen & Stahl, 1984; Allen & Allen, 1986; Miller et al., 1987; Allen & Allen, unpublished data). The mycorrhizal fungi thrive in these associations often infecting over 80% of the root length.

Late Seral Plants

Late seral plants of shrub–grassland include such shrub species as *Artemisia tridentata, Chrysothamnus nauseosus,* and several species of *Atriplex.* These all appear to be facultatively mycorrhizal as are the mid-seral grass species, but inoculation experiments have shown them in general to have larger growth responses than the grasses. For instance, inoculated *Atriplex canascens* had a 20% biomass increase and 15% survival increase in field transplants (Aldon, 1975). Glasshouse seedlings of *Artemisia tridentata* and *Chrysothamnus nauseosus* had up to a 100% biomass increase, depending on the species of mycorrhizal fungus used (Lindsey, 1984). When some mycorrhizal fungal species were used, there was no biomass change, or even a decrease compared to the uninoculated plants. A field transplant experiment on populations of *Artemisia tridentata* from Reno, Nevada, and San Diego County, California, using different species of mycorrhizal fungi, showed that the different species of fungi caused different shrub growth rates (Fig. 15.2). Interestingly, the local fungi did not necessarily cause the largest growth in the local plant population. For instance, *Acaulospora elegans* inoculum from San Diego produced the greatest growth response in the Reno plants at the Reno site (Fig. 15.2A), while inoculum from Reno produced the greatest growth response in the San Diego plants at the Reno transplant site (Fig. 15.2B). These studies indicate that growth is not the most important factor in measuring the fitness of long-lived plants such as shrubs, and that we must measure such factors as survival and reproduction to understand the long-term response to mycorrhizal fungi.

Rhizosphere Nutrient Dynamics: Whole Root and Mycelial Reactions

The presence of mycorrhizal fungi in the rhizosphere greatly affects the nutrient cycling. Fungi produce enzymes and low-molecular-weight organic acids that interact with soil organic compounds resulting in the increased nutrient availability for symbiotic plants. Many fungi are capable of producing a variety of low-molecular-weight organic acids, including oxalic acid, citric acid, and malic acid (Sollins et al., 1981). These acids are capable of chemical weathering as well as mineral liberation (Jurinak et al., 1986). These processes are important in succes-

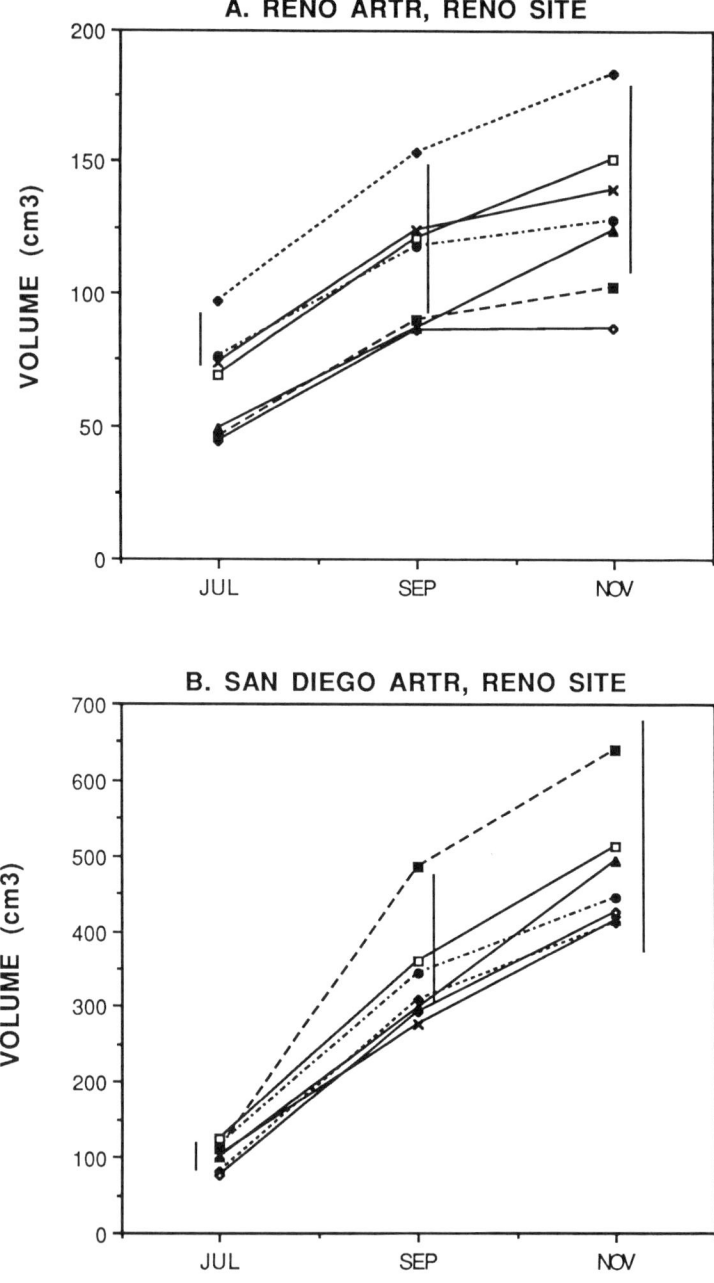

Figure 15.2. Shrub volume of *Artemisia tridentata* on three sample dates during 1990 in transplant gardens near Reno, Nevada and eastern San Diego County, California. Each of the two populations at both transplant sites was either nonmycorrhizal (NM) or inoculated with one of six mycorrhizal treatments: SCUT, *Scutellospora calospora* from Reno;

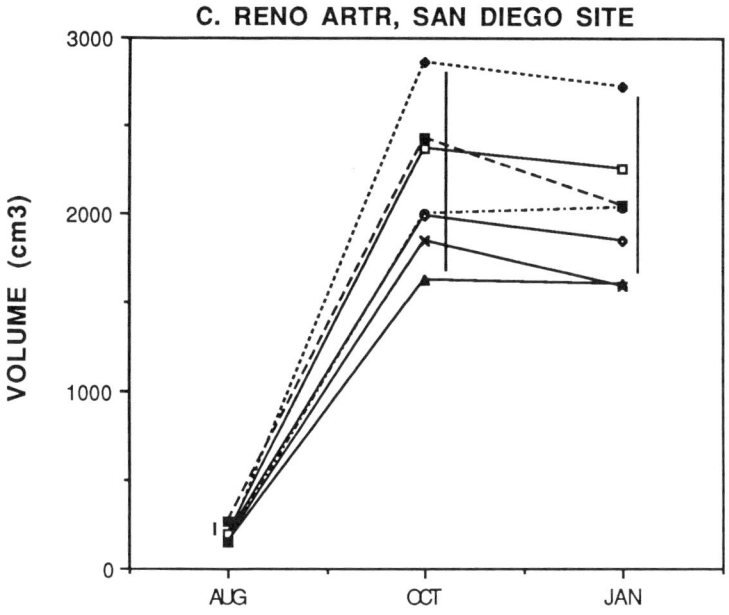

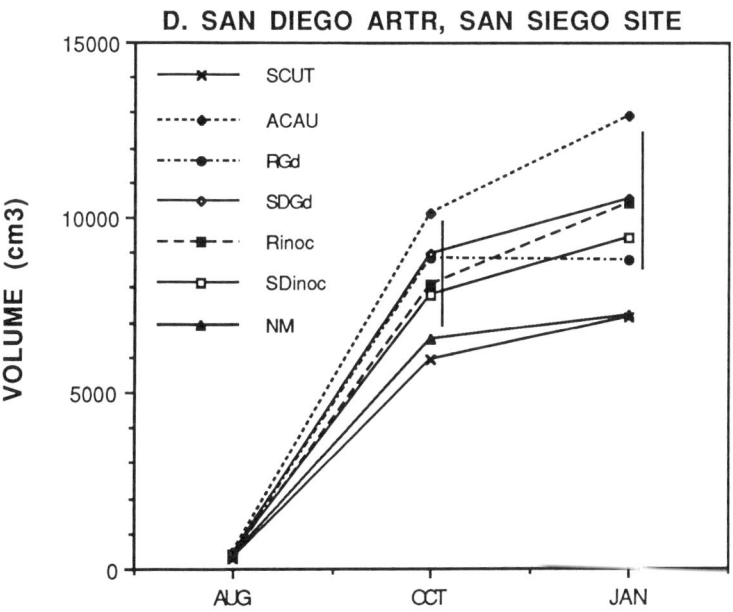

(*continued*) ACAU, *Acaulospora elegans* from San Diego; RGd, Reno *Glomus deserticola;* SDGd, San Diego *Glomus deserticola;* Rinoc, Reno whole soil inoculum; SDinoc, San Diego whole soil inoculum. The whole soil inocula included all the VA mycorrhizal species normally present at each field site. The other inocula were single species.

sion because returning the bound nutrients to forms that can be utilized by plants may affect the rate of succession.

During phosphorous cycling, for example, a disturbance will cause phosphates to become bound to inorganic ions such as Ca^{2+}. In this form they are unavailable to plants. Vegetation that reestablishes in the disturbed area must be able to survive with low phosphorus or have a mechanism to free up this phosphate and allow phosphorus cycling to continue. Two hypothesized mechanisms that free the phosphate are the production of oxalic acid as well as the increase in soil carbon dioxide level by mycorrhizal fungi (Knight et al., 1989). The carbon dioxide is responsible for liberating the bound phosphates from clay micelles. In the presence of aluminum, iron, or calcium ions, these phosphates will once again become bound and unavailable to plants. Oxalic acid has a higher affinity for these inorganic ions than the phosphates, and will therefore bind to the ions leaving the phosphorus free for plant use.

The role of oxalates with regard to classical phosphorus cycle models can be seen in Figure 15.3. The plant or mycorrhizal fungus produces the oxalate, the oxalate becomes bound to calcium aluminum or iron ions, releasing usable phosphates, and then the oxalate is degraded by soil microbes. In addition, mycorrhizal fungi produce increased root–fungal respiration as well as elevated CO_2 concentrations (Knight et al., 1989). This increases their effect on the phosphorous cycling due to the ability of CO_2 to weather phosphate minerals in calcareous soils. In this way the presence of mycorrhizal fungi can reestablish phosphorous cycling, which will also increase the rates at which plants can recolonize.

Plants as well as fungi produce oxalates, and interestingly a number of plant species that are nonmycotrophic produce high levels of oxalate as well. These include members of the Chenopodiaceae such as *Salsola kali* and *Halogeton glomeratus,* and members of the Polygonaceae and Oxalidaceae that have little or no VA mycorrhizal infection in the genera that have been observed (Harley & Harley, 1987). We hypothesize that these plants produce oxalate, rather than associate with VA mycorrhizal fungi, as a mechanism to take up soil P. Tissue levels of oxalate can be quite high, up to 7% in *S. kali* and even higher in *H. glomeratus* (Kingsbury, 1964; Hageman et al., 1988). Preliminary data show that soil oxalates were as high under *S. kali* (533 μg oxalate/g soil) as under mycorrhizal *Agropyron dasystachyum* (520 μg/g), while nonmycorrhizal rhizospere soil of *A. dasystachyum* had lower oxalate (437 μg/g) (L. Dudley, J. Jurinak, E. B. Allen, & M. F. Allen, unpublished data). Another line of preliminary evidence shows that bicarbonate extractable soil P under *S. kali* was twice as high as in soil with no plant growth (J. Cannon, unpublished data). However, in this latter study the concentration of soil oxalate was not higher under the *S. kali,* possibly due to rapid microbial degradation of oxalate (see section below).

We suggest that these oxalate responses at the rhizosphere level can be interpreted to produce results at the organismal and ecosystem succession level as depicted

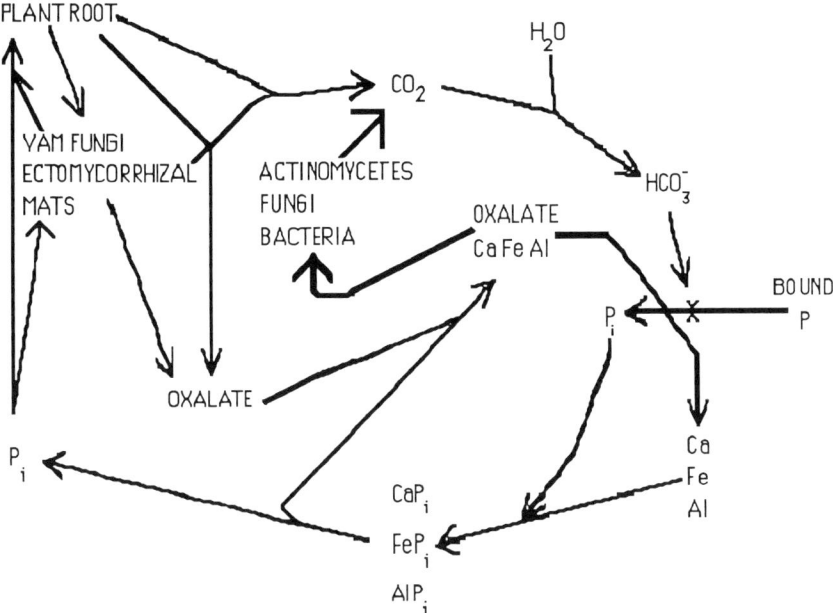

Figure 15.3. The proposed role of oxalates in rhizosphere phosphate cycling. Modified from Allen (1991).

in Figure 15.1. The earliest colonizers in disturbed shrub–grassland of the western United States include a large proportion of oxalate–producing, nonmycotrophic annuals (Allen & Knight, 1984), where they may persist for long time periods if the soil has little or no mycorrhizal inoculum (Allen & Allen, 1980). The production of oxalate by the annuals makes P available to later seral plants at a time when mycorrhizal inoculum is still low. The next seral stage is dominated by grasses, and begins to establish in soil that has available P. Next the mycorrhizal fungi can establish, as the facultatively mycorrhizal grasses provide a host for the fungus. Finally, the later seral shrubs establish. We hypothesize that the nonmycotrophic annual colonizers are actually "nurse plants" that facilitate the process of succession, at least in some situations because they produce oxalate to liberate soil P.

Individual Root Responses to Mycorrhizal Fungi: Tissue Responses

In an effort to better understand the initial responses of roots to infection by VA mycorrhizal fungi, we grew mycotrophic and nonmycotrophic seedlings in root observation chambers (Allen et al., 1989). The two 12 × 30-cm plates of glass were attached to one another by a layer of silicon adhesive around three sides of the perimeter, with the thickness of the silicon making a 5-mm gap between the plates, the thickness of the soil layer. Drainage holes were left through the silicone layer in the bottom (12 cm edge). The chambers were filled with inoculum soil collected from our Boden Field research site near Warner Springs, California. This soil contained several VA mycorrhizal fungal species within the genera *Acaulospora, Glomus,* and *Scutellospora*. Chambers were watered and allowed to stand for 7 days to initiate hyphal development in the soil. Overall, three replicates of each of 10 plant species were seeded into root observation chambers. The plant species tested in this study are listed in Table 15.3. The plates were covered with foil to keep the root systems in the dark. Chambers were tilted at a slight angle to allow some of the roots to grow along the lower plate where they could be observed (Finlay & Read, 1986). Plants were grown at 25°C under artificial light. Roots were scanned for mycorrhizal activity using autofluorescence as described by Ames et al. (1982), as modified by Allen et al. (1989). Specifically, root systems were scanned for autofluorescence using a Leitz Laborlux II microscope with a violet-blue excitation filter and a long pass suppression filter. In addition, roots were scanned for browning or other external reactions using a dissecting microscope.

Utilization of this root chamber method allowed us to demonstrate that the external hyphal architecture of VA mycorrhizal fungi is even more variable and complex within the soil matrix than previously thought. The complexity of these external hyphae appears to be based primarily on the unique function of each hyphal architectural type observed. Since internal spread of the fungus appears limited (e.g., Allen, 1982), the first major function of the external hyphae is to initiate root colonization. The diversity of architectural forms involved in the infection process was surprising (Friese & Allen, 1991). The structure of the infection unit near the root tended to vary depending on the origin of the hypha(e). Structural details of the infection unit produced by hyphae emerging from spores

Table 15.3. *Browning and Autofluorescence Responses of Various Plants to Invasion by VA Mycorrhizal Fungal Hyphae*

No invasion	*Beta vulgaris*
	Arabadopsis thaliana
	Brassica napus
Invasion: No fluorescence	*Agropyron dasystachum*
	Brassica nigra
Invasion: Fluorescence with slow browning	
No VAM structures	*Chenopodium album*
	Halogeton glomeratus
With VAM structures	*Artemisia tridentata*
Invasion: Fluorescence with rapid browning and root death	
	Salsola kali
	Atriplex rosea

and root fragments were most interesting. Single points of infection developing into spore-derived infection networks (SIN) are a striking illustration of the potential of an individual germ tube. This type of infection network was also observed by Gemma and Koske (1988) in sterile root organ culture experiments. Shortly after the fungal germ tube achieves its initial point of root penetration and infection, it responds by developing numerous secondary branches to initiate further points of infection. In this case, root contact in a soil system clearly triggers secondary branching in VA mycorrhizal fungal germ tubes. The second type of infection network seen entering roots were those found emerging from root fragments. Numerous hyphae were seen emerging from root fragments from the inoculum soil, and intertwining into unusual chord-like structures. Importantly, vesicles in the root fragments were apparently not required for initiating new infections as has been hypothesized for other systems (Bierman & Lindermann, 1983). On approaching a viable root, these hyphal chords began to unwind, resulting in a series of root-derived infection networks (RIN). Once again it appears that there is some recognition phenomenon by the mycorrhizal hyphae to the proximity of a root. At this time, however, it is not known why this distinctive chord-like architecture forms.

Once these different VA mycorrhizal hyphal architectures came into contact with the root, a wide range of responses was detected in relation to the plant species in question. These plant responses are summarized in Table 15.3. *Atriplex rosea* and *Salsola kali* roots rapidly changed from a healthy white translucence to a withered dark brown within the first 10 days of observation. Root browning in *A. rosea* and *S. kali* was preceded by VA mycorrhizal fungal hyphal penetration and subsequent autofluorescence. Vesicles and arbuscules were found in the brown segments of the roots, but not in healthy root segments. Often, *A. rosea* and *S. kali* roots were infected at several points along a single root. In *S. kali* 3 of 12 seedlings died, showing signs of wilt despite adequate soil moisture. Two of 8 *A. rosea* seedlings were lost, and the remaining plants showed signs of stress with the tips of leaves curling, turning brown, and occasional leaf loss.

Chenopodium album and *Halogeton glomeratus* roots began discoloring about the sixth day of observation. *C. album* roots changed to an overall shade of light brown. The major portions of the roots were discolored, with only the new roots remaining translucent. In *C. album,* root collapse, fluorescence, and vesicle and arbuscule formation were also minor occurrences. After 20 days, the tips of some leaves in *C. album* began to brown, but no leaf loss or plant mortality was seen. *H. glomeratus* discolored unevenly, with patches of dark browning along with light discoloring and healthy appearing roots. Brown and discolored roots occurred infrequently. Root collapse, fluorescence, vesicle formation, and arbuscule development were seen in moderate amounts. Plants appeared healthy throughout the observation period.

After 20 days no VA mycorrhizal hyphal penetration, or any of the characteristic browning reactions, had occurred in the Brassicaceae. The plates containing

Brassica napus and *Arabidopsis thaliana* were dismantled after the 20 days of inactivity. *B. nigra* and *B. vulgaris* were maintained for another 25 days. In the end, *B. vulgaris* was free from VA mycorrhizal fungi, showing no signs of autofluorescence or browning. *B. nigra* did become slightly infected by VA mycorrhizal fungi after 45 days. No browning or root collapse occurred, but examination under the fluorescence microscope revealed hyphae penetrating the cortical cells and the development of vesicles and arbuscules.

The grass *Agropyron dasystachyum* experienced no browning, root collapse, or autofluorescence when contacted by VA mycorrhizal fungi. After 20 days hyphae, arbuscules, and vesicles were present. With *Artemisia tridentata* hyphal penetration and arbuscule and vesicle development were slow. Over time, browning and autofluorescence set in, but no root collapse was observed.

Preliminary root chamber results demonstrate a wide range of plant responses to infection by VA mycorrhizal fungi. This range in plant response can be classified on the basis of a compatible or an incompatible reaction to fungal invasion. A compatible reaction can result in two possible outcomes, the first being normal mycorrhizal development and root expansion (as was seen in *Agropyron dasystachum*). A compatible reaction can also lead to a slow browning response as was seen in *Artemisia tridentata,* and result in the loss of any internal mycorrhizal structures. Incompatible root reactions to VA mycorrhizal fungal infection can lead to several possible outcomes, the most severe being death of the fungal endophyte. It is also possible that the plant could actively attempt to reject the fungus, but the mycorrhizae are still capable of limited spread within the root segment. A more typical incompatibility reaction is rapid root browning. This root browning can result in host root death (as was seen in *S. kali*), or survival of the root as seen in *Chenopodium album*. Thus, the degree to which the host rejects or accepts fungal invasion can possibly determine the fate (fitness and survival) of the plant, and the success of the fungus.

Molecular Responses by a Nonmycotrophic Plant to Mycorrhizal Fungi

There appears to be an interesting set of tissue reactions that characterizes the interactions of VA mycorrhizal fungi and both host and nonhost plants. Indeed, the death of individual seedlings and root segments of the *S. kali* indicated that it was not the fungal invasion per se that caused the tissue death but the reaction of individual cells in those tissues that turned brown and senesced when the hyphae were present. Consequently, we began our experiments to assess what the fundamental molecular reactions of the plant to the presence of the mycorrhizal fungi might be, and, in turn, how these reactions may control the basic formation of a mycorrhizal symbiosis. The following discussion presents our initial results and an overview of the reactions that may cause the observed responses.

Molecular and Biochemical Analysis of Inducible Enzymes in
Plant–Fungal Interactions

Plants respond to fungal infection by inducing the synthesis of a variety of enzymes including those involved in lignin and phytoalexin biosynthesis and lytic enzymes such as β-1,3-glucanase and chitinase. Perhaps the best characterized defense-related pathway is the phenylpropanoid pathway. This central pathway of plant secondary metabolism is responsible for the production of lignin, anthocyanins, signal molecules involved in *Agrobacterium* infection and *Rhizobium* symbiosis, and, in legumes, isoflavonoid phytoalexins. The phenylpropanoid biosynthetic pathway is an attractive model system for studies on gene regulation in host–microbe interactions since several of the genes code for enzymes that are rapidly and coordinately induced by microbial infection. Over the past 15 years, several key enzymes in this pathway have been purified to homogeneity, facilitating the production of antibodies to the purified proteins and isolation of cDNA and genomic clones for genes encoding these enzymes (Hahlbrock & Scheel, 1989). Recently, a variety of transgenic plants expressing these genes have been constructed. This section will briefly review assay conditions for inducible enzymes involved in lignification as a host response to fungal infection and list probes available for Northern blot analysis and *in situ* hybridization. The application of this technology to mycorrhizal studies will open many new experimental pathways to deepen understanding of the molecular mechanisms involved in the establishment of the plant–mycorrhizal interaction.

Control of Lignin Biosynthesis and Its Possible Relation to
VA Mycorrhizal Establishment

Lignin is universally associated with vascular tissue in higher plants (Grisebach, 1981) and is deposited on the secondary walls of tracheary elements late in the xylogenic process (Esau, 1965). Lignin is also deposited in nonvascular tissue in response to infection (Lewis & Yamamoto, 1990; Vance et al., 1980). Our results showing autofluorescence and browning responses of *S. kali* to VA mycorrhizae (see above) indicated that the deposition of lignin (or a related polyphenolic compound) may be important in this interaction as well. The complex three-dimensional structure of lignin is generated by peroxidase-mediated polymerization of the phenylpropanoid monomers 4-coumaryl alcohol, coniferyl alcohol, and sinapyl alcohol. Lignin monomers are derived via the phenylpropanoid pathway that begins with the *trans*-elimination of ammonia from L-phenylalanine by the enzyme phenylalanine ammonia lyase (PAL; EC 4.3.1.5) to form *trans*-cinnamic acid (Hahlbrock & Grisebach, 1979). Cinnamic acid is then converted by hydroxylation and *o*-methylation to 4-coumaric, ferulic, or sinapic acids, which are esterified to acetyl-CoA by 4-coumarate CoA ligase (4CL). This represents a central branchpoint in the pathway, one branch leading to anthocyanins, flavonoids, and isoflavonoids via chalcone synthase and chalcone isomerase, and the other specifically to lignin via cinnamyl-CoA-reductase (CCR) and cinnamyl alcohol dehydrogenase (CAD; EC 1.1.1.195). Thus, while PAL is the first enzyme in a pathway leading to a multitude of products, CAD is the terminal enzyme in the production of lignin-specific monomers.

Lignification has long been associated with active plant defense response to fungal infection (Hijwegen, 1963). Induced lignification appears to be particularly important as a resistance mechanism in the family Gramineae. Several experiments with *Puccinia graminis* f. sp. *tritici* (stem rust of wheat) have correlated lignification of hypersensitively reacting wheat cells to stem rust resistance. Hypersensitive cells fluoresce yellow under UV light and stain positively with phloroglucinol/HCl (Beardmore et al., 1983), indicating the presence of lignin-like polymers. Feeding radioactive lignin precursors results in incorporation of insoluble wall-bound phenolics in incompatible interactions (Fuchs et al., 1967). Furthermore, enzymes involved in lignin biosynthesis in wheat leaves (PAL, 4CL, CAD, and peroxidase) showed increased activity in incompatible interactions at the time of the resistance response to *P. graminis* f. sp. *tritici* while the same activities dropped in the compatible interaction (e.g., Moerschbacher et al., 1988). Another study on lignin biosynthetic enzymes has shown concomitant increases in peroxidase activity and lignin content of elicited cell suspension cultures of castor bean (Bruce & West, 1989). Because of the universal occurrence of lignin in xylem and the ease of assaying PAL activity, many studies have also attempted to correlate PAL activity with lignin biosynthesis and xylogenesis in culture (Fukuda & Komanine, 1982; Kuboi & Yamada, 1978; Rhodes et al., 1976; Rubery & Fosket, 1969).

As an initial attempt to study the molecular mechanisms of the interaction between *S. kali* and VA mycorrhizae, we measured PAL activity over time in roots of *S. kali* seedlings grown in a natural field soil containing *Glomus* and *Gigaspora* species and in the same soil autoclaved to kill the fungi. Figure 15.4 shows PAL activity (assayed as described in Bolwell, 1985) after various times of growth. Although there are obvious fluctuations in PAL activity in both control and inoculated roots, a steady increase in PAL activity in the inoculated roots is evident between 15 and 20 days. In all cases during this period, PAL activity is higher in inoculated than in control treatments. It is during this window of 15 to 20 days that the autofluorescence and root-browning observed in *S. kali* roots infected with VA mycorrhizae are generally seen. Fluctuations in PAL activity in control plants were also observed by Moerschbacher et al. (1990) in their study of lignin biosynthetic enzymes in wheat stem rust. Resistant and susceptible plants both showed a peak of PAL activity 1 to 2 days after inoculation with *P. graminis* f. sp. *tritici* uredospores, with PAL activity from the susceptible genotype exceeding that of the resistant plant. PAL activity then dropped in both genotypes followed by a rapid increase in PAL activity in the resistant genotype only, concomitant with the hypersensitive response in the resistant plants.

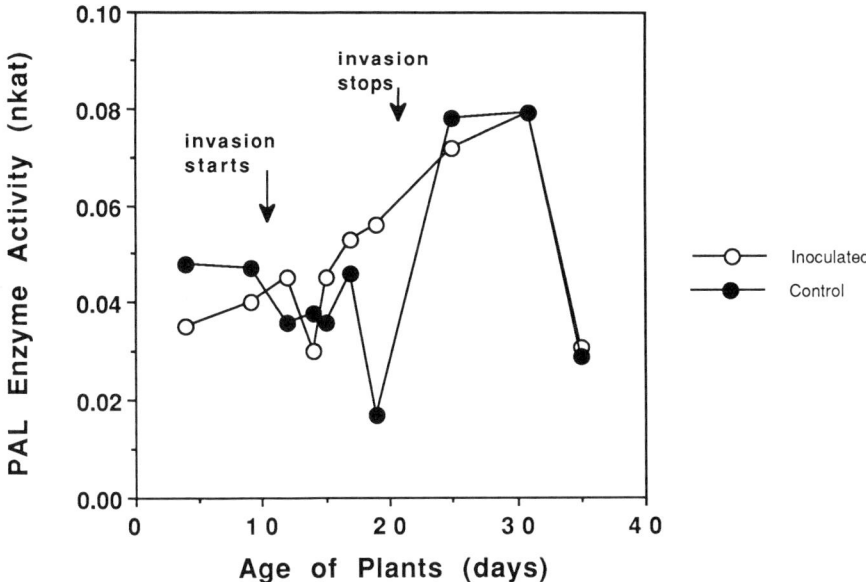

Figure 15.4. Time course of phenylalanine ammonia-lyase activity in *Salsola kali*.

Several aspects of our initial study of enzyme activity in *S. kali* are less than ideal. First, whole roots were ground for enzyme extraction while only a few layers of cells are autofluorescing in response to fungal infection. Thus, we are trying to measure a small change in activity against a high background of PAL activity in noninfected cells. Second, PAL is the first enzyme in a pathway leading to multiple products besides lignin and is inducible by wounding, light, and developmental signals in addition to fungal infection. Assaying the lignin-specific enzymes CCR and CAD may show a difference between the two treatments more clearly. Third, the autoclaved soil, while devoid of living inoculum, contained fungal wall breakdown products that in other systems have been shown to elicit the phenylpropanoid pathway. Because of this, we also assayed PAL activity at selected time points from *S. kali* plants grown in sterile sand and found in each case no significant difference in activity when compared to plants grown in autoclaved soil. Finally, the amount of inoculum in the soil may not have been optimal to give the most easily measurable PAL signal. We will repeat the experiment using single spore isolates of the fungi, employ a sterile synthetic soil mix, monitor autofluorescence with photomicrographs daily, and assay 4CL, CCR, CAD, and peroxidase in addition to PAL.

Even when enzyme data show a marked difference between resistant and susceptible plants, this provides only correlative evidence of the involvement of these enzymes in the establishment of an incompatible or compatible interaction. To establish a casual role, it is necessary to eliminate or inhibit the enzyme activity. Moerschbacher et al. (1990) used three competitive inhibitors of PAL and two highly specific irreversible suicide inhibitors of CAD to eliminate the activity of these enzymes in wheat plants resistant to stem rust. Treatment with any of these inhibitors leads to reduced lignification and increased fungal growth in the resistant plants, providing evidence for a casual link between lignification and disease resistance. The application of these inhibitors to the *S. kali* system should be informative in terms of the relationship between autofluorescence and the establishment of an incompatible interaction with VA mycorrhizae.

Localizing the Defense Response to Single Cells

The approach of assaying the activity of biosynthetic enzymes in the lignin pathway, while valuable, does not have the sensitivity to pinpoint the cell types that gives rise to the increased activity. The increased activity of these enzymes in response to fungal infection has been shown to be due, in large part, to *de novo* protein synthesis resulting from transcriptional activation of the genes encoding the enzymes. This fact, coupled with the availability of cloned sequences for many of the genes involved in the lignin pathway, will allow *in situ* RNA hybridization to be employed to visualize specific cell types that are inducing lignin synthesis in response to fungal attack. This method has been used successfully by Schmelzer et al. (1989) in a study of nonhost resistance of parsley to *Phytophythora megasperma* f. sp. *glycinea*. Radioactive RNA probes were synthesized from cDNAs for parsley PAL and 4CL, partially hydrolyzed to allow entry in tissue sections, and incubated with sections of parsley tissue inoculated with *P. megasperma* f. sp. *glycinea*. The induction of PAL and 4CL mRNA was shown to be sharply localized to the cells in and around the infection center that had undergone hypersensitive cell death in response to attempted infection. Another powerful approach for single cell localization is to use transgenic plants expressing chimeric constructs of defense gene promoters fused to a reporter gene such as β-glucuronidase (GUS). GUS activity can be assayed in a single cell by histochemical techniques (Jefferson, 1987). Liang et al. (1989) fused the promoter for bean PAL to GUS and moved the construct into tobacco by *Agrobacterium*-mediated transformation (Horsch et al., 1988). GUS activity (driven by the PAL promoter) was expressed in a tissue-specific manner in developing xylem and was wound inducible. Other defense gene promoters that have been transferred to tobacco include chitinase (Roby et al., 1990) and chalcone synthase (Schmid et al., 1990). The availability of such transgenic plants should be of immediate value in analyzing the interaction of both ecto- and endomycorrhizae with tobacco.

In summary, these preliminary results indicate that there is a molecular interaction between VA mycorrhizal fungi and nonhost plants that may trigger the observed reaction whereby the *S. kali* root segments senesce, and, in some cases, the entire seedling dies. These preliminary results, of course, should be followed by extensive experimentation using both mycotrophic and nonmycotrophic plants. Nevertheless, they represent an exciting challenge to understanding the fundamental nature of the mycorrhizal symbiosis and an exciting direction in plant science.

Synthesis: A Hierarchical Model of the Role of VA Mycorrhizae in Succession in Semiarid Shrub–Steppe

Observations on the successional patterns in semiarid habitats indicated that VA mycorrhizae play a major role in the composition of the plant communities that subsequently affected the various ecosystem processes critical to forming a stable ecosystem. Our initial hypothesis was that the pattern could be explained simply by invoking the enhanced growth of mycotrophic species caused by mycorrhizal formation that would increase competitiveness. Our subsequent research over the past 15 years have shown that as we studied each hierarchical level, a new set of factors appeared that generated new hypotheses, and played a major role in understanding this process. We have outlined the various stages in Figure 15.5.

Generally, mechanisms are assessed at one scale that cause patterns at the next higher scale (e.g., O'Neill et al., 1986). For instance, ecosystem processes are highly dependent on the community composition. This composition depends on the presence of mycorrhizal fungi and the growth responses of the plants to the fungus. Importantly, the growth of the plants depends on how individual tissues respond to individual invasions. These responses are dictated by the molecular interactions between the invasion of individual hyphae and the molecular recognition features that determine whether the interaction is compatible or incompatible (Fig. 15.5). This series represents a rather unique data set that directly relates processes of interest at the ecosystem level to molecular interactions occurring at the interface of two cells from different organisms.

We believe that explicit recognition of the differing scales of interaction between mycorrhizal fungi and plants is the research challenge to understanding how mycorrhizae affect processes important to both anthropogenic and natural landscapes. When these features are better understood at a wide range of scales, we will be better able to manage lands for improved resource conservation. Clearly, our attempts, along with others working in the cold desert regions of western North America, have been to understand the mechanisms of succession as a means not only of advancing the theories of our pet organisms (we all love fungi), but also to minimize the environmental damage to this susceptible landscape and to improve our ability to restore those lands disturbed to provide natural resources. It is only with this basic understanding that human and environmental needs will be met.

Acknowledgments

We thank Allen Hall, Joe Cannon, and Leslie Hickson for assistance with some of the previously unpublished experiments reported here. Various stages of the

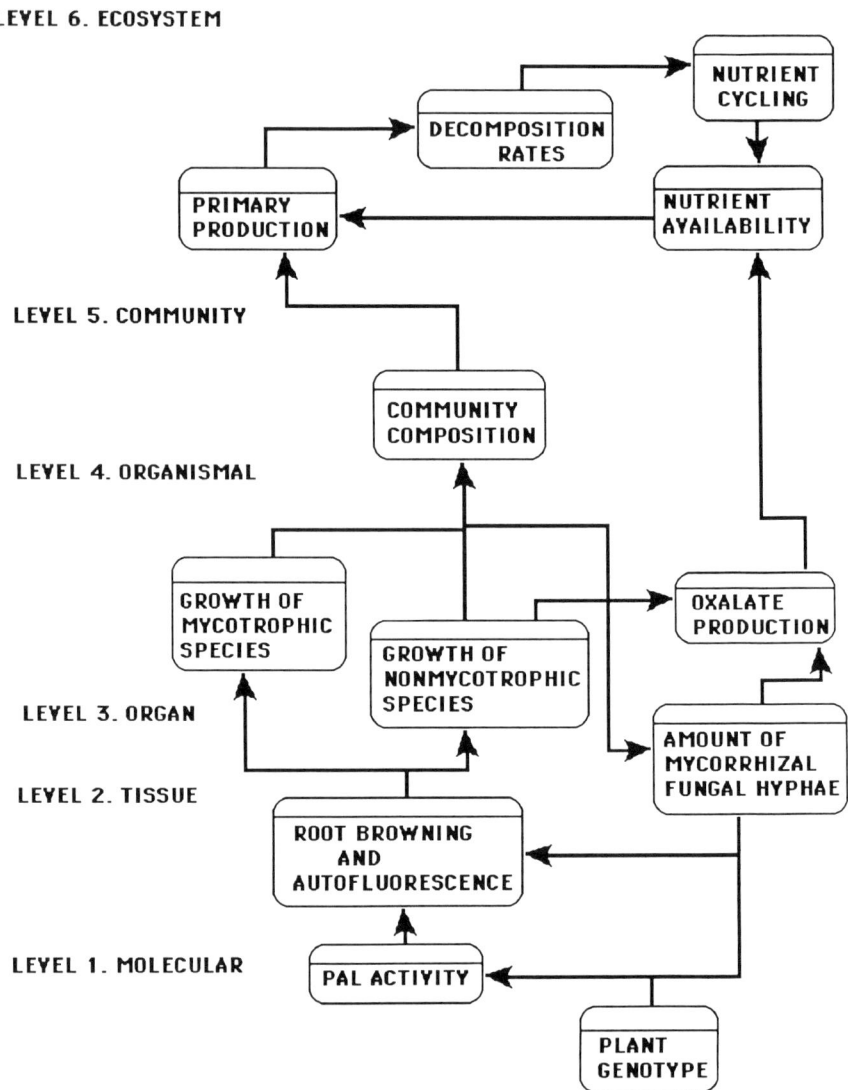

Figure 15.5. Hierarchical linkages of reactions among VA mycorrhizal fungi, nonmycotrophic annual plants, and mycotrophic plants ranging from molecular interactions to the processes regulating succession in a western semiarid shrub–steppe ecosystem.

research were supported by grants from the United States Department of Agriculture Competitive Grants Program and the National Science Foundation Ecology Program and Ecosystems Program.

References

Aldon, E. F. (1975). Endomycorrhizae enhance survival and growth of fourwing saltbush on coalmine spoils. *USDA Forest Service Research Note R.M.* **294**.

Allen, E. B. (1984). VA mycorrhizae and colonizing annuals: Implications for growth, competition, and succession. In: *VA Mycorrhizae and Reclamation of Arid and Semiarid Lands* (Ed. by M. F. Allen & S. E. Williams), pp. 41–51. University of Wyoming Agricultural Experiment Station Science Report SA1261, Laramie, WY.

Allen, E. B. (1988). Some trajectories of succession in Wyoming sagebrush-grassland: Implications for restoration. In: *The Reconstruction of Disturbed Arid lands. An Ecological Approach.* (Ed. by E. B. Allen), pp. 89–112. Westview Press, Boulder, CO.

Allen, E. B. (1989). The restoration of disturbed arid landscapes with special reference to mycorrhizal fungi. *Journal of Arid Environments* **17**, 279–286.

Allen, E. B., & Allen, M. F. (1980). Natural re-establishment of vesicular-arbuscular mycorrhizae following stripmine reclamation in Wyoming. *Journal of Applied Ecology* **17**, 139–147.

Allen, E. B., & Allen, M. F. (1984). Competition between plants of different successional stages: Mycorrhizae as regulators. *Canadian Journal of Botany* **62**, 2625–2629.

Allen, E. B., & Allen, M. F. (1986). Water relations of xeric grasses in the field: Interactions of mycorrhizae and competition. *New Phytologist* **104**, 559–571.

Allen, E. B., & Allen, M. F. (1988). Facilitation of succession by the nonmycotrophic colonizer *Salsola kali* (Chenopodiaceae) on a harsh site: Effects of mycorrhizal fungi. *American Journal of Botany* **75**, 257–266.

Allen, E. B., & Allen, M. F. (1990). The mediation of competition by mycorrhizae in successional and patchy environments. In: *Perspectives on Plant Competition* (Ed. by J. B. Grace & G. D. Tilman), pp. 367–389. Academic Press, New York.

Allen, E. B., & D. H. Knight. (1984). The effects of introduced annuals on secondary succession in sagebrush grassland, Wyoming. *The Southwestern Naturalist* **29**, 407–421.

Allen, M. F. (1982). Influences of vesicular-arbuscular mycorrhizae on water movement through *Bouteloua gracilis* (H.B.K.) Lag ex Steud. *New Phytologist* **91**, 191–196.

Allen, M. F. (1983). Formation of vesicular-arbuscular mycorrhizae in *Atriplex gardneri* (Chenopodiaceae): Seasonal response in a cold desert. *Mycologia* **75**, 773–776.

Allen, M. F. (1988). Re-establishment VA of mycorrhizae following severe disturbance: Comparative patch dynamics of a shrub desert and a subalpine vulcano. *Proceedings of the Royal Society of Edinburgh* **94B**, 63–71.

Allen, M. F. (1991). *The Ecology of Mycorrhizae*. Cambridge University Press, New York.

Allen, M. F., & E. B. Allen. (1990). Carbon source of VA mycorrhizal fungi associated with Chenopodiaceae from a semi-arid steppe. *Ecology* **71**, 2019–2021.

Allen, M. F., Allen, E. B., & Stahl, P. D. (1984). Differential niche response of *Bouteloua gracilis* and *Pascopyrum smithii* to VA mycorrhizae. *Bulletin of the Torrey Botanical Club* **111**, 361–325.

Allen, M. F., Allen, E. B., & West, N. E. (1987). Influence of parasitic and mutualistic fungi on *Artemisia tridentata* during high precipitation years. *Bulletin of the Torrey Botanical Club* **114**, 272–279.

Allen, M. F., Allen, E. B., & Friese, C. F. (1989). Responses of the non-mycotrophic plant *Salsola kali* to invasion by vesicular-arbuscular mycorrhizal fungi. *New Phytologist* **111**, 45–49.

Ames, R. N., Ingham, E. R., & Reid, C. P. P. (1982). Ultraviolet-induced autofluorescence of arbuscular mycorrhizal root infections: An alternative to clearing and staining methods for assessing infections. *Canadian Journal of Microbiology* **28**, 351–355.

Beardmore, J., Ride, J. P., & Granger, J. W. (1983). Cellular lignification as a factor in the hypersensitive resistance of wheat to stem rust. *Physiological Plant Pathology* **22**, 209–220.

Benjamin, P. K., & Allen, E. B. (1987). The influence of VA mycorrhizal fungi on competition between plants of different successional stages in sagebrush-grassland. In: *Mycorrhizae in the Next Decade, Practical Applications and Research Priorities* (Ed. by D. M. Sylvia, L. L. Hung, & J. H. Graham), p. 144. IFAS, University of Florida, Gainesville, FL.

Biermann, B., & Lindermann, R. G. (1983). Use of vesicular-arbuscular mycorrhizal roots, intraradical vesicles and extraradical vesicles as inoculum. *New Phytologist* **95**, 97–105.

Billings, D. W. (1989). Bromus tectorum, a biotic cause of ecosystem impoverishment in the Great Basin. In: *The Earth in Transition. Patterns and Processes of Biotic Impoverishment* (Ed. by G. M. Woodwell), pp. 301–322. Cambridge University Press, Cambridge.

Bolwell, G. P. (1985). Tissue cultures for studies on vascular differentiation. In: *Plant Cell Culture: A Practical Approach* (Ed. by R. A. Dixon), pp. 107–126. IRL Press, Oxford.

Bruce, R. J., & West, C. A. (1989). Elicitation of lignin biosynthesis and isoperoxidase activity by pectin fragments in suspension cultures of castor bean. *Plant Physiology* **91**, 889–897.

Carpenter, A. T., & Allen, M. F. (1988). Responses of *Hedysarum boreale* to mycorrhizas and *Rhizobium* plant and soil nutrient changes. *New Phytologist* **109**, 125–132.

Duce, D. H. (1987). Effects of vesicular-arbuscular mycorrhizae on *Agropyron smithii* grown under drought stress and their influence on organic phosphate mineralization. M.S. Thesis, Logan, UT.

Esau, K. (1965). *Vascular Differentiation in Plants*. Holt, Rinehart & Winston, New York.

Finlay, R. D., & Read, D. J. (1986). The structure and function of the vegetative mycelium of ectomycorrhizal plants. I. Translocation of 14C-labelled carbon between plants interconnected by a common mycelium. *New Phytologist* **103**, 143–156.

Friese, C. F. (1991). Ph.D. Dissertation, Utah State University, Logan, UT.

Friese, C. F., & Allen, M. F. (1991). The spread of VA mycorrhizal fungal hyphae in the soil: Inoculum type and external hyphal architecture. *Mycologia* **83**, 409–418.

Fuchs, A., Rohringer, R., & Samborski, D. J. (1967). Metabolism of aromatic compounds in healthy and rust-infected primary leaves of wheat. *Canadian Journal of Botany* **4**, 2137–2154.

Fukuda, H., & Komanine, A. (1982). Lignin synthesis and its related enzymes as markers of tracheary-element differentiation in single cells isolated from the mesophyll of *Zinnia elegans*. *Planta* **155**, 423–430.

Gemma, J. N., & Koske, R. E. (1988). Pre-infection interactions between roots and the mycorrhizal fungus *Gigaspora gigantea:* Chemotropism of germ-tubes and root growth response. *Transactions of the British Mycological Society* **91**, 123–132.

Grisebach, H. (1981). Lignins. In: *The Biochemistry of Plants*, Vol. 7 (Ed. by E. E. Conn), pp. 457–478. Academic Press, New York.

Hageman, J. H., Fowler, J. L., Suzikida, M., Salas, V., & LeCaptain, R. (1988). Analysis of Russina thistle (*Salsola* species) selections for factors affecting forage nutritional value. *Journal of Range Management* **41**, 155–158.

Hahlbrock, K., & Grisebach. H. (1979). Enzymic controls in the biosynthesis of lignin and flavonoids. *Annual Review of Plant Physiology* **30**, 105–130.

Hahlbrock, H., & Scheel, D. (1989). Physiology and molecular biology of phenylpropanoid metabolism. *Annual Review of Plant Physiology and Molecular Biology* **40**, 347–369.

Harley, J. L., & Harley, E. L. (1987). A checklist of mycorrhiza in the British flora. *New Phytologist* **105**, 1–102.

Hijwegen, T. (1963). Lignification, a possible mechanism of active resistance against pathogens. *Netherlands Journal of Plant Pathology* **69**, 314–317.

Horsch, R. B., Fry, J., Hofmann, N., Neidermeyer, J., Rogers, S. G., & Fraley, R. T. (1988). Leaf disc transformation. In: *Plant Molecular Biology Manual* (Ed. by S. B. Gelvin & R. A. Schilperoort), pp. A5/1–A5/9. Kluwer Press, Boston.

Janos, D. P. (1980). Mycorrhizae influence tropical succession. *Biotropica* **12**, 56–64.

Jefferson, R. A. (1987). Assaying chimeric genes in plants: The GUS gene fusion system. *Plant Molecular Biology Reports* **5**, 387–405.

Jurinak, J. J., Dudley, L. M., Allen, M. F., & Knight, W. G. (1986). The role of calcium oxalate in the availability of phosphorus in soils of semiarid regions: A thermodynamic study. *Soil Science* **142**, 255–261.

Kingsbury, J. M. (1964). *Poisonous Plants of the United States and Canada*. Prentice-Hall, Englewood Cliffs, NJ.

Knight, W. G., Allen, M. F., Jurinak, J. J., & Dudley, L. M. (1989). Elevated CO_2 and solution phosphorus in soil with vesicular-arbuscular mycorrhizal *Agropyron smithii* Rydb. *Soil Science Society of America Journal* **53**, 1075–1082.

Kuboi, T., & Yamada, Y. (1978). Regulation of the enzyme activities related to lignin synthesis in cell aggregates of tobacco cell culture. *Biochimica Biophysica Acta* **542**, 181–190.

Lewis, N. G., & Yamamoto, E. (1990). Lignin: Occurrence, biogenesis and biodegradation. *Annual Review of Plant Physiology and Plant Molecular Biology* **41**, 455–496.

Liang, X., Dron, M., Schmid, J., Dixon, R. A., & Lamb, C. J. (1989). Developmental and environmental regulation of a phenylalanine ammonia-lyase-β-glucuronidase gene fusion in transgenic tobacco plants. *Proceedings of the National Academy of Sciences U.S.A.* **86**, 9284–9288.

Lindsey, D. L. (1984). The role of vesicular-arbuscular mycorrhizae in shrub establishment. In: *VA Mycorrhizae and Reclamation of Arid and Semiarid Lands* (Ed. by S. E. Williams & M. F. Allen), pp. 53–68. University of Wyoming Agricultural Experiment Station, Laramie, WY.

Marx, D. H. (1975). Mycorrhizae and establishment of trees on strip mined lands. *The Ohio Journal of Science* **75**, 288–297.

Miller, R. M. (1979). Some occurrences of vesicular-arbuscular mycorrhiza in natural and disturbed ecosystems of the Red Desert. *Canadian Journal of Botany* **57**, 619–623.

Miller, R. M., Jarstfer, A. G. & Pilai, J. K. (1987). Biomass allocation in an *Agropyron smithii-Glomus* symbiosis. *American Journal of Botany* **74**, 114–122.

Moerschbacher, B. M., Noll, U., Berenike, E., & Reisener, H. J. (1988). Lignin biosynthetic enzymes in stem rust infected, resistant and susceptible near-isogenic wheat lines. *Physiological and Molecular Plant Pathology* **33**, 33–46.

Moerschbacher, B. M., Noll, U., Gorrichon, L., & Reisener, H. J. (1990). Specific inhibition of lignification breaks hypersensitive resistance of wheat to stem rust. *Plant Physiology* **93**, 465–470.

Nelson, L. L. (1991). Restoration of *Stipa pulchra* grasslands: Effects of mycorrhizae and competition from *Avena barbata*. M.S. Thesis. San Diego State University, San Diego, CA.

Newman, E. I., & Reddell, P. (1987). The distribution of mycorrhizas among families of vascular plants. *New Phytologist* **106**, 745–751.

Nicolson, T. H. (1960). Mycorrhizae in the Gramineae. II. Development in different habitats particularly sand dunes. *Transactions of the British Mycological Society* **43**, 132–145.

O'Neill, R. V., DeAngelis, D. L., Waide, J. B., & Allen, T. F. H. (1986). *A Hierarchical Concept of Ecosystems*. Princeton University Press, Princeton, NJ.

Odum, E. P. (1969). The strategy of ecosystem development. *Science* **164**, 262–270.

Reeves, F. B., Wagner, D. W., Moorman, T., & Kiel, J. (1979). The role of endomycorrhizae in revegetation practices in the semi-arid west. I. A comparison of incidence of mycorrhizae in severly disturbed vs. natural environments. *American Journal of Botany* **66**, 1–13.

Rhodes, M. J., Hill, C. R., & Wooltorton, L. S. (1976). Activity of enzymes involved in lignin biosynthesis in swede root disks. *Phytochemistry* **15**, 707–710.

Roby, D., Broglie, K., Cressman, R., Biddle, P., Shet, I., & Broglie, R. (1990). Activation of a bean chitinase promoter in transgenic tobacco plants by phytopathogenic fungi. *Plant Cell* **2,** 999–1007.

Rubery, P. H., & Fosket, D. E. (1969). Changes in phenylalanine ammonia-lyase activity during xylem differentiation in *Coleus* and soybean. *Planta* **87,** 54–62.

Schmid, J., Doerner, P., Clouse, S. D., Dixon, R. A., & Lamb, C. J. (1990). Developmental and environmental regulation of a bean chalcone synthase promoter in transgenic tobacco. *Plant Cell* **2,** 619–631.

Schmelzer, E., Kruger-Lebus, S. & Halbrock, K. (1989). Temporal and spatial patterns of gene expression around sites of attempted fungal infection in parsley leaves. *Plant Cell* **1,** 993–1001.

Schramm, J. E. (1966). Plant colonization studies on black wastes from anthracite mining in Pennsylvania. *American Philosophical Society* **56,** 1–94.

Sollins, P., Cromack, K. Jr., Fogel, R., & Li, C. Y. (1981). Role of low-molecular-weight organic acids in the inorganic nutrition of fungi and higher plants. In: *The Fungal Community. Its Organization and Role in the Ecosystem* (Ed. by D. T. Wicklow & G. C. Carroll), pp. 607–619. Marcel Dekker, New York.

Stahl, E. (1900). Der Sinn der mycorrhizenbildung. *Jahrbuecher fur Wissenschaftliche Botanik* **34,** 539–668.

Vance, C. P., Kirk, T. K., & Sherwood, R. T. (1980). Lignification as a mechanism of disease resistance. *Annual Review of Phytopathology* **18,** 259–288.

Waaland, M. E., & Allen, E. B. (1987). Relationships between VA mycorrhizal fungi and plant cover following surface mining in Wyoming. *Journal of Range Management* **40,** 271–276.

Warner, N. J. (1985). M.S. Thesis. Utah State University, Logan, UT.

West, N. E., & Durham, S. (1991). Seed rain on and near a coal stripmine in southwestern Wyoming. In: *Proceeding of the Annual Meeting of the American Society for Surface Mining and Reclamation* (Ed. by W. Oaks & J. Bowden), pp. 573–581.

Williams, S. E., Wollum, A. G., & Aldon, E. F. (1974). Growth of *Atriplex canescens* improved by formation of vesicular-arbuscular mycorrhizae. *Proceedings of the Soil Science Society of America* **38,** 962–965.

Organism Index

Abies 135, 387, 389, 409
Abies amabilis 146, 431
Abies grandis 392, 409
Abies lasiocarpa 392
Abies procera 392
Acarina 334, 335, 338
Acaulospora 166, 172, 502
Acaulospora elegans 497, 499
Acaulospora laevis 15, 18, 204, 211,
    216, 217, 221, 226, 228
Acaulospora trappei 217
Adenostoma 320
Afzelia 409
Agaricus bisporus 265
Agrobacterium 474, 505, 508
Agrobacterium rhizogenes 43
Agrobacterium tumefaciens 40, 44
Agropyron 496
Agropyron dasystachyum 318, 322, 500,
    502, 504
Agropyron smithii 314, 317, 492, 494
Aizoaceae 360
alder (= Alnus)
alfalfa (= Medicago sativa)
Allium cepa 83, 88, 89, 138, 144
Allium porrum 21, 44, 46, 48, 51, 79,
    84, 85, 87–89, 91, 93, 138, 139,
    220, 221, 344, 345, 346
Allomyces macrogynus 22
Alnus 363, 372, 389, 399, 402
Alnus crispa 69, 76
Alnus rubra 315–317, 389, 390, 401

Alpova 71, 386
Alpova diplophloeus 69, 76, 317
Amanita 105, 107, 262, 278, 279, 283,
    389, 409
Amanita fulva, 264, 265
Amanita muscaria 51, 262, 264, 270,
    271, 274, 275, 278, 283, 284,
    288, 314, 388, 406, 407
Amanita rubescens 264, 265
Amaranthaceae 360
Ambrosia artemisiifolia 317
Amherstieae 173
Amillaria mellia 277
Ammophila breviligulata 450–452
Amphimena 263
Anabaena 173, 177
Andropogon 497
Andropogon gerardii 318
ant 340
Aphelenchusavenae 343
aphids 340
apple 118
Arabadopsis thaliana 502, 504
Arabis hirsuta 308
Arbutus 380, 383, 391, 392, 403
Arbutus menziesii 391, 392, 408, 409
Arbutus unedo 392, 393
Arctostaphylos 320, 380, 383, 391, 392,
    403, 408
Arctostaphylos uva-ursi 391, 408
Arctostaphylus viscida 408
Armillaria mellea 136, 394

Artemisia tridentata 120, 492, 493, 497, 498, 502, 504
Ascomycete 137
Ascomycotina 360
aspen (= Populus tremuloides)
Aspergillus 474
Atriplex 448, 494, 497
Atriplex canescens 497
Atriplex gardneri 492, 493
Atriplex rosea 317, 322, 492, 495, 502, 503
Austrogautieria 386
Avena barbata 496
Azospirillum 178
Azotobacter 169, 172, 177–178, 179
Azotobacter chroococcum 474

Bacillus 169, 474
Bacillus cereus 475
Bacillus subtilis 475
Bacillus thuringensis 476
barley 75
Basidiomycete 137
Basidiomycotina 103, 252, 360
bean (= Phaseolus)
beech (= Fagus)
beet (= Beta vulgaris)
beetles 340
Beta vulgaris 38, 502, 504
Betula 114, 249, 250, 252–258, 260–263, 265, 269, 270–277, 315, 385, 387, 389, 391, 400, 403, 405, 409
Betula alleghaniensis 47
Betula pendula 252, 314, 391
Betula pubenscens 252, 255, 265
Betula verrucosa 392
Betulaceae 359
birch (= Betula)
Boletaceae 262, 263, 386
Boletinus cavipes 142
Boletus 105, 107, 265, 393
Boletus edulis 366, 387
Boletus leptospermi 387
Boletus mirablis 397
Botrytis fabae 77

Bouteloua gracilis 49, 317, 350, 492, 495, 502, 503
Brassica 52, 381
Brassica napus 15, 502, 504
Brassica nigra 317, 495, 502, 504
Brassica olercea 18
Brassicaceae 491, 494, 495, 503
Brauniellula 386
Bromus tectorum 396, 496

Cantharellus 263
Cantharellus cibarius 264, 265, 396
carrot (= Daucus carrota)
Caryophylaceae 360
Castanea 389, 409
Casuarina 430
Ceanothus 320
Ceasalpiniodeae 386, 409
Cedrus 399
Cenococcum 396
Cenococcum cenococcum 283
Cenococcum ectomycorrhizae 309
Cenococcum geophilum 75, 146, 262, 275, 279, 280, 428, 429, 383, 388, 391, 393, 395–397, 408
Centrosema pubescens 145
Ceratocystis 165
Cericiccum geophilum 106
Cesalpinoideae 173
Cestanopsis 408
Chalara elegans 92
cheatgrass (= Bromus tectorum)
Chenopodiaceae 52, 360, 448, 491, 494, 495, 496, 500
Chenopodium album 502–504
Chenopodium quinona 18, 20, 39
Chloridium paucisporum 266, 392
Chondragaster 407
Chroogomphus 386
Chrysothamnus nauseasus 497
Citrus 38
Clavaria argillaceae 393
Clavaria oronornsis 393
clover (= Trifolium)
Cochliobolus sativus 182
Coleoptera 340
Collembola 338, 340

Comelinaceae 360
Convolvulus sepium 8
Coriolus versicolor 312, 394
corn (= Zea mays)
Corticum bicolor 169
Cortinarius 253, 255, 256, 262, 263, 265, 278, 385, 396
Cortinarius bulbosus 262
Cortinarius debilitus 262
Corylus avellana 72
crimson clover 51
Cruciferae 52, 360
Cryptostigmata 334, 335
cucumber 140
Cyanobacteria 173, 177
Cycadaceae 173
Cycadales 177
Cyperaceae 360, 399

Daucus carrota 15, 19
Detariae 173
Diplopoda 340
Dipterocarpaceae 386
Dipterocarpus grandiflorus 390
douglas fir (= Psuedotsuga menziesii)
Dryopteris 383

earthworm 340
Elaphomyces 263
Elaphomyces muricatus 262, 393
Empetraceae 393
Enchytraeidae 338
Endogonaceae 381
Entoloma sericeum 314
Epacridaceae 393
Equisetum 381
Ericaceae 391–393
Ericales 359, 383, 393
Ericoidaae 393
Erwinia caratovora 182
Eucalyptus 47, 48, 260, 267, 285, 286, 320, 363, 383–387, 399, 401–407, 430
Eucalyptus globus 265
Eucalyptus pilularis 141
Euphorbiaeceae 409

Fagaceae 359, 383, 389, 403, 408
Fagus 70, 115, 116, 401
Fagus sylvatica 104, 114, 282
Fava bean 52
fir (= Abies)
Folsomia candida 339–345
Formicidae 340
Frankia 173, 176–177
Fumariaceae 360
Fusarium 167, 182, 475
Fusarium moniliforme 181
Fusarium oxysporum 42, 180, 181, 288
Fusarium solani 46

Gaemannomyces 182
Galium 383
Geranium robertianum 346
Gigaspora 166, 172, 506
Gigaspora calospora 216, 309
Gigaspora decipiens 210, 206, 217
Gigaspora gigantea 5, 17, 18, 22, 38, 43, 269
Gigaspora margarita 7, 15, 16, 38, 43, 52, 168, 226, 228, 340, 343
Gigaspora rosea 340
Gilbertiodendron 407
Ginkgo bilboa 51, 79, 85, 91
Glomeales 136
Glomus 48, 50, 52, 53, 94, 144, 166, 172, 179, 343, 395, 502, 506
Glomus aggregatum 10
Glomus caledonium 8, 11, 23, 79, 204, 208, 216–218, 232, 345
Glomus calospora 144
Glomus clarum 48, 145, 232
Glomus decipiens 208, 211, 231
Glomus deserticola 41, 499
Glomus etunicatum 10, 18, 216
Glomus etunicatus 343
Glomus fasciculatum 10, 38, 41, 47, 79, 140, 144, 204, 207, 210, 216, 220, 221, 225, 228, 309, 340, 345, 346, 474
Glomus geosporm 208, 232, 496
Glomus intraradices 10, 18, 53, 476
Glomus invermaium 10

Glomus macrocarpum 10, 80, 216, 220, 314
Glomus microcarpum 10, 216
Glomus monosporum 208, 216–218, 224, 225, 227
Glomus mosseae 8, 10, 11, 14, 16, 18, 41, 53, 138, 166, 167, 169, 172, 174, 182, 204, 211, 216, 220, 231, 232, 310, 340
Glomus occultum 348, 496
Glomus E3 89, 343, 344
Glomus tenue 204–208, 211, 216, 224–228, 231, 345, 382
Glomus versiforme 15, 21, 79, 80, 82
Glycine max 47, 50, 53, 83, 139
Gomphidiaceae 386
Gomphidius 385, 386
Gomphogaster 386
Gramineae 182, 348, 349
grass (= Graminae)
Gymnascella dankaliensis 393
Gymnosperms 173, 177

Halogeton 448
Hebeloma 38, 106, 181, 252, 257–262, 267, 274, 280–289, 385, 396
Hebeloma crustuliniforme 253, 260, 262, 264, 265, 270, 282, 288, 310, 314, 337, 388, 407, 410, 411, 428, 429
Hebeloma cylindrosporum 266, 281, 282
Hebeloma fragilipes 253
Hebeloma leucosarx 253, 262
Hebeloma mesophaeum 253, 265, 269
Hebeloma sacchariolens 253, 262, 274, 275, 279, 280, 282–286, 288
Hebeloma subsaponaceum 269, 271, 282
Hebeloma vaccinum 253
Hebeloma velutipes 255, 256
Hedysarum boreale 493
Helianthemum 390
Holcus lanatus 318, 348, 349, 382
Halogeton glomeratus 492, 495, 500, 502, 503
Hordeum jubatum 322
Hydnangium 406, 407
Hydnum 262, 263

Hymenogaster 406, 407
Hymenogaster tener 253
Hymenoscyphus ericae 125, 137, 146, 393
Hymenoscyphus monotropue 394
Hynansium carneum 407
Hysterangium 105, 407
Hysterangium setchellii 337

Inocybe 256, 259, 260, 262, 274, 280, 287, 396
Inocybe geophila 262
Inocybe lacera 260, 262
Inocybe lanuginella 253, 257, 262
Inocybe petiginosa 253
Isopoda 340

jack pine (= Pinus banksiana)
Juncaceae 360
Juniperus 320

Keteleeria 400
Klebsiella pneumonia 41
Kobresia 399
Koeloria cristata 497
Kohlrabi 38

Laccaria 180, 181, 262, 267, 269, 274, 275, 278, 280, 282, 287, 384, 389, 404, 407, 409
Laccaria amethystina 16, 278
Laccaria bicolor 16, 266, 282, 425, 396
Laccaria laccata 41, 49, 50, 75, 76, 77, 180, 253, 262, 265, 281, 314, 388, 396, 403, 405, 407, 410, 411, 428, 429, 433
Laccaria proxima 110, 111, 253, 262, 263, 270, 274, 275, 278, 284
Laccaria tortilis 253, 262
Lactarius 105, 253, 256, 262, 263, 265, 269, 274, 275, 278, 287, 409
Lactarius blennius 262
Lactarius deterrimus 106
Lactarius glyciosmus 253, 277, 278
Lactarius helvus 264
Lactarius pubescens 253, 255, 257, 259,

262, 263, 269, 271, 274, 275,
    277, 284, 286, 288
Lactarius rufus 146, 262, 278, 337
Lactarius spinulesus 253, 262
Lactarius subdulcis 115
Lactarius tabidus 275, 278
Lactarius torminosus 263
Lactarius turpis 262, 264, 265
Lactarius victus 262
Lactuca 383
larch (= Larix)
Larix 114, 116, 267, 385, 387, 389, 392,
    399, 400, 406, 409
Leccinum 253, 264–266, 277–279, 386,
    392
Leccinum aurantiacum 16
Leccinum scabrum 253, 262, 264, 274,
    275, 284, 387
Leccinum versipelli 253, 288
leek (= Allium porrum)
Leguminosae 94, 173, 177, 409
Lepidium 448
Lepidoptera 340
Leptospermum 387
Liquidambar styraciflua 310
Lithocarpus 408
Lithocarpus densiflora 392, 408, 409
lodgepole pine (= Pinus contorta)
Lolium perenne 138, 318
Lumbricidae 340
Lycopersicon esculentum 16, 19, 182,
    265, 269

Marasmius androsageus 335
Medicago sativa 53, 169, 174, 340
Melaleuca uncinata 394
Meloidogyne 182
Messtigmata 338
Mesophellia 407
millipedes 340
Mimosoideae 173
mite 334, 335
Monotropa 393
Monotropa hypopitys 311, 312
Monotropa uniflora 394
Monotropoideae 393
morning glory (= Convolvulus sepium)

moth larvae 340
Muciturbo 267
Muciturbo reticulatus 267
Mycelium radicis myrtillis 393
Mycenagalopus 335
Myrica cerifere 448
Myrtaceae 359, 385
Myxotrichum setosum 393

Neea lactivirens 389
nematode (= Nematoda)
Nematoda 338–340, 343
Neurospora 480
Norway spruce (= Picea abies)
Nostoc 173, 177
Nothofagus 386
Nyctaginaceae 360, 389

oak (= Quercus)
onion (= Alliuum cepa)
Onychiurus ambulans 345, 346
Onychiurus armatus 337
Onychiurus encarpatus 340
Onychiurus folsomi 340
Onychiurus latus 335
Onychiurus subtenuis 335
Oribatida 338
orchid (= Orchidaceae)
Orchidaceae 359, 394
Oryzopsis hymenoides 120
Oxalidaceae 500
Oxidiodeadron cercalis 393
Oxidiodeadron griseum 393
Oxidiodeadron maius 393
Oxidiodeadron rhodogenum 393

Papilonodeae 173
Parasponia (Ulmaceae) 174
Paxillus 105, 106, 107, 279, 284, 285
Paxillus involutus 42, 76, 77, 110, 138,
    149, 181, 256, 262, 264, 265,
    279, 283, 284, 285, 289, 290,
    314, 391
pea 52, 53, 87
pearl millet 53
Pemphigus piceae 340
Penicullium 474

Phaseolus 17, 50, 78
Phaseolous vulgari 17
Phialophora finlandia 266, 392
Phialophora fortinii 266
Phytolaccaceae 360
Phytophthora 182, 475
Phytophthora cinnamomi 288
Phytophthora glycinea 508
Picea 76, 106, 114, 266, 267, 389, 391, 396, 399, 409
Picea abies 16, 18, 40, 51, 392
Picea engelmannii 392
Picea excelsa 309
Picea glauca 392
Picea pungens 392
Picea sitchensis 111, 260, 262, 381, 392
Piceirhiza nigra 106
Piloderma cibarius 16
Piloderma croceum 18
Pinaceae 50, 73, 75, 76, 110, 111, 114, 116, 181, 251, 260, 262, 265–269, 275, 320, 359, 366, 383–392, 398, 400–406, 409, 411, 428, 430
pine (= Pinaceae)
Pinus banksiana 266, 384, 262, 263, 403, 404
Pinus cembra 411
Pinus contorta 52, 53, 141, 142, 275, 307–310
Pinus monticola 386
Pinus pinastar 266
Pinus ponderosa 319, 403, 409, 410
Pinus radiata 22, 113, 141, 169, 265, 267, 275, 285, 286, 405–407, 428
Pinus resinosa 152
Pinus strobus 141, 147
Pinus sylvestris 16, 49, 50, 105, 111, 137, 142, 145, 303, 387
Pinus taeda 141
Pisolithus 78, 106, 107, 181, 282, 283, 407
Pisolithus tinctorius 43, 46, 77, 138, 141, 275, 279–283, 286, 289, 337, 387, 388, 395–397, 403, 405, 407, 411, 428

Pisonia 71
Plantago lanceolata 321
Polygonaceae 360, 399, 500
Polygonum 399
ponderosa pine (= Pinus ponderosa)
poplar (= Populus)
Populus 367, 316, 320, 363, 389, 409
Populus tremuloides 16
Portulacaceae 360
Proisotoma minuta 340
Protura 338
Pseudogymnoascus roseus 393
Pseudomonas 169, 179, 474
Pseudomonas fluorescens 41, 169, 474, 475, 476
Pseudomonas putida 474, 475
Pseudomonas syringae 474
Pseudotsuga 380, 390, 399, 400
Pseudotsuga menziesii 41, 53, 106, 116, 146, 268, 309, 314–319, 336, 340, 385–387, 390, 392, 395, 401, 403, 404, 406–411, 428, 432
Pterospora 393
Puccinia graminis 506
Pyrola 393
Pythium 182

Quercus 320, 387, 389, 399, 408, 409, 428
Quercus chrysolepis 408
Quercus garryana 408
Quercus robur 391

rape plants (= Brassica napus)
red alder (= Alnus rubra)
red spruce 409
red yeast 8
Rhizina undulata 267
Rhizobia 40, 44, 52, 94
Rhizobium 139, 140, 176, 177, 179, 382, 493, 505
Rhizobium inoculants 471, 474
Rhizobium japonicum 140
Rhizobium meliloti 174
Rhizoctonia 182, 394
Rhizoctonia cerealis 394

Rhizoctonia solani 136, 181, 394
Rhizopogon 50, 54, 76, 77
Rhizopogon 107, 263, 367, 385–387, 390, 394, 396, 400, 403, 404, 406, 409–411
Rhizopogon luteus 22, 41, 141, 169, 265, 275, 285, 405–407
Rhizopogon occidentalis 53
Rhizopogon parksii 406, 407
Rhizopogon roseolus 77, 137
Rhizopogon rubescens 403, 406, 407
Rhizopogon vinicolor 314, 386, 394, 403, 407, 428
Rhododendron 393
Rhodotorula 165
Rhodotorula glutinus 8
rose 75
Roseofractum 253, 254, 262, 275
Russulaceae 263
Russula 105, 253, 255, 256, 262, 265, 389, 409
Russula atropurpurea 253
Russula betularum 253, 255
Russula cyanoxantha 262
Russula grisea 253, 255, 262
Russula nitidia 264, 265
Russula versicolor 253, 255
ryegrass (= Lolium perenne)

Salicaceae 359
Salix 110, 316, 320, 399
Salix rotundifolia 314
Salmonella 44
Salsola kali 52, 94, 118, 119, 317, 448, 489, 492–496, 500–508
Sarcodes 393
Saxifraga 381
Schizophyllum commune 279
Sclerocystis 172, 395
Scleroderma 106, 263
Scleroderma citrinum 262, 289, 391
Sclerotium 475
Sclerotinium 182
Scutellospora 502
Scutellospora calospora 201, 210, 211, 216–220, 489, 499
sea oats (= Uniola paniculata)

Sebacina vermifera 394
Serpula lacrymans 106, 277
Setchelliogaster 407
Shorea leprosulans 390
Shorea maxwelliana 390
Shorea obtusa 390
Shorea parvifolia 392
sitka spruce (Picea sitchensis)
Sorghum vulgare 23, 221
soybean (= Glycine max)
Sphaerosporella brunnea 267, 287, 384, 392, 396
springtails 340
spruce (= Picea)
Stipa comata 497
Stipa pulchra 497
Streptomyces cinnamomeous 41
Streptomyces orientalis 41, 167
Striga asiatica 6, 9
subterranian clover (= Trifolium subterraneum)
sudan grass 23, 38, 221
sugar maple 269
Suillus 106, 107, 312, 367, 385, 386, 387, 390, 396, 403, 405, 406
Suillus bovinus 73, 105, 107, 110, 137, 138, 145, 275, 387
Suillus cavipes 387, 399, 400
Suillus granulatus 141, 265, 266, 283, 405, 406
Suillus grevillei 387, 406
Suillus luteus 169, 262, 265, 304, 405, 406

tabacco 92
Taxus 83
Thelephora 256, 259, 260, 274, 285, 384, 404
Thelephora terrestris 106, 110, 111, 169, 253, 256, 259, 260, 264, 265, 280, 281, 285, 286, 317, 319, 388, 396, 410
Thielaviopsis 182
Thielaviopsis basicola 9
tomato (= Lycoperscicon esculentum)
Tricharina melanosporum 266
Tricharina mikolae 266, 267

Trichoderma 169
Trichoderma aureoviride 167
Trichoderma hamatum 476
Trichoderma harzianum 476
Tricholoma matsutake 105
Trifolium 138, 144
Trifolium pratense 345, 348, 349
Trifolium repens 38, 145, 221, 318
Trifolium subterraneum 144, 220
Trisetum 381
Tsuga heterophylla 384
Triticum aestivum 18, 39, 53, 309
Truncocolumella 386, 396
Truncocolumella citrina 393, 394
Tsuga 396, 397
Tsuga heterophylla 279, 404
Tuber albidum 71
Tuber melanosporum 72, 76
Tydeidae 338

Uapaca 409
Uniola paniculata 118, 450, 451

Uromyces appendiculatus 43
Uromyces viciafabae 44
Uromyces vignae 46
Urticaceae 360

Vacciniodaae 393
Vaccinium 138
Verticillium 167, 182
Vicia faba 139
Villosuli 386, 403
Vitis vinifera 79

western hemlock (= Tsuga heterophylla)
wheat (= Triticum aestivum)
white clover (= Trifolium repens)
Wilcoxia 391, 392
Wilcoxia mikolae 391
Wilcoxia rehmii 391
woodlice 340

Zea mays 17, 18, 73, 221, 396
Zygomycotina 360

Subject Index

absorbing hyphae (see hyphae, absorbing)
absorptive nutrition 79, 147
achlorophyllous plant 6, 9, 311, 383, 393, 394
actinomycetes 151, 152, 166–168, 170, 173, 176, 279, 280, 287
actinorrhizal 176, 177
active transport 73, 75, 89, 90
aeration 426, 448
aerial germ tube 38
agglutins 42, 72, 106
aggregates 151, 451–458
agriculture 41, 173, 200, 219, 226, 235, 252, 271, 358, 438, 439, 468–471, 473, 475, 476, 481
algae 151, 283, 398, 451
allelopathy 7
alpine 67, 350
altitude 427
aluminum (Al) 500, 501
amensalism 178
amides 149
amino acids 9, 16, 23, 39, 48, 50, 115, 125, 149, 168, 303, 308
ammonium (see nitrogen)
anastomosis 305, 309–312, 342, 476
animal 241, 337, 343, 397, 399–402
annual 396, 401, 402, 494, 496, 501, 510
antagonism 40–42, 475
antagonistic-interactions 249, 287–289, 495

antibiotics 41, 42, 135, 152, 169, 181, 183, 287, 288, 475
antibodies 87, 92, 472, 475, 480, 481, 505
ants 340, 491, 495
apoplast 67–69, 71, 82, 85–87, 89–91
appressorium 43, 44, 53, 78
arbuscule 45–52, 78–92, 138, 139, 175, 439, 472, 489, 490, 503, 504
arbutoid 66, 68
arbutoid mycorrhiza 320, 363, 383, 391–395, 397, 403, 408
ascomycete 124, 363, 392–395, 398
ascospores 266, 267
asexual 395, 396, 398, 477
assimilates 73, 113, 114, 125, 135, 138–142, 147, 148, 150, 287, 288
ATPase 49, 50, 73–75, 89, 90
autoflourescence 502–508, 510
autotroph 105, 110, 135
auxins 169, 174, 387
avirulence genes 53

bacteria 6, 40–44, 49, 53, 151, 165–173, 178, 179, 182, 264, 265, 271, 280, 333, 451, 453, 474, 475
basidiomes 252
biocontrol agents 475, 476
biological control 40, 41, 164, 167, 179–181
biomass 113, 116, 118, 135, 139–149,

151, 318, 319, 335–337, 339, 347, 400, 401, 431, 432, 452
biomass estimates, fungal 143–147
biome 318, 495
biotechnology 440, 468–470, 472, 481
bottleneck 432
brown earths 250, 256, 262, 284, 285

C - selected (combative) 263, 289
Ca (see Calcium)
Calcium (Ca) 44, 70, 83, 90, 125, 173, 176, 178, 282, 500, 501
canopy 310, 493
carbohydrates 18, 39, 42, 50, 65, 68, 73, 136–138, 141, 142, 150, 152, 174, 183, 210, 259, 301, 302, 305, 308, 311, 383, 424, 426
carbon ^{14}C-labeled 107, 110, 119, 142, 178, 304, 308, 310, 311
carbon (C) 8, 39, 47, 68, 103, 105, 107, 110, 113, 114, 116, 119, 124, 125, 139, 142, 149, 150, 152, 164, 287, 302–305, 307, 308, 311, 312, 323, 360, 382, 384, 386, 394, 396, 408, 409, 427, 431
carbon allocation (partitioning) 135, 138–141, 149, 176, 314, 336
carbon compounds 135, 137, 148, 149, 151, 152
carbon cost (loss, drain) 139, 140, 142, 149, 150, 164, 342
carbon exchange (transfer) 23, 45, 73, 82, 135, 150, 152, 278
carbon fixation (assimilation) 16, 139, 140, 149, 152
carbon flow (movement) 17, 47, 137, 141, 147, 148, 150, 152, 153, 304, 305, 308
carbon gain 4, 308, 311
carbon incorporation 137, 138, 140
carbon sink 135, 137, 138, 141, 142, 149
carbon translocation 135, 137, 141, 152
carbon, CO_2 5, 7, 14, 38, 39, 107, 139, 140–142, 151, 168, 500
carbonicolous 267

catabolite repression 136
cDNA 472, 473, 505, 508
cell 21, 39, 42–46, 48–52, 67–69, 73–76, 78, 81, 83–92, 94, 124, 163, 167, 168, 170, 180, 203, 391, 393, 491, 504, 506–509
cellulose 46, 136, 287, 288
Chaparral 320
chemical defenses 20, 153
chemical trigger 17, 24, 472
chitin 48, 52, 72, 75, 81, 82, 87, 93, 94, 143, 145, 166, 173, 478
chitnase 42, 51, 52, 82, 125
chlamydospores 79, 266, 267, 395, 398
chlorine 70, 83
clay 453, 454
clearcutting 261, 279, 309, 317, 391, 408
climate 397, 401, 412, 431, 439, 445
coal deposits 256, 273, 280, 289
coenocytic 15, 20, 69
coevolve 54
collembola 334, 335, 337–340, 343–346, 348
colonizing 38, 40, 41, 43–45, 47, 51–53, 67, 78, 82, 85, 87, 92–94, 105, 107, 116, 118, 119, 125, 149, 152, 164, 166, 169, 172–179, 181–183, 201, 219, 231, 274–277, 340, 359, 363, 382, 383, 386, 391–395, 401, 402, 404, 406, 407, 428, 430, 442, 443, 445, 447–450, 469, 491, 496
commensalism 496
communication 40, 54, 67, 177
community 107, 119, 120, 235, 249, 250, 263, 301, 302, 305, 309, 320, 334, 335, 348, 358, 360, 372, 382, 387–398, 400, 402, 403, 409–412, 439–441, 444–448, 458, 459, 490–494, 509, 510
community composition 315, 412
community development 135, 149, 358, 409
community structure 252
compatible 18, 94, 110, 119, 142, 171,

183, 358, 359, 380–383, 385, 387, 389, 390, 391, 401, 403, 405–409, 412
competition 41, 105, 150, 153, 165, 166, 169, 174, 180, 201, 208, 210, 217, 218, 220, 221, 228, 229, 249, 250, 283, 287, 288, 301, 302, 304, 308, 311–319, 321, 323, 335, 341, 380, 383, 404, 408, 410, 440, 442, 473, 488, 489, 494
competition theory 302, 308
conidia 266, 395
controlled hyphal growth 210
cooperate 254
copper 125, 173, 176
cost 468, 469, 471, 476, 481
Cretaceous 399
crop 153, 271, 454–456, 469, 471, 473, 476
culture 7, 396
cyanide 41, 42

decompose 304, 440, 457
decomposer populations 135, 148, 335
decomposition 124, 135, 139, 152, 153, 304, 401, 444, 454
desert 390, 442, 492, 494, 509
dialysate 166
differentiation 42–44, 106, 107, 111, 116–120, 470–472
dikaryon 266–268
dipterocarp 366, 372, 386, 389, 400
disaccharides 136
disease 469, 475, 476
disomycete 394
dispersal 105, 263, 277, 340, 341, 397–402, 407, 409, 412, 444
distribution 103–105, 125, 136, 255–258, 260, 263, 397, 399, 409, 432, 452, 456
disturbance 104, 256, 261, 263, 310, 316, 358, 402–404, 408, 411, 412, 427, 441–444, 446–449, 453, 489, 491, 492
diversity (see plant or fungal)
DNA 432, 433, 472, 477, 479, 480

DNA probe 143, 477
dominance 249, 256, 275, 281, 283, 285, 451
dormancy 215, 218, 275, 286, 287, 288, 289, 451
drought 176, 314, 387, 401, 407, 439, 448, 469, 494, 497

E-strain fungus (see ectendomycorrhiza) 343, 344
ecological crunch 124
ecological specificity 380, 382, 386, 388, 389, 391, 406, 407, 410
ecological strategy theory 249, 289, 290
ecosystem 118, 120, 126, 171, 334, 336, 341, 358, 359, 380, 381, 402–404, 408, 410–412, 424, 427, 439, 441, 443, 445, 449, 451, 490–493, 500, 509, 510
ecosystem-soil 134, 135, 139, 148, 152, 164
ectendomycorrhiza 68, 266, 267, 279, 286, 343, 344, 358, 363, 383, 391, 392, 395–397, 404
ectomycorrhizins 91, 441, 458, 508
elicitor 51, 52, 78, 89
endomycorrhizins 92, 441, 508
energy 26, 116, 119, 146, 149, 152, 153, 178, 287, 309, 311, 396–398, 424, 450
enzymatic 89, 90, 428
enzymes 42, 45–52, 71, 75, 76, 78, 82, 83, 87, 89–93, 125, 135–137, 148, 152, 164, 171, 175, 176, 180, 183, 387, 472, 478, 497, 505–508
epidermis 44–46, 68, 73, 76, 83, 85, 309, 383, 393
ericoid mycelium 45, 124, 125
ericoid mycorrhiza 45, 53, 66, 68, 78, 103, 124, 125, 136, 146, 147, 149, 152, 303, 307, 320, 358, 363, 380, 383, 392, 393, 395, 397, 399, 441, 458
erosion 150, 173, 397, 450–452
evolution 1, 4, 280, 359, 398–400
exotic 384, 405–407, 411, 428, 430

exudates 5, 8, 14, 23, 24, 38, 39, 42, 46, 135, 139, 140, 151, 152, 163, 168, 170–172, 174, 181, 182, 203, 216, 277, 472

F horizon 116
facultative (see mycotrophic)
faeces 401
fairy ring 104, 105
fatty acids 9, 46
Fe (see iron)
fertilization (fertility) 358, 427, 429, 474, 476
filimentous fungi 11
fire 401, 403
fitness 301, 315, 410, 497, 504
flavonoids 40
fluoroscein diacetate (FDA) 143, 145
food web 334
foredune 450, 451, 489
forest 105–107, 111, 116, 125, 126, 146, 147, 152, 261, 264, 266, 278, 288, 289, 309, 310, 315, 385, 388–391, 396–412, 424–432, 441, 442
forestry 173, 249, 290, 358, 425, 430–432, 439, 469
fossil 398, 399
fruiting bodies 103, 105–107, 113, 115, 135–139, 142, 146, 249–256, 260, 278, 279, 359, 385, 395, 397, 399, 426, 427, 432
fungal aggression 221
fungal diversity 150, 250, 253, 335, 359, 384, 389–391, 400, 401, 403, 405, 411, 427, 429, 431
fungi - early stage 106, 107, 249, 275, 277, 278, 284, 289, 384, 400, 403–405, 427
fungi - late stage 106, 249, 275, 277, 278, 284, 289, 290, 403, 404, 427, 430
fungistasis 165, 166, 168, 182

gene 19, 52, 53, 94, 431, 433, 472, 473, 477, 479–481, 505, 508

gene expression 40, 42, 44, 49, 50, 87, 92, 94
generalists 264, 411
genetic diversity 477
genotype 52, 250, 449
germ tube 43, 122, 470, 503
germination 11, 40, 41, 45, 78, 81, 119, 165–169, 180, 380, 393, 394, 405, 447, 451, 480
germ tubes 13, 37–39, 471, 472, 477, 480
global 401, 412, 431
glucose 42, 136, 270, 288, 304, 305, 386
grassland 67, 118, 150, 318, 334, 336, 338, 348, 350, 402, 441, 449, 501
grazing 151, 335–350
green revolution 358
growth 49–51, 53, 67, 105–107, 110, 111, 113, 116, 125, 126, 164, 165, 167, 168, 173, 175, 177, 179, 183, 210, 267, 269, 271, 274, 281, 283, 285–288, 290, 301, 321, 323, 341, 342, 345, 380, 384, 387, 394–397, 407, 408, 410, 424–426, 428, 430, 440, 441, 447, 450–452, 454, 457, 458, 469, 470, 472, 474, 478, 496
guilds 119, 358, 385, 389–391, 409–412, 441

Hartig net 22, 26, 45, 49, 54, 69, 71–76, 91, 105, 309, 388, 391–392, 425
haustorium 46, 86, 90
heathland 124, 125
heavy metal 167, 430
helicoidal organization 79, 81
helicoidal texture 80
heterogeneous 426
heterothallism 267
heterotroph 110, 138
hierarchical 489, 491, 509, 510
hormones 22, 25, 69, 90, 141, 151, 167, 169, 170, 171, 174, 177–179, 183, 474

horticulture 468, 469, 471, 473, 475, 476, 481
humus 124, 125, 145, 147, 383
hybrids 477, 478
hypersensitivity 52, 53
hypertrophic 119
hyphae, absorbing 113, 120, 323, 491
hyphae, extramatrical 151, 216, 303–305, 307, 309–311, 336
hyphae, extraradical 71, 78, 80, 82, 85, 336, 440, 443, 451–456, 458
hyphae, runner 18, 120, 122, 208, 323, 324
hyphae, thick-walled 118, 122
hyphae, external 120, 216, 217, 303, 347, 440, 443, 451–455, 502
hyphal aggregates 278
hyphal architecture 14, 207, 502
hyphal bridges 120, 122, 290
hyphal connection 149, 150, 235, 301, 302, 308, 310, 311, 317, 321–324, 426
hyphal density 107
hyphal extension 270
hyphal interference 288
hyphal length 110, 111, 113, 118, 123, 143–145, 208, 455–458
hyphal network 277
hyphal transport 301–303, 305, 308, 310, 321, 324
hyphal wefts 275
hypogeous fungi 105, 267, 397–401, 404, 407

immobilize 148, 152
incompatible 142, 182, 312, 314, 316, 385, 388, 405, 406, 410
individual 107, 110, 114, 116, 118, 119, 428
interfacial material 48, 73, 86–88, 90
infection 20, 44–46, 78, 135, 140, 144, 146, 182, 277–279, 282, 400, 445, 446, 451
infection fans 16
infectivity 203, 206, 220, 221, 225, 232, 277, 279, 447, 448
inhibition 7, 41, 49, 206, 319

inoculum 10, 114, 118, 232, 236, 249, 250, 252, 256, 261, 263, 264, 268, 275, 278–288, 320, 391, 401–403, 410, 411, 424, 429, 430, 439–451, 459, 468–473, 490, 491, 494–503, 507
inoculum density 204, 211, 227, 232, 301, 302, 341
inoculum potential 110, 118, 119, 222, 230, 383, 402, 409
inorganic 449, 452, 492
insect 348, 476
interface 50, 67, 73, 75, 76, 86
intraradical 81, 82, 85, 440
invasion 78, 263, 507, 509
invertebrate 338
involving layer 72, 75, 76, 86
iron (Fe) 41, 176, 179, 500, 501
isolvaleric acid 265
isozyme analysis 48, 51, 339, 432

landscape 358, 412, 440, 442, 444, 509
leaf hydration 139
leaf mass 139
leaf mortality 497
lectin 42, 44, 76, 82
light 147
lignase 125
lignification 47, 52, 85, 94, 505–508
lignin 136, 149, 183, 287, 288, 304, 505–508
lipid 83, 125, 138, 179
litter 105, 114–116, 124, 125, 152, 426, 429, 444, 493
litter decomposer 105, 444

M factor 16, 38
magnesium 142
mammals 340, 397, 407
mannitol 137, 138, 304, 305
mantle 41, 69, 72, 114, 170, 267, 285, 287, 304, 309, 391, 425, 426
membrane leakiness model 24
metabolites 16, 380, 381, 384, 472
methyl bromide 281, 282
microbial populations 134, 151, 152, 163–168, 171, 179–184, 277

530 / Subject Index

microcosms 119, 348, 387, 395, 408
microinjection 479, 480
microrhizotrons 6, 104
microsites 114, 426
mine spoils 281, 283, 289, 402, 430
mineral nutrients 66, 78, 79, 105, 110, 115, 137, 147–150, 153, 172, 290, 424
mineral nutrition 141, 171
mineralization 148, 173, 492
mining 150, 441, 442, 444, 448, 454, 455, 489
minirhizotron 104
mites 333, 334, 338–340, 343
models 90, 91, 183, 238, 443, 500, 505
moisture 396, 400, 401, 444, 445, 447–449
molecular biology 39, 44, 48, 54, 432, 510
molecular interaction 19, 488, 491, 508, 509
monokaryons 261, 262, 266, 267
monosaccharides 136
monotropoid mycorrhizae 66, 68, 303, 304, 311, 383, 393, 394
morphology 67, 69, 76, 78, 79, 83, 164, 178, 254, 382, 393, 455
mRNA's 39, 92, 472
mucilages 42
mutualism 67, 102, 110, 126, 335, 358, 359, 398, 402, 496
mycelial aggregate 277
mycelial slurries 262
mycelium 73, 81, 102–107, 110, 113–116, 118–120, 123–126, 135, 138–150, 153, 165, 167–170, 172, 204, 206, 216, 220, 249, 254, 258, 261–268, 271, 275, 277–279, 285, 288, 305, 340–343, 385, 387, 396, 403, 404, 429, 458, 459, 497
mycelium, external 102, 124, 137, 138–150, 143, 145–150, 172, 340–343, 346, 347, 382, 387
mycelium, extramatrical 114, 135, 145
mycelium, network 110, 114, 118, 119, 124, 458, 459

mycocosms 114, 327
mycophagy 139, 151, 397, 401
mycorrhizae, facultative 301, 302, 323, 360, 381, 384, 402, 412, 442
mycorrhizae, obligate 360, 381, 384, 402, 405, 412, 442
mycorrhizins 47, 51, 90–92, 439–459
mycorrhizoplane 152
mycorrhizosphere 41, 135, 151, 164, 170–172, 183
mycotrophy 359, 360, 488, 489, 491, 494, 495, 496, 509, 510
mycotrophy, facultative 287, 301, 302, 318, 323, 363, 381, 442
mycotrophy, non 18, 118, 119, 308, 317, 318, 321, 322, 363, 381, 442, 488, 489, 491–495, 500, 501, 504, 509, 510
mycotrophy, obligate 318, 363, 381, 382, 384, 442, 472, 473, 481

N (see nitrogen)
nematode 8, 181, 182, 333, 334, 338, 339, 343, 346
net primary production (or NPP) 113, 146, 336, 431, 510
niche 302, 315, 397, 401, 473
nitrate (see nitrogen)
nitrogen 8, 39, 47, 50, 82, 103, 115, 116, 124, 125, 142, 149, 177, 179, 270, 303, 310, 314, 319, 425, 427, 492, 493
nitrogen cycle 148
nitrogen, ammonium 50, 116, 125, 147, 149, 177, 179, 303, 310
nitrogen, fixation 139, 170, 172, 173, 175–179, 474, 475
nitrogen, organic 115, 116, 148, 152, 173, 303
nitrogen, uptake 171, 318, 425
nodules 139, 172–177, 343
non-mycorrhizal 39, 47, 51, 138, 139, 140, 141, 171, 318, 322, 360, 402, 441–443, 448
nonmycotrophy (see mycotrophy, non)
nutrient availability 135, 164, 427
nutrient cycling 148, 151, 152, 164, 171,

179, 290, 334, 345, 348, 431,
439, 443, 456–458, 488, 489,
492, 497, 510
nutrient exchange (transfer) 47, 54, 66,
67, 78, 82, 87, 90–92, 94, 103,
116, 307, 321–323, 382, 387
nutrient uptake 142, 387, 407, 410, 429,
431, 473

obligate (see mycotrophy, obligately)
obligate symbionts (see symbionts - obligate)
old growth forest 317, 403, 430
operculate discomycetes 266
orchid mycorrhiza 66, 68, 136, 170, 363,
394, 395, 397
organic acids 39, 41, 137, 455, 497
organic horizons 116
organic matter 111, 116, 119, 149, 151,
152, 168, 171, 271, 303, 397,
398, 400, 401, 452–455, 493
organic nutrients 137
outplanting 428, 429, 430, 451, 452
oxalate 42, 490, 500, 501, 510
oxidase 125

P (see phosphate)
palatability 337, 339
parasitism 67, 76, 94, 105, 136, 163,
166, 311, 341, 359, 394, 475
parent tree 249, 278
patch 107, 116, 442
pathogenic fungi 8, 210
pathogens 40, 42, 43, 46, 49–54, 66, 67,
86, 89, 92, 94, 163, 164, 167,
179, 181–183, 277, 286, 359,
366, 380, 381, 397, 401, 469,
470, 475
PCR (see polymerase chain reaction)
peat, vermiculate 262, 284
pectin 46, 49, 136, 287, 288, 478
penetration 4, 11, 21, 25, 42–46, 50–52,
54, 76, 81, 83, 86–90, 118, 119,
124, 125, 170, 177, 180, 203,
381, 383, 393, 402, 425, 503
peptidolytic 148
pH 90, 125, 219, 387

phenolics 18, 21, 40, 44, 46, 50, 53, 69,
75, 85, 90, 125, 149, 181, 183,
381, 388, 402, 506
phenology 314, 315, 494, 497
phosphatase 48, 75, 83, 89, 125, 314,
425, 428
phosphate 15, 18, 23, 39, 41, 43, 47,
53, 69, 70, 73, 75, 82, 83, 103,
107, 114, 116, 123, 144, 147,
149, 166, 170, 172, 173, 175–
179, 182, 201, 211, 218, 219,
221, 261, 262, 270, 290, 303,
304, 307–309, 315, 318, 319,
321–323, 387, 425, 427, 439,
469, 473, 474, 476, 478, 488,
492, 493, 500, 501
phosphate translocation 142
phosphate, uptake 70, 71, 73, 90, 200,
343–349, 425
phosphorus ^{32}P-labeled 107, 110, 137,
138, 140–142, 305, 308, 321,
322
photosynthate 7, 114, 137–141, 150,
182, 249, 404
photosynthate translocation 137
photosynthesis 124, 135, 141, 174, 249,
303, 304, 398, 407, 411, 497
photosynthetic capacity (rate) 135, 139,
141, 174, 284
physical trigger 472
physiology, fungus 6, 19, 22, 26, 82,
106, 138, 165, 200, 220, 383,
400
physiology, plant 164, 170, 178, 207,
315, 317, 439, 440
plant composition 170
plant diversity 120, 358, 363, 386, 411,
427, 439
plasma membrane 39, 49, 67, 87, 89–91,
180
plasmalemma 45, 48, 49, 74, 75, 82, 84,
86, 87, 89, 90, 180
plasmid 479
plasmogamy 266
polymerase chain reaction (PCR) 394,
432, 477
polypeptides 433

polyphosphate granules 69, 70
population 105, 106, 119, 165, 334, 339, 399, 401, 424, 431, 442–445, 447
potassium 44
production (see net primary production)
production, fungal 440, 451
production, plant 113, 314, 358, 439, 440, 493
propagules 165, 204, 211, 215, 217, 218, 227, 228, 236, 282, 380, 383, 395, 397, 401–403, 406, 440–442, 445–449, 458
protease 39, 149
protein 40, 47, 50, 75, 87, 90–92, 303
proteinase 125
proteolysis 115, 116, 125
proteolytic 116, 125, 148, 149, 152
protoplast 477, 478, 480
pseudomycorrhizae 266, 359
pure culture 115, 124, 125, 136, 137, 165, 366, 380, 382, 385, 386, 388, 389, 391, 406

r&K selection 234, 249, 261–263, 289
reclamation 439, 443, 449, 469, 489, 496
recombinant DNA 478–481
recycling 116, 135, 140, 152, 341
reforestation 424, 425, 429
regeneration 135, 150, 153, 428, 479, 480
reproduction, fungal 231, 384, 395, 477
reproduction, plant 441, 458
resource acquisition 201, 495
respiration, belowground 113, 140, 336
respiration, fungal 113, 139, 140, 142, 143, 146, 304, 500
respiration, plant 139, 141, 151, 176, 336
restoration 439–444, 450–459
restriction fragment length polymorphisms 432, 477
RFLP's (see restriction fragment length polymorphisms)
rizomorphs 70, 106, 110, 277, 323, 387, 395, 426

rhizoplane 37, 42, 142, 172
rhizosphere 6, 12, 23, 37, 40, 41, 136, 151, 163–65, 171, 172, 174, 176, 178, 179, 182, 183, 215, 230, 271, 274, 287, 309, 341, 358, 380, 382, 384, 404, 405, 408, 412, 431, 432, 441, 453, 474, 475, 497, 500, 501
rhizosphere effect 163, 341
RNA 472, 508
rodent 401
root architecture 251
root browning 52, 502–504, 506, 510
root cortical tissue 470, 491
root distribution 13, 59, 256–259
root hair 44, 69, 75, 203, 401
root infection 107, 110, 111, 114, 203, 210, 424
root interception 210
root structure 69, 73, 432
root system 91, 107, 114, 119, 135, 141, 164, 177, 249, 253, 256, 260, 261, 269, 275, 277, 281–286, 288, 290, 312, 401, 404, 450, 455
root-soil interface 163–165, 183
root/shoot ratio 179
ruderal 249, 261, 263, 289

S - selected (stress tolerant) 263, 289, 290
sand dunes 118, 290, 440, 489
saprobic 431
saprophytic 69, 70, 119, 136, 148, 152, 163, 164, 169, 267, 277, 279, 287, 304, 335, 363, 384, 394, 451–454, 472, 478, 493
scale 440, 488, 489, 491, 492, 509
secondary compounds, 151, 183
secretions 45, 163
seed 118, 119, 170, 393, 394, 400, 401, 409, 447, 491, 493, 496
seedling 110, 114, 116, 118, 139, 141, 142, 150, 169, 181, 249, 250, 256, 257–284, 286–290, 303, 308–310, 314, 315, 321, 385, 387, 388, 390–392, 394, 403–

410, 424–430, 432, 442, 444,
  448, 450, 451, 458, 495, 496,
  501, 503, 504, 506, 508
sequences - age related 261, 284
sequences - temporal 252, 253, 256, 260
serral 318, 320, 442, 493–497, 501
sheath 68, 69, 71, 75, 104–106, 114–
  116, 146, 147, 177, 180, 181,
  279, 286, 341, 393, 400
siderophores 41, 42, 171, 172, 179, 182,
  475
sink strength 141, 149, 302, 322, 323
sodium 43
soil aggregates 150, 153, 440, 441, 444,
  452, 453, 455, 456
soil reclamation 173
soil structure 171, 219, 250, 251, 439,
  443, 452, 455, 457, 458
somatic incompatibility 105
source-sink 103, 135, 174, 176, 311,
  321, 322
spatial 252, 253, 256, 260, 312–315,
  335, 337–339, 431
specificity 40, 53, 110, 118, 119, 149,
  150, 314–317, 359, 360, 366,
  380–382, 385–391, 394, 398–
  405, 410–412
specificity 53, 314, 316, 317
spheroplast 478
spore-derived infection network (SIN)
  503
spores 37, 38, 41, 42, 45, 78, 79, 81,
  118, 122, 138, 165–168, 180,
  249, 264–268, 274, 275, 278,
  282, 289, 336–340, 385, 386,
  395–401, 405, 429, 447, 448,
  450, 451, 470–478, 480, 491,
  502, 503, 507
sporocarp 366, 382–391, 395, 397, 403,
  404, 407, 426
stand 105, 106, 389, 401, 405, 407, 409,
  424, 426, 427, 429, 431, 432
starch 136
steppes 402, 491, 494, 495, 509, 510
stomata 43, 314
stomatal resistance 124, 314
strain 411, 425, 470, 474–478

stress 425, 469
suberin 84, 85, 94
succession 250, 256, 260, 290, 314–317,
  319, 350, 358, 388, 402–405,
  408–410, 412, 426, 427, 440,
  442, 447–451, 489–497, 500,
  501, 509, 510
sucrose 8, 43, 136
sugars 39, 43, 46, 48–50, 53, 68, 69,
  73, 75, 76, 78, 81, 82, 249, 261,
  262, 269, 277
sulfur 83
surface area 110, 114, 122, 440
survivorship 124, 150, 235, 316
symbionts 47, 107, 135, 136, 139, 140,
  163, 180, 285, 288, 359, 372,
  381, 382, 397, 398, 401, 430,
  442, 474, 491
symbionts-facultative 136, 363
symbionts-mutualistic 94, 180, 303
symbionts-obligate 135, 136, 165, 363,
  470, 472, 481
symbiotic 45, 92, 125, 149, 170, 173,
  176, 282, 311, 424, 433, 472, 497
symplast 68, 69, 71, 85, 86, 91
synergism 14, 40, 41, 51, 52, 167, 172,
  177, 178, 223, 408, 410

tannic acid 149
tannin 125, 400
teleomorph 266
temperate 441, 451
temperature 7, 8, 111, 210, 215, 218,
  226, 228, 230, 271–273, 289,
  314, 396, 400, 401, 431, 443,
  447
tillering 450, 497
trace elements 469, 473
transformations 477–480
transpiration 123, 124
trehalose 136–138, 304, 305
Triassic 398
trophic 67, 386, 402, 428 430, 440, 442,
  451
truffles 340, 359, 390, 397, 426
tundra 336
turnover 147, 148, 152

534 / Subject Index

vacuoles 69, 73, 83, 92
vector 442, 479, 480
vegetative phase 263, 349
vesicles 48, 398, 439, 503, 504
vigor 250
vitamins 474
volatile 5, 6, 9, 17, 25, 38, 265, 269

water 45, 106, 123, 124, 150, 174, 201, 218, 303, 305, 314, 323, 397, 425, 440, 441, 451, 452

water potential 124
waterlogged soils 6, 14, 111
weed 402, 448, 449, 491, 492, 494
wind 397, 400, 402, 450–452, 493
woodlands 249, 250, 252, 263, 271, 273, 278, 279, 284, 289, 336

xylogenesis 505, 506

zinc 125, 167, 173, 176

DATE DUE

DUE DATE SUBJECT TO CHANGE
IF A RECALL IS REQUESTED

SEP 2 3 2005

Rtnd - LS   JUN 1 6 2005